Lethbridge Community College Lib
C0-AVT-102

Delmar's Inspection & Maintenance Series

A TECHNICIAN'S GUIDE TO

AUTOMOTIVE EMISSIONS SYSTEMS

BY LARRY CARLEY

Contributing Editor
Rick Escalambre

Delmar Publishers

I(T)P™ An International Thomson Publishing Company

Albany • Bonn • Boston • Cincinnati • Detroit • London • Madrid • Melbourne
Mexico City • New York • Pacific Grove • Paris • San Francisco • Singapore • Tokyo
Toronto • Washington

NOTICE TO THE READER

Publisher does not warrant or guarantee any of the projects described herein or perform any independent analysis in connection with any of the project information contained herein. Publisher does not assume, and expressly disclaims, any obligation to obtain and include information other than that provided to it by the manufacturer.

The reader is expressly warned to consider and adopt all safety precautions that might be indicated by the activities herein and to avoid all potential hazards. By following the instructions contained herein, the reader willingly assumes all risks in connection s with such instructions.

The publisher makes no representation or warranties of any kind, including but not limited to, the warranties of fitness for particular purpose or merchantability, nor are any such representations implied with respect to the material set forth herein, and the publisher takes no responsibility with respect to such material. The publisher shall not be liable for any special, consequential, or exemplary damages resulting, in whole or part, from the readers' use of, or reliance upon, this material.

Cover Design:

Delmar Staff:
Publisher: Susan Simpfenderfer
Assistant Publisher: Dale R. Bennie
Administrative Editor: Vernon Anthony
Production Manager: Cheri Plasse
Marketing Manager: Lisa Reale

COPYRIGHT ® 1995
By Delmar Publishers
a division of International Thomson Publishing Inc.

The ITP logo is a trademark under license

Printed in the United States of America

For more information, contact:

Delmar Publishers
3 Columbia Circle, Box 15015
Albany, New York 12212-5015
1-800-347-7707

International Thomson Publishing Europe
Berkshire House 168 - 173
High Holborn
London WC1V7AA
England

Thomas Nelson Australia
102 Dodds Street
south Melbourne, 3205
Victoria, Australia

Nelson Canada
1120 Birchmount Road
Scarborough, Ontario
Canada M1K5G4

International Thomson Editores
Campos Eliseos 385, Piso 7
Col Polanco
11560 Mexico D F Mexico

International Thomson Publishing GmbH
Konigswinterer Strasse 418
53227 Bonn
Germany

International Thomson Publishing Asia
221 Henderson Road
#05 - 10 Henderson Building
Singapore 0315

International Thomson Publishing - Japan
Hirakawacho Kyowa Building, 3F
2-2-1 Hirakawacho
Chiyoad-ku, Tokyo 102
Japan

All rights reserved. No part of this work covered by the copyright hereon may be reproduced or used in any form or by any means–graphic, electronic, or mechanical, including photocopying, recording, taping, or information storage and retrieval systems–without the written permission of the publisher.

1 2 3 4 5 6 7 8 9 10 X X X 01 00 99 98 97 96 95

Library of Congress Cataloging-in-Publication Data

Carley, Larry W.
Automotive Emissions Systems / by Larry Carley.
p. cm
Includes index.
ISBN 0-8273-7048-2: Soft Cover – ISBN 0-8273-7135-7: Shrinkwrapped
1. Automobiles--Pollution control devices. I. Title.
TL214.p6C3697 1994
629.25'28--dc20 9422901
CIP

Contents

Preface

It seems that today's automotive service technicians must be instant experts at almost everything! They have to know basic electricity and understand fuel injection, carburetion and the fundamentals of four-stroke combustion, and use a scan tool to pull trouble codes and other diagnostic data out of a vehicle's onboard computer system. Technicians must be able to relate such things as compression, valve time, and abnormal combustion problems to everyday driveability problems. Most importantly, today's technicians must be able to analyze, diagnose, and repair complex emissions and driveability problems to their customers' satisfaction.

Because of rapidly changing technology, today's technicians must constantly strive to improve their knowledge and skills. They have to live up to higher consumer expectations and make sure their customers' vehicles comply with more stringent emission laws. This is a lot for technicians to master. No training courses or books can provide technicians with all the conceivable tools needed to keep up-to-date. This reference book, however, can, and does, give today's technicians a comprehensive overview of all the essentials.

A Technician's Guide to Automotive Emissions Systems, the first of Delmar's Inspection and Maintenance Series, covers everything from the basics of electricity, ignition, fuel delivery, and engine operation to emissions systems, computerized engine controls, electronic fuel injection, advanced exhaust analysis, testing and diagnosis, and driveability issues. There are several other books and courses available that focus on many of these subjects, but no other resource can match this book's overall content or completeness.

A Technician's Guide to Automotive Emissions Systems is filled with real-world information that makes it especially useful as a self-study guide for working professionals or students.

CHAPTER 1

Introduction

This technician's training manual is a reference supplement for Automotive Emmissions Training especially for I/M training programs. The material that each chapter covers goes with the corresponding classroom learning module. You will be required to read the appropriate background chapter prior to beginning each learning module.

Most of the material in this book is fairly basic and should already be familiar to you. Even so, you should still read and review the material to refresh your memory in case you've forgotten something. You may unclear about what a particular system or component does, how it functions, what sort of problems might affect its operation, and so on. If you're not sure about something you've read, make a note of it, and discuss it with your instructor in the classroom.

WHY I/M TRAINING PROGRAMS?

The purpose of this technician's training manual is to train automotive technicians in the fundamentals of inspection and repair. You're taking this program so you can become a qualified emissions technician. And to accomplish that goal, you must understand the basics of electricity, carburetion, ignition, fuel injection, computerized engine controls, emissions, emissions testing, and troubleshooting. The 18 modules in this book are designed to cover each of the basic topics that you need to do this.

The need for enhanced inspection and maintainance technician training came about because of changes in the Clean Air Act. These changes require certain areas to implement more rigorous vehicle emission inspection and testing programs. Such programs have become necessary to reduce urban air pollution from motor vehicles.

WHY THINGS ARE THE WAY THEY ARE TODAY

The automobile that created our mobile society and sprawling cities has become a major source of environmental concern (Fig. 1-1). Over half the total car-

Fig. 1-1 Motor vehicles are a major source of urban air pollution.

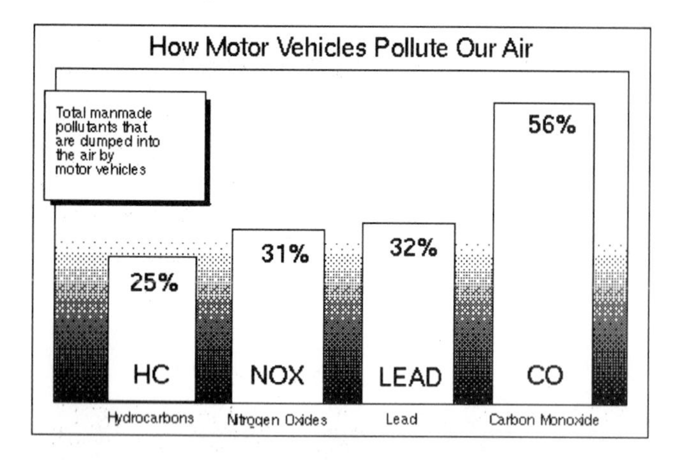

Fig. 1-2 How motor vehicles pollute our air.

bon monoxide, a quarter of the hydrocarbon, and a third of the nitrogen oxides that pollute many urban areas, are attributed to motor vehicles (Fig. 1-2). Consequently, automotive emissions are coming under closer scrutiny to improve urban air quality.

Today's new cars are much cleaner from an emissions standpoint than their predecessors (Fig. 1-3). Pre-1963 vehicles were the worst polluters because they lacked any emission controls whatsoever. In 1963, positive crankcase ventilation (PCV) was added. This was done to recycle the blowby vapors in the crankcase back into the intake manifold, so they could be reburned. This virtually eliminated crankcase emissions as a source of air pollution. Sealed fuel systems and charcoal canisters, which were added in 1971, reduced evaporative emissions to zero. In 1973, exhaust gas recirculation (EGR) was added to reduce oxides of nitrogen (NOX) in the exhaust. An even greater change appeared in 1975 when the catalytic converter (which required unleaded gasoline) was introduced (Fig. 1-4). The converter greatly reduced unburned hydrocarbons (HC) and carbon monoxide (CO) levels in the exhaust, by "reburning" the pollutants. The switch to unleaded gas was necessary because lead poisoned the catalyst. The change to unleaded fuel also eliminated lead as an exhaust pollutant (which has been linked to lead poisoning and various learning deficiencies in urban children). In 1981, onboard computers for closed-loop running, and three-way oxidation/reduction converters, were added to most cars. These computers cut CO and NOX emissions by another 50% (Fig. 1-5).

The tailpipe emissions from late model cars (1981 and later), with computerized engine controls and three-way catalytic converters, are only a fraction of the older pre-emissions controlled cars. Today's cars produce on average 96% less HC and CO, and 76% less NOX than their pre-emission counterparts (Fig. 1-3).

Additional reductions are being phased in, which began in 1994 and continue through 1998. The federal standards up through 1993, allowed no more than 0.41 grams per mile (gpm) of HC, 3.4 gpm of CO, and 1.0 gpm of NOX for the first 50,000 miles. The new standards slash these limits by almost half: CO is cut to 0.25 gpm, and HC and NOX drop to 0.4 g/m.

EMISSION REQUIREMENTS FOR PASSENGER CARS
(IN GRAMS PER MILE)

MODEL	HYDROCARBON (HC)		CARBON MONOXIDE (CO)		OXIDES OF NITROGEN (NO_x)	
YEAR	FEDERAL	CALIFORNIA	FEDERAL	CALIFORNIA	FEDERAL	CALIFORNIA
1978	1.5	0.41	15.0	9.0	2.0	1.5
1979	0.41	0.41	15.0	9.0	2.0	1.5
1980	0.41	0.39	7.0	9.0	2.0	1.0
1981	0.41	0.39	3.4	7.0	1.0	0.7
1982	0.41	0.39	3.4	7.0	1.0	0.4
1983	0.41	0.39	3.4	7.0	1.0	0.4
1984	0.41	0.39	3.4	7.0	1.0	0.4
1985	0.41	0.39	3.4	7.0	1.0	0.7
1986	0.41	0.39	3.4	7.0	1.0	0.7
1987	0.41	0.39	3.4	7.0	1.0	0.7
1988	0.41	0.39	3.4	7.0	1.0	0.7
1989	0.41	0.39	3.4	7.0	1.0	0.4
1990	0.41	0.39	3.4	7.0	1.0	0.4
1960	10.6*		8.4*		4.1*	

*Typical values before controls

Fig. 1-3 Federal and California Emission Standards Compared to Uncontrolled 1960's Vehicles.

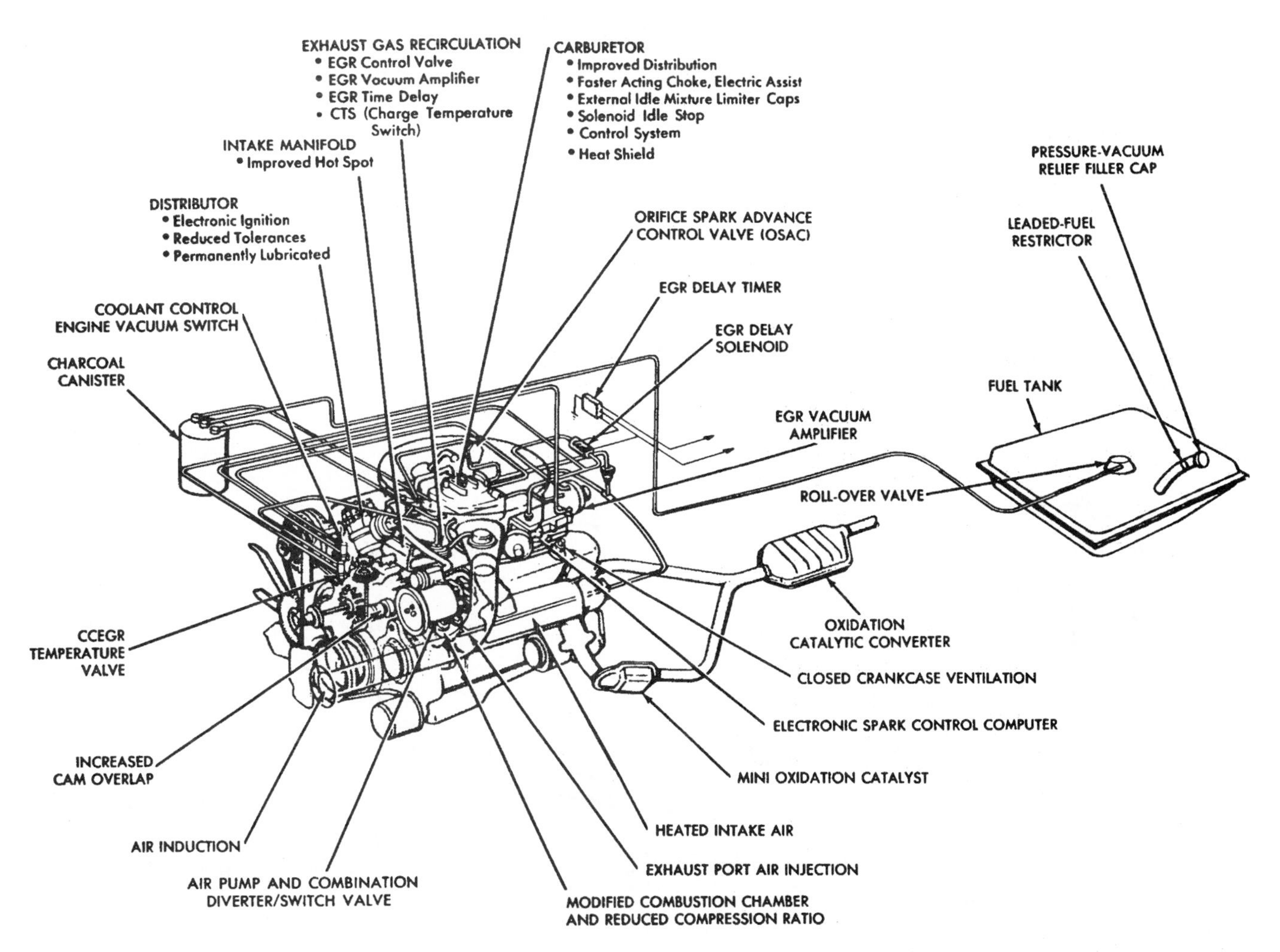

Fig. 1-4 The first generation of automotive emission controls were mostly add-ons, but made significant reductions in pollution (courtesy Chrysler).

What do all these numbers really mean? It means a 1975 to 1979 model car puts out roughly as much HC and CO pollution as four new cars. A 1972 to 1975 model year car produces as much of these pollutants as seven new cars. A 1968 to 1971 model year car produces pollution equivalent to 10 new cars. A 1963 or earlier pre-emissions controlled car pumps as much pollution into the atmosphere as 25 or more new cars!

These comparisons assume that a vehicle produces no more pollution that it did when new. However, this is usually NOT the case once an engine has accumulated a lot of miles. It's not unusual to find high mileage engines in older vehicles (1980 and earlier) that are belching out the pollution equivalent of 100 new cars! It's also not unusual to find relatively new cars producing excessive emission because of mechanical, fuel, or ignition problems. Vehicle inspection programs are necessary to make sure a vehicle's emissions are within acceptable limits. Vehicle inspection programs check that all the major systems that affect emissions on that vehicle are functioning within acceptable norms. It's also just as important to be able to fix the vehicles that fail an emissions test, so they can be brought back into emissions compliance.

HOW INSPECTION LAWS CAME TO BE

Today's emission problems, and the legislation that has subsequently been adopted to address those problems, had their origins in California. Los Angeles has always had an air pollution problem because of its unique geography and urban sprawl. Warm air over the Los Angeles basin forms an "in-

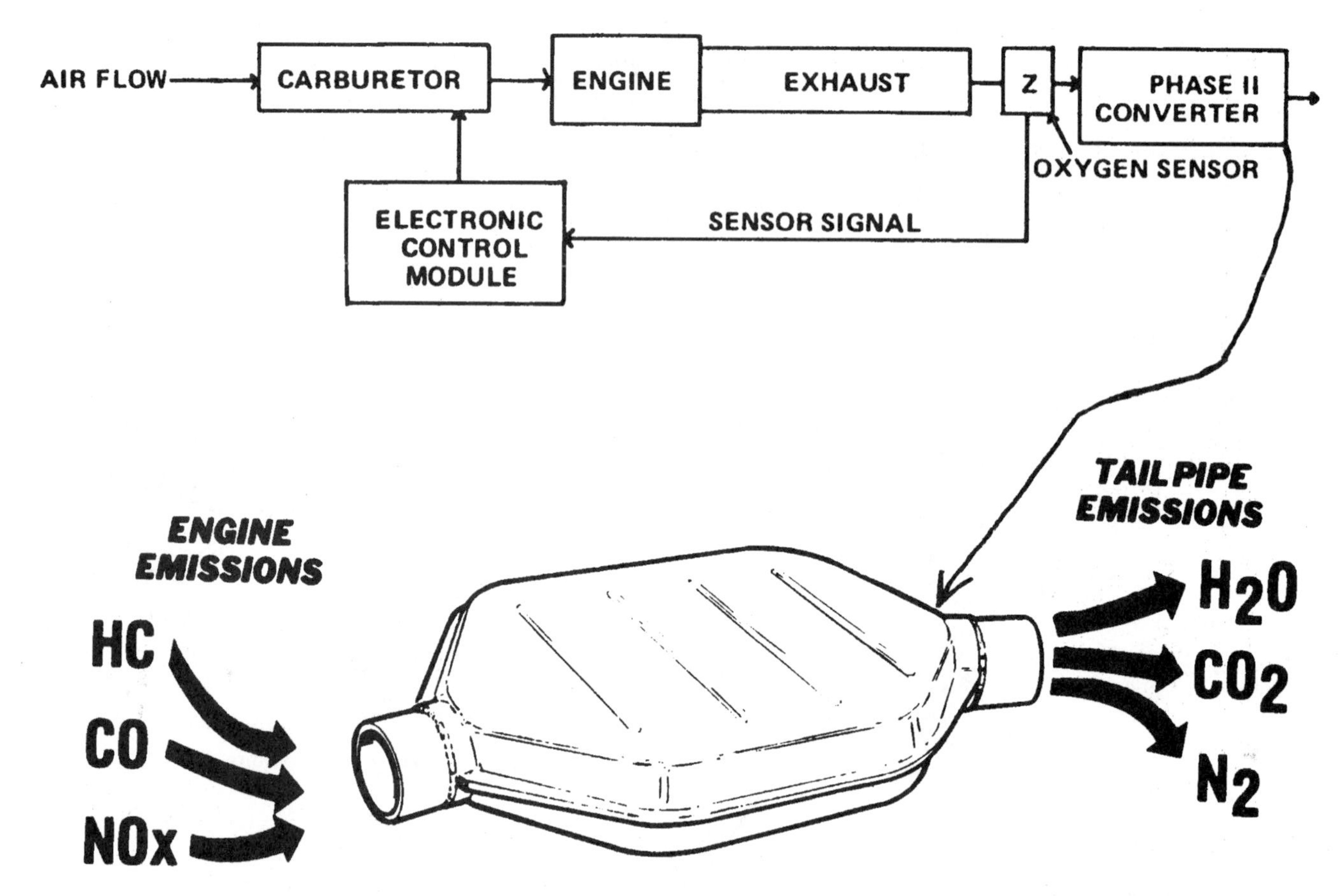

Fig. 1-5 The introduction of feedback carburetion and computerized engine controls made possible further reductions in pollution (courtesy Sun Electric *and* AC Delco.).

version layer," which traps pollution underneath (Fig. 1-6). The surrounding mountains prevent the wind from blowing the pollution away, so a huge dome of stagnant, polluted air hangs over Los Angeles like a cloud.

What many people don't know is that the problem existed long before Los Angeles became a large city. Native Americans lived in the area long before the first missionaries and settlers arrived. The native Americans were familiar with the perpetual haze created by natural sources of pollution.

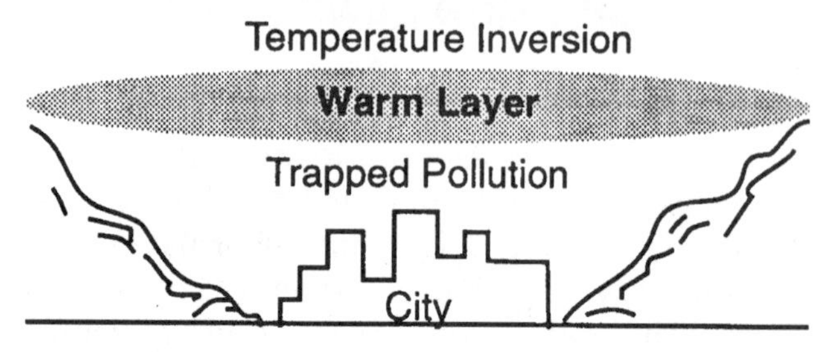

Fig. 1-6 Temperature inversions trap polluted air.

The environmental movement, and the passage of the first Clean Air Act in 1970, prompted Californians to create the California Air Resources Board (CARB). CARB's purpose was to implement more stringent clean air standards. Many Californians believed these were necessary to combat the growing air pollution problem in the L.A. basin.

CARB established model legislation for future federal emission standards. CARB required tougher emission standards than those in the other 49 states. Certain engines and drivetrains, that were offered in the rest of the country, were not available in California because of the different requirements. California consumers also found they had to pay higher prices for new cars because of the added pollution equipment.

The situation in California created a serious problem for the auto makers. They had to build cars capable of meeting one set of standards in California, and another set of standards in the other 49 states. But they soon turned it to their advantage.

California's standards were one to two years ahead of the rest of the country. Therefore, it gave the car companies an opportunity to get the bugs out of their new emission control systems before the equipment was installed on the rest of their cars. This meant Californians sometimes had to deal with unproven and troublesome emission control systems.

One of the approaches CARB undertook, to ensure clean air for the Los Angeles basin, was to implement a program of periodic motor vehicle inspection (PMVI). The idea behind PMVI (or inspection and maintenance (I/M) as it is also called), is to inspect vehicles once a year. This inspection makes sure their emission control equipment is in good working condition. This is accomplished by subjecting the vehicle to a tailpipe emissions test and an underhood inspection. The tailpipe test certifies that the vehicle's exhaust emissions are within the limits set by law. The underhood and/or vehicle inspection verifies that the pollution control equipment has not been tampered with or disconnected.

SIDE-STEPPING THE LAW

Restrictions in the 1970 Clean Air Act made it illegal for professional mechanics to remove, disconnect, or tamper with factory emission controls. But the law placed no such restrictions on the vehicle owner. Consequently, emissions tampering was rampant. Keep in mind that many of the add-on emission systems of the early and mid-1970s did nothing to improve performance or fuel economy. Many people simply unhooked all the "pollution junk" to make their cars run better, which often made their cars run worse. Fuel switching was also common because of the price differential between unleaded and leaded regular gasoline.

In an effort to discourage tampering, the Environmental Protection Agency asked the auto makers to make their emission controls more tamper resistant. The auto makers did this, much to the chagrin of tune-up technicians, who then found it more difficult than ever to make simple idle mixture or choke adjustments.

In California, the annual inspection effectively discouraged tampering. It was now difficult for individuals to sneak vehicles, with modified or missing emission controls, past the state inspectors. Inspection programs in other states, though not as tough as those in California, had a similar effect. The incidence of tampering and fuel switching decreased as motorists got the word that tampering could be expensive. Missing emission control components had to be replaced. Those components that were unhooked, or made inoperative, had to be reconnected or fixed.

During this period, a lot of aftermarket performance parts were being installed on street-driven vehicles to improve performance. Carburetors, intake manifold, camshafts, air cleaners, exhaust headers, etc., for instance, were not designed with emissions in mind. To skirt California's laws, many of the manufacturers of such products labeled their products "For off-road use only."

The Clean Air Act was revised in 1990. The Act made it illegal for anyone (including the vehicle's owner) to remove, disconnect, or tamper with any emissions control component on any vehicle that's driven on a public road. Off-road-only performance parts are still available for true race car applications, but are no longer street legal anywhere in the U.S. The law requires that any aftermarket performance parts that are installed on a street-driven vehicle must be "emissions certified." This means they've been tested and shown not to increase emissions. The test data must be submitted to CARB for review. If approved, the product is issued an "exemption order" (EO) number which makes it "emissions legal."

THE ROAD TO COMPUTERIZATION

In 1980 in California, and 1981 in the remaining 49 states, the first generation of computerized cars arrived with "feedback" engine management systems and electronic carburetion (Fig. 1-5). Fuel, ignition and emissions control functions became more integrated than ever before. Most motorists found the new computer cars were "too complex" to work on. The motorists feared they would hopelessly screw things up, or damage the expensive electronic componentry. At the same time, professional automotive mechanics were starting to be called "technicians." The technician title was more appropriate because of the increasing complexity of the vehicles they worked upon, and the skills needed to diagnose and repair the new systems.

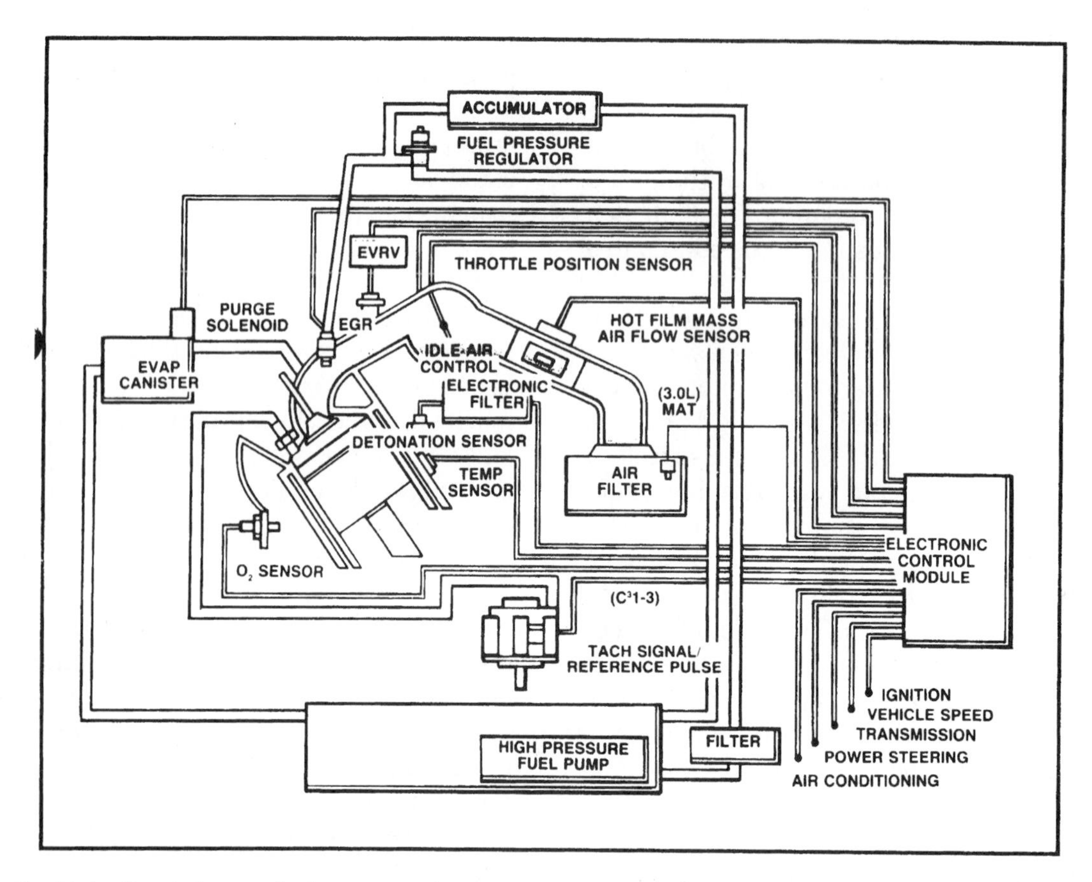

Fig. 1-7 Fuel injection takes emissions control a step further by totally integrating the fuel, ignition and emissions control functions under the ECM (courtesy GM).

In the mid-1980s, carburetors were giving way to electronic fuel injection (Fig. 1-7). Fuel injection proved to be a welcome improvement over carburetion in terms of performance, driveability, fuel economy and emissions. First came the relatively simple throttle body injection (TBI) systems. These systems were followed by much better (and more complex and expensive) multipoint injection (MPI) systems.

Today we have computerized engine and drivetrain management, electronic multipoint fuel injection and sophisticated emissions control strategies. These systems actually result in the best of all worlds: low emissions, good fuel economy, AND excellent performance and driveability. Engine management systems diversity and the complexity of the technology has made it necessary to keep up-to-date with changing technology. That's why emissions diagnosis and repair is becoming a specialty unto itself. It takes a substantial amount of training and expertise, as well as access to up-to-date service information and special tools, to perform emissions diagnostic work.

OBD II

In an effort to make emissions diagnosis easier, the government has mandated a standardized approach called "onboard diagnosis" (OBD) for identifying and troubleshooting emissions performance problems. The OBD rules require a standard diagnostic connector on all 1995 model year vehicles. Therefore, a common scan tool can be used to retrieve trouble codes from the vehicle's computer. The codes correspond to a specific fault that the computer has detected. The code or codes can then be used to begin a troubleshooting procedure. This eventually allows you to pinpoint the exact cause of the problem—or so the theory goes.

Trouble codes and onboard diagnostics have been around since the introduction of the first computerized cars in 1980. However, they haven't necessarily made troubleshooting emissions and performance problems much easier. This are better than nothing but they can also be misleading. A trouble code is not an end in itself, but a beginning. You should follow an exact step-by-step diagnostic procedure to rule out all the other various possibilities. Then you may make the correct diagnosis and identify what's really causing the problem. The new OBD II codes will hopefully improve the situation.

ABOUT EMISSIONS WARRANTIES

The original Clean Air Act required vehicle manufacturers to offer a 5 year/50,000 mile emissions warranty on all new vehicles starting in 1981. This would minimize the repairs costs consumers would have to bear on the new emission controlled vehicles.

There are actually two separate federally-mandated emission warranties, one for performance and the other for defects.

The Performance Warranty kicks in when ALL of the following stipulations are met:

- The vehicle is an '81 or later model ('82 for high-altitude).
- The vehicle has failed an EPA-approved state or local pollution test.
- The state or local government requires that repairs be made to enable the vehicle to pass the test. To put it another way, the owner will have to bear a penalty (a fine, the cost of repairs, the loss of the right to legally use the car) because of the test failure.
- The vehicle has been operated and maintained in accordance with the instructions in the owner's manual (that includes using unleaded gas only in catalyst-equipped conveyances).

For the first two years or 24,000 miles of service, the automaker is obliged to pay an authorized dealer to make ANY repairs necessary to get the car through smog inspection. This would not cover internal engine repairs, but such repairs would likely be covered under an extended powertrain warranty.

Up to five years or 50,000 miles (which ever comes first), the vehicle manufacturer is also liable for the cost of repairing or replacing any emission control device.

The 5/50 Defect Warranty applies to the repair of emission control-related parts that have failed (the vehicle does not have to fail an emissions test, however, for the defects warranty to apply).

Starting with the 1995 model year, the federal emissions warranty is extended to 8 years and 80,000 miles on the engine computer and catalytic converter. This warranty applies to any individual component costing more than $200 to replace. At the same time, the defect warranty is rolled back to 2 years and 24,000 miles. But many new vehicles now have 3 year/36,000 mile or longer bumper-to-bumper warranties that cover virtually everything on the car.

In addition to the federal emission warranties, California has its own specific warranty requirements.

The whole issue of warranties can be confusing, and is always subject to change. So always refer to the vehicle manufacturer's warranty terms and list to determine which parts are covered and for how long.

Generally speaking, the following systems and components are covered under all emissions warranties:

- Exhaust Gas Recirculation: valve, spacer plate, solenoid, thermal vacuum switch, backpressure transducer, sensors and switches used to control flow.
- Early Fuel Evaporation: valve, thermal vacuum switch.
- Air Injection System: air pump, anti-backfire valve or deceleration valve, diverter, bypass, or gulp valve, reed valve.
- Exhaust Gas Conversion System: catalytic converter, thermal reactor, oxygen sensor, dual-walled exhaust pipe.
- Positive Crankcase Ventilation System: valve, solenoid.
- Evaporative Emission Control System: purge valve, purge solenoid, fuel filler cap, vapor storage canister and filter.
- Fuel Metering System: electronic control module, deceleration controls, fuel injectors, fuel injection units and fuel rails developed for feedback EFI or TBI. Also air flow meter, module, or mixture control unit, mixture control solenoid, diaphragm or other fuel metering components that achieve closed-loop operation.

Electric choke, altitude compensator sensor, other feedback control sensors, switches, and valves, and thermostatic air cleaner completes this list.

- Ignition systems: electronic ignition, electronic spark advance, timing advance/retard systems.
- Miscellaneous Components: hoses, gaskets, brackets, clamps, and other accessories used in the above systems.
- Items such as spark plugs and oxygen sensors that have a recommended replacement interval of less than 5/50 are only warranted up to their normal replacement interval.

A few miscellaneous notes on emissions warranties:

- No matter how many owners a car may have had, the emission warranties still apply up to their full time and mileage limits.
- Proper maintenance records and receipts must be kept.
- A claim can't be disallowed because a non-OEM part was installed providing it is an approved equivalent, labeled "Certified to EPA Standards."

THE FUTURE OF EMISSION CONTROL

Have we reached a plateau in automotive exhaust emissions? Hardly. The landscape is constantly changing. California (and possibly some other states as well) will require 2% of all vehicles that are sold in that state to be "zero emission vehicles" (ZEVs) by 1998. The ZEVs will increas to 5% by 2001, and 10% by 2003. ZEVs will probably be electrically-powered because most fuels can't meet a zero emission requirement.

A certain percentage of California vehicles will also be "ultra-low emission vehicles" (ULEVs). These will be a combination of super-clean gasoline-powered vehicles, as well as those that run on alternative fuels such as propane or natural gas.

Low emission vehicle (LEV) legislation is also being considered at the federal level for areas that fail to meet federal air quality standards (primarily the Northeastern states). Implementation of the federal LEV rules would begin in 2001, and would roughly parallel the California requirements.

One change that's already taken place is the introduction of "reformulated" gasoline in many major cities that don't meet federal clean air standards. Reformulated gasoline typically contains alcohols to provide extra oxygen for a slightly leaner, slightly cleaner burning mixture. In older cars, reformulated gasoline can make a measurable improvement in reducing carbon monoxide emissions.

Another change is the phase-in of the new enhanced I/M 240 inspection program. The new inspection program is being implemented primarily in cities with the dirtiest air. The new test is much more comprehensive than a simple idle emissions test. It simulates the same urban driving test cycle that's used by the car makers to certify new vehicles for emissions compliance. It's called I/M 240 because the test is 240 seconds in duration. It involves "loaded mode" testing. This means the vehicle is run at various speeds on a dynamometer to check emissions under various driving conditions. The new test also measures NOX emissions for the first time. The test includes a check of the charcoal canister and fuel system to make sure fuel tank vapors are being properly contained.

CHAPTER 2

Fundamentals of Electricity and Electronics

To understand electrical systems and circuits, you first have to know something about electrons and electricity. In the simplest of terms, electricity is the flow of electrons. Where do electrons come from? Atoms. All matter is composed of atoms. Atoms make up the molecules from which everything else is made. Electrons orbit the nucleus of the atom like miniature planets circling the sun (Fig. 2-1). Only so many electrons can share the same orbit at the same time, and once an orbit is filled, the excess electrons are forced to circle at a higher orbit further from the nucleus. The electrons that are furthest out are not held as strongly as those closer. So they can be stripped away fairly easily and shared with adjacent atoms—but it depends on the number of electrons in the outermost orbit (Fig. 2-2).

Metal atoms have only one to three electrons in their outermost orbits, which means they can supply and share electrons easily. That's why metals are good "conductors" of electricity.

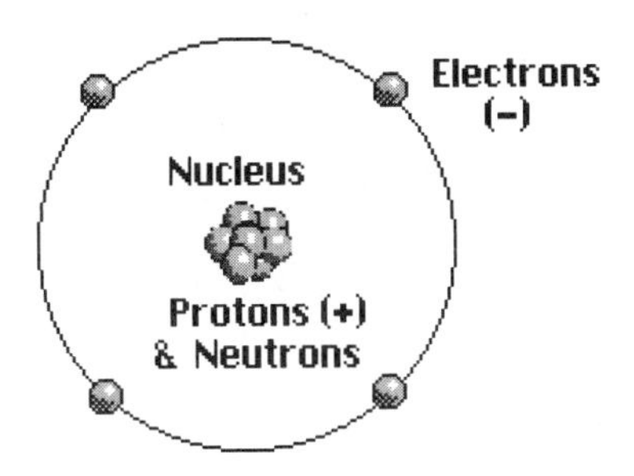

Fig. 2-1 Electrons orbit the nucleus of atoms.

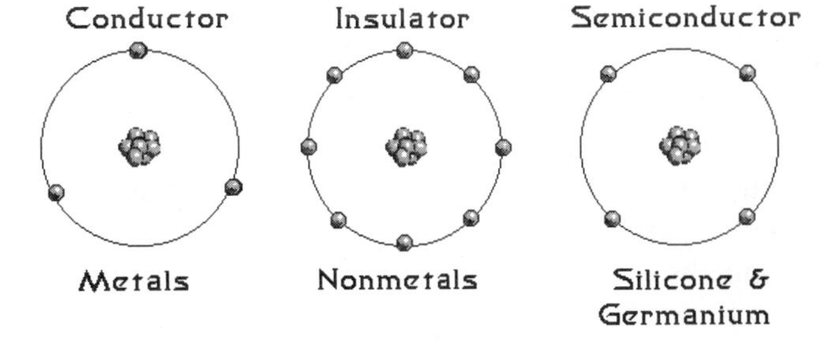

Fig. 2-2 Whether a material is a conductor, insulator, or semiconductor depends on the number of electrons in the atom's outermost orbit (valence ring). Metals have one to three electrons in the outermost orbit, insulators five or more, and semiconductors exactly four.

Materials made up of atoms that have more than four electrons in the outermost orbit do not conduct electricity well because the electrons are held strongly in place. These materials make good "insulators" (rubbers, plastic, glass) because they do not conduct electricity.

Materials that have exactly four electrons in the outermost orbits of their atoms conduct electricity only under certain conditions. These materials also make good insulators. But when crystals of these materials are treated or "doped" with certain other atoms, they become "semiconductors" and are used for solid-state electronic components such as transistors and diodes.

CURRENT FLOW

Copper is a good conductor of electricity because it's a metal and has lots of easily shared electrons (Fig. 2-3). But having a ready supply of electrons doesn't make an electrical current, unless there's a driving force to move the electrons. One of two things is needed to make that happen: either a moving magnetic field to push the electrons through the wire, or a surplus of electrons at one end of the wire and a shortage of electrons at the other. Either will make the electrons move and create an electric current. To keep the current going, there must also be a return path so the electrons can continue to circulate.

Electrical repulsion and attraction are what actually push the electrons through the wire. Like charges repel one another, while unlike charges attract. Negative repels negative, but negative and positive attract. When a bunch of electrons are crowded together, they try to push each other apart because all are negatively charged. And when there's a shortage of electrons, the positively charged atoms that have lost electrons attract new electrons to replace the ones that were lost. So electrons always flow from a region of high concentration to one with low concentration. In other words, electrical current flows from negative to positive.

Electrons move through a conductor like a wave. The positively charged atoms are locked in place by each other, so they can't move towards the electrons. Electrons are extremely small compared to atoms, so the electrons easily circulate through an interlocking grid of atoms. As they move, they're momentarily captured by one atom and then another, knocking adjacent electrons loose and creating a ripple-like current. And although each individual electron moves only a zillionth of an inch, the effect of the push multiplied billions of times over is felt throughout the wire almost instantly. The result is a net flow of electrons and an electrical current.

VOLTS, AMPS AND OHMS

This brings us to the basic measurements of electricity:

- Voltage is the difference in electrical potential between two points, or the amount of "push" that makes the electrons flow. It's also called the "electromotive force" (EMF). It's like the pressure that forces compressed air through a hose. But instead of being measured in pounds per square inch, voltage is measured in units called "volts."
- Current is the amount or volume of electrons that flow through a conductor or a circuit. It is a measure of volume, specified in units called "amperes," or "amps" for short. The analogy with an air hose would be the number of cubic feet per minute of air passing through the hose. One amp is equal to 6.3 million trillion electrons (6.3 with 18 zeros after it) flowing past a point in one second! That's a lot of electrons, but a relatively small current in many automotive circuits. A starter, for example, can draw several hundred amps when cranking the engine.
- Resistance is the opposition to the flow of current, or the restriction that impedes the flow of electrons. Resistance is measured in units called "ohms." The flow of air though a hose can be reduced by pinching it, by reducing the diameter of the hose, or by holding your finger

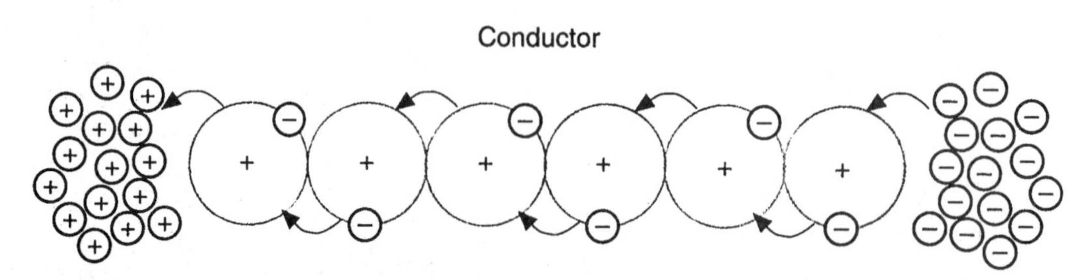

Fig. 2-3 Electric current flows when electrons jump from atom to atom in a conductor (reprinted, by permission, from Hollembeak, *Automotive Electricity and Electronics*. Copyright 1994 by Delmar Publishers Inc.).

over the outlet. Likewise, current flow through a wire can be slowed or controlled by adding resistance. Resistance can be created by altering the composition of the material, by decreasing the size of the conductor or wire (smaller wire has more resistance than larger wire), or by adding heat (heat increases resistance).

Ohms Law

One volt equals the amount of force needed to push a one-amp current through a circuit with a resistance of one ohm. They call this "Ohm's Law" after the scientist who first figured it out (Fig. 2-4). It can be expressed in various ways:

$$\text{AMPS} = \frac{\text{VOLTS}}{\text{OHMS}} \text{ or OHMS} = \frac{\text{VOLTS}}{\text{AMPS}} \text{ or VOLTS} = \text{AMPS} \times \text{PHMS}$$

Understanding Ohm's Law and the relationships between volts, ohms, and amps is the key to understanding electrical currents and circuits. Ohm's Law explains why high resistance in a circuit chokes off the current and causes a voltage drop. It also explains why an electrical short can cause a wire to rapidly overheat and burn because of a runaway current.

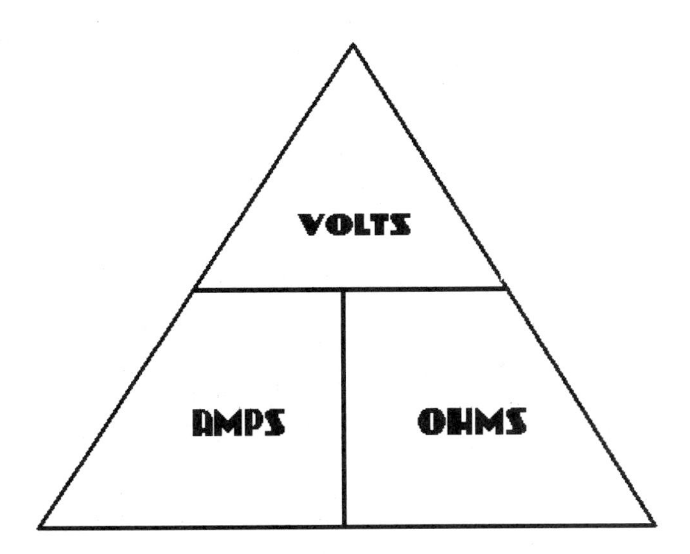

Fig. 2-4 Ohm's Law allows you to figure the amps, ohms, or volts in a particular circuit if you know two out of the three values.

ELECTRICAL CIRCUITS

An electrical circuit is basically a route or path through which electrons flow. As we said earlier, it must form a complete loop so the current will continue to flow. The electrons need a return path back to their source (the battery or alternator). Otherwise, they have no place to go.

There are essentially two kinds of electrical circuits:

- A "series" circuit is one in which all the circuit elements are connected end-to-end in chain-like fashion (Fig. 2-5). The current has only one path to follow, so the amount of current passing through it will be the same throughout. The total resistance in a series circuit is equal to the sum of the individual resistances within each circuit element. If one element in a series circuit goes bad, continuity is broken, and the entire circuit goes dead because the current can't complete its journey through the circuit.
- A "parallel" circuit is one in which circuit elements are connected next to, or parallel to one another. This creates multiple branches or pathways through which current can flow. The resistance in any given branch will determine the voltage drop and current flow through that branch and that branch alone. One of the advantages of a parallel circuit is that the various

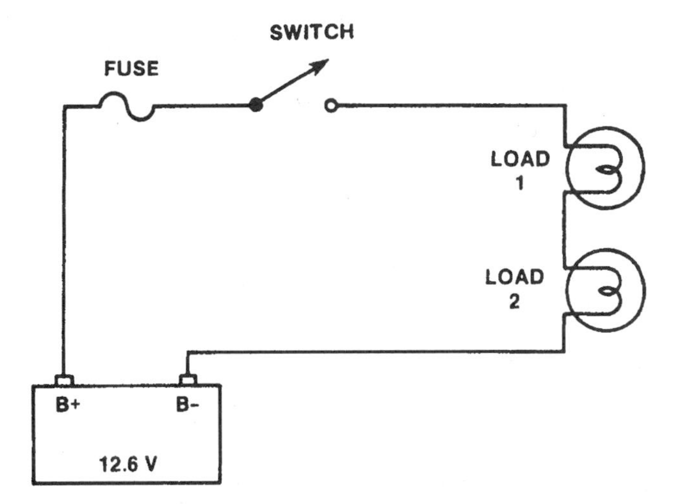

Fig. 2-5 A typical series circuit with two light bulbs in series with a switch (reprinted, by permission, from Santini, *Automotive Electricity and Electronics, 2E.* Copyright 1992 by Delmar Publishers Inc.).

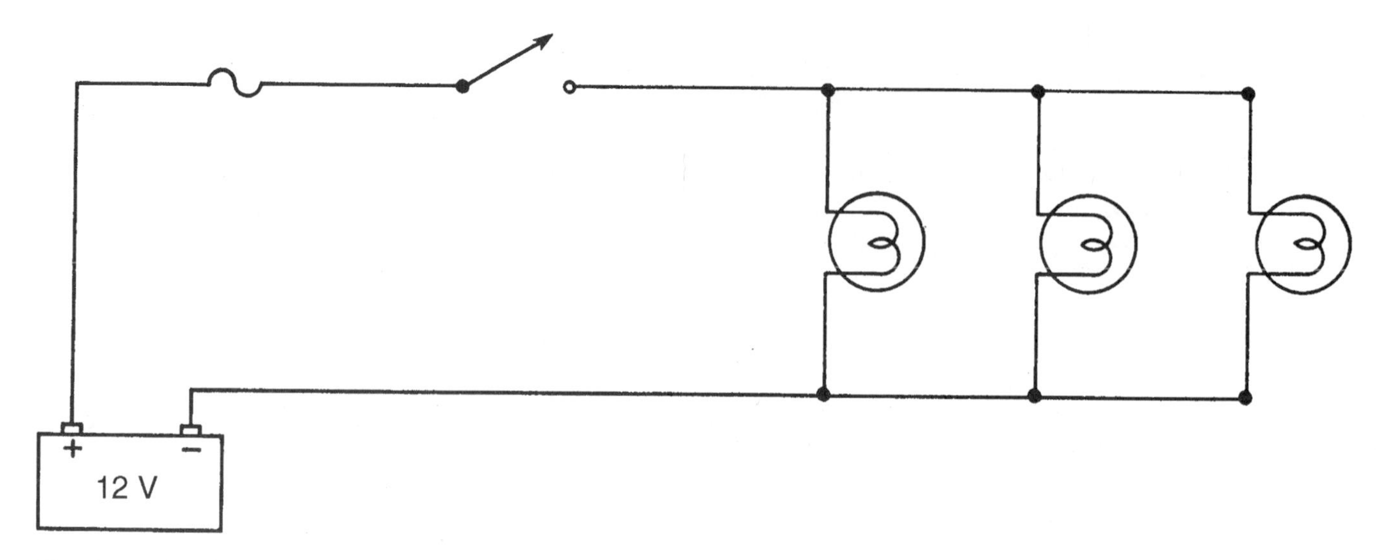

Fig. 2-6 The three bulbs in this circuit are wired in parallel to one another, and in series with the switch (reprinted, by permission, from Santini, *Automotive Electricity and Electronics, 2E.* Copyright 1992 by Delmar Publishers Inc.).

segments or pathways of the circuit can operate independently of one another (Fig. 2-6). If one element goes open (breaks continuity), it won't disrupt the function of the other.

Some circuits combine elements of both a series and parallel circuit. These would be called a "series-parallel" circuit (Fig. 2-7). In this type of circuit, part of the circuit might have loads in series, while in another part the loads would be parallel. Figuring the amount of resistance in a series-parallel circuit is more complicated. You first have to calculate the equivalent series loads of the parallel branches, then add it to the load in the series portion of the circuit. In other words, if the parallel part of the circuit had 3 ohms of resistance, and the series portion had 1 ohm of resistance, the total resistance for the entire circuit would be 4 ohms (3 + 1).

Voltage Drop

A voltage drop occurs when current flows through a component in a circuit. The resistance created by

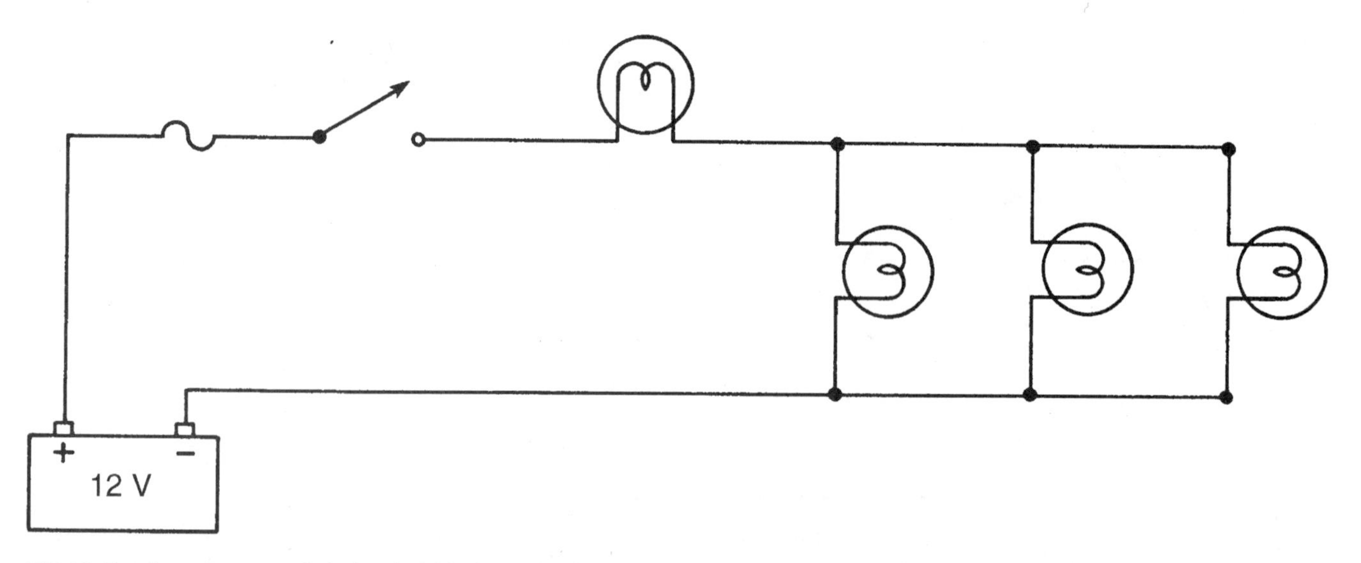

Fig. 2-7 A series-parallel circuit with three bulbs wired parallel to one another, and in series with the fourth bulb and switch (reprinted, by permission, from Santini, *Automotive Electricity and Electronics, 2E.* Copyright 1992 by Delmar Publishers Inc.).

the device produces a corresponding drop in voltage. This drop can be calculated using Ohm's Law if you know the resistance of the component and current flow.

VOLTAGE DROP = RESISTANCE × CURRENT

In the shop environment, voltage drop is measured with a voltmeter (see Chapter 3). The voltmeter's leads are connected on either side of the circuit component or connection that's being tested (Fig. 2-8). If a connection is loose or corroded, it will create resistance in the circuit and restrict the flow of current causing an excessive voltage drop. As a rule of thumb, a voltage drop of more than one-tenth volt (0.1v) across any connection means trouble.

Measuring voltage drop is an effective means of quickly pinpointing circuit problems such as loose or corroded connectors, wires, switches, etc., because it doesn't require you to disassemble anything prior to testing.

Types of Current in a Circuit

Most automotive electrical circuits operate on "Direct Current" (DC). In a DC circuit, the polarity of the voltage and current do not change—as opposed to "Alternating Current" (AC) circuits, where they do.

In a DC circuit, positive (+) is always positive, and negative (-) is always negative (Fig. 2-9). Positive or "hot" wires may be (but not always) color-coded red, while negative or "ground" wires may be

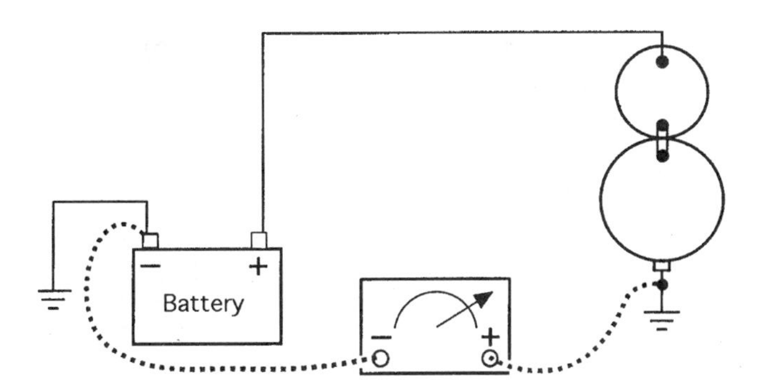

Fig. 2-8 Checking voltage drop across a circuit or connection is a good way to find resistance (reprinted, by permission, from Hollembeak, *Automotive Electricity and Electronics.* Copyright 1994 by Delmar Publishers Inc.).

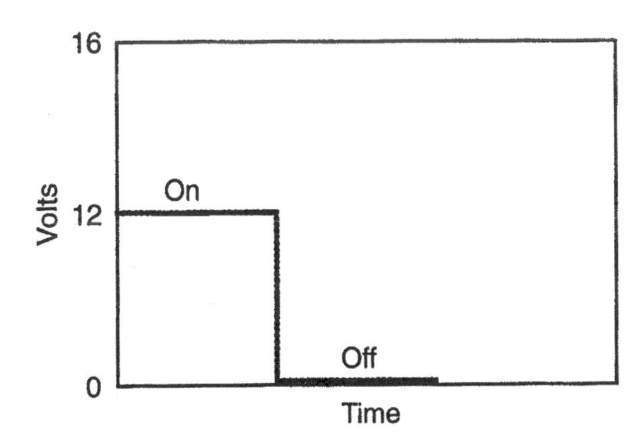

Fig. 2-9 Direct current (DC) always flows in the same direction and remains constant throughout the circuit (reprinted, by permission, from Hollembeak, *Automotive Electricity and Electronics.* Copyright 1994 by Delmar Publishers Inc.).

coded black. (Remember that electrons actually flow from negative to positive!)

All modern automotive electrical circuits are "negative ground" meaning the body is connected to the battery's negative terminal. In some antique vehicles and older British cars, the body is connected to the positive battery cable, creating a "positive ground" electrical system.

DC current is used to drive the starter, solenoids, fuel injectors, relays, idle speed control motor, fuel pump, and most other electrical and electronic components on the vehicle. Most sensors also operate on direct current.

The alternator, however, is an exception. It uses direct current to produce an alternating current charging voltage. But before the alternating current leaves the alternator, it is converted back to direct current again by the alternator's diodes. (See Chapter 4 for additional information about the operation of the charging system.)

In an AC circuit, voltage and current do not remain constant. An AC current reverses direction and goes from positive to negative and back to positive again in a cyclic fashion. If an AC current is plotted on a graph or viewed on an oscilloscope, it forms a "sine wave" pattern (a DC current would form a straight line). The AC wave starts at zero volts, increases to its maximum positive value, then falls back to zero and reverses to its maximum negative value.

Think of the difference this way: a DC current flows one-way (negative to positive), while an AC current surges back and forth (Fig. 2-10).

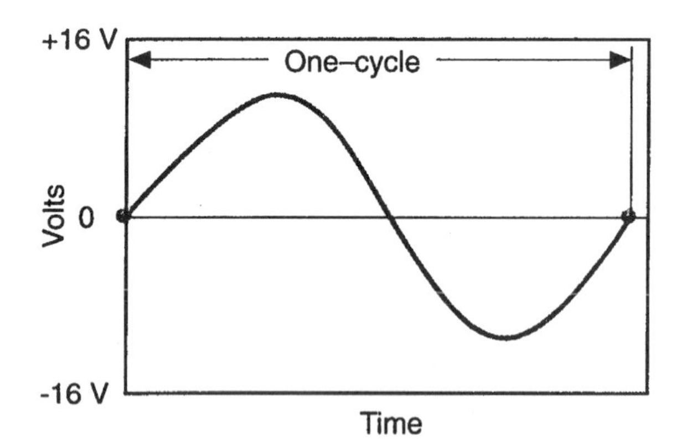

Fig. 2-10 Alternating current (AC) reverses direction with every cycle and does not remain constant (reprinted, by permission, from Hollembeak, *Automotive Electricity and Electronics.* Copyright 1994 by Delmar Publishers Inc.).

BASIC ELECTRONICS

The fundamental difference between an electrical system circuit and an electronic circuit is the amount of electrons that flow through the circuit. Electrical currents are generally measured in amps, while electronic currents are measured in milliamps (.001 amps).

Electronic circuits are also based on "solid-state" components such as "transistors" and "diodes." They're called solid-state components because they have no moving parts. Solid-state electronics has revolutionized the world, and along with it automotive ignition, fuel, and emissions-control systems. Electronics is now used to regulate everything from antilock braking and traction control to automatic transmissions and "smart" suspensions.

To understand how solid-state electronic devices work, we have to go down to the atomic level. As we described earlier, the electrical properties of a material are determined by the number of electrons in the outermost shell (valence ring) of its atoms. Metal atoms have three or less electrons in the outer shell, making them good conductors of electricity. Insulators, such as glass, rubber, and most plastics, have five or more electrons in the outer shell. This inhibits the flow of electricity. But some materials have exactly four electrons in the outermost shell. This gives them unique electrical properties that allow the material to either act like an insulator or a conductor under certain conditions (as when doped with other atoms). These materials are called "semiconductors," and are the basis of solid state electronic devices.

Transistors

The age of solid-state electronics actually began with the invention of the vacuum tube, not the transistor. The glass vacuum tubes once used in radios and later televisions were really the first electronic switching devices. They used electronics rather than mechanical contacts to reroute the flow of electrons through a circuit. Vacuum tubes are no longer used today, because they consume too much power, generate too much heat, and are bulky and fragile. They also tended to fail frequently. But they laid the groundwork for the solid-state revolution that was to follow.

In 1948, Bell Laboratories made what was to become one of the key inventions of this century: the transistor. A transistor is essentially a switch that reroutes an electrical current according to a voltage input. A transistor performs the same switching function as a vacuum tube, except that it does not require a heated filament or magnetic field to switch the circuit path. The switching function is accomplished by changing the conductivity of the junction inside the transistor.

A transistor is made by sandwiching three layers of silicone or germanium crystals that have been doped to create opposite electrical properties (Fig. 2-11). When certain trace impurities are added to either silicone or germanium crystals during their manufacture, it alters the material's electrical properties. Remember what we said about semiconduc-

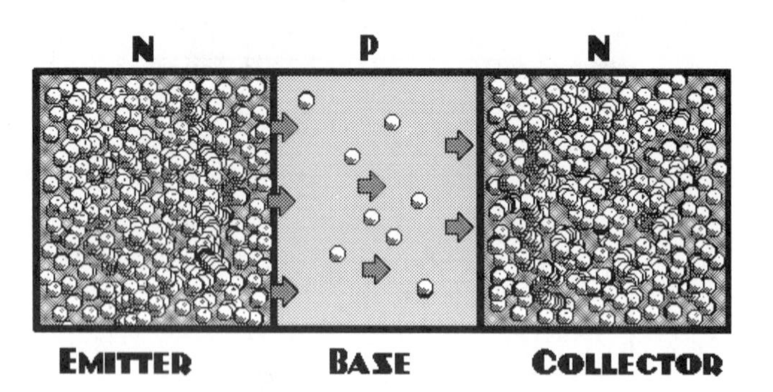

Fig. 2-11 A transistor consists of three layers of alternating semiconductor material sandwiched together. The base layer serves as the "switch" that either passes or blocks the flow of current from the emitter to the collector.

tors having four electrons in the outermost ring? In a pure crystal of either material, the outer rings of the atoms link up in such a way that there aren't any "holes" left over for extra electrons to move about. (A hole is a space in an outer orbit for an electron to orbit.) Pure silicone and germanium are actually good insulators. But when the crystals are doped with trace "impurities," the neat arrangement of electrons is upset. The trace impurities (only about one atom in 10 million!) add just enough extra electrons (or create just enough empty holes) so that the material will conduct a current when a voltage is applied. This transforms them into semiconductors.

Doping creates one of two basic types of semiconductor materials. One is "Negative" or "N-type." To create an N-type semiconductor material, the crystal is doped with atoms (such as phosphorus) that have five or more electrons in their outer valence ring. The extra electron doesn't fit into the neat arrangement within the crystal, so a surplus of electrons is created that gives the crystal a negative charge. The other type of semiconductor material is the "Positive" or "P-type." Positive semiconductors are doped with atoms (such as boron, indium, or aluminum) that have three or fewer electrons in the outer valence ring. The missing electrons leave holes in the crystal lattice into which electrons can move when a voltage is applied. Thus P-type semiconductors have a positive charge.

So how does all this work together to make a transistor act like a switch? The transistor consists of three layers of alternating semiconductor material: either N-P-N or P-N-P. Each layer is connected to its own electrical lead. The "in"-lead is called the "emitter." The "out"-lead is called the "collector." And the center control lead is called the "base." The middle base layer performs the switching function by either allowing current to pass from the emitter or in-lead to the collector or out-lead.

Current cannot move from the emitter side to the collector side within the transistor unless voltage is also supplied to the middle base layer. The opposite charge of the center base layer creates a boundary across which current can't pass because there aren't enough extra electrons (N-type) or holes (P-type) to carry it. But when an outside voltage is applied to the base layer, it reverses the charge and makes it conductive. The outside voltage provides the extra electrons or holes depending

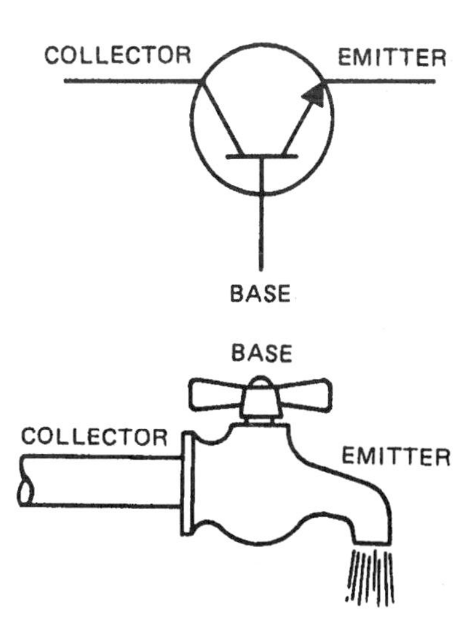

Fig. 2-12 Think of a transistor as a kind of electrical faucet (reprinted, by permission, from Herman, *Electronics for Industrial Electricians,* second edition, Copyright 1990 by Delmar Publishers Inc.).

on the type of material (N-type or P-type) needed to carry the current. The switch "closes," and the transistor passes current to the opposite side (Fig. 2-12).

Thus the on/off switching function of a transistor is controlled by the application of voltage, rather than opening a set of mechanical contacts as in a conventional switch.

The same type of transistor can also be made to act like a variable resistor by varying the current to the middle base layer. As the current increases to the base lead, the transistor passes more and more current to the collector or out-lead. A typical application for a variable resistor transistor might be the power transistor for the blower motor in an automatic climate control system.

Diodes

A diode is a type of "one-way" switch or filter that allows current to flow in one direction only. A diode is made by sandwiching P-type and N-type silicone crystals back to back. The P-N junction will only pass electrons when an outside voltage pushes the holes in the P-material towards the extra electrons in the N-material. This happens when the applied voltage polarity is in the same direction as the P-N junction (positive on the P-side, negative on the N-side). This is called "forward bias." When this happens, the junction conducts current.

But when current is applied in the opposite di-

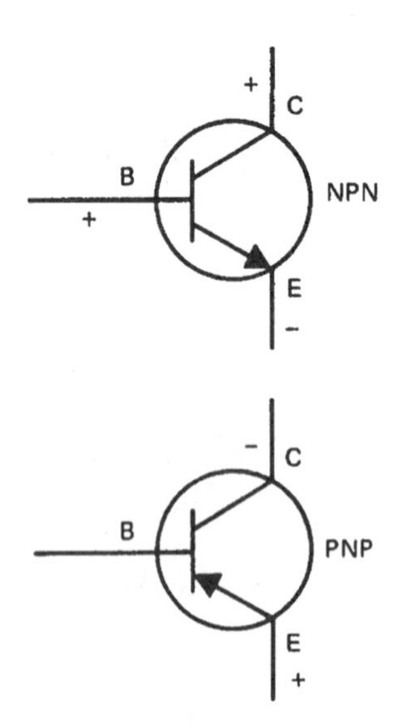

Fig. 2-13 Schematic symbols of transistors (reprinted, by permission, from Herman, *Electronics for Industrial Electricians,* second edition, Copyright 1990 by Delmar Publishers Inc.).

rection ("reverse bias"), the reversed polarity pulls the electrons away from the P-N junction on the N-side, and the holes away from the junction on the P side (Fig. 2-13). This leaves nothing to carry the current across the junction boundary. So the diode effectively blocks the passage of current.

Of course nothing is perfect, and diodes are no exception. If enough reverse bias voltage is applied, it will punch through the diode and flow in the opposite direction, destroying the diode in the process (Fig. 2-14). That why alternator diodes that convert alternating current (AC) to direct current (DC) are usually damaged when the battery is hooked up backwards, or when someone uses a battery

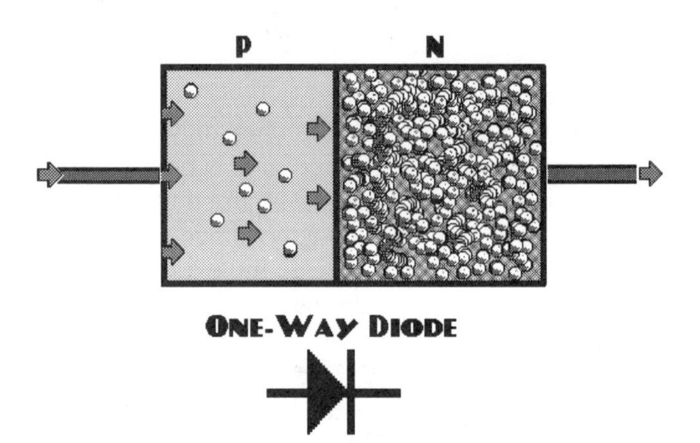

Fig. 2-14 Diodes consist of a P-N or N-P junction that allows current to flow only one way.

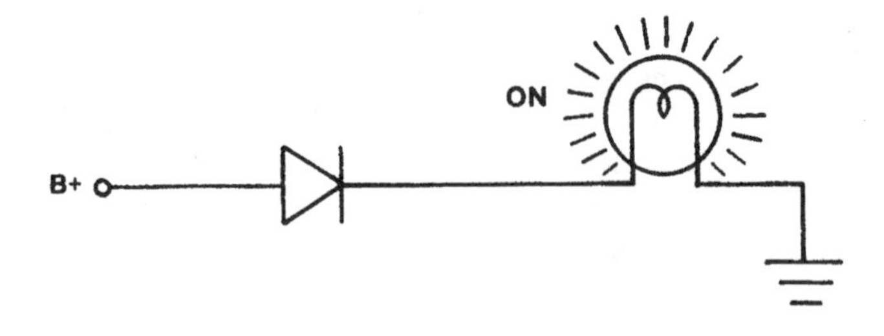

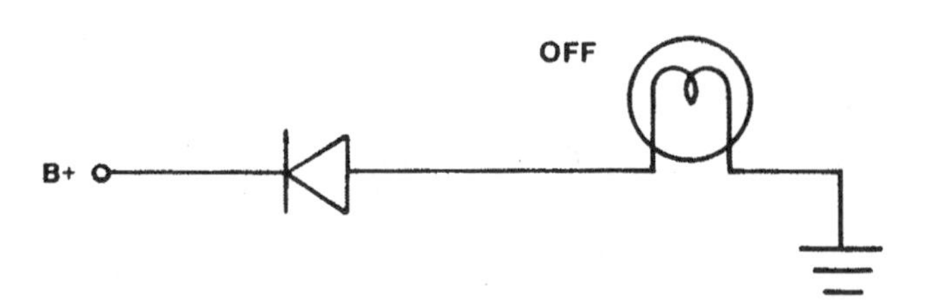

Fig. 2-15 On top, the diode is allowing current to pass through the circuit and illuminate the bulb. On the bottom, the direction of the diode is reversed and prevents the flow of current (reprinted, by permission, from Santini, *Automotive Electricity and Electronics, 2E.* Copyright 1992 by Delmar Publishers Inc.).

charger that puts out too much voltage. The voltage at which a diode breaks down is called the "peak inverse voltage." Exceed it and the diode is ruined.

The same precautions apply to transistors. All semiconductors are designed to handle limited current loads, even the large power transistors used in ignition modules and distributorless ignition systems. If the maximum current load is exceeded, it usually ruins the component.

Zener Diodes

"Zener" diodes are a special type of diode designed to flow backwards under certain circumstances. The semiconductor material is more heavily doped so the P-N junction will allow reverse current flow without damage when a certain voltage level is exceeded (Fig. 2-15). These are sometimes called "avalanche" diodes because they don't pass reverse current until a certain voltage is achieved. Then they open up all at once. Zener diodes are used in voltage regulators to provide voltage overload protection.

LED

Another special type of diode is the "Light Emitting Diode" (LED). The crystal in this type of diode glows red when current is applied, much like the filament in a light bulb. But since there's no filament to burn out, LEDs tend to be long lived. LEDs also require less current than conventional bulbs. LEDs are used for digital displays in some test instruments, and in some CHMSL (center high-mounted stop light) brake lights. LEDs are also used to trigger ignition and/or injector pulses in some engines (such as Chevrolet's Opti-Spark Ignition System or Nissan's ECCS system).

INTEGRATED CIRCUITS

As solid-state electronics evolved, components were miniaturized so more and more components could be crammed onto smaller and smaller circuit boards. The "integrated circuit" (IC) allowed a complete electronic circuit consisting of transistors, diodes, and resistors to be formed on a single silicon wafer or "chip."

Invented by Jack Kilby of Texas Instruments in 1958, ICs are made by plating and etching multiple layers of P-type and N-type material over one another. Thus forms interconnecting circuits on a tiny wafer that may be no larger than a small letter "o" on this page! IC technology has allowed engineers to pack the electronic equivalent of 6,000 lbs. of spaghetti into a 1-lb. box!

Under a microscope, these interconnecting circuit elements resemble a complex street map (Fig. 2-16). Keep in mind, however, that it's a three-dimensional map. The little "roads" that crisscross each other are several layers deep with numerous underlying interconnections, like the overpasses and access ramps on an intercity expressway. But the layers are extremely thin, only a few microns thick at most.

These integrated circuits may be combined like building blocks to create even larger more complicated chips. For example, "very large-scale integration" (VLSI) refers to ICs that contain 50,000 or more such building blocks.

The silicon chip that contains the integrated circuits may be packaged several different ways. One is to mount the chip on a flat ceramic or metal plate so it can be installed on a printed circuit board. This is done by soldering the IC's leads to the circuit board.

Another type of packaging is to encapsulate the chip with plastic. The surrounding plastic not only supports the chip but protects it against corrosion. An example would be Ford's Thick Film Integrated (TFI) ignition module.

PROM Chips

Replaceable IC's found inside electronic control modules (ECM's) and other electronic devices are usually square or rectangular in shape. They have anywhere from 16 to 28 pins in rows underneath like

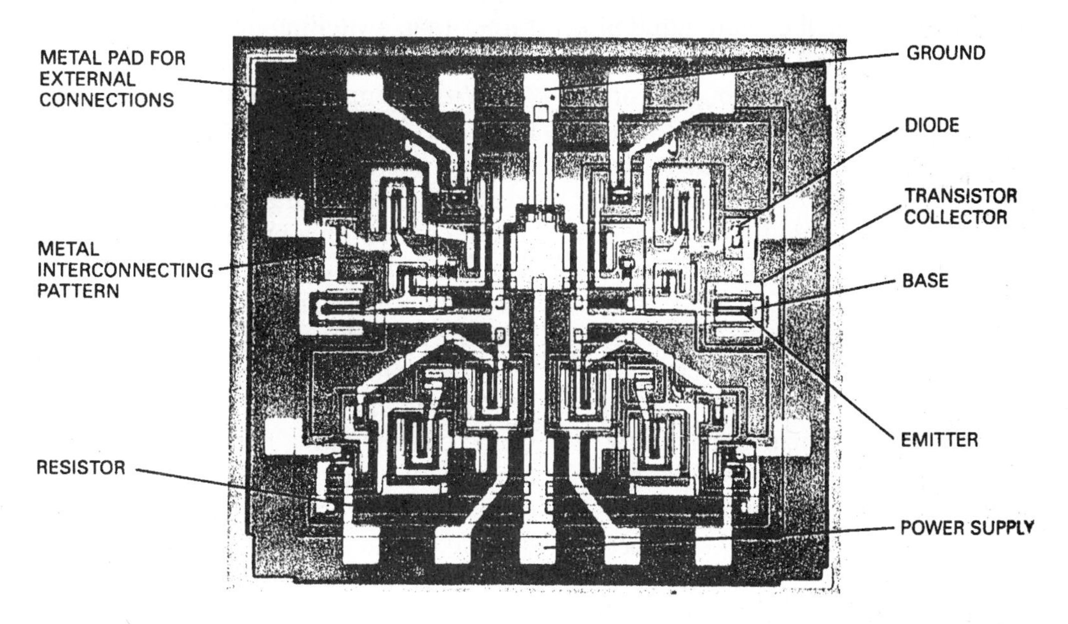

Fig. 2-16 An integrated circuit consists of various circuit elements etched on a silicon wafer (courtesy Texas Instruments).

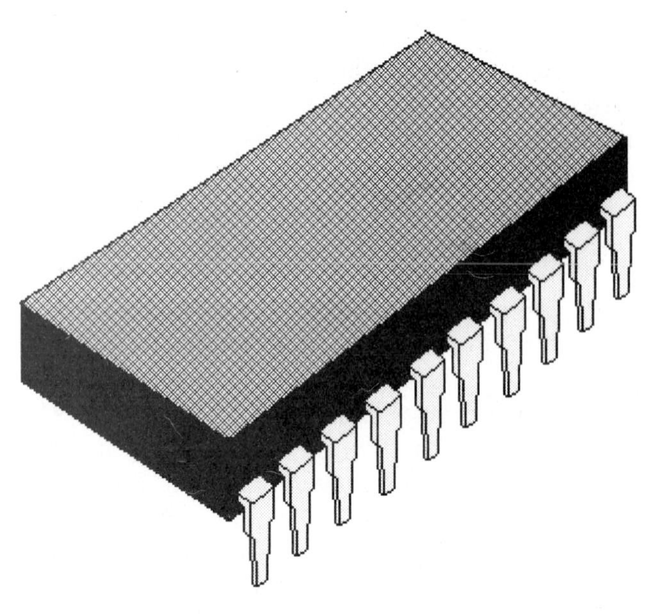

Fig. 2-17 Integrated circuits such as this PROM (Program Read Only Memory) chip can be programmed with the instructions to control other circuit functions.

the legs of a caterpillar (Fig. 2-17). The pins allow the IC to be plugged into a circuit board. This eliminates the need for soldering and allows the IC to be removed and replaced if necessary. Examples of this type of IC include the "Program Read Only Memory" (PROM) chips and Electronically Erasable Program Read Only Memory (EEPROM) chips used in General Motors ECMs and other automotive computers. The PROMs or EEPROMs are used to program the engine's fuel and ignition curves as well as other emission functions. The PROM or EEPROM chip calibrates the engine control computer for a specific vehicle application. This is necessary because different accessories or drivetrain combinations can affect a vehicle's emission performance. An engine with an automatic transmission, for example, may require a different spark curve than a manual transmission.

PROM chips cannot be reprogrammed, but they can be replaced with a new PROM that contains different or updated instructions. Replacing a PROM with one that contains updated information is sometimes necessary to cure a specific driveability or emissions problem. If a vehicle manufacturer believes a problem is due to the computer's calibration, it may issue a Technical Service Bulletin (TSB) describing the problem and announcing the availability of a new or updated replacement PROM. The PROM can then be obtained from the new car dealer and installed in the vehicle's computer.

With EEPROMS, physical replacement isn't necessary to recalibrate a computer. An EEPROM can be reprogrammed by downloading updated instructions through the computer's diagnostic hookup. To prevent "unauthorized" tampering with the EEPROM's instructions, special access codes are required. These are only available to new car dealers at this time.

IC RELIABILITY

One of the major advantages of using ICs rather than conventional circuit boards with soldered individual circuit components is reduced power consumption. An IC can do the same job on a fraction of the amperage required to run a similar large-scale circuit. But because the circuit elements in an IC have been reduced to such tiny proportions, there's a limit to how much voltage and current they can safely handle. The limit for most ICs is 20 volts or less, and current ratings are measured in milliamps.

The reason for the power limitation is heat. The more current an electronic circuit carries, the more heat it generates. If the heat is concentrated in a tiny device such as an IC, it can produce temperatures high enough to damage or destroy the chip. That's why the IC's role in most electronic devices is usually limited to information processing or controlling rather than switching or handling the voltages and currents that make things happen.

With an ignition module or a voltage regulator, the control output from the IC is used to switch a power transistor on and off. The power transistor can safely handle larger voltages and currents, but the IC cannot. This brings us to the question of reliability. As long as an IC is operated within its voltage and current limits, it is generally more reliable than a large-scale electronic circuit with individual soldered components.

The main reason why most electrical and electronic circuits fail is because of breaks in connections between circuit components rather than the outright failure of individual components. The overlapping circuit elements within an IC are deposited on top of one another by a plating and etching process, so there's little chance of anything wiggling or vibrating loose. And if the IC is encapsulated in plastic, the circuits are protected against environmental contamination and corrosion as well. But that doesn't make ICs immune to trouble because the pin connectors that link the chip to the outside world are the weak point.

Soldered pin connections can and do break loose, causing opens that result in circuit failures. The pins on the push-in variety of ICs can also be bent or broken during installation. That's why GM recommends using a special tool to remove and install PROM chips in their ECMs. The IC pins must also fit tightly in the circuit board receptacle and be corrosion-free.

ICs are most vulnerable, however, to damaging voltage spikes. Because the overlapping circuit layers on the chip are so thin, it doesn't take much of an overload to destroy a connection. And once a chip is damaged, it's history.

Voltage spikes can occur when an electrical connection is broken while a circuit is still hot. An example would be unplugging an engine sensor or wiring connector while the ignition is on. The sudden break in the circuit causes the voltage to momentarily surge, creating a spike that may be as high as 50 or 60 volts in a circuit that normally sees only 5 or 12 volts. The surge occurs because the electrons want to keep flowing as the circuit is being broken. They pile up and try to push their way across the gap, creating a transient voltage spike that can fry a chip. The car makers "harden" their electronic circuits by building in voltage overload protection, but safeguards may not be enough to protect the chips under all circumstances. So never disconnect or connect any wiring connector or electronic component when the ignition is on because some components receive current directly from the battery.

Voltage surges and spikes can also be created by arc and MIG welding equipment. As a precaution, the battery should be disconnected prior to welding. If welds are being made in close proximity to an electronic control module or other device, the module or device should also be unplugged for additional protection.

Potentially damaging voltage overloads can also be created by electrostatic discharges. Sliding across a vinyl seat can build up a static electrical charge of thousands of volts. The actual amperage of such charges isn't much, but the voltage can be high enough to produce a visible (and sometimes painful) shock when you touch a conductive surface. If the thing you touch happens to carry the voltage back to a computer chip, the spark can literally punch a hole right through the microscopic circuit layers of the IC, leaving it permanently maimed.

When handling any type of chip-based electronic module, therefore, take care to avoid electrostatic discharges that might damage the circuitry. You can minimize the danger by "grounding" yourself with an anti-static wrist strap, by placing modules and computers on non-conductive, anti-static mats, and/or by wearing cotton-fiber clothing.

Testing and Diagnosis

There's no practical way for a technician to test or repair faulty integrated circuit chips inside a control module. Your only option is to replace the module if it isn't working correctly. (Unless, of course, the problem is really module calibration and there is an updated PROM chip for the module, or new programming for an EEPROM.) But first you have to isolate the problem to the module by following a detailed, step-by-step diagnostic procedure (diagnosing driveability problems is covered in Chapter 19).

With transistors and diodes that are not part of an integrated circuit, it is possible to replace individual circuit components in some instances (the diodes in an alternator, for example). But in most instances, the labor to do so is too time-consuming, so the entire unit is replaced if a circuit component fails.

CHAPTER 3

Understanding Electrical Circuits and Wiring Schematics

To diagnose and repair complex driveability and emissions problems on today's vehicles, you have to understand electrical circuits, know how to troubleshoot and repair electrical problems, and read wiring diagrams. With that in mind, let's take a closer look at circuits.

CURRENT PATH

Since all electrical circuits require a continuous path for current to flow, most automotive electrical systems use the steel body for the "ground" path. The wiring makes up the other half of the circuit. This approach eliminates the need for two wires to many components. A side indicator light, for example, needs only one wire because it is grounded to the body. Other components may be electrically connected to the vehicle by two or more wires, but the ground wire may only run a short distance before terminating at a common ground point.

It's important to remember that the ground side of a circuit is just as important as the "hot" or positive side of the circuit. If the current can't flow from the battery, it won't flow. We tend to think of the positive side of the battery as being the hot side that supplies the current. It's actually the negative side of the battery that provides the electrons that flow through the ground path and return to the positive side.

As we said in the last chapter, the body on all modern vehicles is connected to the negative battery terminal. The wiring is connected through the ignition switch and fuse box (or power distribution center) to the positive battery terminal. As stated above, we tend to think of current flowing from the positive "hot" wire to a component. It actually flows to the component through the ground (negative to positive). Electric current flows from negative to positive because the negatively charged electrons leave the battery through the negative terminal, travel through the body to the various circuits and components, then return to the battery through the so-called "hot" wires in the electrical system.

It may sound confusing, but a "hot" wire in a negative ground electrical system is really a return wire that completes a circuit. A "hot" wire always sparks when grounded because it completes a direct path back to the battery. It will also show battery voltage when checked with a grounded voltmeter (Fig. 3-1).

Opens

When a circuit is not complete (no continuity), it is said to be "open". Electricity obviously can't flow through an open circuit because there's no return path back to the power source. An open can be created intentionally by using a switch, circuit breaker, or relay to turn a circuit off. Or it may be unintentional due to a broken wire, or a loose or corroded connection.

Opens can be detected a variety of ways:

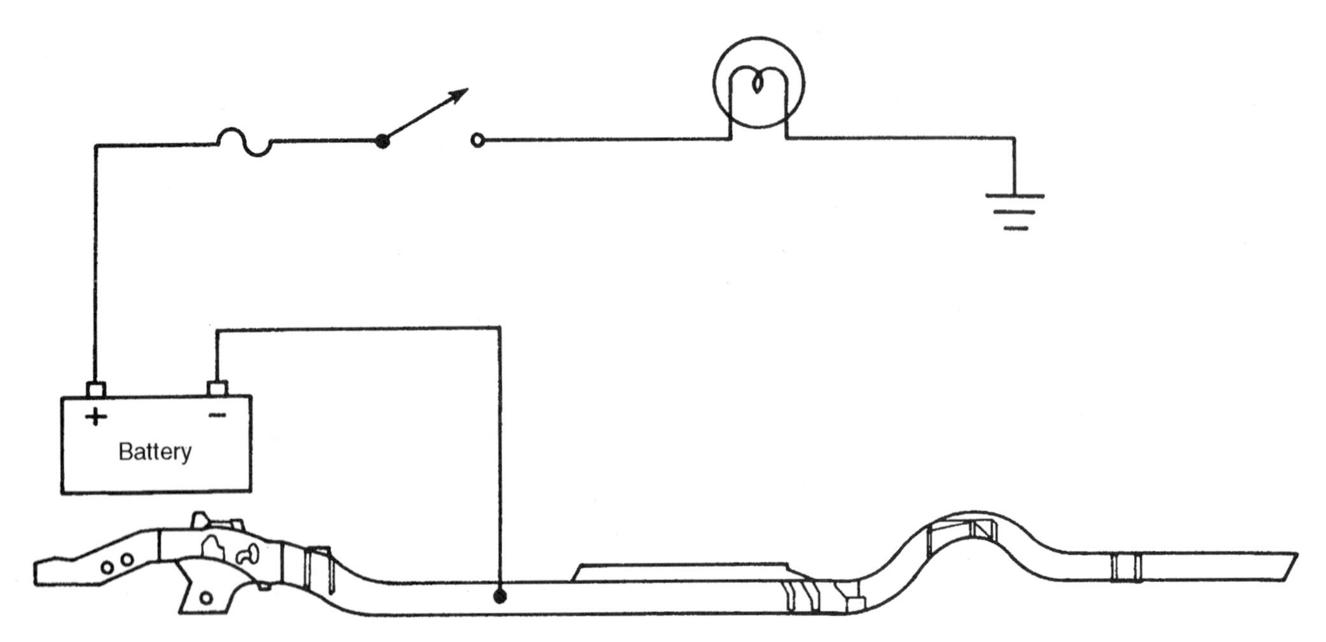

Fig. 3-1 The vehicle's frame or unibody serves as the ground for many electrical circuits (reprinted, by permission, from Santini, *Automotive Electricity and Electronics, 2E.* Copyright 1992 by Delmar Publishers Inc.).

- Checking wiring continuity with a continuity tester (no continuity would indicate an open).

  Note: A simple continuity tester can be made by connecting a low-wattage 12-volt bulb or buzzer to a pair of test leads. Adding a battery will make it self-powered.
- Measuring resistance with an ohmmeter. Infinite resistance would indicate an open.

  Note: Never use an ohmmeter to measure a "live" circuit as doing so may damage your meter.

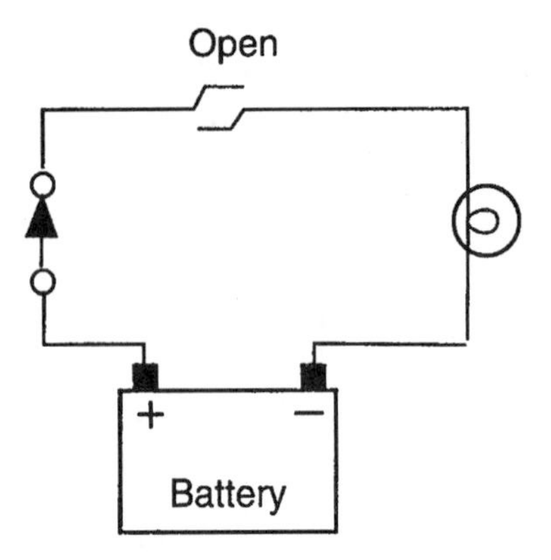

Fig. 3-2 An open anywhere in a circuit stops the flow of electrical current (reprinted, by permission, from Hollembeak, *Automotive Electricity and Electronics.* Copyright 1994 by Delmar Publishers Inc.).

- Checking a "hot" circuit for voltage with a voltmeter. No voltage would indicate an open.

  Note: When diagnosing electronic circuits, computer modules, sensors, etc., a high impedance (10 megohm) digital voltmeter or multimeter must be used to protect electronic components against damaging voltage overloads.
- Measuring current flow with an ammeter. No current flow would indicate an open.
- Visually inspecting a wiring circuit for obvious damage, breaks, or loose or corroded connections (Fig 3-2).

Shorts

A "short" occurs when a portion of an electrical circuit is bypassed unintentionally (Fig. 3-3). It's called a short because it creates a shorter return path for the current to follow (Fig. 3-4). An example of a short would be a break in the insulation on a wire touching metal.

When a short occurs, two things happen. The first is that electricity always prefers the path of least resistance. This is usually via the short rather than through the electrical circuit. Second, the absence of resistance allows a runaway current that can quickly overheat and melt the wire, possibly starting a fire. To guard against such mishaps, various

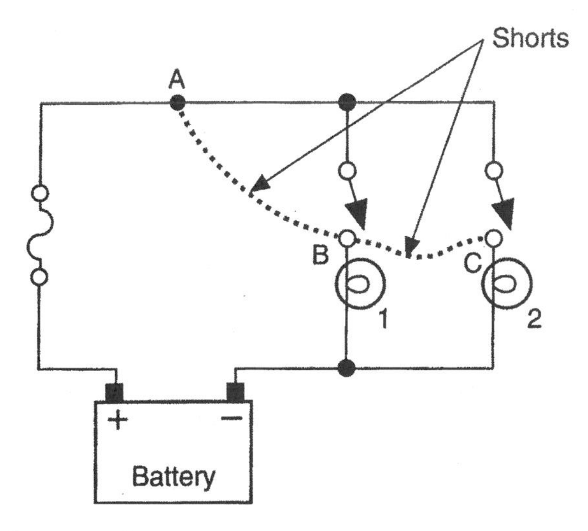

Fig. 3-3 Shorts anywhere in an electrical circuit can allow the current to take a "short cut" along a different path (reprinted, by permission, from Hollembeak, *Automotive Electricity and Electronics*. Copyright 1994 by Delmar Publishers Inc.).

types of safety devices are used to protect electrical circuits.

Shorts can be detected by:

- Checking between the circuit and ground and/or adjacent circuits for continuity with a continuity tester or ohmmeter. Continuity or a measurable resistance reading other than infinity would indicate a short.
- Checking for the presence of voltage in adjacent wiring circuits that should not have voltage with a voltmeter. A voltage reading in a circuit that is supposed to be "off" would indicate a short.

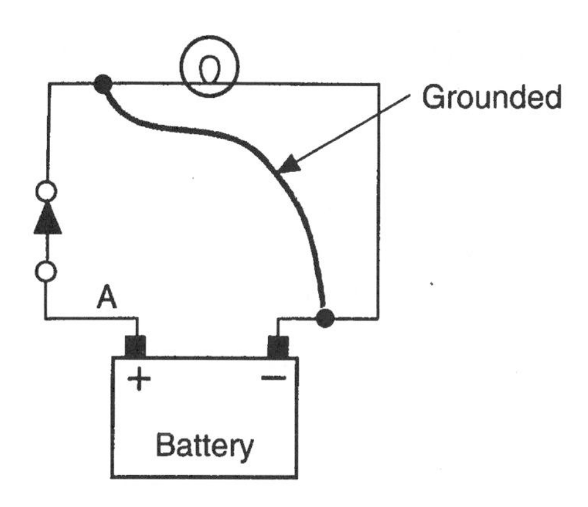

Fig. 3-4 A short to ground allows the current to bypass the normal load in the circuit (reprinted, by permission, from Hollembeak, *Automotive Electricity and Electronics*. Copyright 1994 by Delmar Publishers Inc.).

- Inspecting the fuse box or power distribution center for blown fuses, or the wiring circuit for failed "fusible links."
- Inspecting wiring for cracked, burned, melted, or bare wiring.
- Checking for misrouted wires or reversed connectors.

CIRCUIT PROTECTION

To protect electrical circuits from shorts or overloads, any of the following devices may be used:

- A "fuse" is the most simple form of protection (Fig. 3-5). A fuse contains a piece of wire that melts at low temperature. When the flow of current through a circuit approaches the limit of the fuse, the fuse wire melts. This opens the circuit and stops the flow of electricity. The amp rating of any given fuse is always clearly marked on the fuse. It is determined by the vehicle manufacturer according to the size of wiring used and the electrical loads the circuit is designed to handle. Fuses are usually located in a fuse box under the dash (Fig. 3-6) and/or a power distribution center under the hood. Some devices or accessories—particularly those installed after the vehicle was manufactured, such as auxiliary lights, radios, alarm systems, cellular phones, aftermarket air conditioning, etc.—may also have an "in-line" fuse in the wiring for protection.

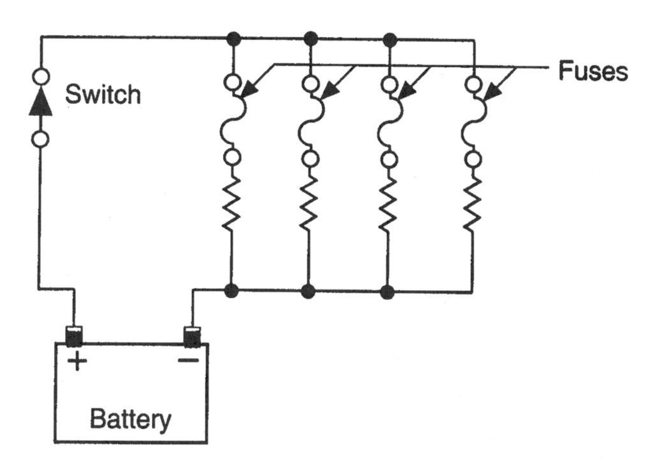

Fig. 3-5 Fuses are usually in series with the individual circuits they protect (reprinted, by permission, from Hollembeak, *Automotive Electricity and Electronics*. Copyright 1994 by Delmar Publishers Inc.).

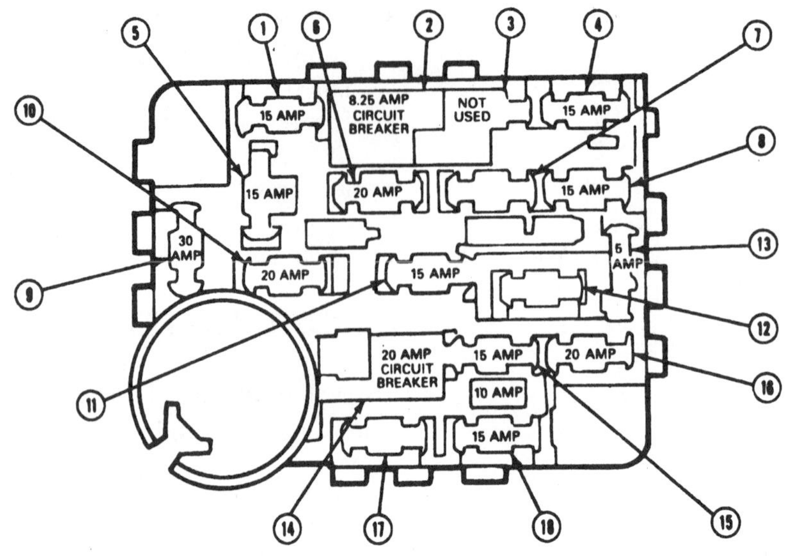

Cavity Number	Fuse Rating	Color	Circuit Protected
1	15 Amp	Light Blue	Stoplamps, Hazard Warning Lamps, Speed Control
2	8.25 Amp Circuit Breaker		Windshield Wiper, Windshield Washer Pump, Interval Wiper, Washer Fluid Level Indicator
3	Spare		Not Used
4	15 Amp	Light Blue	Tail Lamps, Parking Lamps, Side Marker Lamps, Instrument Cluster Illumination Lamps, License Lamps
5	15 Amp	Light Blue	Turn Signal Lamps, Backup Lamps, Fluids Module, Rear Window Defroster Relay
6	20 Amp	Yellow	A/C Clutch, Luggage Compartment Lid Release, Speed Control Module, Clock/Radio Display, A/C Throttle Positioner, Day/Night Illumination Relay
7	Spare		Not Used
8	15 Amp	Light Blue	Courtesy Lamps, Key Warning Buzzer, Radio, Power Mirror
9	30 Amp	Light Green	Heater Blower Motor
10	20 Amp	Yellow	Flash-to-Pass, Low Oil Warning Relay
11	15 Amp	Light Blue	Radio, Tape Player, Premium Sound, Graphic Equalizer
12	Spare		Not Used
13	5 Amp	Tan	Instrument Cluster Illumination Lamps, Radio, Climate Control, Ash Receptacle Lamps, "PRNDL" Lamp
14	20 Amp Circuit Breaker		Power Windows
15	15 Amp	Light Blue	Fog Lamps
16	20 Amp	Yellow	Horn, Cigar Lighter
17	Spare		Not Used
18	15 Amp	Light Blue	Warning Indicator Lamps, Throttle Solenoid Positioner, Low Fuel Module, Dual Timer Buzzer, Tachometer, Engine Idle Track Relay, Fluid Module/Display

K8719-F

The fuses and circuit breakers are color-coded by amp rating.

The locations and values of the fuses and circuit breakers not contained in the panels are shown in the following chart.

Circuit	Circuit Protection and Rating	Location
Headlamps and High Beam Indicator	22 Amp. CB	Integral with Lighting Switch
Heated Rear Window	16 GA Fuse Link	Engine Compartment
Power Windows, Power Seat, Power Door Locks	20 Amp. CB	Starter Motor Relay
Load Circuit	Fuse Link	In Harness
Engine Compartment Lamp	Fuse Link	In Harness
Convertible Top	25 Amp. CB	Lower Instrument Panel-Reinforcement

CK4356-J

Fig. 3-6 Fuses and circuit breakers (courtesy Ford Motor Corp.).

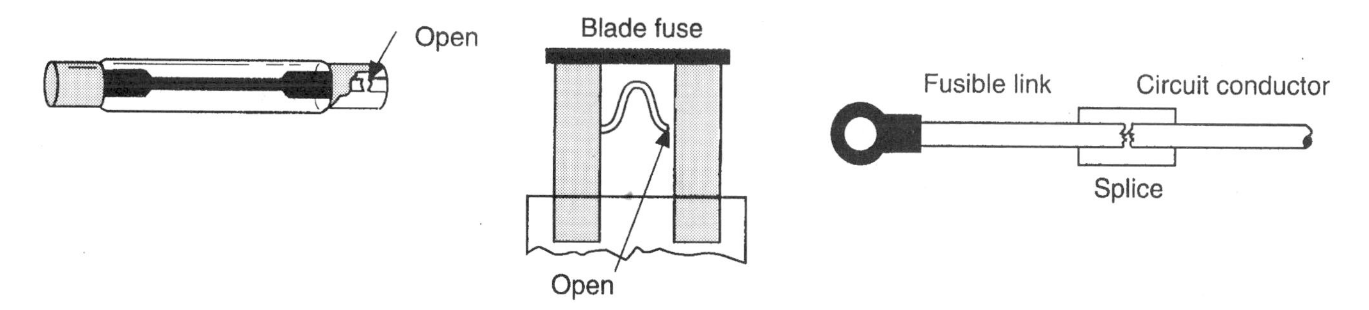

Fig. 3-7 When a fuse is overloaded, a metal link inside melts and opens the circuit, protecting the circuit against damage (reprinted, by permission, from Hollembeak, *Automotive Electricity and Electronics.* Copyright 1994 by Delmar Publishers Inc.).

Never replace a fuse with one of greater capacity. Always use a fuse with the same rating as the original—or as specified on the fuse panel in a shop manual (in case someone else installed the wrong fuse).

If a fuse is blown, find out why before you replace it (Figs. 3-7 and 3-8). Occasional overloads that result from unusual operating conditions can sometimes cause a fuse to fail. But in most instances, fuses don't blow unless there's a short or other electrical problem in the wiring or a component. Replacing the fuse may restore power to the circuit temporarily, but unless the underlying problem is diagnosed and repaired, the fuse will likely fail again.

- A "fusible link" is a length of special wire designed to melt (like a fuse) when a circuit is overloaded. Fusible links are sometimes used in ignition circuits and other circuits that carry high amperage. The location of a fusible link will be specified in a vehicle's wiring diagram. You can tell if a fusible link has burned out by noting the condition of the insulating tape wrapped around it. If the tape appears to be blistered, the wire inside has burned out. Most late-model vehicles now use replaceable fuses rather than fusible links for circuit protection. If the vehicle has an underhood power distribution center, chances are it does not have fusible links in the wiring (Fig. 3-9).
- A "circuit breaker" consists of a bimetallic switch that breaks the circuit when an overload occurs. When the amount of current flowing through the circuit exceeds the built-in limit of the circuit breaker, the bimetallic switch heats up and pulls opens the contact points. Once the current stops, the circuit breaker begins to cool off and eventually recloses. If the circuit breaker opened because of a temporary overload, the circuit will resume functioning normally. But if

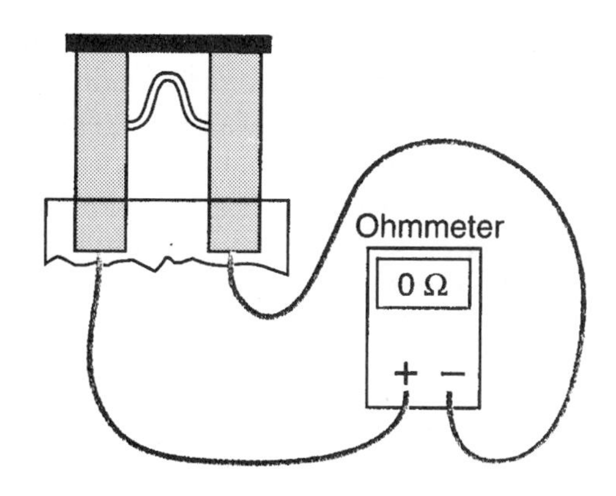

Fig. 3-8 An ohmmeter can be used to test a fuse. A good fuse will have zero resistance (reprinted, by permission, from Hollembeak, *Automotive Electricity and Electronics.* Copyright 1994 by Delmar Publishers Inc.).

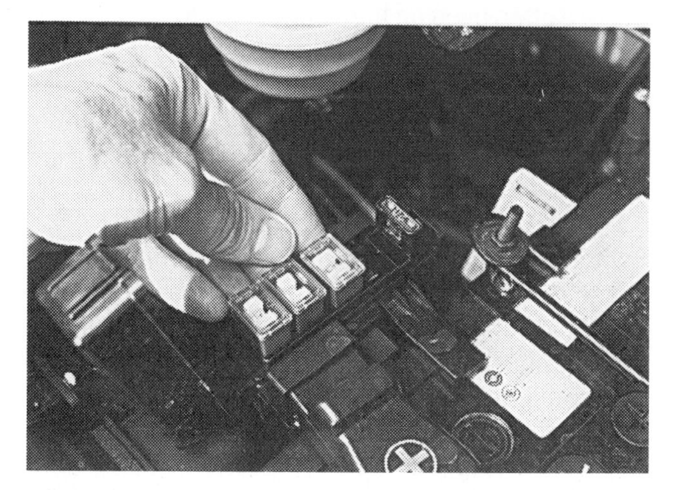

Fig. 3-9 Fusible links on most late-model cars have been replaced by fuses in a "power distribution center" in the engine compartment.

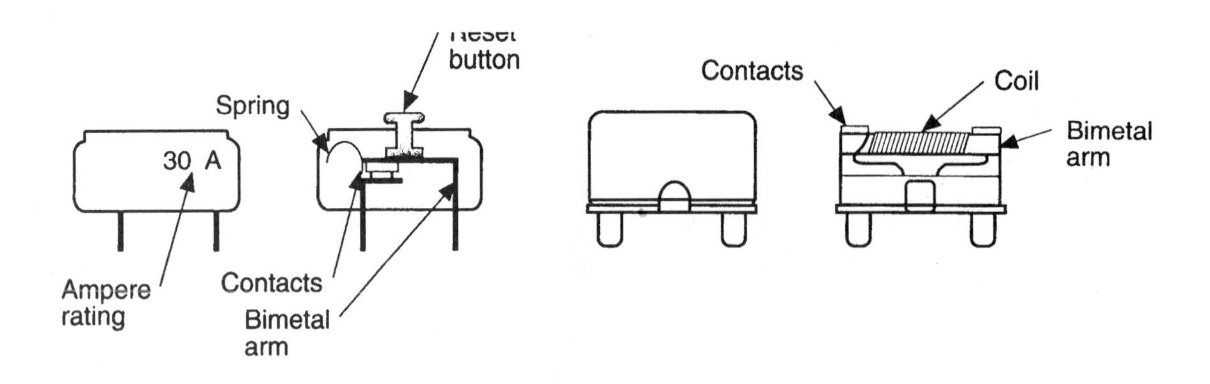

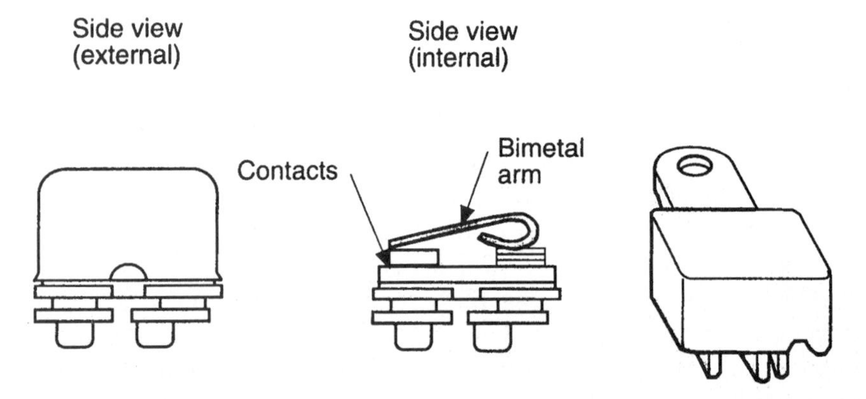

Fig. 3-10 Some circuit breakers (such as those at the top) are "noncycling" and must be manually reset (A) or removed or disconnected from the power supply (B). Others (bottom) will reset automatically once the bimetal arm cools down (reprinted, by permission, from Hollembeak, *Automotive Electricity and Electronics*. Copyright 1994 by Delmar Publishers Inc.).

the overload persists, the circuit breaker will continue to cycle on and off. Circuit breakers are often used on high-amp-load circuits such as the headlights and air conditioner (Fig. 3-10).

SWITCHES

Switches are an important electrical component because they turn the current on or off to various circuits and components including the ignition circuit, fuel pump, starter motor, blower motor, A/C compressor, lights, and other electrical accessories.

Switches may also function as status or position sensors to keep the engine-control module or other modules informed about events that are taking place. The brake-pedal switch, for example, not only turns on the brake lights when the pedal is depressed but also signals the ECM that the vehicle is braking. This, in turn, may cause the ECM to disengage the torque converter lockup clutch, lean the fuel mixture, adjust spark timing, and so on. If the vehicle has antilock brakes, it may signal the ABS-control module to be prepared to modulate brake pressure if the wheels start to slip. Some other "special" switches that have such dual roles include the park/neutral switch and/or transmission gear position switch, wide-open throttle switch and/or idle switch on the throttle linkage, and brake-pressure-differential switch.

A switch may be normally open (NO) or normally closed (NC) depending on its function. The switch itself may have a lever, pushbutton, push-pull, or rotary knob that has two, three, or multiple positions. A simple two-position toggle switch, for example, is either on or off. A three-position toggle switch, on the other hand, is usually off in the middle position, but turns on one of two different circuits depending on which way it's flipped.

Some pushbutton switches are "momentary" switches because they only remain on as long as the button is held in (an electronic radio channel selection button, for example). Others have a detent that holds the button in (on) after it's been pushed. Pushing it a second time releases the button.

Some switches are also multi-function switches, such as the ignition switch (Fig. 3-11), combination turn-signal/headlight-dimmer/cruise-control switch found on many late-model cars and trucks, or a combination windshield-wiper/interval-wiper/washer switch.

The important thing to remember about switches is that problems within the switch itself, such as worn, dirty, or corroded contacts, can create problems for the circuit the switch controls. A faulty neutral/park safety switch, for example, can cause a no-start by preventing current from flowing through the ignition circuit. An open inertial safety switch on a fuel-injected car that's been tripped because of a hard jolt or accident will prevent the engine from starting because it won't pass current to the fuel-pump relay. A worn ignition switch may cause intermittent stalling and/or starting problems.

If an electrical problem can be isolated to a bad switch, then the only way to cure the problem is to replace the switch. Switch problems can be isolated by using jumper wires to bypass the switch. If the problem goes away when the switch is bypassed, then the switch is at fault and needs to be replaced.

The switches for the various electrical circuits and accessories come in a variety of configurations. Each of has its own type of symbol on a wiring schematic. Some of the basic types you should know include:

- Single Pole Single Throw (SPST). This type of simple switch has only one set of contacts. "Single throw" means the switch is either on or off. Throwing the switch one way turns it on, the other way turns it off.
- Single Pole Double Throw (SPDT). This type of switch has a double set of contacts. When the

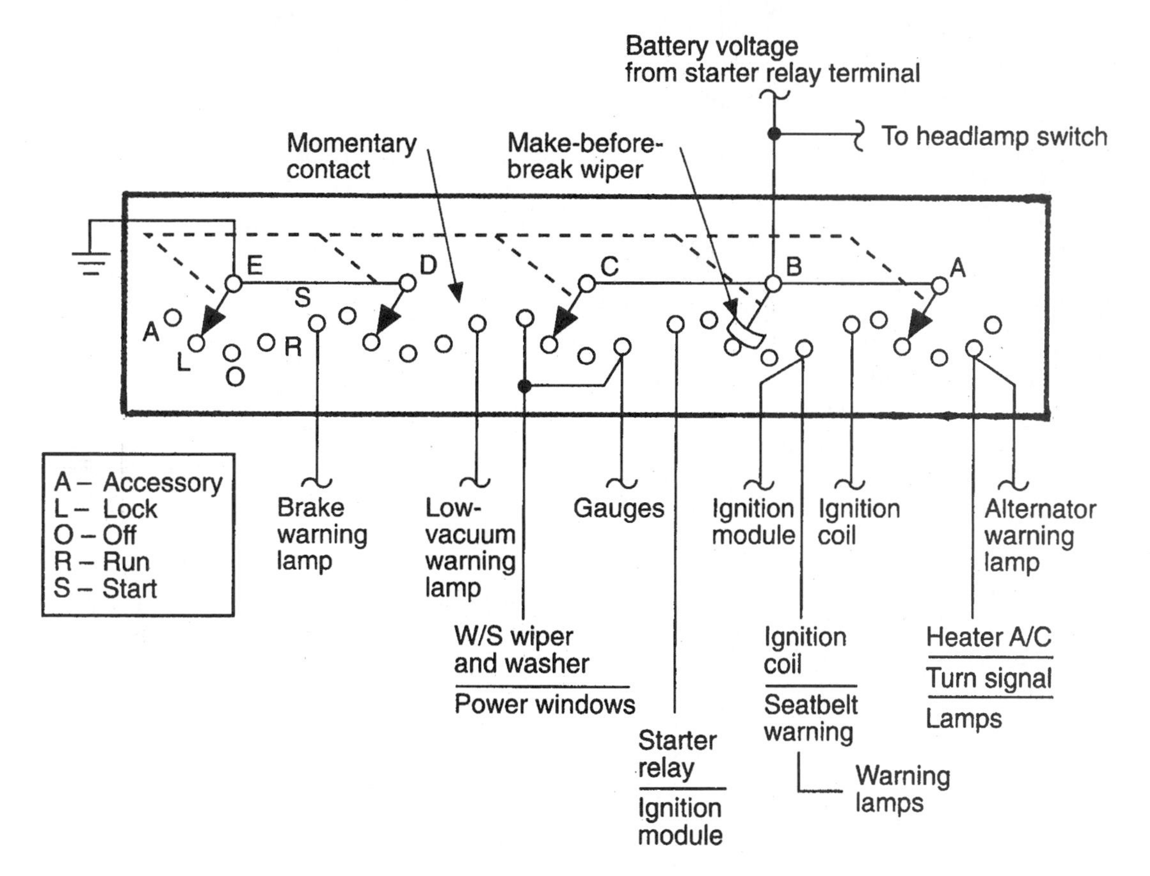

Fig. 3-11 A ignition switch is a good example of a multi-function switch (reprinted, by permission, from Hollembeak, *Automotive Electricity and Electronics.* Copyright 1994 by Delmar Publishers Inc.).

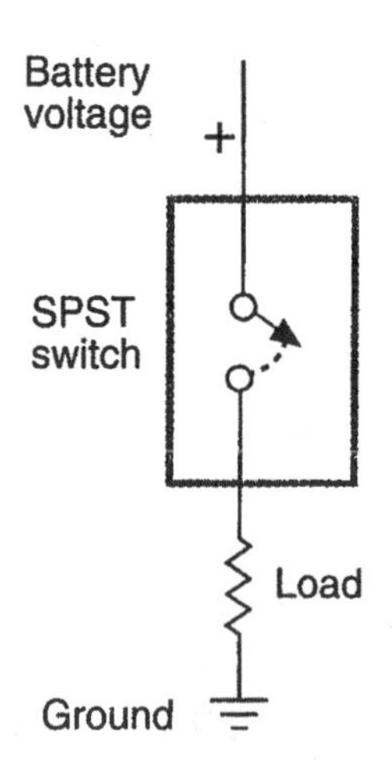

Fig. 3-12 A single pole single throw (SPST) switch (reprinted, by permission, from Hollembeak, *Automotive Electricity and Electronics.* Copyright 1994 by Delmar Publishers Inc.).

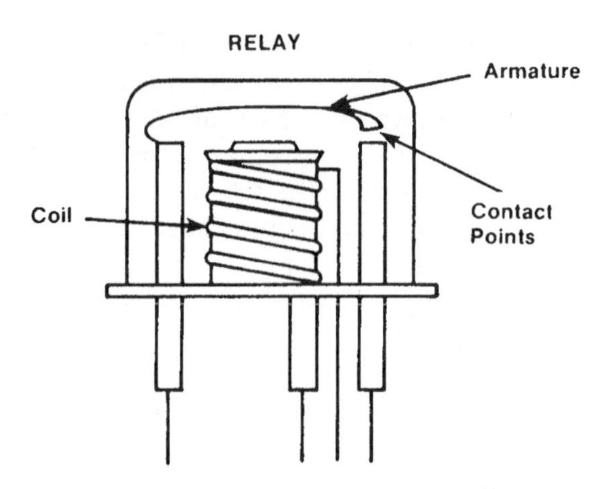

Fig. 3-13 A relay uses an electromagnet to close a set of contact points (reprinted, by permission, from Santini, *Automotive Electricity and Electronics, 2E.* Copyright 1992 by Delmar Publishers Inc.).

switch is thrown one way (Fig. 3-12), it opens the first set of contacts and closes the other. Reversing the position of the switch closes the first set of contacts and opens the second.

- Double Pole Double Throw (DPDT). This switch is like two single pole double throw switches in one. It works exactly the same as the SPST switch, except that it has two sets of contacts that open or close simultaneously when the switch is thrown.

Relays

A relay is nothing more than an electrically actuated switch that's used to turn a circuit or component on and off (Fig. 3-13). When voltage is supplied to the relay, a magnet pulls a set of contact point close and the relay routes power to the circuit or component (Fig. 3-14). When power to the relay's control circuit is cut off, the relay releases its grip on the contact points, the points open, and power is discontinued to the circuit or component. Some relays, on the other hand, are normally closed so energizing them opens the circuit.

Relays are used to power the electric fuel pump, fuel injectors, electric cooling fan, A/C compressor, headlights, horn, rear-window defroster and just about every other power accessory that pulls a high-amp load.

For example, when the ignition key is turned on, power may be routed to the fuel-pump relay. The ignition circuit is the control circuit in this case. The voltage from the ignition circuit flows through a small coil (actually an electromagnet) inside the relay. The magnetic field created by the coil pulls a set of contact points shut. This completes the main relay circuit and turns the fuel pump on. When the key is turned off, the relay points reopen, breaking the circuit to the pump. This shuts the pump off.

Why use a relay to turn a circuit on and off? Because relays are designed to handle higher current loads. Thus, a relatively small control current can be used to energize a relay that in turn provides a much higher current to a circuit or component.

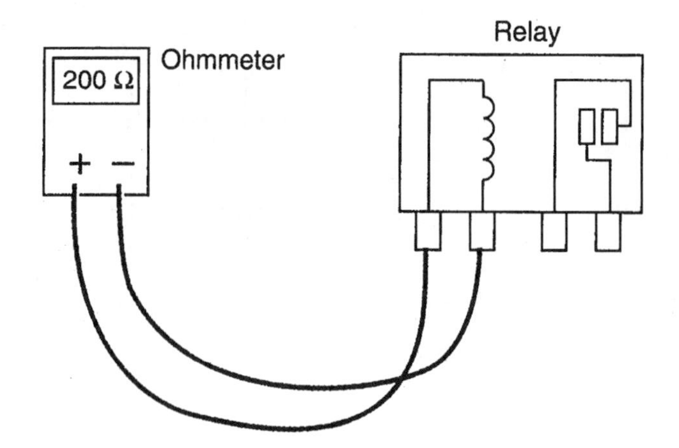

Fig. 3-14 The coil inside a relay can be tested with an ohmmeter. If the coil is open or resistance exceeds specifications, the coil is defective and the relay needs to be replaced (reprinted, by permission, from Hollembeak, *Automotive Electricity and Electronics.* Copyright 1994 by Delmar Publishers Inc.).

One way to tell if a relay is working or not is to listen for an audible click when power is applied. If voltage is reaching the relay, but there's no click, either the relay is defective or not grounded properly. Relays usually fail because the wiring connectors to the internal coil break or the contact points wear out or stick (Fig. 3-15).

Something else to note about relays: *they're seldom located anywhere near the device they operate.* So, you usually have to hunt them down by referring to a wiring diagram or component location chart. Relays may be scattered throughout a vehicle: under the hood, under the dash, under a seat, in the trunk, or behind a kick panel. Most relays are also not identified with anything other than an original equipment manufacture (OEM) part number, so make sure you have the correct relay for the application if it needs to be replaced.

Solenoids

Like relays, solenoids are also electrical switches or valves (Fig. 3-16). They can be used for a variety of purposes. Some solenoids (like the starter solenoid) are used to operate an electrical circuit like a relay. Some (like ported vacuum switches for exhaust gas recirculation (EGR) valves, or the charcoal canister purge valve) are used to open or block vacuum circuits. Some (like those in antilock brake systems or electronically controlled automatic transmissions) are used to open or block hydraulic

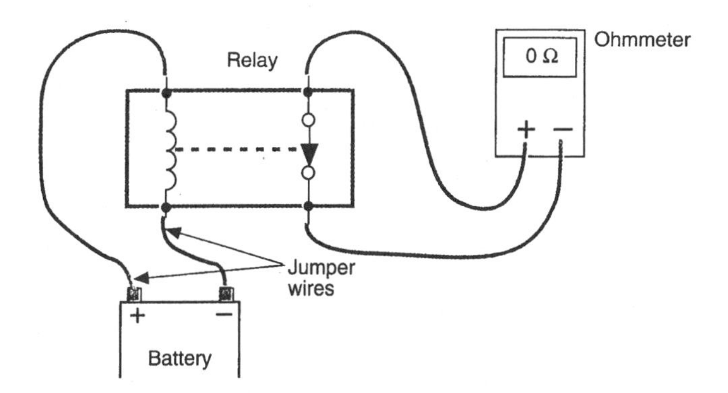

Fig. 3-15 The operation of a relay can be checked by applying voltage to the coil to see if it closes the contact points. An ohmmeter connected to the input and output terminals of the contact points should go from infinity (no voltage applied) to zero resistance (voltage applied) if the relay is good (reprinted, by permission, from Hollembeak, *Automotive Electricity and Electronics.* Copyright 1994 by Delmar Publishers Inc.).

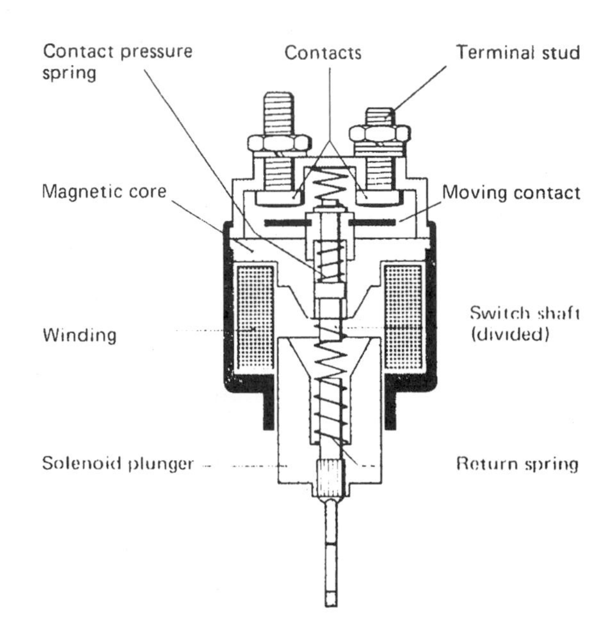

Fig. 3-16 A solenoid consists of a coil with a sliding pole piece or plunger that moves when voltage is applied to the coil (courtesy Bosch).

circuits. Others (such as mixture control solenoids in electronic carburetors) are used to open and close fuel metering circuits. An EGR valve solenoid may open and close a vacuum line to the valve that prevents EGR under certain operating conditions.

Fuel injectors are another device that contain a solenoid. The solenoid in the top of the injector is connected to the injector valve. When the solenoid is energized, it pulls opens the valve and allows pressurized fuel to spray out the injector nozzle.

Like relays, solenoids contain an electromagnetic coil. Except for starter solenoids, which are essentially relays, most solenoids move a "pole piece" when energized, instead of closing a set of contact points. The pole piece is nothing more than a small piece of iron that moves when pulled upon by a magnetic field. When the solenoid is energized, it pulls the pole piece up. When the solenoid's control voltage is cut, the spring-loaded pole piece snaps back to its original rest position.

With a fuel injector, the pole piece is connected to the pintle valve that opens the injector nozzle. With a ported vacuum switch, the pole piece uncovers or blocks vacuum passageways as it moves up and down.

Solenoids can be cycled on and off very rapidly. The buzzing that a fuel injector makes is nothing more than the rapid cycling of the solenoid and pintle valve. With antilock brake systems, the ABS solenoids alternately hold, release, and reapply brake pressure in the brake lines. These solenoids can cycle on and off 4 to 10 times a second depending on the system.

Another feature about solenoids is that the pole piece doesn't necessarily have to move all the way from its rest position to a fully retracted position. By varying the voltage to the solenoid, the pole piece can be slid partially open. This ability enables a solenoid to provide multiple operating positions, rather than just on or off. Bosch, for example, uses a three-position solenoid in its antilock brake systems.

How do you check a solenoid? Usually by measuring its resistance with an ohmmeter. If the reading isn't within specs, the solenoid coil is either open or shorted. You can also listen for a click when the solenoid is energized. No click or movement means it needs to be replaced.

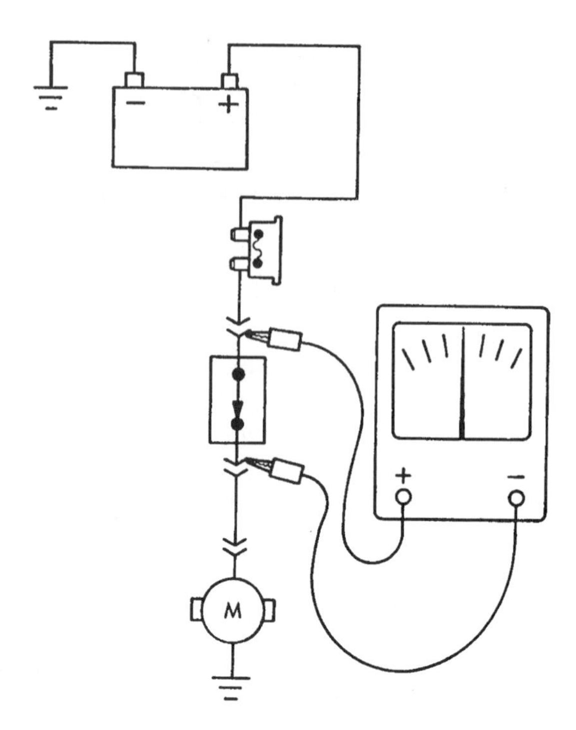

Fig. 3-17 Measuring the voltage drop across circuit connections is a good way to find loose, corroded, or damaged connectors (courtesy Chrysler).

VOLTAGE REQUIREMENTS

Every electrical device requires a minimum voltage to operate. A light bulb will glow with reduced brilliance as the voltage drops. But for some components there is a minimum threshold voltage below which the component may not work properly or at all. A starter motor, for example, will crank the engine more slowly if battery voltage is low, but may not engage or crank at all if the voltage is below a certain threshold (usually less than 10 volts). Minimum threshold voltage is especially critical for such components as solenoids (which need a certain amount of voltage to overcome spring resistance), relays, timers, buzzers, horns, fuel injectors (which are solenoids, too), and most electronics (the ignition module, engine computer, ABS control module, radio, etc.).

Checking the load point for full battery voltage will tell you whether or not sufficient voltage is reaching the component in question. Voltage checks should be made with a digital voltmeter or multimeter. A reading within a few tenths of a volt of base battery voltage usually means the circuit is okay.

Voltage Drop

Low circuit voltage will result if there's excessive resistance at any point in the circuit (Fig. 3-17). Low voltage usually indicates a weak or corroded connector, a faulty switch or relay, or poor ground connection. To find the point of high resistance, a voltmeter can be used to perform a "voltage drop test" at various points throughout the circuit. A drop of more than one-tenth (0.1 v) across any connector, switch, ground, or relay indicates a problem.

To test the amount of voltage drop across a wiring connector, for example, one of the voltmeter's test leads is attached to the wire on one side of the connector. The other test lead is attached to the wire on the opposite side of the connector. The test lead connections will probably require backprobing the connector (if room permits) or inserting pins through the insulation of each wire to make contact. If the voltmeter reads 0.1 volts or less, the connection is okay. But a greater reading indicates excessive resistance in the connector. If the reading is high, the connector should be opened up, inspected, cleaned, reconnected and retested. If it still shows excessive resistance, replacement is recommended.

Sometimes undersized wiring can cause low voltage. If someone has rewired a circuit or added an accessory (driving lights, stereo system, a rear-window defogger, etc.), the wires may not be a heavy enough gauge to carry the load. The higher the amp load, the larger the required gauge size for the wiring. The following list includes recommended wire gauge sizes:

Wire size	Amp Capacity
18	6
16	8
14	15
12	20
10	30
8	40
6	50

CONTINUITY

Every electrical circuit requires a complete circuit to operate. Voltage to the load won't do any good unless there is also a complete ground path to the battery. The ground path in the case of all metal-bodied cars is the body itself. In plastic-bodied cars, a separate ground wire is needed to link the load to the chassis. In either case, a poor ground connection has the same effect as an open switch. The circuit isn't complete so current doesn't flow.

To check wiring continuity, all that's needed is an ohmmeter or a self-powered test light (Fig. 3-18 and 3-19). An ohmmeter is probably the better of the two because it displays the exact amount of resistance between any two test points. A test light, on the other hand, will glow when there's continuity, but the intensity of the bulb may vary depending on the amount of resistance in the circuit. A trained eye can usually detect the difference, but an ohmmeter is more exact.

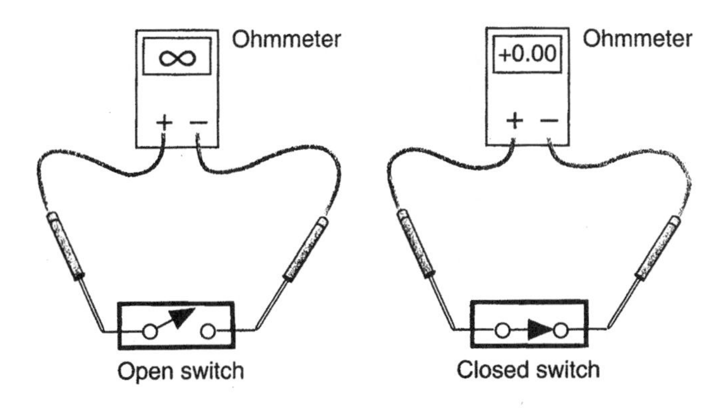

Fig. 3-18 The continuity (as well as the resistance) of a switch can be checked with an ohmmeter. The meter should read infinity when the switch is open, and zero when the switch is closed (reprinted, by permission, from Hollembeak, *Automotive Electricity and Electronics*. Copyright 1994 by Delmar Publishers Inc.).

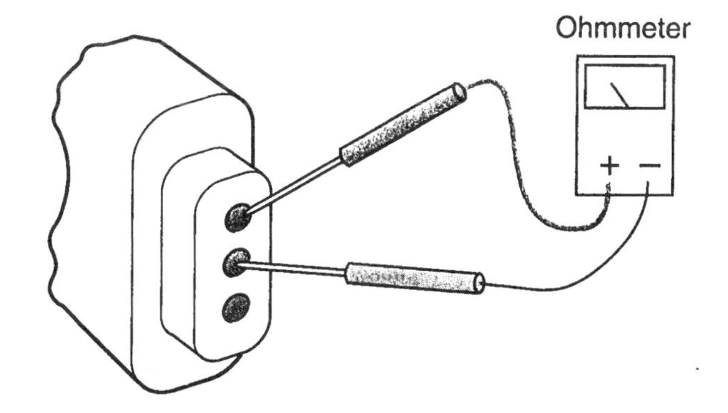

Fig. 3-19 Using an ohmmeter to check the continuity of various wiring circuits and connectors is a must to find electrical problems (reprinted, by permission, from Hollembeak, *Automotive Electricity and Electronics*. Copyright 1994 by Delmar Publishers Inc.).

An ohmmeter should never be used to check resistance in a "live" circuit. Doing so can damage the ohmmeter. Before testing the circuit's resistance or continuity, therefore, the circuit should be isolated by disconnecting it from its power source. Pulling the circuit's fuse will usually do the trick.

Ohmmeters are great for measuring resistance and checking continuity in normal electrical circuits. But care must be used when working on electronic components. An ohmmeter works by applying a small voltage through its test leads. This voltage may be enough to damage sensitive electronic components such as computer chips. Do not attempt to directly probe any module connectors or terminals unless a diagnostic procedure specifically states to do so.

DETECTING ON-OFF DIGITAL VOLTAGE SIGNALS

In some circuits (sensors, fuel injectors, mixture control solenoids, digital EGR valves, etc.), on-off digital signals passing through the wires either carry vital information or provide power to drive a particular device. In either case, it sometimes becomes necessary to check one of these circuits to see if the signal is getting through properly. A high-impedance digital voltmeter isn't much help because it won't react quickly enough to give you an accurate voltage reading. An ordinary analog voltmeter shouldn't be used because of possible danger to the circuit. So a special tool called a "logic probe" is needed.

A logic probe contains three colored LEDs: the red one indicates high voltage, the green one low voltage, and the yellow one whether or not a pulse is present. The probe is externally powered so all it does is monitor what's happening inside the circuit. When the probe's test leads are attached to the circuit (usually by backprobing a connector), the LEDs indicate what's happening in the circuit.

If a signal is present, the yellow LED should be on. If the yellow LED is not on, there is no on-off signal. The voltage may be staying high (red LED on) or low (green LED on).

The relative brightness of the red and green LEDs can also indicate the duration or width of the signal pulse, and whether or not the pulse is changing.

COLOR CODE	COLOR	STANDARD TRACER COLOR	COLOR CODE	COLOR	STANDARD TRACER CODE
BL	BLUE	WT	OR	ORANGE	BK
BK	BLACK	WT	PK	PINK	BK OR WT
BR	BROWN	WT	RD	RED	WT
DB	DARK BLUE	WT	TN	TAN	WT
DG	DARK GREEN	WT	VT	VIOLET	WT
GY	GRAY	BK	WT	WHITE	BK
LB	LIGHT BLUE	BK	YL	YELLOW	BK
LG	LIGHT GREEN	BK	*	WITH TRACER	

Fig. 3-20 Vehicle manufacturers have their own wiring color codes, so refer to their key when tracing wires (courtesy Chrysler).

CIRCUIT	FUNCTION
A	Battery Feed
B	Brake Controls
C	Climate Controls
D	Diagnostic Circuits
E	Dimming Illumination Circuits
F	Fused Circuits (Secondary Feed)
G	Monitoring Circuits (Gauges)
H	Open
I	Not Used
J	Open
K	Powertrain Control Module
L	Exterior Lighting
M	Interior Lighting
N	ESA Module
O	Not Used
P	Power Option (Battery Feed)
Q	Power Options (Battery Feed)
R	Passive Restraint
S	Suspension/Steering
T	Transmission/Transaxle/Transfer Case
U	Open
V	Speed Control, Washer/Wiper
W	Open
X	Audio Systems
Y	Open
Z	Grounds

Fig. 3-21 Vehicle manufacturers may also use letter codes to designate various circuits by function (courtesy Chrysler).

TRACING WIRES

Tracing wires is an essential element of checking continuity and finding wiring problems, but isn't as easy as it looks. A circuit wire will sometimes change color after passing through a connector, switch, or relay. That's why you should always refer to a detailed and accurate wiring diagram when doing electrical work. Working without a wiring diagram is like working in the dark.

Wires may be solid color with or without a stripe of a second color. On most wiring schematics, wire colors are indicated by an alphabetical code (Fig. 3-20). One or two letters used alone usually indicates a standard wire color. Two or more letters separated by a dash or slash indicates a striped wire. The first letter(s) indicates the basic wire color, and the second letter(s) the color of the stripe.

Abbreviations for color codes vary somewhat from one vehicle manufacturer to another. Always refer to the manufacturer's color identification chart for an explanation of the particular code.

Chrysler, for example, identifies gray wires with "GY" on its wiring schematics, while Toyota calls them "GR." Chrysler would identify a light blue

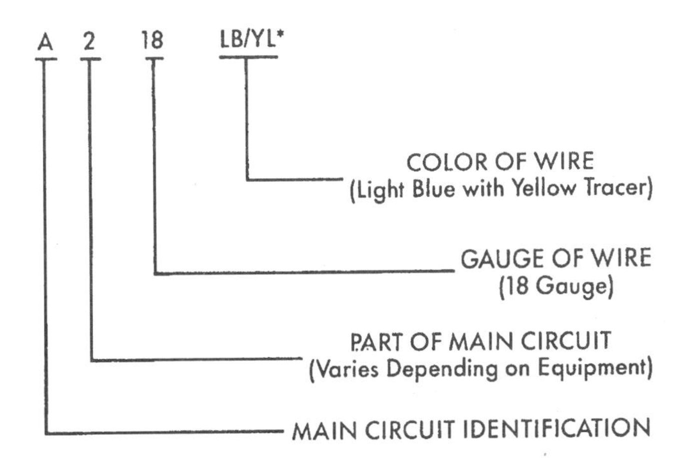

Fig. 3-22 Wires in a wiring schematic will usually be identified a number of ways (courtesy Chrysler).

wire with a white tracer as "LB/WT," while Toyota would label it "L-W."

Reading Wiring Schematics

Rule number one for troubleshooting any type of electrical problem on a vehicle today is this: *always refer to a wiring schematic to identify the circuit components.* Besides identifying the color of the wires in the circuit and how they're connected, the schematic will also show you all the components in the circuit. There may be a "hidden" relay, fusible link, connector, or ground connection you're not aware of that may be causing the problem. So always refer to the schematic for the basic layout of the circuit.

When tracing wires on a schematic, circuit identification numbers (Fig. 3-21), terminal identification numbers, connector numbers, and other information will help you figure out how the circuit

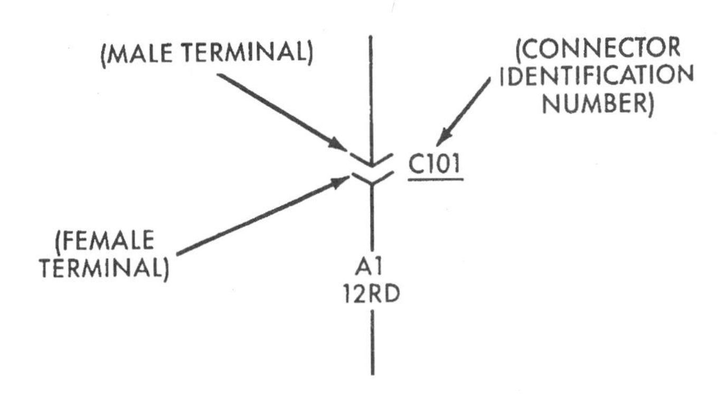

Fig. 3-23 Connector identification information may include the type of connector (male or female) as well as a numeric code (courtesy Chrysler).

is wired (Fig. 3-22). The schematic will also show you which circuits share common ground points, fuses, switches, or other devices. This kind of information can help you isolate wiring problems more quickly when more than one circuit is affected.

Once you've studied the wiring schematic, you may then have to refer to assembly wiring diagrams (often found in the body repair section of a shop manual) for wiring routing and connector location information (Fig. 3-23 through 25).

FINDING WIRING FAULTS

Now that we've covered the basics, what's the best way of finding a wiring fault? It depends on the problem.

If you're confronted with a "dead" circuit, the first thing to look for is voltage at the load point. Use a 12-volt test light or voltmeter to check for battery voltage at the component in question. If there's voltage, the problem is either a bad ground or the component itself. Check the ground connection with an ohmmeter. No resistance problems? Then the problem is in the component and not the wiring.

If you don't find battery voltage at the load point, however, the problem is on the hot side of the circuit. Inspect the fuse, circuit breaker, or fusible link that protects the circuit first. If okay, try bypassing the switch, relay, or solenoid that supplies power to the circuit. Still no voltage? Then check wiring continuity between the various links in the circuit for opens or shorts to ground. Eventually you'll find the link that's open or shorted. It may be a bad connector or wire. Sometimes the bad component will be obvious, but many times not.

The most difficult kinds of electrical problem to troubleshoot are intermittent ones. Everything works fine in the shop, but as soon as the customer gets his vehicle back, it acts up again. Intermittent opens or shorts can be caused by thermal expansion and contraction (something gets hot and opens or shorts out). Intermittents can also be caused by vibration, chaffing, or fatigue failure (frayed or worn insulation that allows intermittent contact with the body or other wires; cracked, broken, or loose wires, connectors, etc.).

To duplicate the conditions that cause an intermittent short or open, it may be necessary to wig-

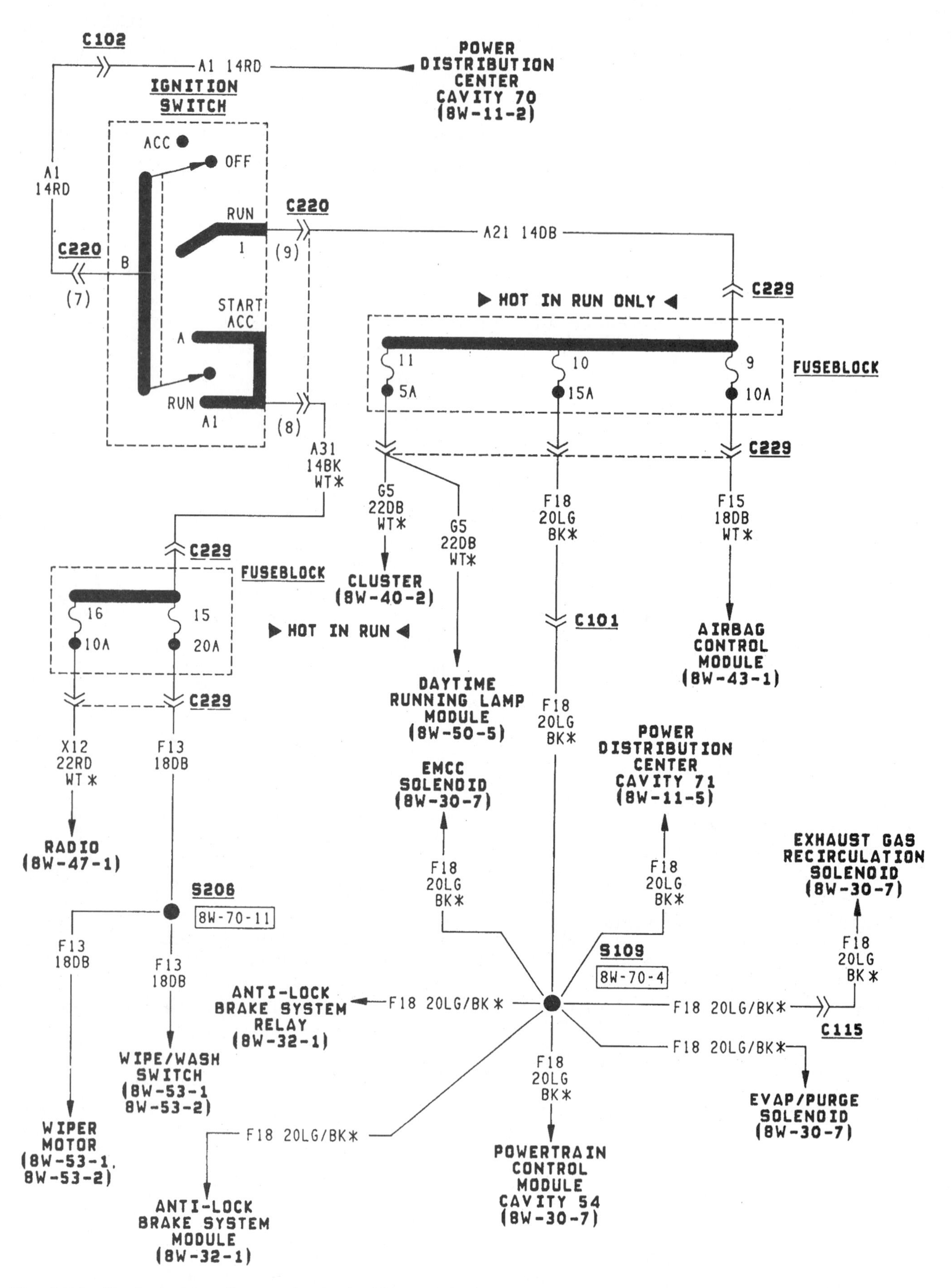

Fig. 3-24 Typical wiring diagram for the ignition switch, fuseblock, and related circuits (courtesy Chrysler).

LEGEND OF SYMBOLS USED ON WIRING DIAGRAMS			
+	POSITIVE		BY-DIRECTIONAL ZENER DIODE
−	NEGATIVE		MOTOR
	GROUND		ARMATURE AND BRUSHES
	FUSE	C100	CONNECTOR IDENTIFICATION
	GANG FUSES WITH BUSS BAR		MALE CONNECTOR
	CIRCUIT BREAKER		FEMALE CONNECTOR
	CAPACITOR		DENOTES WIRE CONTINUES ELSEWHERE
Ω	OHMS		DENOTES WIRE GOES TO ONE OF TWO CIRCUITS
	RESISTOR		SPLICE
	VARIABLE RESISTOR	S100	SPLICE IDENTIFICATION
	SERIES RESISTOR		THERMAL ELEMENT
	COIL	TIMER	TIMER
	STEP UP COIL		MULTIPLE CONNECTOR
	OPEN CONTACT		OPTIONAL — WIRING WITH / WIRING WITHOUT
	CLOSED CONTACT		"Y" WINDINGS
	CLOSED SWITCH	88:88	DIGITAL READOUT
	OPEN SWITCH		SINGLE FILAMENT LAMP
	CLOSED GANGED SWITCH		DUAL FILAMENT LAMP
	OPEN GANGED SWITCH		L.E.D. — LIGHT EMITTING DIODE
	TWO POLE SINGLE THROW SWITCH		THERMISTOR
	PRESSURE SWITCH		GAUGE
	SOLENOID SWITCH		SENSOR
	MERCURY SWITCH		FUEL INJECTOR
	DIODE OR RECTIFIER		948W-192

Fig. 3-25 Symbols commonly used in wiring diagrams (courtesy Chrysler).

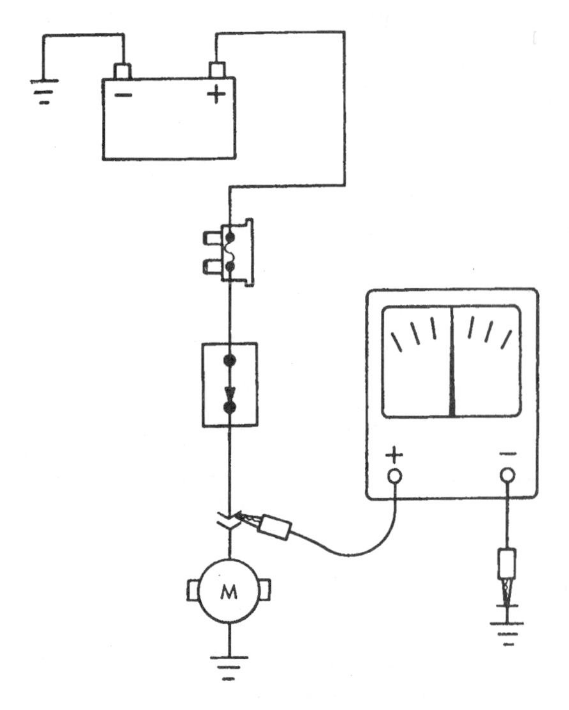

Fig. 3-26 When troubleshooting wiring problems, checking for the presence of voltage at the load point is a good place to start (courtesy Chrysler).

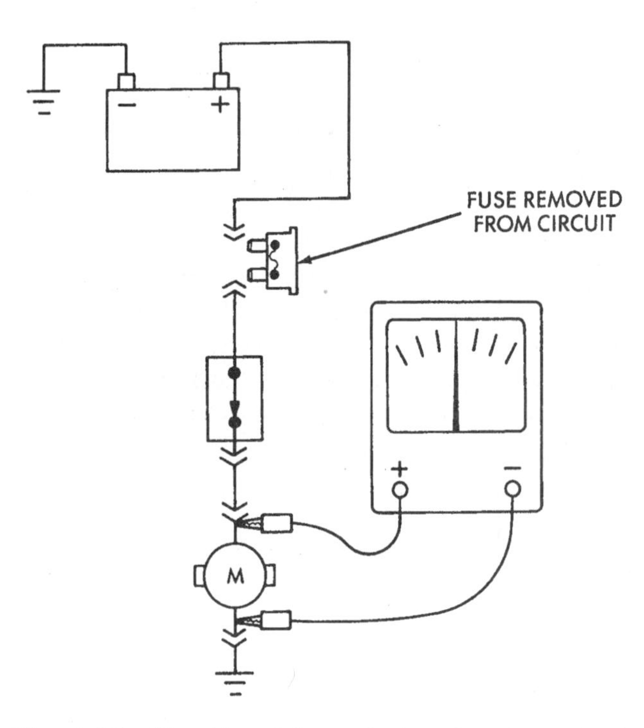

Fig. 3-27 Checking the continuity of various circuit components, wiring, and connectors is often necessary to pinpoint a wiring problem (courtesy Chrysler).

gle wires, heat up (or in some cases chill) suspicious connectors, wires, or other components.

Environmental factors can also play havoc with electrical systems. Extreme humidity, road splash (especially slush from heavily salted roads), or water leaks in the cowl area, windshield, firewall, or underbody may cause a temporary short. Look for obvious signs of corrosion or stains from water leakage.

Electromagnetic interference (EMI) is another factor that's sometimes overlooked as a source of trouble in electronic circuits. When current passes through a wire, it creates a magnetic field around the wire that can induce a current in adjacent wires if the wires are routed too closely together (Fig. 3-26). This kind of "crosstalk" can produce false signals in computer sensor circuits that confuse the computer. The result may be a performance problem and/or the setting of "false" trouble codes.

WIRING REPAIRS

Most wiring problems occur at harness connectors and terminals. These components are the point where the wiring is most vulnerable to environmental contamination, vibration, and mechanical stress. Wiring connectors and terminals sometimes work loose or are damaged by careless handling, but rust and corrosion are usually the main culprits (Fig. 3-27).

Though it is sometimes possible to clean a dirty wiring connector, the best cure is often replacement. Once the protective plating on the metal has been breached by corrosion, no amount of cleaning can fully restore it. Coating it with protective dielectric grease can help keep moisture out, but eventually corrosion will penetrate the connector and disrupt the circuit.

Damaged wiring is less common, but vibration, heat, and physical abuse can damage insulation and break wires. If a circuit has shorted out, the wiring should be carefully inspected for heat damage (discolored, melted, or burned insulation). Any damaged sections should be replaced.

When replacing wiring, always use the same gauge wire as the original. Make sure the wires are routed the same as before, and if possible, use the same color wire as the original. This will help prevent confusion in the event future repairs are needed.

If replacing a fusible link, do not use ordinary

wire. Ordinary wire provides no overload protection and may create a potential fire hazard!

When splicing wires don't just twist them together and wrap electrical tape around the connection. Use a solderless crimp-on connector. Or twist the wires together, solder them, and use shrink-wrap electrical insulation tubing to seal the repair (Figs. 3-28 & 3-29).

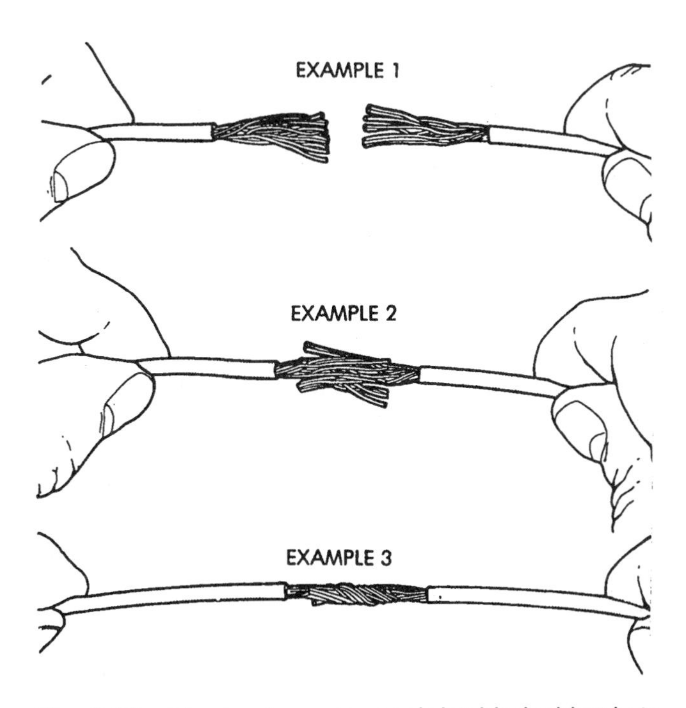

Fig. 3-28 The best way to repair braided wiring is to intermesh the wires as shown, prior to soldering (courtesy Chrysler).

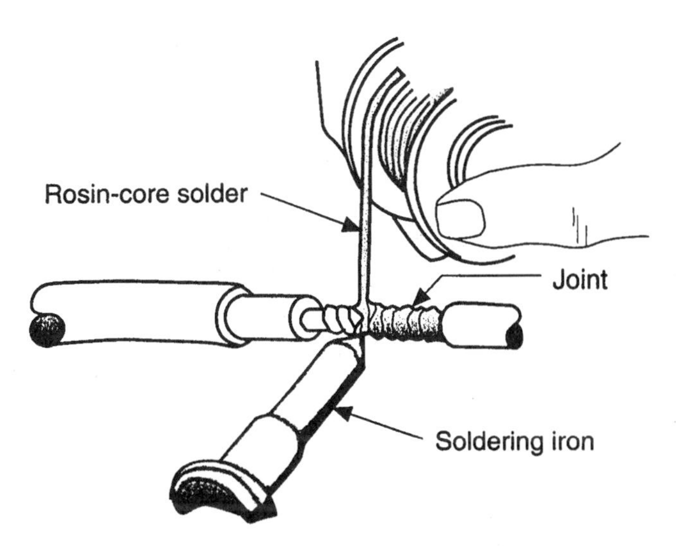

Fig. 3-29 After wires have been joined, they should be soldered to assure good electrical contact. The soldered wires should then be covered with shrink tubing or electrical tape (reprinted, by permission, from Hollembeak, *Automotive Electricity and Electronics.* Copyright 1994 by Delmar Publishers Inc.).

CHAPTER 4

Battery, Charging & Starting System

HOW A BATTERY WORKS

All batteries use chemical reactions to create and store electrical energy. When two dissimilar metals are placed in an acid solution, they create an electrochemical potential called voltage. This is the driving force that makes electrons flow through a circuit.

The two dissimilar metals in the case of automobile storage batteries are lead and lead peroxide. The positive plates contain lead peroxide (a soft, dark brown material). The negative plates are made of finely ground lead (called "sponge lead") which is gray in color. The positive and negative plates are sandwiched together in pairs, and separated by a nonconductive layer of paper, plastic, or fiberglass.

The acid solution is a mixture of 25% water and 75% sulfuric acid (H_2SO_4). Battery acid is often referred to as the "electrolyte" because it allows charged particles (called "ions") to migrate between the plates when current is drawn from the battery. The chemistry created by the different layers of lead and acid produces a surplus of free electrons at the negative terminal, and a shortage of electrons at the positive terminal. When a conductive path is connected between the two battery terminals, it completes the circuit. Stored chemical energy is converted into electrical energy, and power is drawn from the battery.

A 12-volt battery has 6 "cells", each of which contains 9 to 13 positive and negative plates. (The greater the number of plates, the greater the storage capacity of the battery.) Each cell produces 2.1 volts. The cells are connected together in series, so when all 6 cells are added together, total battery voltage is actually 12.6 volts.

As a battery discharges, both the positive and negative plates take on sulfate (the SO_4 part of the battery acid). This lowers the concentration of the remaining sulfate in solution. (The concentration can be measured on batteries with removable caps or tops by checking the "specific gravity" of the solution with a hydrometer.) At the same time, the two "dissimilar" metals become less dissimilar. The result is a gradual decrease in voltage output as the battery discharges. In other words, as the battery runs down, both positive and negative plates become saturated with sulfate. This makes them more alike and less able to continue the chemical reaction. Voltage and current output drop and eventually the battery runs dead.

If not returned back to solution by recharging, sulfate forms a barrier on the surface of the battery plates. This interferes with the battery's ability to make electricity. The accumulation of sulfate over time, can be due to chronic undercharging (defective voltage regulator, weak alternator, slipping fan belt, insufficient driving time between starts to recharge the battery, etc.). Sulfate buildup can render a battery useless. A sulfated battery will resist accepting a charge. But slow charging at perhaps 6 to 8 amps for up to 24 hours can often

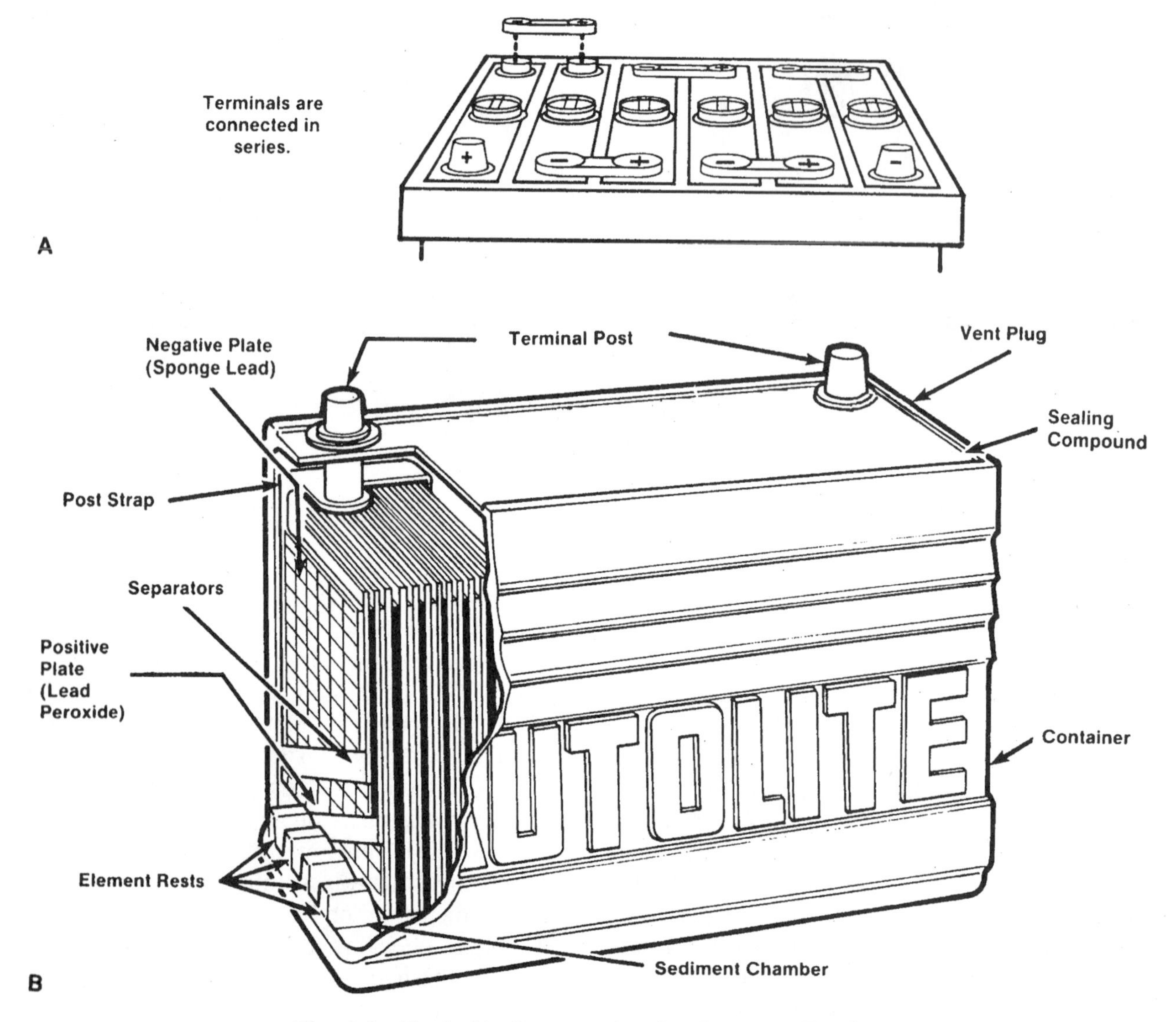

Fig. 4-1 Typical battery construction (courtesy Ford).

restore heavily sulfated plates to a usable condition (Fig. 4-1).

To recharge a battery, the chemical reaction has to be pushed in the opposite direction. A battery charger or alternator forces more voltage into the battery than it makes. This breaks the sulfate away from both plate surfaces and causes it to go back into solution. As the battery becomes recharged, the increase in sulfate that's gone back into solution can be checked by reading the specific gravity (the density of the electrolyte) with a hydrometer.

If water has to be added to a battery, always use pure distilled water. Ordinary tap water contains dissolved minerals and salts which can interact adversely with the acid and plates to shorten the life of the battery.

Adding acid to a discharged or old battery will not renew it or bring it back to life. The only way to restore or recharge a battery is to push the sulfate back out of the plates by reversing the current flow through the battery. If it won't accept a charge, the battery needs to be replaced.

For maximum life, automotive batteries should be kept at or near full charge. But this isn't always easy because it takes a lot longer to put amps back into a battery than it does to take them out. If an engine draws 225 amps when it is cranked, and is cranked for several minutes in an attempt to start it, it will take about half an hour to fully recharge the battery at a rate of 75 amps. Since this is close to the maximum output of many alternators, it may take 20 to 30 miles of driving to replace those amps.

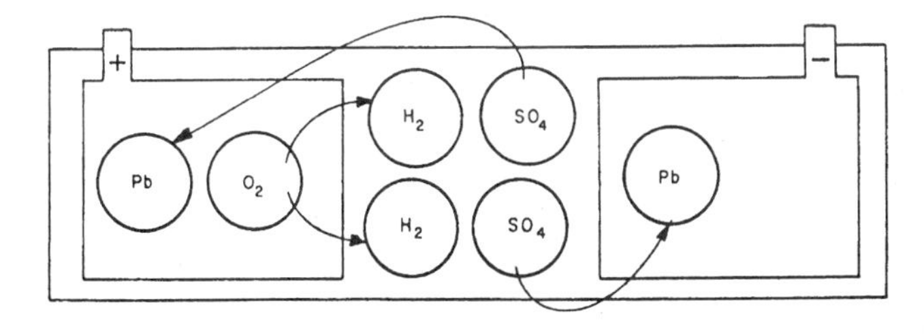

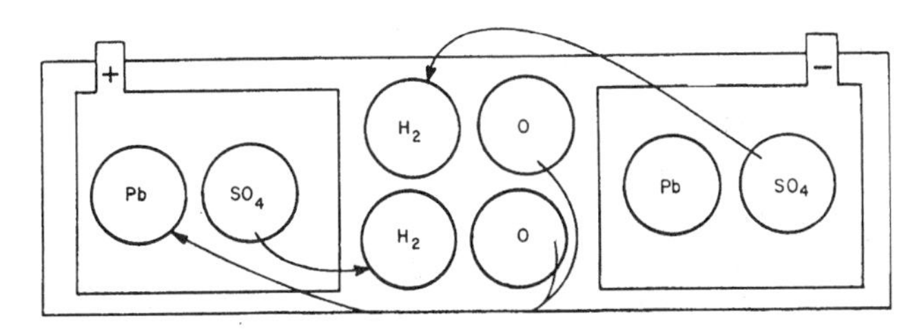

Fig. 4-2 The movement of sulfate and hydrogen in and out of solution during discharge (top) and recharge (bottom) (reprinted, by permission, from Santini, *Automotive Electricity and Electronics, 2E.* Copyright 1992 by Delmar Publishers Inc.).

It's important to note that recharging a battery causes "gassing" (Fig. 4-2). The reverse flow of current through the electrolyte breaks some of the water down into hydrogen and oxygen gas. Gassing is especially noticeable as a battery nears full charge, or is being charged at a faster rate than it can handle. Overcharging is particularly bad in this respect because it can literally boil a sizable quantity of water out of a battery. If too much water is lost, the exposed cell plates will dry out and become useless, reducing the battery's storage capacity. A faulty voltage regulator that pumps too much current into a battery can cause excessive water loss and premature battery failure. Overcharging can also create excessive heat (Fig. 4-3). Heat can warp cell plates, damage the insulating separators between the plates, and cause internal shorts and dead cells.

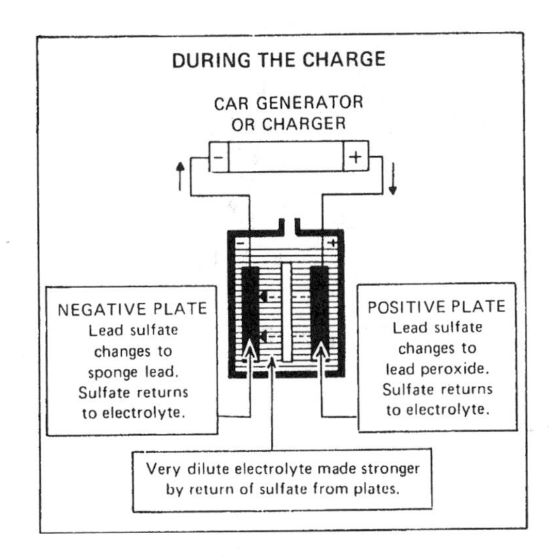

Fig. 4-3 What happens during the charging cycle (courtesy Battery Council International).

Gassing can be dangerous because of the explosive gases given off (both hydrogen and oxygen). *Hydrogen is extremely explosive when mixed with oxygen so always treat the battery with respect.* Never get an open flame or spark near a battery. Many a battery has blown up because someone used a cigarette lighter for illumination when trying to check the water level inside. Do not smoke near a battery. Make sure a charger is off when making or breaking booster connections with the battery terminals.

SAFE JUMP STARTING PROCEDURE

If you find yourself in a situation where an engine has to be jump started, never make the final jumper connection directly to the battery itself (Fig. 4-4). Connect the positive jumper cable to the positive battery terminal, but make the final negative jumper connection to a good ground away from the battery. The final jumper connection usually produces a spark. So making the connection away from the battery lessens the odds of touching off any hydrogen and oxygen that may be in the battery.

MAINTENANCE-FREE BATTERIES

To help reduce the gassing problem and thus the periodic need for water, the "maintenance-free" battery was invented. Water consumption in these types of batteries has been reduced or eliminated by changing the alloy of the lead plates. In some designs, the amount of antimony is reduced to 3.4% or less to reduce gassing. In most, however, calcium is used. Strontium and selenium are two other alloys that can also reduce gassing. But because calcium has established itself as the dominant technology, most battery manufacturers have generally stayed with calcium.

Besides reducing gassing to the point where

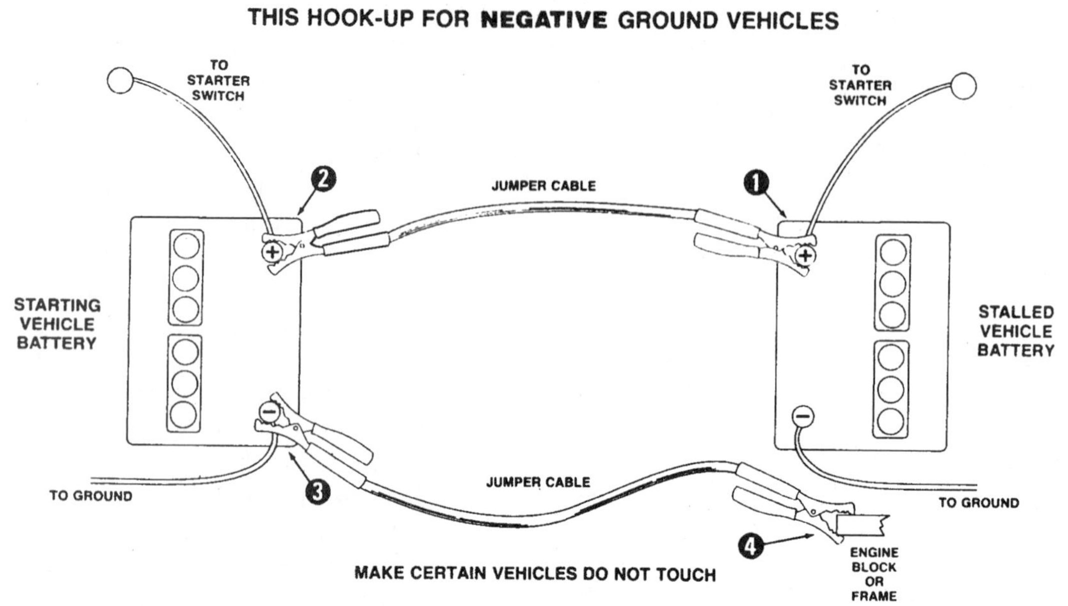

Fig. 4-4 Using this procedure to jump start a vehicle with a dead battery will keep any sparks away from the dead battery (courtesy Battery Council International).

adding water is no longer required, calcium is more resistant to overcharging. And its self-discharge rate is 20–30% lower than that of a conventional lead-antimony alloy battery. The tradeoff, however, is that calcium grids are not as well suited for deep cycling (being completely discharged and re-charged). If a calcium battery is run completely dead several times, its ability to deliver current can be diminished with each subsequent recharge.

In some batteries a "hybrid" approach is used. Antimony is used for the positive grids and calcium for the negative grids. Taking the antimony out of the negative grids reduces gassing while retaining some of the deep cycling performance of antimony-lead. Calcium batteries perform best in applications where the battery is kept at or near full charge, and where the vehicle is driven frequently. Maintenance-free calcium batteries typically require a slightly higher charging voltage (14.0 volts or higher). General Motors, for example, sets most of their late-model regulators to charge at 14.8 volts to keep their "Freedom II" lead-calcium battery properly charged. (The actual charging voltage will vary with temperature.) Ford and Chrysler use regulator settings in the 14.3-volt range.

Some maintenance-free batteries have sealed tops (can't be opened to add water or check the electrolyte). But sealed-top batteries are still vented so hydrogen gas can escape. Therefore, use the same safety precautions when jumping one of these batteries as you would with any other.

RESERVE BATTERIES

A special type of aftermarket battery that features a small "reserve" battery in addition to the regular battery has been available for several years. This type of battery is designed for motorists who forget to turn their headlights off. A switch atop the battery allows the motorist to switch to the reserve plates, which provide enough cranking amps to start the engine. This type of battery is essentially two batteries in the same case. So service and test procedures are essentially the same as with a conventional battery, except that each part needs to be tested independently.

RECOMBINATION BATTERIES

A more recent battery design that does not use a liquid electrolyte is the "recombination" battery. It uses separators that hold a gel-type material. The separators are placed between the grid plates and have very low electrical resistance. The result is a battery that uses no water, won't spill if cracked, is corrosion-free, and can be installed in virtually any position (even upside down!).

Performancewise, recombination batteries produce about 0.6 more volts (13.2 volts total) than standard lead-acid batteries, which provides additional power for cranking. Recombination batteries are also capable of withstanding deep cycling without damage, unlike conventional lead-acid passenger car batteries. Most also carry a comparatively high cold-cranking amp rating (800 amps or more), and can last up to several times longer than a conventional battery.

STATE OF CHARGE

As a battery discharges, the amount of sulfate ions in the acid solution decreases (Fig. 4-5). The concentration of acid becomes less and less and the percentage of water goes up. Measuring the "specific gravity" of the electrolyte with a hydrometer is a way to tell how much acid is left. From that, you can deduce the battery's state of charge.

Specific gravity is the weight of a given volume of liquid, divided by the weight of an equal part of water. Pure water has a specific gravity of 1.000. A fully charged battery with its electrolyte at maximum acid concentration should produce a hydrometer reading in the 1.260 to 1.280 range at 80 degrees F. Any battery less than 75% charged should be recharged. This is especially important when testing batteries because a low state of charge can make an otherwise good battery appear to be defective.

The following chart shows typical hydrometer readings:

Hydrometer reading (80 degrees F.)	State of charge
1.260 or higher	100%
1.225	75%
1.190	50%
1.155	25%
1.120 or less	Discharged

Specific gravity readings must be adjusted for temperature changes. As a liquid heats up, it becomes less dense. Since the concentration of acid doesn't change, you have to compensate to get an accurate reading. Most hydrometers have a thermometer that shows you how much to add or subtract from the reading depending on the temperature.

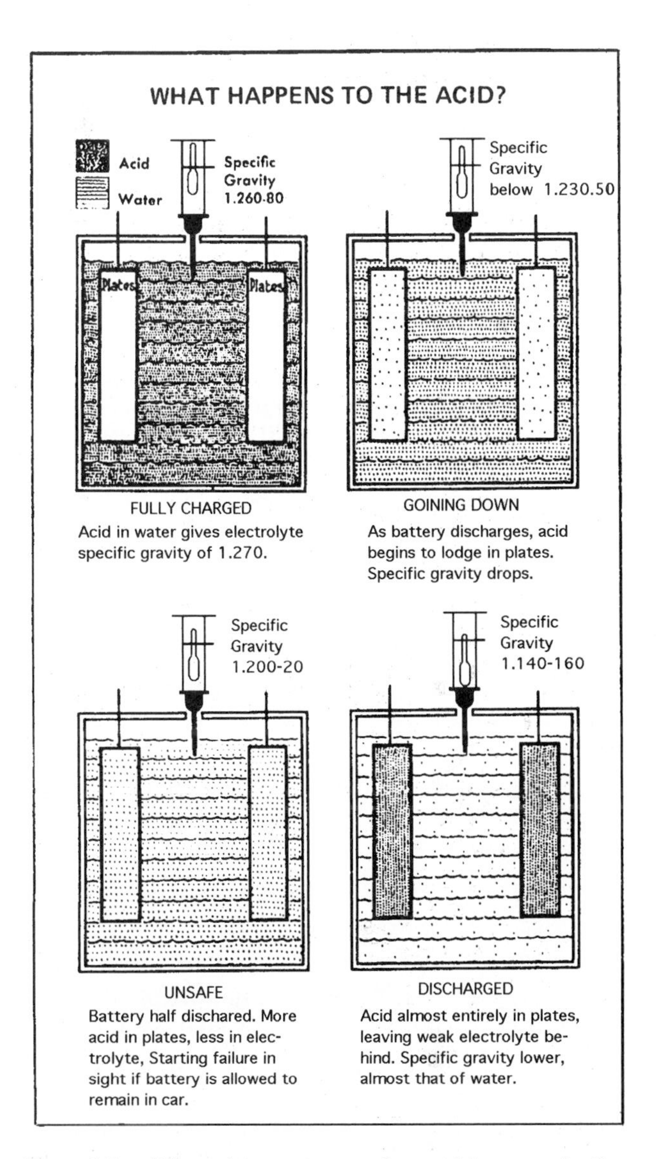

Fig. 4-5 What happens to the acid as a battery charges and discharges (courtesy Battery Council International).

As a rule of thumb, subtract 4 gravity points (0.004) for each 10 degrees F. that the battery temperature is below 80 degrees F.

If specific gravity readings vary by more than 50 points between any two cells, the battery is bad and needs to be replaced.

On sealed-top, maintenance-free batteries, a built-in hydrometer or charge indicator is sometimes included:

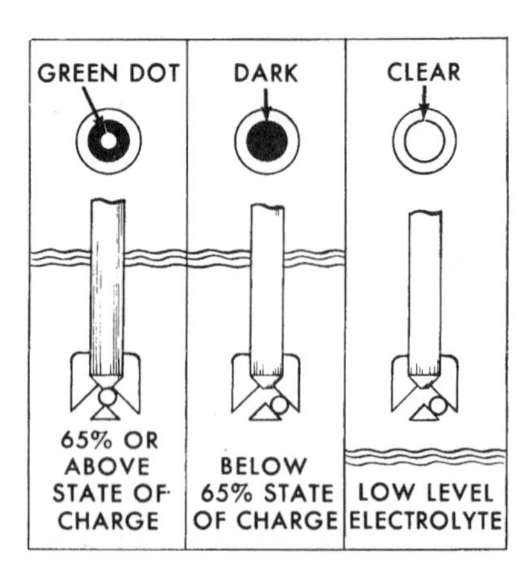

Fig. 4-6 How to read a built-in hydrometer or charge indicator on a sealed-top battery (courtesy Delco Remy).

- A colored dot (Fig. 4-6) (usually green) visible in the indicator means the battery is at least 75% charged and is okay for further testing if necessary.
- A dark indicator (no colored dot) means the battery is less than 75% charged and must be recharged before further testing. (**Note**: on some Chrysler OEM batteries, a red dot indicates a charge of less than 50%).
- A clear or yellow indicator means the level of electrolyte inside the battery has dropped too far below its normal level to give a reading. If the top cannot be opened to add water, the battery will probably need to be replaced soon, if it doesn't already. Once the water level drops below the top of the cell plates, the plates dry out and lose their ability to hold a charge. Never attempt to jump start or recharge a battery with a low electrolyte level because of the danger of explosion. The slightest spark can ignite the hydrogen and oxygen gas that accumulates in the top of the battery, and a battery that's low on water will likely have a lot of hydrogen and oxygen in its cells.

Using Voltage to Determine a Battery's State of Charge

A battery's state of charge can also be determined by measuring its "open circuit" voltage. This is the rest or base voltage of the battery with no external loads placed upon it. If the battery has been used to crank the engine or subjected to a load test, you must wait at least 10 minutes before checking open circuit voltage for an accurate reading.

Open circuit voltage	State of charge
12.6v	100%
12.4v	75%
12.2v	50%
12.0v	25%
11.7 or less	Discharged

BATTERY RATINGS

When it comes to battery ratings, it's a real numbers game. The number most manufacturers tout today is cold cranking amps (CCA), or the battery's ability to deliver a sustained amp output at zero degrees Fahrenheit without dropping below 7.2 volts.

There's a rule of thumb that says a vehicle's battery should have a CCA rating equal to, or greater than the engine's displacement in cubic inches (Fig. 4-7). In other words, for every cubic inch of displacement, the engine needs at least 1 cold cranking amp of battery power for reliable winter starting. Thus a 300-cubic-inch V8 would require a 300 CCA battery. For reliable cold starting in subzero weather, 2 amps per cubic inch is a more realistic figure. For diesel engines, two amps per cubic inch is also needed.

When it comes to cranking power, bigger is definitely better. You'll find batteries on the market today with CCA ratings of up to 1000 amps, though most fall in the 550 to 750 range. But are such batteries really necessary for today's generation of small-displacement V6 and four-cylinder engines? It depends.

One of the advantages of making a small battery with a large CCA rating is that it can be used in a variety of applications. This reduces the number of different batteries that a parts store has to stock to provide broad application coverage. To build a battery with a very high CCA rating, however, the cells have to be redesigned to reduce resistance and increase the active surface area of the materials therein. Thinner plates, more plates per cell, more efficient grid designs, and reconfigured grid connections are all needed to squeeze more starting amps out of a smaller-sized box.

Unfortunately, the kind of changes that enable a battery to deliver a lot of amps in a short period of time reduce the battery's ability to deliver a sustained current over a long period of time. This is called the battery's "Reserve Capacity" rating. It is

Battery Efficiency

Battery cold power ratings established at 0°F

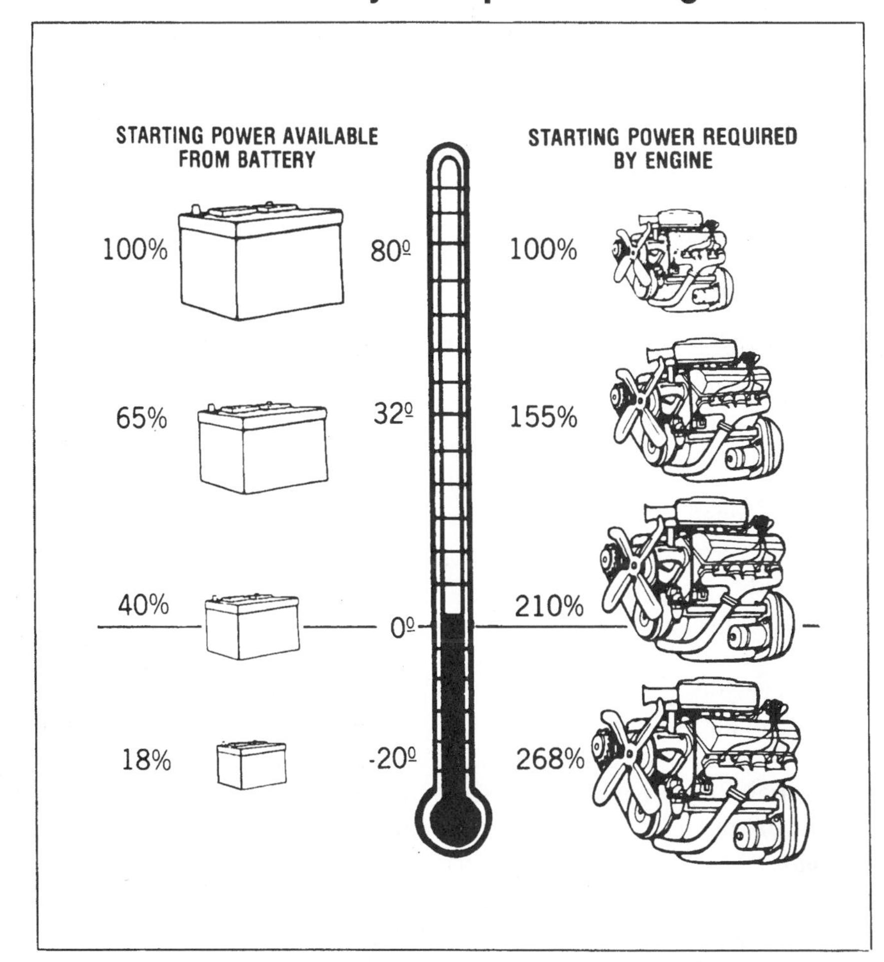

1. Keep batteries clean and dry.
2. Hydrometer testing, open-circuit voltage testing and load testing are methods to check a battery's charge and condition.
3. Batteries can be charged quickly or slowly, alone or in groups. Alternate methods are available.
4. Batteries self discharge whether in storage or installed. When stored, keep batteries in a cool, dry area. When installed, check the battery's charge often to avoid sulfation.
5. Cable connection should be maintained bright and tight and replaced when necessary.
6. Batteries must be securely mounted and carefully handled.
7. Take precautions when working around batteries.
8. A good battery PM program will avoid the most common causes of premature battery failure.

Both a battery's ability to supply current and an engine's starting power requirements are affected by the weather. The colder it gets, the less efficient a battery becomes and the greater the starting power needed to crank an engine. The chart above shows that a battery is 100 percent efficient at 80°F. At 0°F, the same battery is only 40 percent as efficient, but the engine needs over twice as much power to start as it did at 80°F. Starting requirements of diesel engines in cold weather are even greater than those illustrated here.

Fig. 4-7 How temperature affects battery efficiency (courtesy Battery Council International).

a measure of how long a fully charged battery can sustain a drain of 25 amps at 80 degrees Fahrenheit. The higher the reserve capacity rating, the longer the battery can keep the engine and accessories functioning in case of a charging system failure.

What's more important: cold cranking amps or reserve capacity? It depends on the application. Both are important, but a battery that sacrifices too much reserve capacity for brute cranking power might not be the best battery for a vehicle that sits for long periods of time between use or for one that is rarely driven long enough to fully recharge the battery.

The physical dimensions of the battery itself have nothing to do with its suitability for a given application. As long as the battery fits the battery box, has the right kind of terminals (post terminals or side mount), the necessary CCA ratings to handle the engine and accessories, and sufficient reserve capacity to handle sustained current drains or periods of inactivity, any size of battery will do.

BATTERY TESTING

There are several ways to test a battery's condition. A traditional "load test" will show the battery's abil-

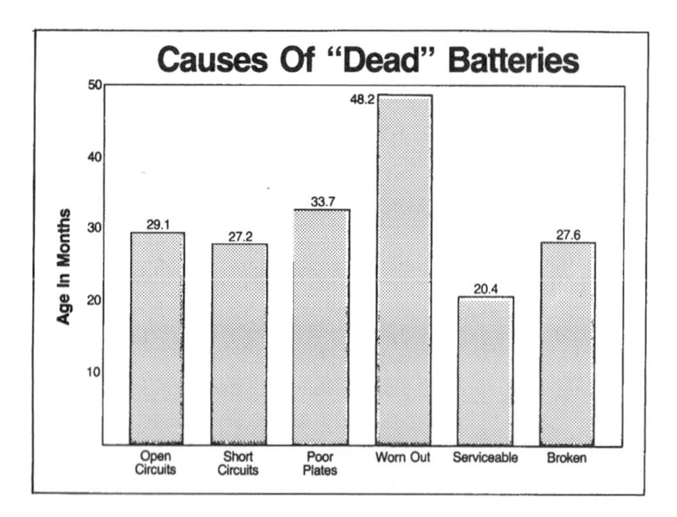

Fig. 4-8 Not all batteries fail because of old age or sulfation (courtesy Battery Council International).

ity to deliver current. The battery should be at least 75% charged before load testing. Otherwise, it may fail the load test. Any battery that fails a load test should be recharged and retested before being condemned.

To load-test, a "carbon pile" is needed to create the appropriate load on the battery. The carbon pile is set according to the battery's CCA (or amp/hour) rating. The carbon pile is usually set to one-half the battery's CCA rating (or three times its amp/hour rating). The load is then applied to the battery for 15 seconds, while the battery's voltage output is observed. If battery voltage remains above 9.6 volts, the battery is good. If less than 9.6 volts, the battery can be recharged and retested, or given a "three-minute charge test."

A three-minute charge test checks for a sulfated battery. Slow charge the battery at 40 amps for three minutes. Then check the voltage across the terminals with the charger on. If the voltage is above 15.5 volts, the battery is not taking a charge. Slow charging for 20 hours may reverse the sulfation condition. If not the battery will have to be replaced.

The newest type of electronic battery testing equipment uses a different approach to determining a battery's condition and state of charge. Instead of applying a large load with a carbon pile and monitoring the drop in battery voltage, the new generation of "smart" tester applies a small current to the battery and measures its internal resistance. This eliminates the need to recharge low batteries prior to testing and generally gives a more reliable indication of a battery's true condition than a traditional load test.

BATTERY REPLACEMENT

One of the cardinal rules when replacing a battery is to always test the performance of the charging system. It doesn't make much sense to install a new battery in a vehicle that has a faulty charging system. In many instances, the problem isn't the battery but (Fig. 4-8) a bad regulator or alternator, or perhaps even a slipping drive belt or corroded battery terminal.

New batteries should always be installed fully charged. Most battery warranty claims (and charging system failures) are the result of someone installing a battery that wasn't fully charged. Alternators are designed to maintain battery charge, not to recharge dead batteries. Using the alternator to recharge a dead battery may overtax it and damage the unit.

For dry-charge batteries (ones stored without acid in them), the battery should be boost-charged at 30 amps after adding acid until the battery reaches full charge. If the battery fails to activate (due to moisture contamination of the plates inside), switch to a slow charge to bring it up to full charge.

The battery cables should always be inspected and cleaned prior to installation, and the starter and ground connections checked to make sure they're also clean and tight (Fig. 4-9).

The battery must be mounted securely with a strap or holddown when installed to protect it against vibration. The grid plates inside many maintenance-free batteries are fairly brittle, and if vibration causes an internal separation, the battery will go dead.

On late-model cars with computers and electronic accessories, temporarily disconnecting the battery for testing or replacement may affect the settings and electronic memory of the radio, climate control, and clock settings. On vehicles with electronic fuel injection, there may be a temporary slight change in performance, because the portion of the computer memory that keeps track of the average fuel calibration over time will be lost. It will soon recalculate the average fuel calibration setting. Until then, there may be a noticeable change in the way the engine runs. The engine computer fault code memory will also be erased when the battery

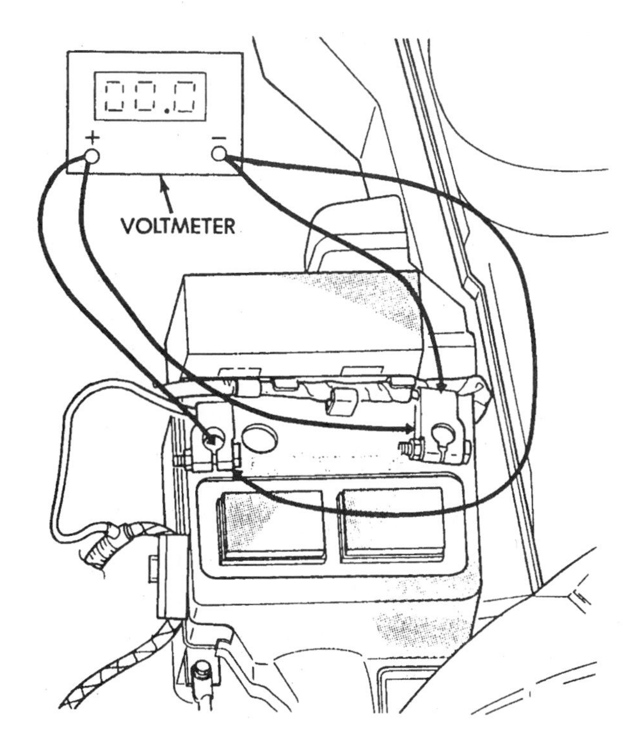

Fig. 4-9 Checking voltage drop across the battery cable connections can find "hidden" corrosion (courtesy Chrysler).

is disconnected. If the vehicle has been experiencing additional performance problems in addition to battery/charging problems, or the "check engine" or "power loss" light on the dash has been on, any fault codes that are stored in memory should be recovered and recorded prior to disconnecting the battery (see Chapter 19).

Aftermarket "memory saving" devices are available to keep the computer supplied with voltage while a battery is being changed. These devices use an ordinary 9-volt household battery and plug into the cigarette lighter.

THE CHARGING SYSTEM

The charging system consists of the alternator, a regulator to control the alternator's charging output, and a charge indicator (either a gauge or warning light). The charging system's job is to maintain the battery in a fully charged condition, and to meet the additional current requirements of the electrical system.

The charging system uses magnetism to make electricity. To understand how this works, we first need to briefly look at the subject of magnetism.

An ordinary "permanent" magnet has two "poles," a "north" pole and a "south" pole. The poles behave something like electrical charges in that like poles repel, while unlike poles attract. Between the two opposing poles is a region called a "magnetic field" which contains invisible lines of magnetic "flux." The lines run from one pole to the other. The stronger the magnet, the greater the number of lines or magnetic flux in the magnetic field.

When electric current flows through a wire, it also creates a magnetic field around the wire. There's a "left-hand" rule that says if you point your left thumb in the direction the current is flowing inside a wire (negative to positive), the "lines of flux" in the magnetic field will be traveling in the same direction as your curled fingers.

If the wire is wrapped around a piece of iron, the iron becomes a powerful "electromagnet" with a very strong magnetic field. The left-hand rule again applies to determining the north pole of an electromagnet.

When a conductive wire is passed through a magnet, the magnetic field acts on the electrons inside the wire, pushing them along and inducing an electrical current in the wire. The more powerful the magnet, the stronger the electrical current produced in the wire. Since electromagnets can be hundreds of times stronger than permanent magnets, the alternator uses very powerful electromagnets to create high-output currents.

Controlling the current output of the alternator, therefore, is basically a matter of controlling the voltage flowing through its electromagnets (called "field coils"). The higher the "field" control voltage, the higher the alternator's output.

Understanding the Alternator

To better understand how the alternator and voltage regulator work, we need to look at how these components evolved. Before alternators became common, a "generator" was used to keep the battery charged. The generator's big limitation was that its output was proportional to speed. The faster the engine turned the generator, the more voltage it produced. At idle and slow speeds, therefore, a generator didn't put out much juice. As a result, the battery could run down. Generators also suffer from another inherent weakness. All the current produced by a generator flows from its armature through its brushes. This creates a lot of heat and arcing, which eats up the brushes rather quickly. Consequently the brushes had to be replaced frequently. Feeding the generator's output through its brushes also put a limit on how much current it could produce.

It soon became apparent that a better means of generating electricity was needed to meet the increased demands of the electrical system. The advent of air conditioning as a common option made a change absolutely essential. The answer was the alternator (refer to Figs. 4-10 through 4-13).

Essentially a generator turned inside out, an alternator produces current by rotating the magnetic field inside a stationary conductor instead of rotating a conductor inside a stationary magnetic field as in a generator. The alternator's rotor carries the magnetic field current, which creates a rotating magnetic field as the rotor turns. As the magnetic poles pass beneath the three stationary stator windings, a three-phased alternating current (AC) is induced in the stator windings. The current is then "rectified" (converted) to direct current (DC) by the alternator's diodes. This keeps the battery charged and meets the demands of the electrical system.

An alternator's output is controlled by switching its field current on, and off. The voltage regulator does this by monitoring battery voltage. When voltage is low, the regulator switches the field current on, causing the alternator to go to full output. When the voltage becomes too high, the regulator shuts off the field current, causing the alternator's output to go to zero.

Constantly cycling the field current on and off many times a second maintains an "average" output that is just right for the demands of the battery and electrical system. In other words, as the voltage regulator increases field current "dwell time" (the ratio of on to off time), alternator output goes up. As it decreases the average dwell time, alternator output drops.

In theory it sounds quite simple, but in practice it can be anything but. The old electromechanical

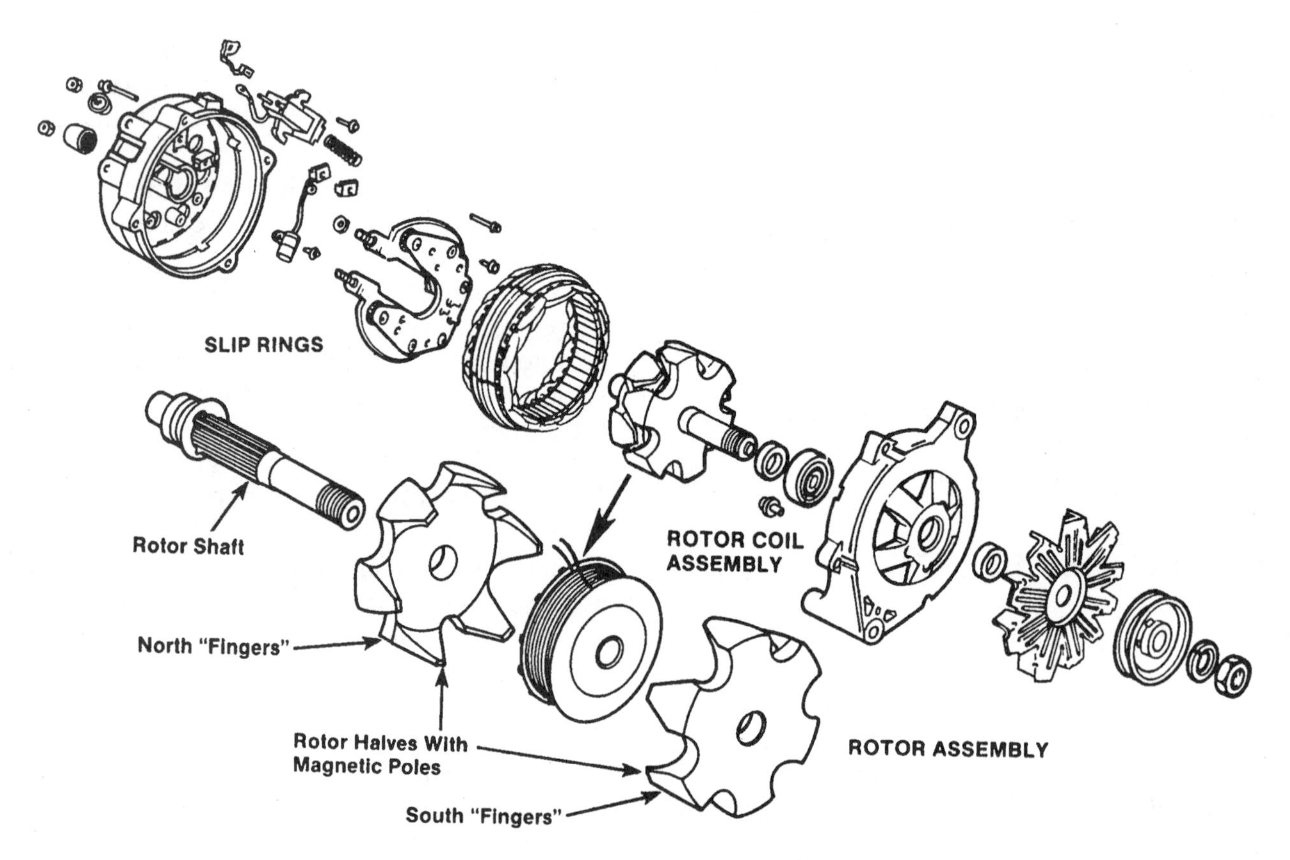

Fig. 4-10 Rotor coil and case assembly (courtesy Ford).

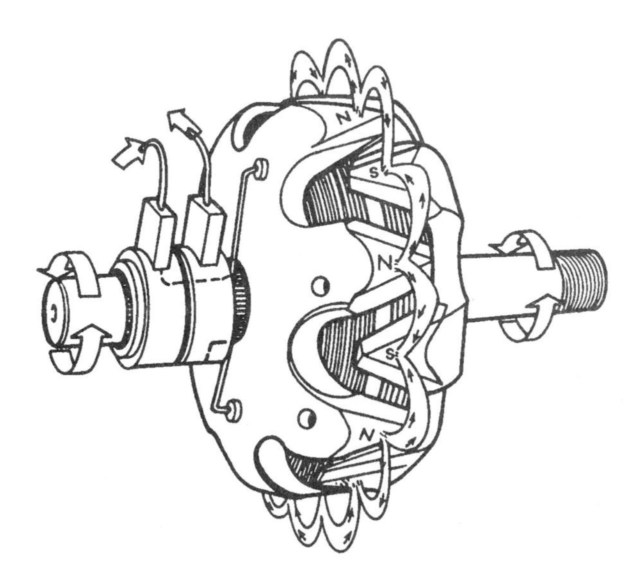

Fig. 4-11 Current flows through the brushes and slip rings to create a magnetic field in the rotor (courtesy Delco Remy).

regulators that used to be part of the charging system used two or three electromagnetic coils, several sets of silver-coated or tungsten contact points, some bimetal springs, wires, and resistors to control voltage output. The regulator control "settings" were determined in part by the air gaps between the various contact points, the spring tension on the contact point armatures, and by the electrical resistance in the wiring. If the settings needed to be adjusted, bending the bimetal spring bracket upwards to decrease spring tension raised the voltage setting, while bending it down lowered the voltage setting.

Electromechanical regulators were trouble prone because of their complexity and vulnerability to misadjustment and wear. The contacts points suffered wear, and if they froze, the charging system could either overcharge or discharge the battery. The average life span of a typical electromechanical regulator was only about 30,000 to 40,000 miles, so it was standard practice to recommend a new regulator if the alternator had to be replaced.

In 1972 the domestic car makers began changing over to solid-state electronic voltage regulators, and by 1974 the electromechanical regulator was obsolete at the OE (original equiptment) level. In comparison to its electromechanical counterpart, the current generation of electronic regulators are much simpler and compact. A typical unit contains a pair of transistors, a Zener diode, some capacitors, diodes, and resistors.

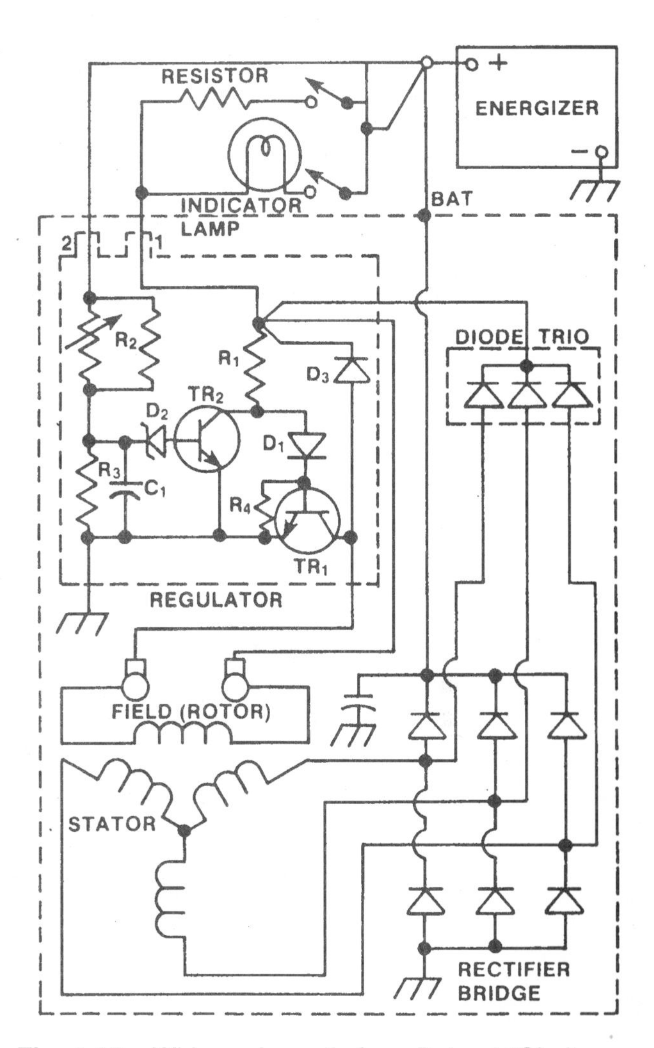

Fig. 4-12 Wiring schematic for a Delco 10SI alternator (courtesy Delco Remy).

ELECTRONIC VOLTAGE REGULATION

A typical charging system uses a large power transistor as the main switch for turning the alternator's field current on and off. The transistor will complete the power circuit to the field coils until it is toggled off by the second transistor which gets its clue from a "Zener" diode.

A Zener diode is an electronic component that suddenly becomes conductive (turns on) when a certain threshold voltage is reached. The voltage threshold of the Zener diode, therefore, determines the reference voltage (see Glossary) for the charging system. It is the functional equivalent of the magnetic coil and spring-loaded armature in the electromechanical voltage regulator. The only difference is that there are no moving contacts in the

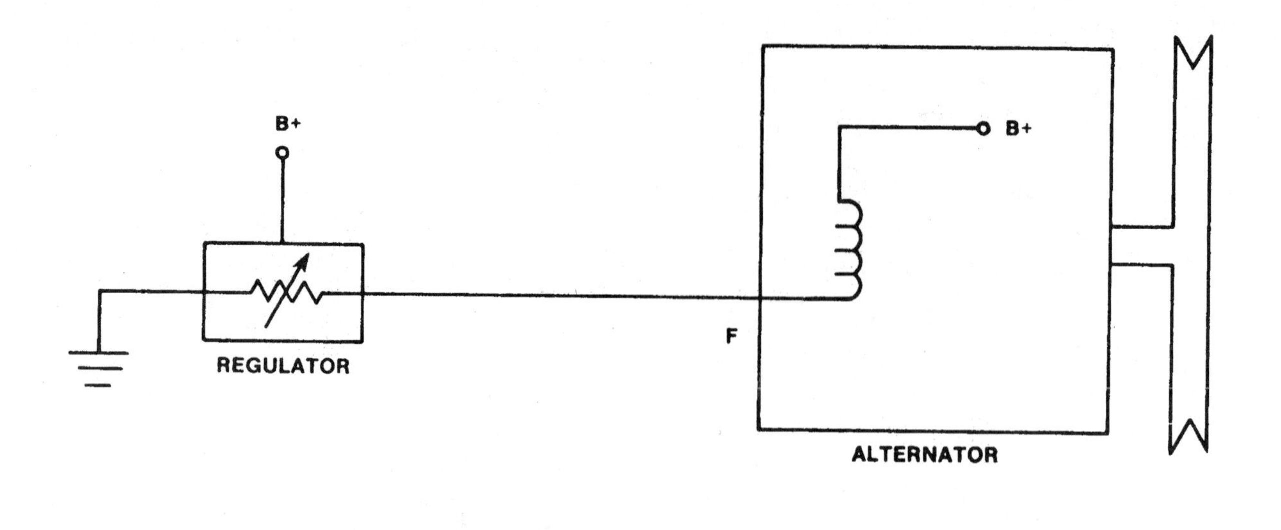

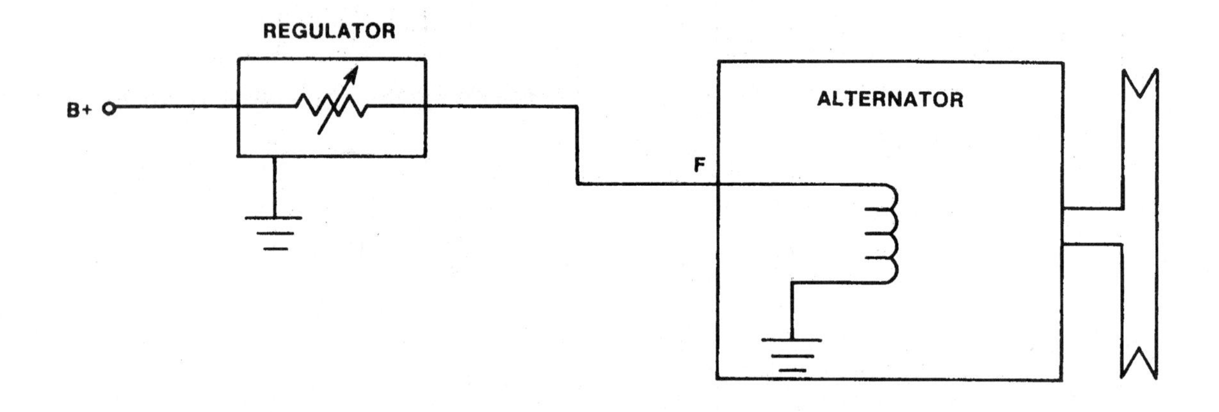

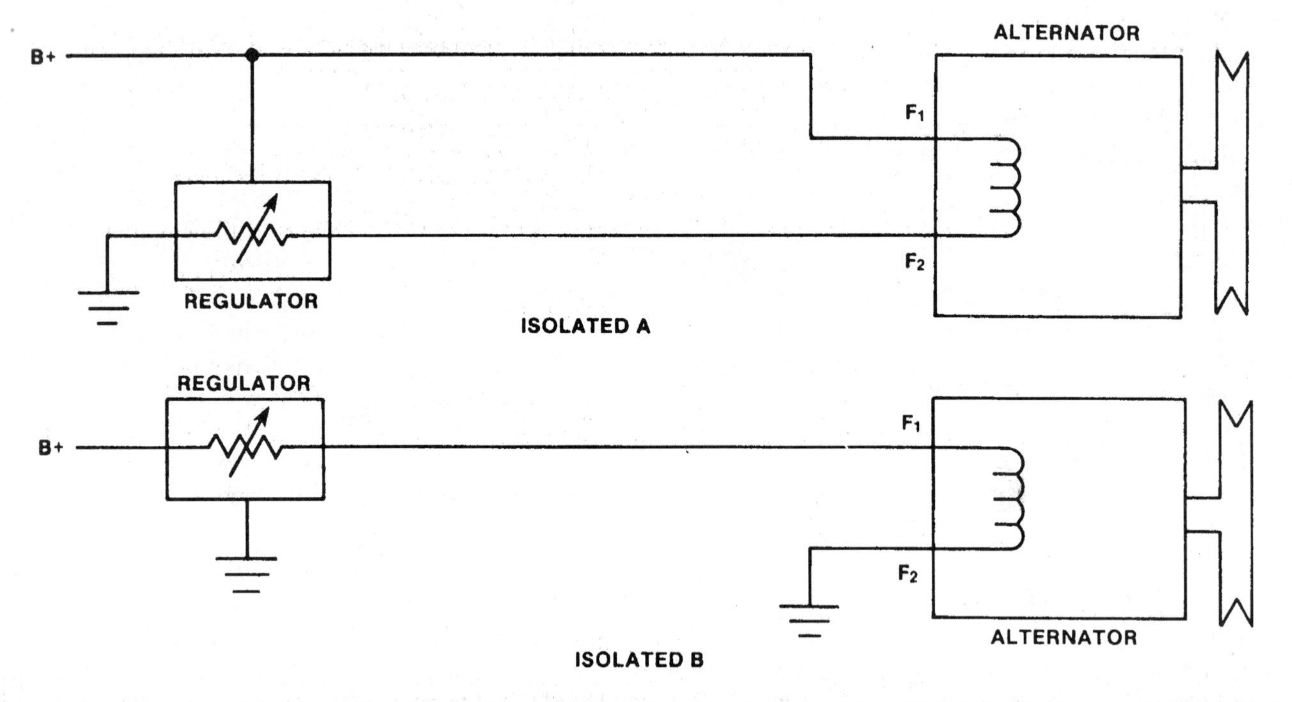

Fig. 4-13 Four types of alternator circuits: (top to bottom) A circuit, B circuit, Isolated A circuit, and Isolated B circuit (reprinted, by permission, from Santini, *Automotive Electricity and Electronics, 2E.* Copyright 1992 by Delmar Publishers Inc.).

solid-state version. So there's no way to mechanically adjust the regulator's built-in setting to increase or decrease charging output.

When alternator output voltage reaches the threshold limit of the Zener diode, the diode passes voltage to the number two transistor which in turn toggles the main power transistor off. The field current is broken and the alternator's output drops to zero. As the charging voltage drops back under the Zener diode's threshold level, the Zener diode turns off. This trips the number two transistor back to its original position, which allows the main power transistor to switch back on. Thus the on and off cycling that regulates alternator output is achieved with no moving contacts, springs, or armatures.

Temperature compensation in the electronic regulator is accomplished electrically by using a "thermistor" (a special kind of resistor that changes conductivity according to temperature) in the Zener diode's supply circuit. As the thermistor changes its resistance with the changing underhood temperature, the threshold voltage of the Zener diode is manipulated to alter charging system output. Thus a GM Delco alternator that is normally set to produce a maximum voltage output of 14.8 to 14.9 volts can actually range from as low as 13.8 to 14.9, depending on ambient temperatures.

The main enemies of the electronics inside the voltage regulator are heat, excessive current, and excessive voltage. A sudden current surge or a voltage spike can literally burn right through a semiconductor and render it useless. When an alternator fails, a lot of nasty things can happen electrically. Depending on the nature of the failure, the voltage regulator may be subjected to serious voltage overloads. The regulator should always be tested, therefore, to make sure it is still good.

On many alternators, the voltage regulator is mounted inside or on the alternator itself. These are called "integral" alternators because they are a part of the alternator.

GENERAL MOTORS "CS" SERIES ALTERNATORS

In the 1986 model year, the first of a new series of "CS" alternators from Delco Remy began appearing on certain General Motors cars. The three most common CS Delcotron units are the CS-121, the CS-130, and the CS-144.

"CS" stands for "Charging System," and the number denotes the outside diameter of the stator. The CS series alternators are high-output units and differ from the earlier SI series alternators in a number of unique ways. One difference is that CS series alternators have a totally different type of voltage regulator. The switching pattern is digital rather than analog. Previously, all Delco voltage regulators used an analog switching pattern. This means that the frequency at which the regulator switches the field current off and on changes as the load and rpm of the alternator change. The CS series, however, switches with a digital pattern. This means the switching frequency of the regulator stays constant at about 400 cycles per second throughout the rpm and load range of the alternator. By varying the on-off time, the average field current is regulated for the correct charging voltage. At high speeds, the on-time may only be 10% and the off-time 90%. At low speeds with high electrical loads, the on-time may be as high as 90%. The regulator also compensates for temperature.

One reason for the change was to make the regulator "interactive" with the onboard computer. Computers can only process a digital signal, so analog signals are not compatible with computer controls. GM wanted load response control in the alternator. When an engine is running at low speed and a heavy electrical load is applied, it can lug down the engine if it reacts too quickly to load. In a small displacement engine, this can cause an objectionable idle shake.

The CS series voltage regulator is a "pulse width modulated" unit. It gradually increases the field current as an electrical load is applied. In other words, it gradually cuts in the alternator more gradually, going from zero output to full output in about 2 1/2 seconds. Providing this gradual transition reduces the kind of idle shake problems that can result when a sudden load is applied.

COMPUTERIZED CONTROLS

Chrysler was the first manufacturer to go to "computerized" or "smart" voltage regulation. It did so in 1985 by eliminating the voltage regulator on certain vehicles and incorporating the necessary voltage regulation circuitry on the engine computer power board.

Computerized voltage regulation has several

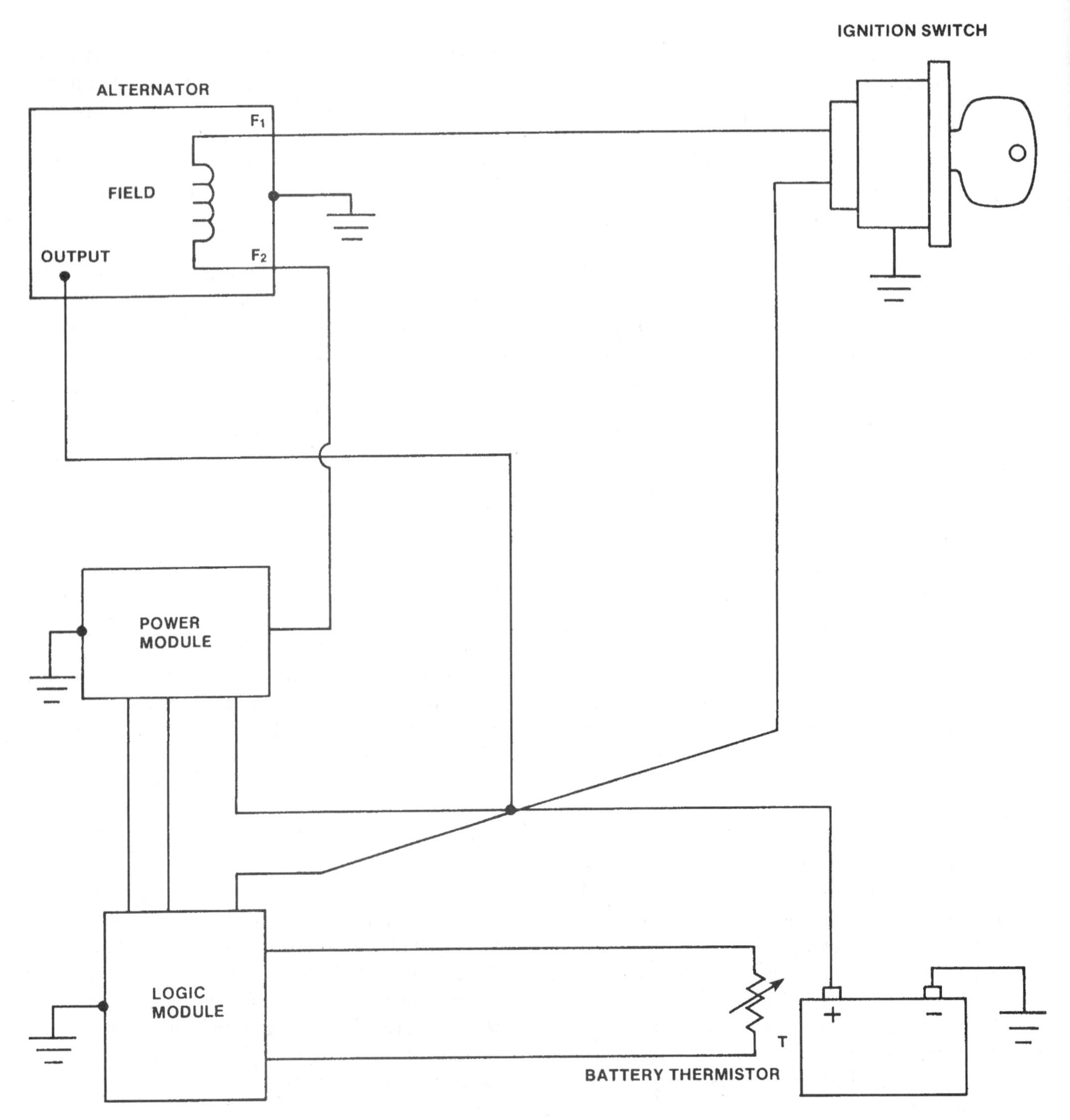

Fig. 4-14 Computerized voltage regulation allows the output of the charging system to be regulated independent of load (reprinted, by permission, from Santini, *Automotive Electricity and Electronics, 2E.* Copyright 1992 by Delmar Publishers Inc.).

important advantages. The main one is that the computer can modify or adjust the charging rate to compensate for such factors as engine speed, internal battery temperature (which may be different from underhood temperature), or the accessory load on the engine (Figs. 4-14). A small four-cylinder engine can really lug down if the regulator runs the alternator at full output at the same time the air conditioner is on. If the engine also has automatic idle speed control, the result can be a surging idle condition that is annoying to the driver.

By integrating the voltage regulator function

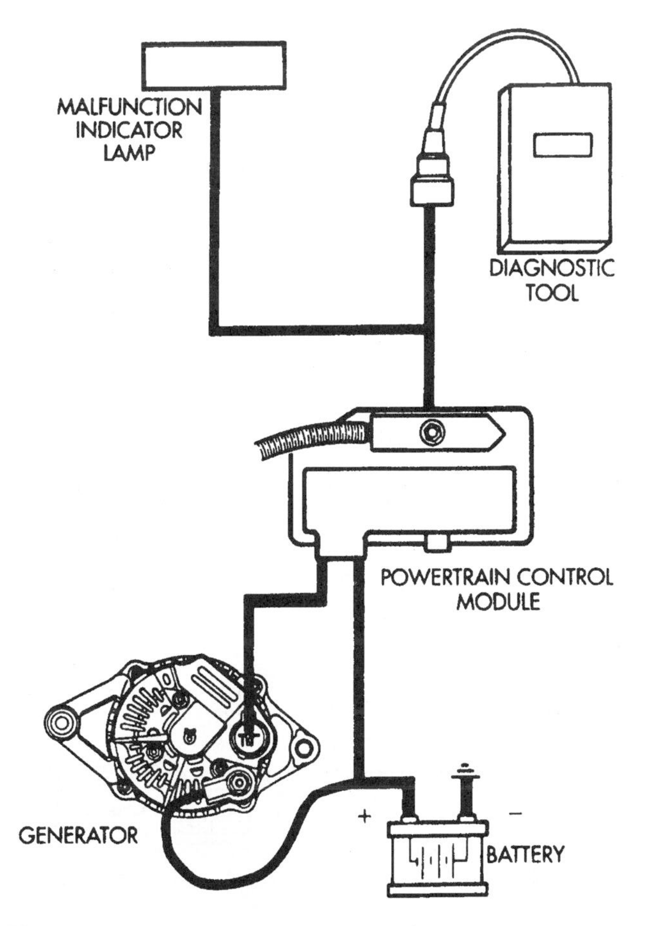

Fig. 4-15 Voltage regulation in Chrysler's charging system is handled by the powertrain control module. Tapping into the system with a scan tool can help identify charging problems (courtesy Chrysler).

into the computer, however, voltage regulation has become more integrated with the powertrain and electrical system. The computer compensates for such variables as engine speed and load to come up with a better schedule for charging the battery and meeting electrical needs. The computer already has the necessary inputs from its various sensors, so it can alter the alternator's charging curve to suit prevailing conditions. Instead of going to full charging output as soon as the engine is started or a load is applied, the computer may delay full charging until the engine can warm up, or the vehicle is off idle and is being driven. Under full throttle acceleration, the computer can also cut out charging temporarily to totally eliminate any parasitic horsepower loss, leaving more power for passing.

Another advantage of computerized voltage regulation is the ability to self-diagnose system problems (Fig. 4-14). On Chrysler's computer system, for example, a "code 41" means excessive or no field current. A "code 46" signals high battery voltage while a "code 47" warns of low voltage. By following the appropriate diagnostic charts in the manual, you can track down the fault and fix the problem.

CHARGING SYSTEM TESTING

A charging system that is working properly should produce a charging voltage of somewhere around 14 volts at idle with the lights and accessories off. When the engine is first started, the charging voltage should rise quickly to about two volts above base battery voltage, then taper off, leveling out at some point within the accepted specifications range for the vehicle.

It's important to note that the exact charging voltage will vary according to the battery's state of charge, the load on the vehicle's electrical system, and temperature. The lower the temperature, the higher will be the charging voltage. The higher the temperature, the lower the charging voltage. On a GM application, for example, the accepted voltage charging range is 13.9 to 14.4 volts at 80 degrees F. At 20 degrees F. below zero, the charging range is 14.9 to 15.8 volts. At 140 degrees F., the charging voltage is 13.0 to 13.6 volts.

Output can also be checked with an adjustable carbon pile, voltmeter, and ammeter. The carbon pile is attached to the battery and adjusted to obtain maximum output while the engine is running at 2000 rpm. Voltage and current outputs are then monitored on the instruments and compared to specs.

If a voltage check of the charging system shows low output, the standard question that follows is always "Is it the alternator or the regulator?" One way to find out is to bypass the voltage regulator using a procedure called "full fielding" the alternator. If the alternator produces the specified voltage or current output after full fielding, the problem is in the regulator (or the wiring), not the alternator.

The exact procedure for full fielding an alternator varies from vehicle to vehicle depending on how the alternator is wired. Always refer to a service manual for the exact procedure. The alternator and/or regulator can be damaged if the wrong test procedure is used. Generally speaking, the regulator is bypassed by connecting a jumper wire between the field (FLD or "F" terminal) and battery positive (BAT) terminal on the alternator (Fig. 4-16).

On older GM applications that have an integral

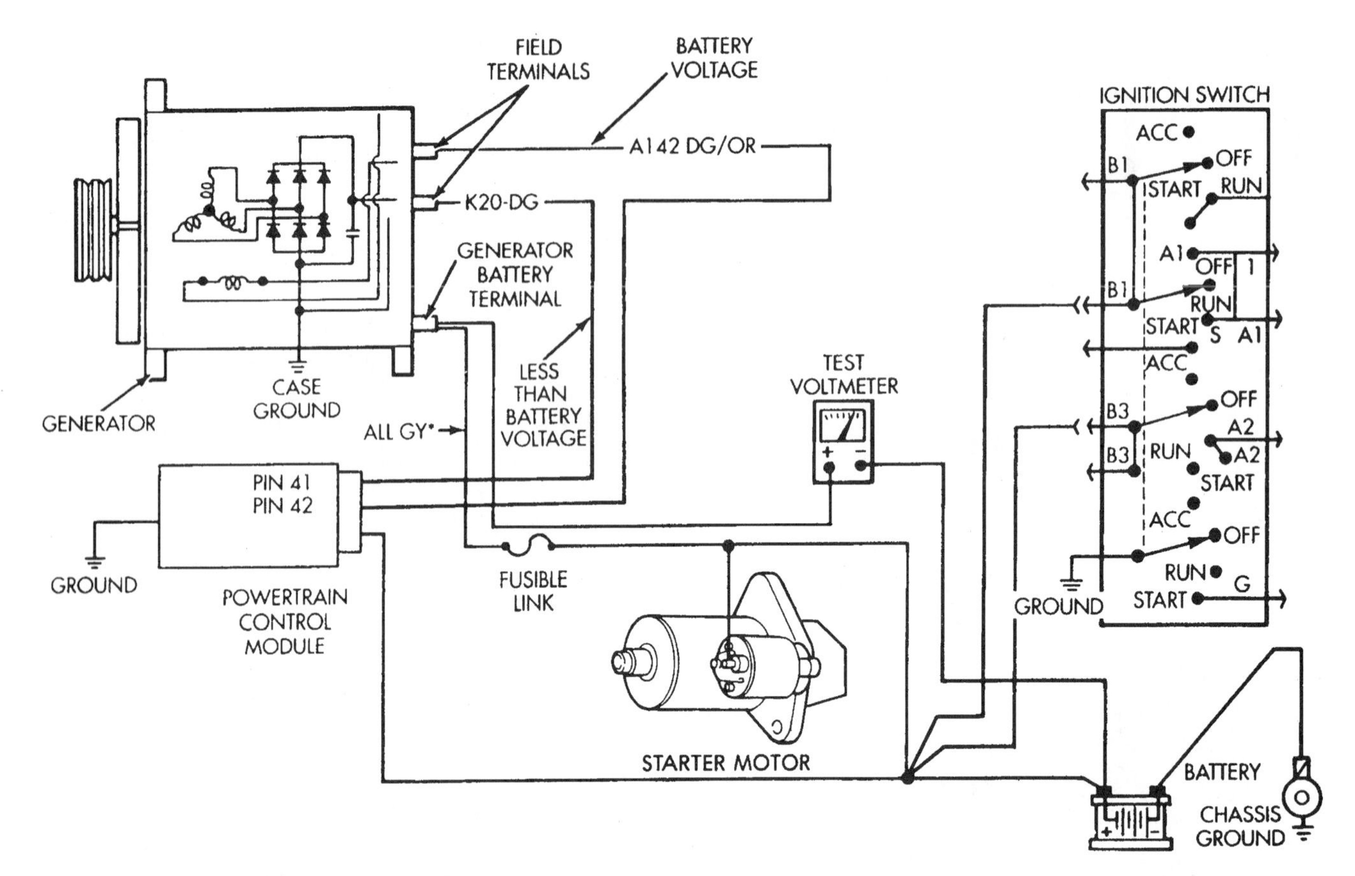

Fig. 4-16 Measuring charging output with a voltmeter (courtesy Chrysler).

regulator SI alternator, inserting the tip of a screwdriver through the D-shaped hole in the back of the alternator full fields the unit. This supplies full field current to the rotor which should cause the alternator (if good) to go to maximum output.

On Chrysler externally regulated alternators, battery voltage must not be applied to the "F" terminal. This system is full fielded by grounding the green wire at the regulator connector.

On externally regulated Ford alternators, the alternator is full-fielded by disconnecting the 4-wire connector from the regulator and jumping across the "A" and "F" terminals.

Either voltage or current output can be compared against the manufacturer's specifications to determine if the alternator is functioning at full capacity. Generally speaking, an alternator's output should fall within 10 amps or 10% of its rated capacity at 2000 rpm.

It's important to follow the manufacturer's full fielding test procedures exactly for several reasons. If only one diode or stator winding is bad, the alternator may still make enough electricity at high rpm to keep the battery charged but not enough at idle or low speed (Fig. 4-17).

If charging output goes up when the regulator is bypassed by full fielding, but falls back to zero when not full fielded, the regulator may have a poor ground connection. This is especially important on Ford and Chrysler systems. Poor or open wiring connections between alternator and regulator can also cause a charging problem.

A slipping drive belt, which is more of a problem with V-belts than serpentine belts, is one of the most common causes of undercharging. A belt that grips at idle or low rpm may slip when the alternator is under load. Checking belt tension and/or looking for glazing or burned streaks on the sides of the belt will tell you if it has been slipping.

THE STARTING SYSTEM

The starting system consists of four main parts: the ignition switch, a starting safety switch, a starter

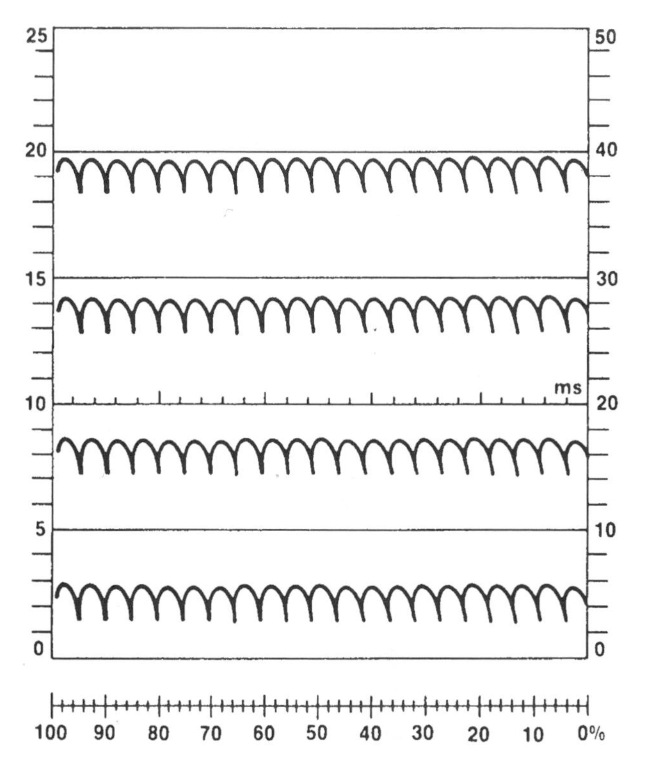

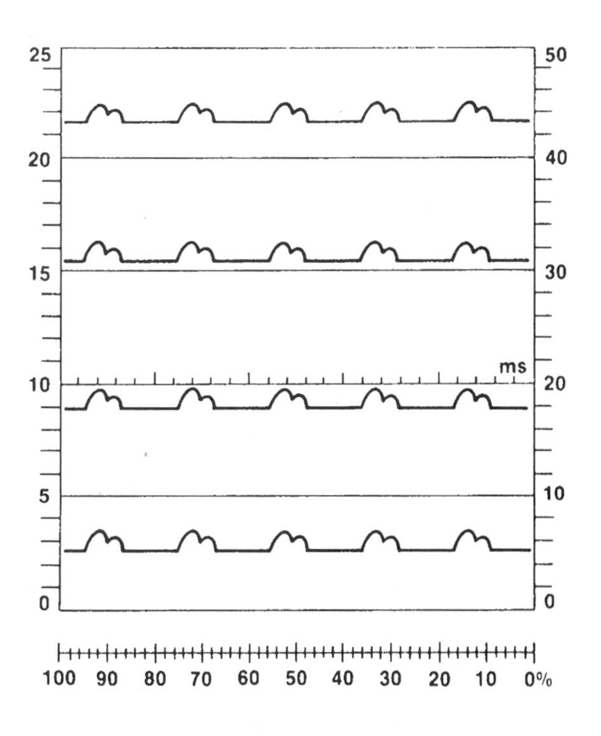

Fig. 4-17 Observing an alternator's output on an oscilloscope can identify charging problems. The pattern at the left shows a typical pattern for an alternator that's functioning properly. The pattern at right results from one open and one shorted diode (courtesy Sun Electric).

relay or solenoid, and the starter motor (refer to Figs. 4-18 through 4-21).

The ignition switch usually has at least four positions: "Off" cuts the power to the ignition system, starter and all accessories; "On" passes current to the ignition and accessories, but not the starter; "Start," which is spring loaded, passes current to the starter relay or solenoid while maintaining power to the ignition circuit; and "ACC" provides power to only the accessories when the engine is off. The ignition switch also has a fifth "lock" position which locks the steering wheel in a fixed position.

A park-neutral safety switch on the transmission shift linkage or a clutch pedal switch is included in the ignition switch circuit to prevent the car from starting in gear. On many vehicles with manual transmissions, the ignition switch will not crank the motor unless the clutch is depressed. In automatic equipped vehicles, the engine will only crank if the transmission is in neutral or park.

A misadjusted safety switch can prevent the starter from working, or it can allow the starter to crank even though the transmission is in gear or the clutch is engaged. If the switch is faulty, loose, or has a wiring problem, however, the starter circuit will be dead and the starter won't crank.

The starter relay or solenoid is a magnetic switch that controls the opening and closing of the starting circuit (Fig. 4-20). A "relay" uses a magnetic coil to pull a hinged armature above it down to close a pair of contact points. A "solenoid," on the other hand, pulls a plunger into the magnetic coil to close the contact points (see Chapter 3).

A relay or solenoid is necessary because the starter requires too many amps for the ignition switch itself to handle. A starter can pull 100 to 200 or more amps when cranking, which means the wires that carry the current must be very thick. By using the ignition switch to close the relay or solenoid switch, the current can go directly to the starter through the relay or solenoid rather than through the ignition switch.

Starters

The starter itself is a high-torque, direct current electric motor. It has a rotating armature, brushes, and a pair of field coils in the housing. The starter

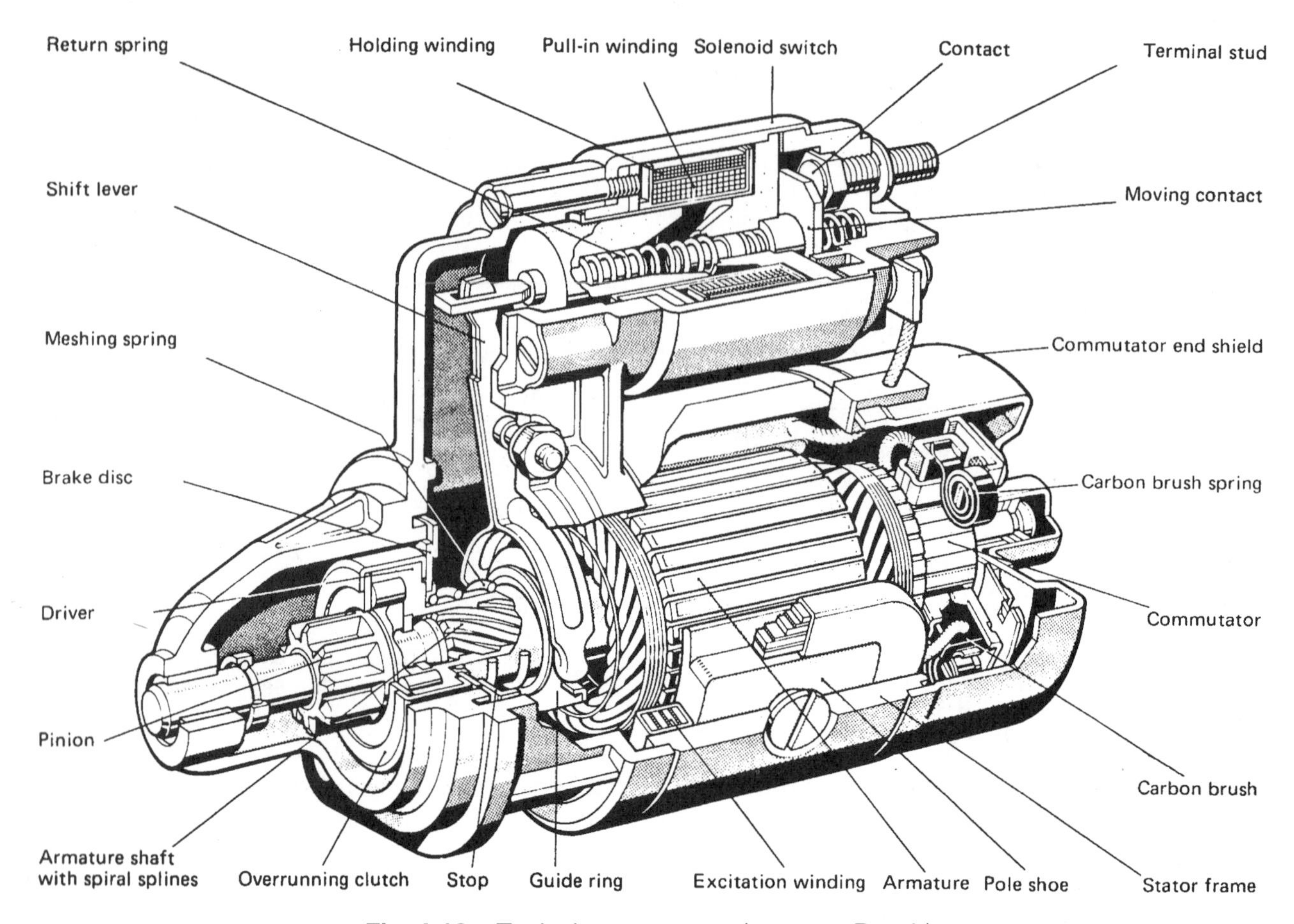

Fig. 4-18 Typical starter motor (courtesy Bosch).

(Fig. 4-18) uses the opposing magnetic forces of the armature and field coils to drive itself. Though most starters are "direct drive," some have an extra set of "reduction gears" to increase cranking torque. Reduction gears also allow a smaller motor to be used.

Another variation in starter design are "permanent magnet" starters (Fig. 4-19). This type of starter has powerful magnets glued to the starter motor case and uses no field coils. Current is delivered directly to the armature through the commutator and brushes. This allows for a lighter, more compact starter. Planetary reduction gears may also be used for increased torque.

The starter shaft has a gear and drive mechanism which engages the flywheel to crank the engine. On starters that have a solenoid mounted on them, the solenoid works a lever to push the starter drive out to engage the flywheel. The drive mechanism is only supposed to engage the flywheel when the starter is cranked, then disengage as soon as the engine starts. An overrunning clutch mechanism in the drive mechanism helps to protect the starter against damage if the driver continues cranking after the engine has started. If the starter fails to disengage, the engine will overrev it and literally blow it apart.

STARTING SYSTEM DIAGNOSIS

Cold weather is hardest on starters, not because of any inherent weakness in the starter itself, but because of changes which occur elsewhere that make the starter's job doubly difficult. Cold temperatures cause the oil inside the engine to thicken, and subzero temperatures can turn oil into molasses. At the same time, lower temperatures inhibit fuel vaporization and slow the chemical reactions inside the battery that create current. As temperatures plunge, fuel ignites less easily and the battery produces fewer amps. Consequently, the starter has to work harder to pump life into a cold engine. Relentless cranking can overheat the starter and damage its windings. And once that happens, the engine won't start no matter how many booster amps are force-fed to it.

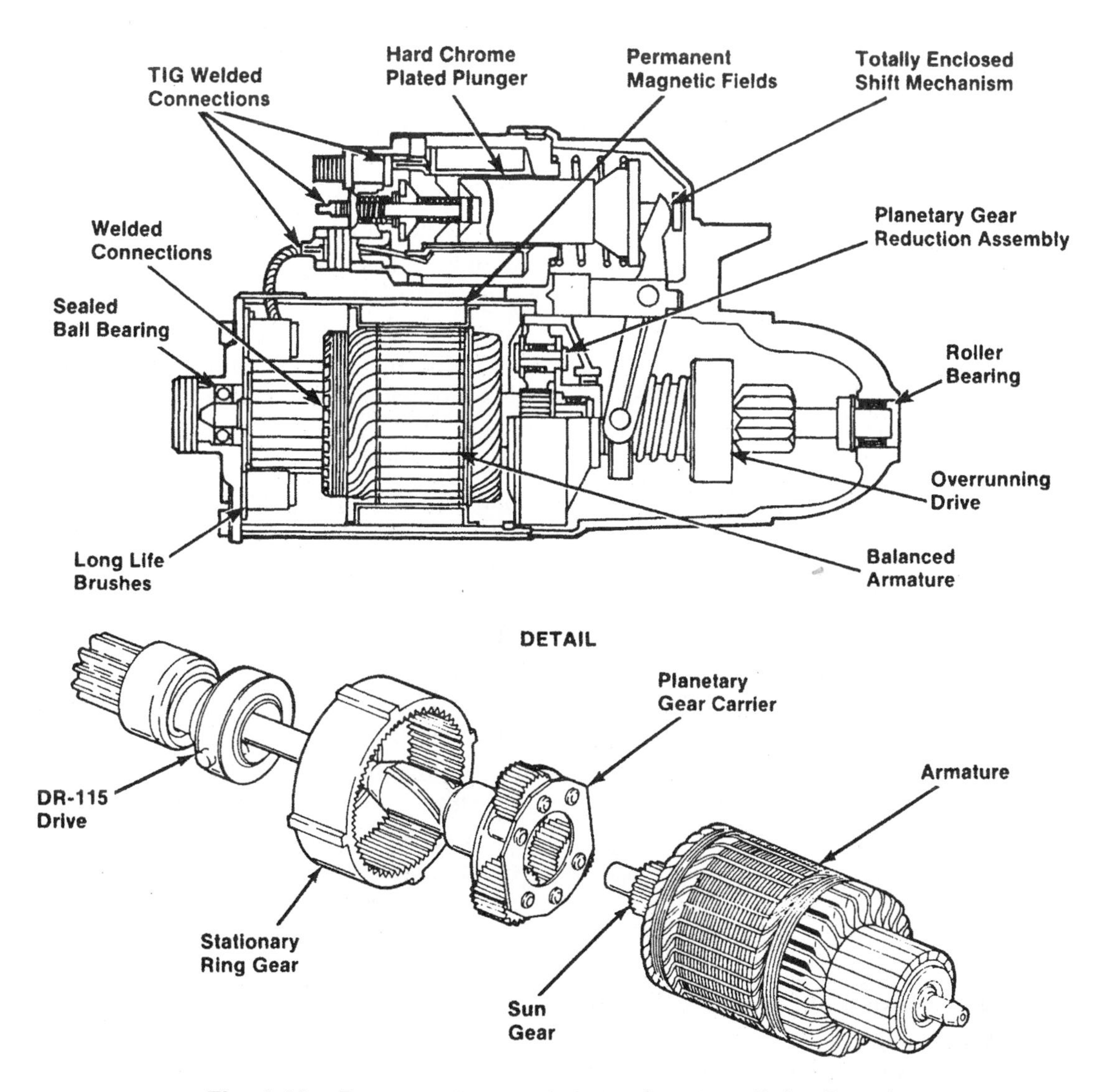

Fig. 4-19 Permanent magnet starter (courtesy Delco Remy).

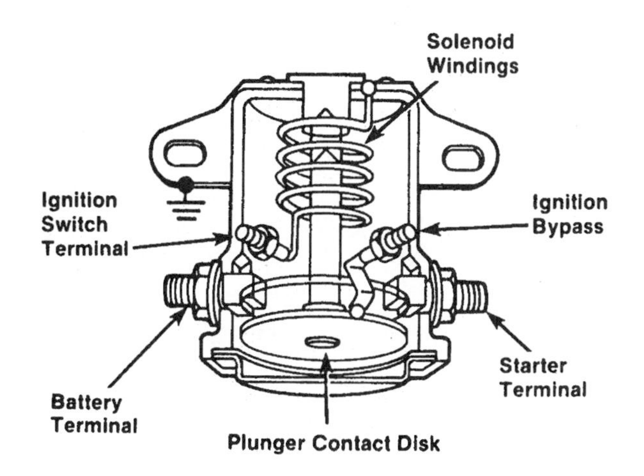

Fig. 4-20 This type of starter relay routes high-amperage current from the battery to the starter motor (courtesy Ford).

The most frequent reason for starter failure is overcranking. Unless the starter is allowed to rest every 30 seconds or so to cool off, continuous cranking can overheat and damage soldered brush connectors, or field or armature wires—especially if the starter is being force-fed more than its usual diet of amps.

Overboosting the battery to coax a dead engine back to life is probably the most common mistake well-intentioned service personnel make on service calls. If an engine won't start with a normal jump, doubling the dosage of amps or volts may do more harm than good. Too many volts can be especially damaging to sensitive electronics. Most electronics

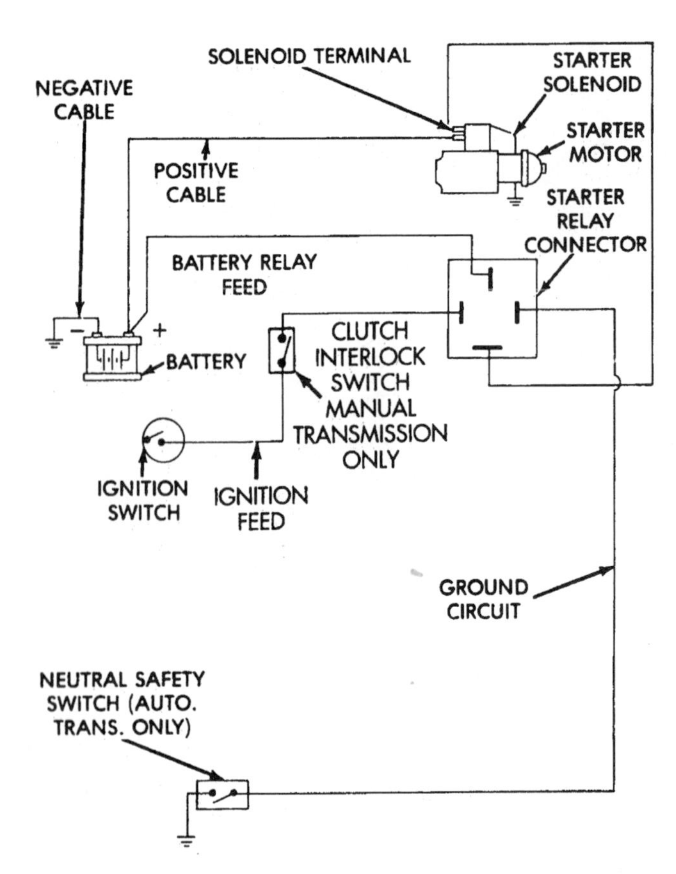

Fig. 4-21 Typical starter motor circuit (courtesy Chrysler).

can withstand brief exposure of up to 18 amps, but any more can be very dangerous.

Starters also wear out. Brushes and bushings take a beating as do the starter drive, solenoid, armature, and field coils. A starter may fail to crank an engine for any of a number of reasons, many of which are not the starter's fault. That's why diagnosis is so important. Yet some technicians don't subscribe to the "diagnose-before-repair" theory of auto repair. They play by hunch, and if the symptoms seem to indicate a bad starter then that's what they replace—only to discover it wasn't the starter after all!

Considering the way that many so-called starter problems are "diagnosed," it's not surprising that nearly 80% of the "defective" starters that are returned under warranty work perfectly when tested. To accurately diagnose starter problems requires not only an understanding of the basics (Fig. 4-21), but also a willingness to investigate all possible causes.

An engine that turns sluggishly when cranked is often an engine that's starved for amps. The first thing to check, therefore, is the battery. Is it fully charged? Can it maintain 9.6 volts throughout a load test? Does it have sufficient amp capacity to crank the engine?

The next items that deserve attention are the battery cables. Are the connections tight and corrosion-free? All it takes is a thin film of oxide between the cable clamp and battery post to effectively choke off the supply of amps from the battery. Using a voltmeter or multimeter to check for voltage drop across the battery cable connections can help you find this kind of problem.

A quick check you can make is to switch on the headlights and note what happens when the engine is cranked. If the lights go out, it usually means a bad battery cable connection can't handle enough current to crank the engine and feed the lights at the same time. Check both battery posts (Fig. 4-22), the solenoid connection, starter connections, and ground cable. Remember, a circuit will only flow as much current as the ground connection can return to the battery. A voltage drop check of the cable connections should show no more than 0.1-volt drop at any point, and no more than 0.5 volts for the circuit on either side of the starter motor.

Cranking problems can be created when undersized battery cables are installed. Some cheap replacement cables have small gauge wire encased in thick insulation. The cable may look normal sized, but the small-diameter wire may not be capa-

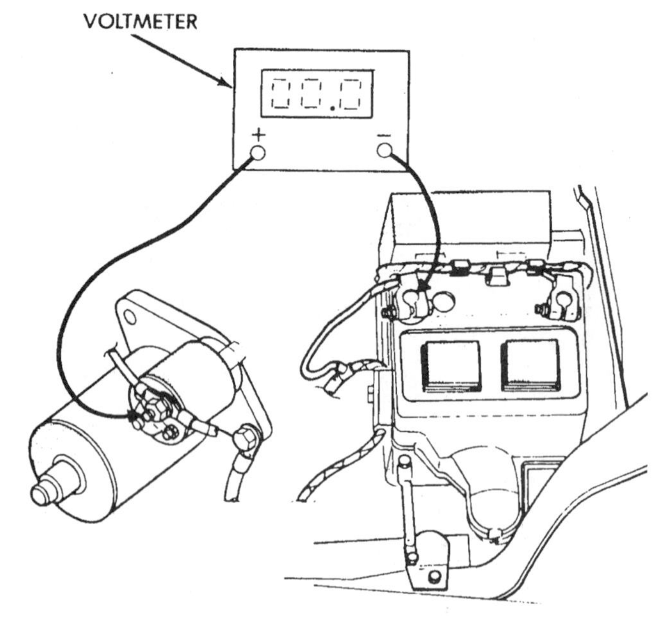

Fig. 4-22 Using a voltmeter to check for voltage drop in the positive battery cable connection (courtesy Chrysler).

ble of carrying an adequate amp load to the starter. During summer months when cranking loads are light, an undersized cable may work fine. But when cold weather arrives and cranking demands increase, the engine may experience hard-starting problems because the battery cables can't carry enough amps.

The cold cranking amp (CCA) capacity of the battery is also extremely important. If it doesn't match the amp requirements of the engine, it won't be able to start the engine in cold weather. Remember, the minimum is 1 CCA per cubic inch of displacement, with 2 per cubic inch being best for reliable subzero starting.

On some applications (GM cars with an "MX high-torque" starter that draws about 100 amps more than a standard starter), a larger battery and/or cables may be necessary for reliable cold-weather starting.

Increased starter drag, especially in starters with reduction gears, can also cause sluggish cranking and hard starting. Worn or damaged gears can increase internal friction in the starter, causing it to draw more amps than usual when it cranks the engine. Excessive or unusual starter noise when cranking might indicate this type of problem.

When a starter draws too much current, it can sap all the juice in the system and not leave enough for the ignition. The engine may crank, but won't start. A starter current draw or load test should be performed to check starter performance. The test only takes a minute or so to run, and it quickly tells you whether or not the starter is drawing too many amps.

Pulling a starter and using a pair of jumper cables to "bench test" it is not a very accurate test. It tells you only whether or not the starter will spin, not how much torque it can deliver, or how many amps it draws. For accurate starter testing, a bench tester that can test the starter under load should be used.

Excessive starter draw can be caused by grounds or shorts in the armature or coil windings. It can also result from increased internal friction due to shaft bushings that bind, or an armature that is rubbing against the housing. The magnets in permanent-magnet starters can sometimes break or separate from the housing, dragging against the armature.

If a starter is noisy, it may be dragging. But noise can also be caused by starter misalignment. A high-pitched whine after starting may signal a starter drive that's hanging up. Variations in the machining of the bellhousing can cause this kind of problem. It may be necessary to shim the starter by placing washers between it and the bellhousing. If the starter teeth are too close, they can bind when the engine gets hot. The recommended clearance is usually about .020 inches. When properly aligned, the drive teeth should engage the flywheel teeth about three-quarters of the way down from the top of the each tooth. To check clearances, you have to push a starter drive out (by inserting a screwdriver through a hole in the bottom of the starter housing) and then use a hooked wire gauge to check the clearance between the teeth.

A loose starter may crank an engine slowly, noisily, or not at all. Loose bolts will make for a weak ground connection. The starter may also flop around, slip, chatter, or fail to engage depending on how loose it is. That's why you're supposed to use lock washers or locking compound on the starter bolts, and torque them to specs.

If the starter appears to be dead, and does absolutely nothing when you turn the key, check the battery voltage. Most starters need a minimum of 10.5 volts to engage and crank. Also check the solenoid and starter terminals for battery voltage when you attempt to crank. If there's no voltage reading, check for voltage at the ignition switch lead to see if a misadjusted or open neutral safety switch is breaking the circuit.

If the headlights continue to shine brightly when you attempt to crank the engine, it usually means voltage isn't reaching the starter. If the lights go dim and there's no cranking, the battery may be low, or the starter may be locked up or suffering from high internal friction. There's also the possibility that the engine may be seized or have a cylinder full of oil, coolant, or gasoline.

A "click" means the solenoid is engaging, but there may not be enough juice to spin the starter. A bad battery cable connection, a poor solenoid ground connection, or high resistance in the solenoid or starter could be the problem. Check the state of charge of the battery, and all cable connections for excessive voltage drops. Then try bypassing the solenoid to see if the starter will spin. If it cranks, the problem is in the solenoid. If it doesn't, the starter is at fault.

Sometimes a starter will spin but won't crank the engine. This is usually an engagement problem due to a weak solenoid or a defective starter drive. A starter drive that is on the verge of failure may engage briefly but then slip. The starter drive has a one-way overrunning clutch mechanism that you can check (and replace if necessary) once the starter is out of the car. The drive should turn freely in one direction but not in the other if good. A bad drive will turn freely in both directions or not at all. If a drive locks up, it can overrev and destroy the starter.

The weak point on Ford sliding-pole starters is the pole shoe that pulls in towards the armature to engage the starter. This starter needs a minimum of 10.5 volts and about 300 to 400 amps to work—otherwise it just clicks. To make matters worse, as the starter wears, the pivot bushing sometimes hangs up and prevents the engagement pole shoe from being pulled down. The starter will spin but it won't engage the flywheel.

A similar problem can happen to GM starters. Sometimes the solenoid is too weak to overcome the force of the return spring. When this happens, the starter won't work. If replacing the solenoid doesn't help, a lighter return spring may be needed.

Missing or damaged flywheel teeth can prevent the starter from engaging and cranking. So always inspect the condition of the flywheel teeth when changing a starter.

Finally, hard-starting problems can be caused by conditions totally unrelated to the starter or battery:

- Oil that's too heavy a viscosity for prevailing temperatures (straight 30W oil at subzero temperatures, for example).
- Engine mechanical problems such as low compression, no compression, hydrolocking, a seized piston or bearing, broken or slipped timing belt, etc. (see Chapter 7)
- Fuel problems caused by such things as a defective or misadjusted choke, a faulty cold-start injector, defective fuel pump relay, plugged fuel line, etc. (see Chapters 8 and 9)
- Ignition problems such as fouled or worn spark plugs, bad plug wires, cracked distributor cap, cracked rotor, weak coil, etc. (see Chapters 5 and 6).

CHAPTER 5

Ignition System Operation

The function of the ignition system is to provide an accurately timed spark of sufficient intensity to ignite the air/fuel mixture inside the engine's cylinders (Fig. 5-1). The amount of voltage needed to fire the spark plugs can range from 5,000 to 25,000 volts or higher.

Why does it take so much voltage? There are three basic reasons:

1. It takes high voltage to overcome the resistance created by the air gap between the spark plug's electrodes. The little gap between the electrodes may not seem like much, but to a tiny electron it's an enormous distance to jump across. The only way the electron can make it is with a big enough push, and that means lots of voltage.

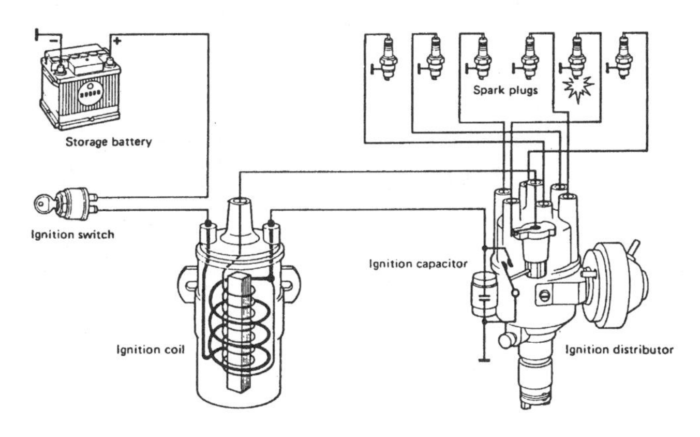

Fig. 5-1 Basic ignition system (courtesy Bosch).

2. Engine compression makes it even harder for the spark to jump the gap. The higher the compression ratio, the greater the voltage required to push the spark through the densely packed fuel mixture.
3. Temperature is a factor. The higher the engine temperature, the greater the firing voltage required to overcome the resistance. The temperature of the center spark plug electrode is also important. The hotter the spark plug center electrode, the less voltage is required to jump the gap. A spark is more easily generated from a hot electrode than a cold one. The long path that heat must follow from the center electrode through the ceramic insulator to the spark plug shell keeps the tip of the plug hot. This makes it easier for the spark to jump from the center electrode to the much cooler ground electrode on the side of the plug tip. If the polarity of the plug were reversed so that the spark had to jump from the cooler ground electrode to the center electrode, up to 40% more voltage would be needed to fire the plug!

Once the spark succeeds in bridging the gap, the firing voltage required to sustain the spark is reduced because the air in the gap is temporarily "ionized." This means the air is made conductive by blasting it with electrons. The bombardment overloads the atoms in the air with extra electrons. This transforms them from electrically neutral to conductive, thus forming a path across which the spark can flow with much less effort.

It only takes a few millionths of a second for the initial spark to occur, but once it does the firing voltage drops to as low as 3,000 volts for the remaining duration of the spark. The actual duration of the spark is measured in hundred-thousandths of a second, and may last for up to 20 degrees of crankshaft rotation at high rpm. Though it may not seem like much time, it is long enough to ignite the compressed air/fuel mixture.

COIL THEORY

The ignition coil generates 25,000 to 40,000 or more volts of electricity from a primary voltage of only 12 volts by "electromagnetic induction." The coil contains two sets of copper wire windings around a central iron core. The outer windings inside the coil are the "primary" windings. These consist of a few hundred turns of relatively heavy-gauge wire. The inner windings are the "secondary" windings, which consist of several thousand turns of small-gauge wire. There may be ten times as many turns of secondary windings as there are primary windings. The iron core in the center of the windings is made of thin strips of iron "laminated" together. The strips of iron sandwiched together help to concentrate the magnetic field that's created when current flows through the coil's primary windings. In essence, the coil is nothing more than an electromagnet. But because it has two sets of windings, it acts like a transformer.

When current flows through the outer primary windings, it creates up a strong magnetic field. This in turn induces a voltage in the secondary windings, but not enough to fire the spark plugs. That doesn't occur until the primary current is suddenly cut off.

The instant the primary voltage is cut, the magnetic field begins to collapse (Fig. 5-2). The stored energy in the field tries to find someplace to go by first trying to keep the current flowing in the primary windings. Since that isn't possible, it creates a powerful voltage surge in the secondary windings (up to 40,000 volts depending on the application). The available voltage quickly exceeds what is needed to bridge the electrode gap at the spark plug, and the plug fires. That's why automotive ig-

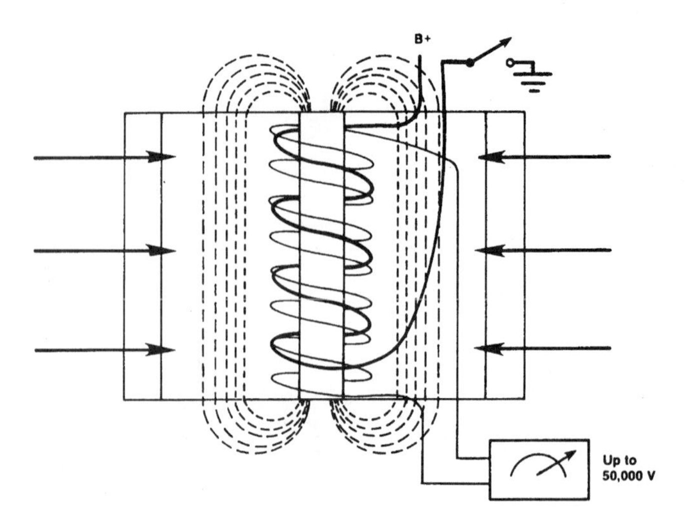

Fig. 5-2 The collapse of the magnetic field that occurs when voltage is cut to the primary windings induces a high-voltage surge in the coil's secondary windings (reprinted, by permission, from Santini, *Automotive Electricity and Electronics, 2E*. Copyright 1992 by Delmar Publishers Inc.).

nition systems are sometimes called "inductive-discharge" ignitions.

Coil Polarity

At the top of the ignition coil are the primary terminals, and the center high-voltage terminal. One primary terminal is marked positive (POS) or (+), and the other negative (NEG) or (-). The positive terminal is connected to the "hot" side of the ignition circuit. The negative terminal is a ground path that connects to either the contact points in the distributor or the ignition module in an electronic ignition system.

Inside the coil, the primary windings form a continuous path from the negative to positive terminal. The high-voltage secondary windings have one end connected to the positive primary terminal, and the other end to the high-voltage center terminal in the top of the coil. This arrangement gives the high-voltage center terminal a "negative polarity." This means that the spark will travel from the coil to the spark plug, where it will jump across the plug gap from the hot center electrode to the cooler side-ground electrode. This lowers the required firing voltage because, as mentioned earlier, a spark requires less voltage to jump from a hot electrode to a cooler one.

If the primary coil leads are accidentally reversed (that is, the positive terminal is attached to the distributor wire and the negative terminal is attached to the "hot" wire), the polarity of the coil will be reversed. The center high-voltage electrode will be positive, causing the coil to draw the spark backwards. The spark now has to jump from the cooler side-ground electrode to the hot center electrode. This increases the required firing voltage by up to 40%. If the coil can't make the extra voltage, the plug may not fire, causing the engine to misfire.

BASIC IGNITION CIRCUIT

The ignition system is actually two systems in one. The "primary" or low-voltage side of the system is the portion that triggers the ignition coil. It includes the battery, ignition switch, the primary resistor, the breaker points or electronic pickup and module, and the primary windings inside the ignition coil. The "secondary" or high-voltage side of the system is the portion that delivers the spark to the spark plugs. The secondary ignition components include the secondary windings in the coil, the distributor cap and rotor, the ignition cables, and the spark plugs.

The ignition systems operates as follows:

1. When the ignition switch is turned on, battery voltage is applied to the primary circuit.
2. A primary resistor (or resistor wire) between the ignition switch and coil may be used to reduce the voltage and limit the amount of current that can flow through the circuit. In point ignition systems, the resistor helped prolong the life of the points. Resistors are not used with most electronic ignition systems.
3. Voltage is applied to the positive terminal on the ignition coil. But nothing happens until the negative coil terminal is grounded to complete the circuit.
4. In the older breaker point ignitions, the circuit was completed when the breaker points closed. But with electronic ignition systems, the circuit isn't completed until the ignition module power transistor switches to ground.
5. When the circuit is complete, current flows through the primary windings inside the coil and creates a magnetic field.
6. When the switching device (breaker points or ignition module) breaks the circuit, the collapsing magnetic field in the coil induces a high-voltage surge in the coil's secondary windings.
7. A high-voltage current passes from the coil to the distributor.
8. The high-voltage surge jumps across the air gap between the tip of the rotor and one of the ignition cable terminals inside the distributor cap. It then passes on to a spark plug through the cable.
9. The current arcs across the electrode gap at the tip of the spark plug, producing a spark and igniting the air/fuel mixture in the cylinder.

BREAKER POINTS

Breaker points have not been used in new car ignition systems since 1975. But understanding how a breaker ignition works makes it much easier to comprehend electronic ignition. Breaker points and an electronic ignition are both switching devices for the coil's primary current. Both perform exactly the same function. Breaker points do the switching mechanically by opening and closing contact points. An electronic ignition does the switching electronically with a transistor triggered by a magnetic pickup coil.

Breaker points were replaced by electronic ignition systems because the latter do not wear or require periodic adjustment or maintenance. There were two main problems with breaker points. (1) Wear caused by the rubbing block against the distributor cam gradually narrowed the point gap. This resulted in dwell and timing changes that had an adverse effect on engine performance. (2) Arcing across the contact points caused them to pit, resulting in increased resistance that reduced voltage to the coil. This could lead to misfiring, higher emissions, poorer fuel economy and performance. A condenser attached to the points helped reduce arcing somewhat to prolong point life. Even so, contact points usually had to be replaced every 12,000 to 15,000 miles as part of a periodic tune-up.

Electronic ignition, on the other hand, has no moving parts to wear out. Nor are there any mechanical contact points to pit or adjust. Once set, electronic ignition maintains consistent performance indefinitely. This virtually eliminates any need for maintenance (except for replacing the spark plugs periodically, and inspecting the distributor cap, rotor and wires).

Point gap and Dwell

On breaker point ignitions, the gap between the points had to be carefully adjusted. This setting determined not only the instant at which the points opened and closed (and thus ignition timing), but also how long the coil received its dose of primary voltage, called "dwell" (Fig. 5-3). Point dwell could be set with a feeler gauge, but a more accurate method was to use a dwell meter. This displayed the actual number of degrees of dwell while the engine was idling.

When used for this purpose, a dwell meter showed the number of degrees of cam rotation (called "cam angle") during which the points remained closed. But dwell is actually the amount of "on" time measured in degrees of distributor rotation that the points are making contact and routing voltage to the coil. Dwell meters are no longer used for tuning engines, except for checking the dwell on a feedback carburetor with a mixture control solenoid (see Chapters 8 and 17).

Decreasing the point gap increased dwell. Thus as the points wore, dwell would increase. Though this actually increased coil saturation and firing voltage, it also retarded ignition timing (about 1 degree for every 1 degree change in dwell). So as the points wore down, engine timing became more and more retarded, which reduced performance and fuel economy.

ELECTRONIC IGNITION

An electronic or "solid-state" ignition system has no breaker points. The coil on-off switching function is performed electronically inside the "ignition module" by a power transistor. The module, in turn, receives a trigger signal from a pickup inside the distributor, which functions essentially as a piston or crankshaft position sensor.

There are basically three kinds of solid-state electronic ignitions: magnetic impulse, Hall effect, and optical (light sensing). Magnetic impulse and Hall effect are most commonly used on passenger car and light truck applications, with optical systems (such as GM's Opti-Spark and Nissan's ECCS) being used in limited applications.

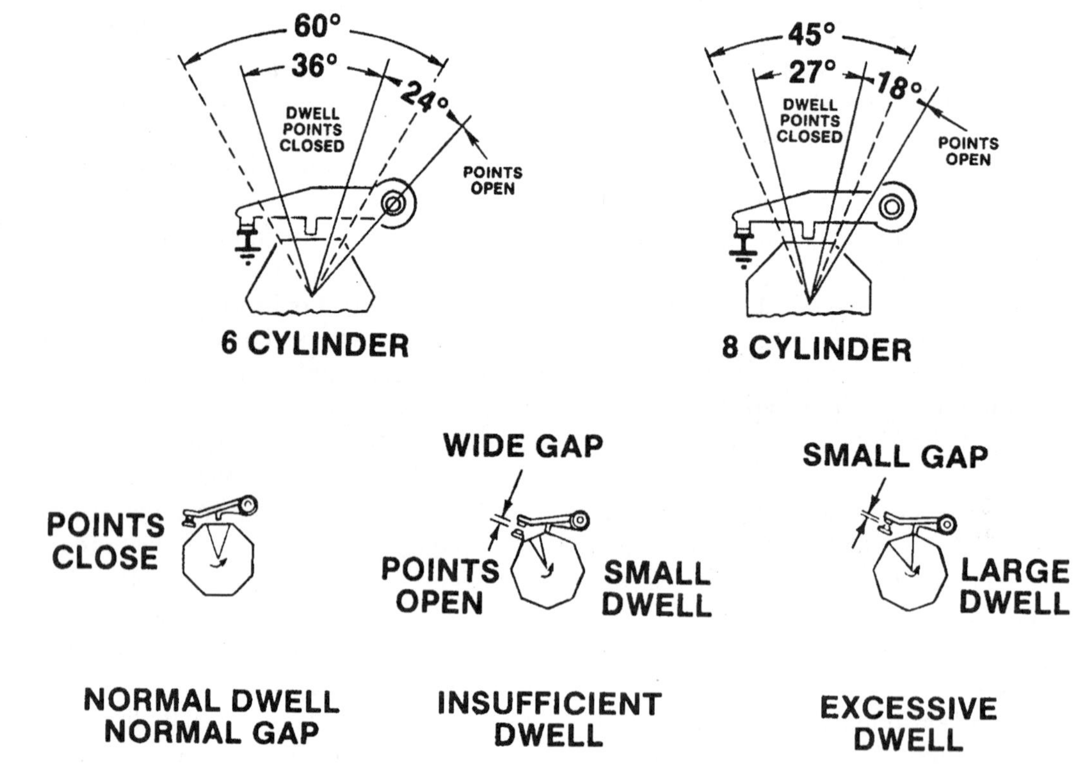

Fig. 5-3 Effect of point gap on dwell (reprinted, by permission, from Santini, *Automotive Electricity and Electronics, 2E.* Copyright 1992 by Delmar Publishers Inc.).

Electronic ignition systems are capable of producing higher firing voltages than breaker point systems. They don't have to deal with the problem of point arcing. Thus higher voltage or full battery input voltage can be used for a hotter spark. (Some systems still use a ballast resistor to reduce voltage, and therefore don't produce as hot a spark as others.) Electronics also perform better at high rpm because there is no danger of point bounce. The rpm capabilities of the ignition system are limited only by coil output and module circuitry.

In addition to switching on and off the primary voltage to the ignition coil, many modules also perform a "current limiting" function. This is found on systems that do not have a primary ballast resistor. When the primary current to the coil is switched on, the magnetic field in the coil builds until the coil is fully saturated. Once full saturation is reached, the module reduces the coil voltage to prevent coil overheating.

Another function that many electronic ignition modules provide is variable dwell. Such modules typically increase coil dwell at higher rpm to make sure the coil is fully saturated before it fires. As engine speed increases, there is less and less time between cylinder firings to achieve complete coil saturation. So by increasing the dwell, the module assures maximum coil output at high speed. If this were not done, the engine might misfire at high rpm due to a weak spark.

Magnetic Pickups

This type of electronic ignition replaces the breaker points with a "magnetic pickup" assembly (called a "stator" by Ford and a "pole piece" by GM) consisting of a permanent magnet, coil, and pole piece (Fig. 5-4). Most systems use a "trigger wheel" (which Ford calls an "armature" and Chrysler calls a "reluctor") in place of a cam on the distributor shaft. The pickup coil is connected to a "control module" which performs the actual on-off switching of the coil primary current.

The trigger wheel, which is the only moving part, is made of anti-magnetic steel to provide a low "reluctance" (low resistance) path for magnetic lines of force. The trigger wheel has one tooth for each cylinder in the engine. As the trigger wheel rotates, each tooth momentarily aligns with the pole piece. As this happens, the lines of magnetic force from the permanent magnet in the pole piece begin to concentrate in the trigger wheel. This increases the strength of the magnetic field, which causes a small alternating current (AC) to be induced in the pickup coil.

As the trigger-wheel tooth moves past the pickup magnet, the magnetic attraction between tooth and pole piece begins to decrease. The magnetic field drops, causing the polarity of the pickup coil to reverse direction, going from positive to negative voltage. This triggers the module's power transistor to

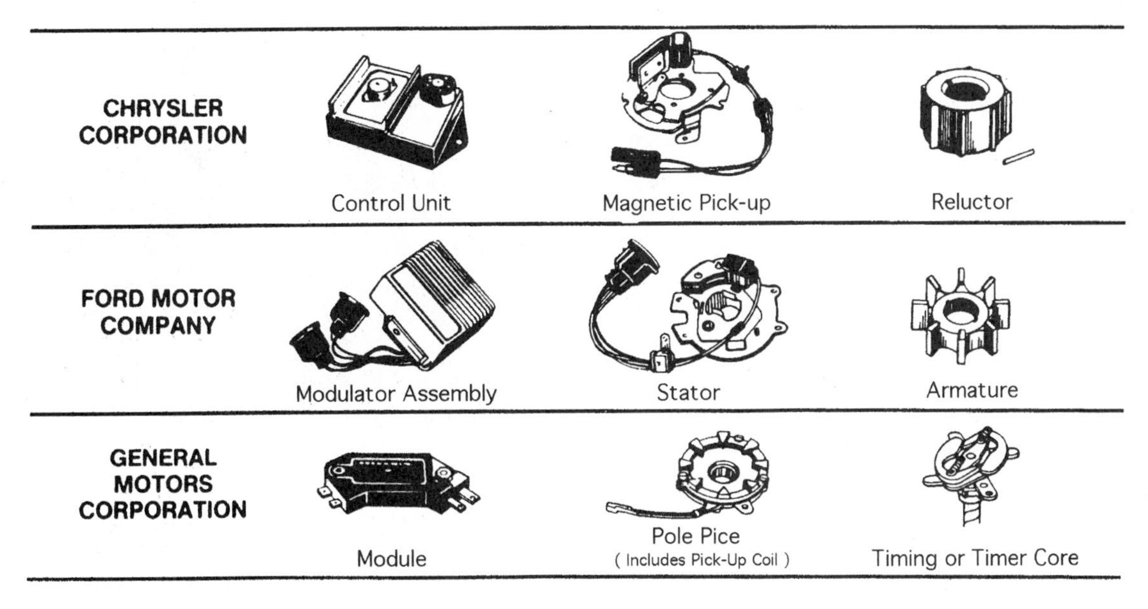

Fig. 5-4 Electronic ignition components (courtesy Borg Warner).

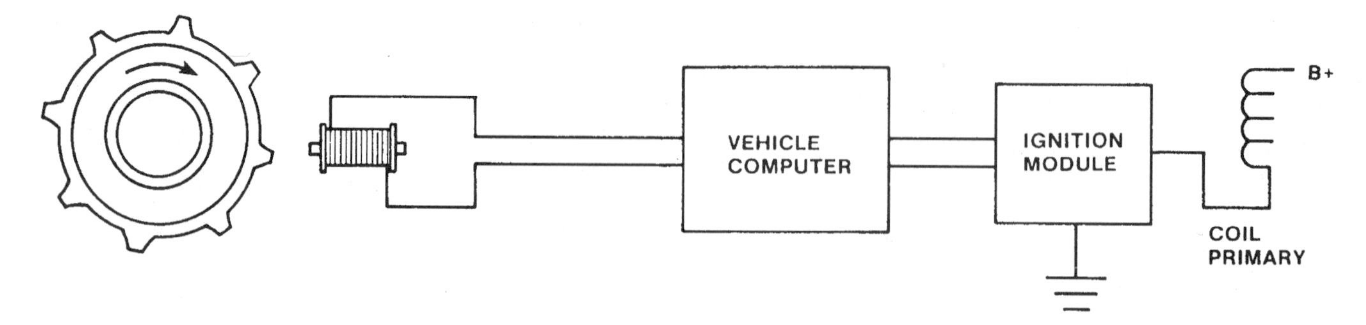

Fig. 5-5 The current induced in the magnetic pickup sends an ignition signal to the engine computer and ignition module (reprinted, by permission, from Santini, *Automotive Electricity and Electronics, 2E.* Copyright 1992 by Delmar Publishers Inc.).

cut the voltage to the ignition coil (Fig. 5-5). The magnetic field collapses in the coil, creating a voltage surge and the coil produces a spark, just as with a breaker point ignition. A timer circuit within the ignition module decides when to turn the coil on again. So dwell is determined electronically rather than mechanically, and is not adjustable. Depending on the application, dwell may remain fixed as engine speed changes, or it may increase or decrease.

Some pickups are designed differently. GM, Mitsubishi, and Hitachi (Datsun/Nissan) pickups, for example, use a circular magnet that wraps all the way around a trigger wheel. There is one protruding notch in the pole piece for each cylinder. All are evenly spaced around the circumference of the ring. As the notches in the pole piece and those on the trigger wheel align, a current is induced which then triggers the ignition module.

On some magnetic systems, the only adjustment required is the air gap between the trigger wheel and pickup coil. If this gap is too wide, it can cause starting problems. As the air gap between the trigger wheel and pickup widens, the field strength between the magnet and the trigger-wheel teeth decreases. This results in a weak signal voltage in the pickup coil. The air gap should be checked with a nonmagnetic (brass) feeler gauge, and adjusted to specs by repositioning the pickup coil. Once set, the gap should not change for the life of the unit.

Pickup coils can short-out or open. If this happens, the trigger signal to the module is lost, and the coil ceases to fire. On GM high energy ignition (HEI) pickup coils (Fig. 5-6), broken wire leads are a common problem. An ohmmeter can be used to test for shorts between the pickup and ground. Connect one ohmmeter lead to the base of the distributor, and the other lead to first one and then the other pickup lead. The ohmmeter should read infinity if the pickup is not shorted. The resistance of the pickup coil itself can be tested by attaching the two ohmmeter leads to the two pick leads. If the pickup coil is not within the manufacturer's specifications, it should be replaced.

In some vehicles (older Chrysler Lean Burn, for example), the distributor contains two pickup coils. One is the "start" pickup while the other is the "run" pickup. For easier starting, the start pickup is offset with respect to the run pickup to reduce spark advance when cranking. When the engine is cranked, the ignition module selects the start pickup for its basic trigger signal. Then, it switches over to the run pickup for normal operation. The same thing

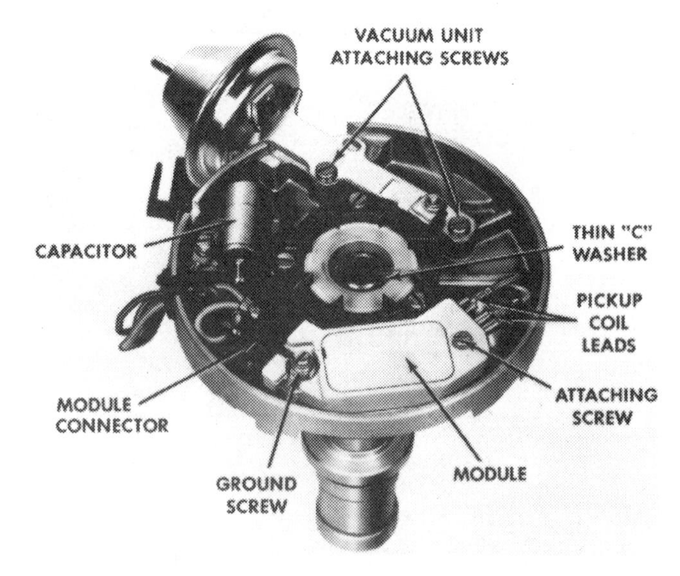

Fig. 5-6 General Motors high energy ignition (HEI) distributor components (courtesy GM).

can be accomplished in other systems by the ignition module itself. In Ford ignition modules, for example, timing is retarded slightly when cranking by a special circuit in the module. Some cheap replacement modules leave out the retard circuit to reduce costs.

The ignition module itself may be mounted in several places: inside the distributor (GM and many imports), on the base of the distributor (Ford TFI modules), on the firewall or inner fender (older Ford Dura-Spark modules and certain Chrysler ignitions), or integrated into the spark control computer (older Chryslers).

When replacing a GM HEI or Ford thick film integrated (TFI) module, it is extremely important to apply a coating of dielectric silicone grease to the base of the module prior to installation. The grease helps to conduct heat away from the module. Forget the grease, and the module may soon fail due to overheating.

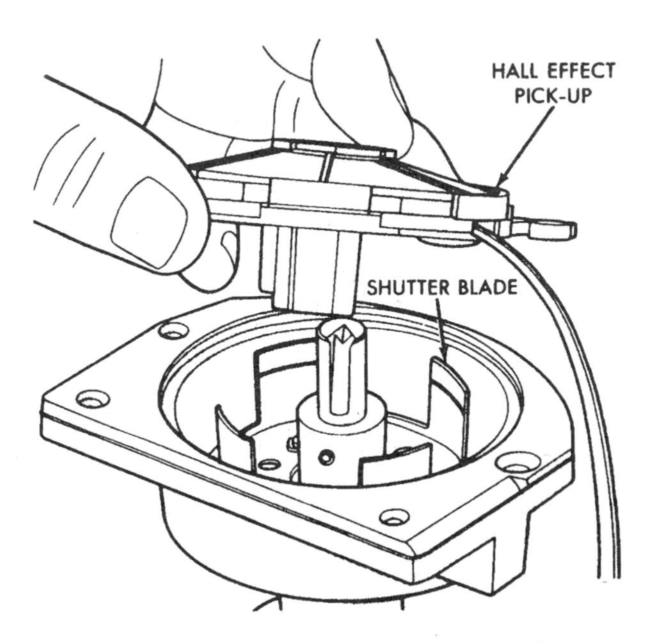

Fig. 5-7 Hall effect distributor pickup (courtesy Chrysler).

Hall Effect

Hall effect electronic ignition can be found on many domestic and import applications. A Hall effect ignition is similar to a magnetic impulse ignition. A signal is generated by a pickup in the distributor, which the ignition module then uses to turn the coil on and off. The difference, however, is that a Hall effect pickup does not generate an AC signal pulse. Rather, it generates a square wave on-off signal that is much more sharply defined. Hall effect sensors also require an input voltage to produce a signal. Finally, the output signal will not vary or diminish with engine speed.

A special kind of trigger wheel is mounted on the distributor shaft under the rotor (Figs. 5-7 and 5-8). The wheel has flat shutter-like blades. There is one blade for each cylinder. As the distributor turns, the blades pass through a magnetic "window" between a sensor element and permanent magnet. When there's nothing in the window, the magnet produces no voltage in the sensor. But when a shutter blades passes through the window, it blocks the magnetic field inducing a current in the sensor that creates a voltage signal. The effect is like trying to sneak a steel plate through an airport metal detector. The passing shutter blade sets off the beeper.

Two problems to watch for with Hall effect ignitions are: (1) a loss of voltage to the sensor that prevents it from working, and (2) loss of ground on the shutter blades which is necessary for a strong signal.

Light Effect

Two electronic ignition systems that use an optical or light-sensing pickup are Nissan's Electronic Concentrated Engine Control System (ECCS), and General Motor's Opti-Spark Ignition.

Nissan's ECCS uses a photo-sensitive "crank-angle sensor" in the distributor to inform the engine computer about both engine rpm and crankshaft po-

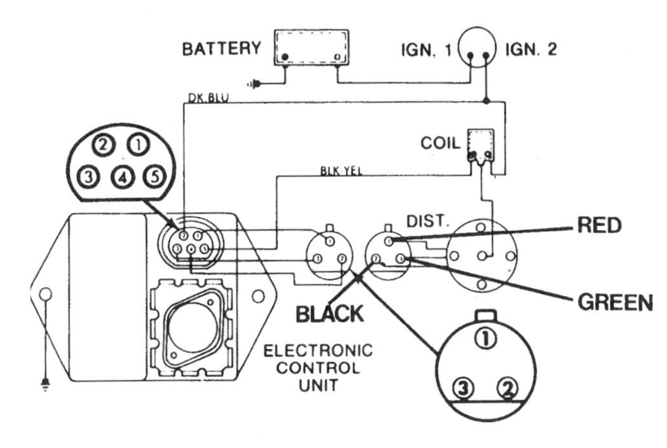

Fig. 5-8 Hall effect electronic ignition wiring diagram (courtesy Chrysler).

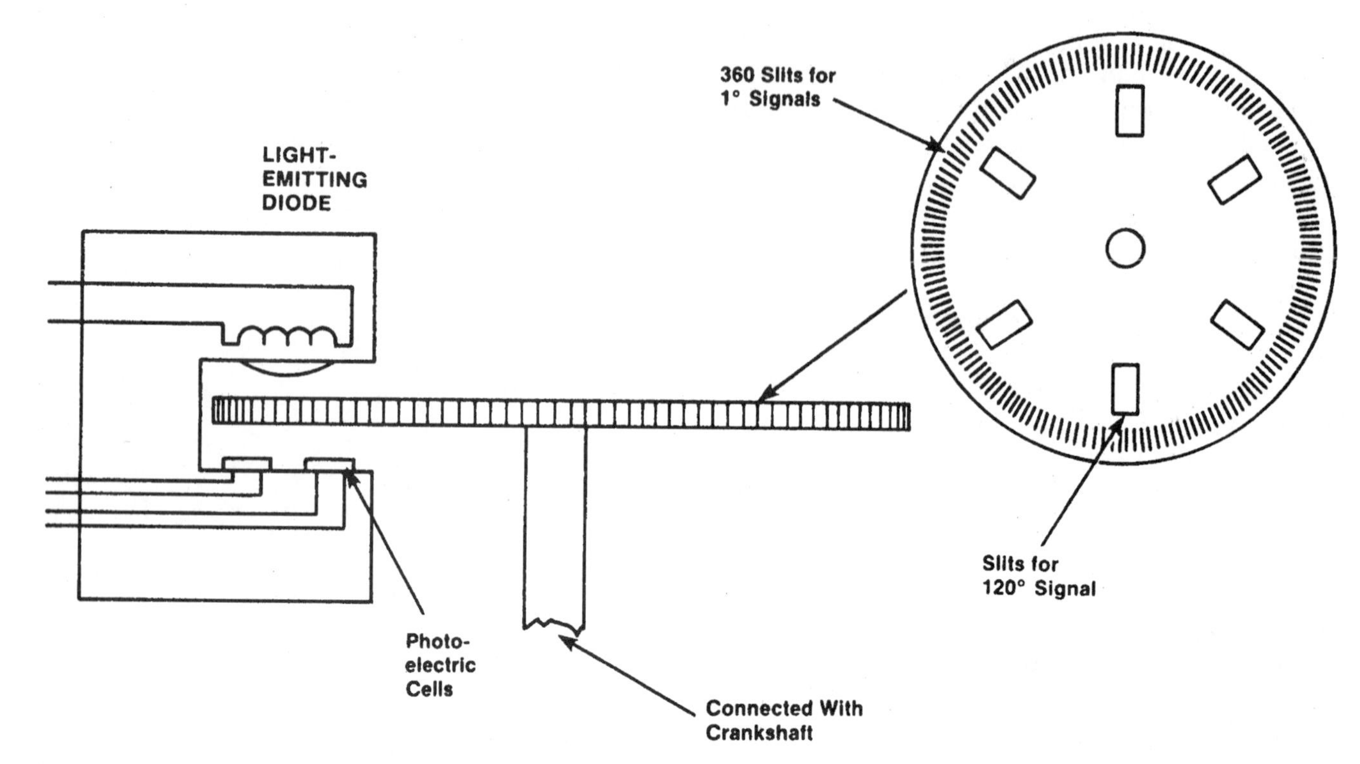

Fig. 5-9 Nissan's ECCS optically triggered ignition system (reprinted, by permission, from Santini, *Automotive Electricity and Electronics, 2E.* Copyright 1992 by Delmar Publishers Inc.).

sition. ECCS first appeared back in 1981 on the 280Z turbo cars, was added to other models in subsequent years, and has been standard on all Nissans since 1987.

The ECCS crank-angle sensor (Fig. 5-9) consists of three Light Emitting Diodes (LEDs), three photo-cell diodes, a slotted rotor plate, and an electronic signal conversion circuit board.

The LEDs are mounted above the rotor plate at various positions. Under them are their corresponding photo-cells. The rotor plate has 360 slits along its outer edge, and 6 additional slits at 60-degree intervals on 6-cylinder engines or 4 slits every 90 degrees for 4-cylinder engines to provide a crankshaft position reference signal. As the distributor shaft turns, the slotted plate creates a shutter effect between the LEDs and photo-cells. This produces an alternating voltage in the photo-cells, in effect switching them on and off with every flash.

The crank-angle sensor actually generates two signals. One signal is generated for every degree of rotation by photo-cells number 1 and 3 as each of the 360 slits in the rotor plate passes by. This is referred to as the "1-degree" signal. The second signal, which keeps the computer informed about the engine's firing order, is referred to as either the "120-degree" signal (6-cylinder engines), or a "180-degree" signal (4-cylinder engines). It is generated by photo-cell number 2 every time a piston reaches top dead center (TDC). The slit that corresponds to TDC for the number 1 cylinder is larger than the others, producing a longer "on" pulse when it passes between the LED and photo-cell.

The alternating voltage created in the photo-cells as they switch on and off doesn't make the kind of nice, crisp, digital signal the engine computer needs. So the wavy signal is "squared off" and converted to a digital on-off signal by the circuit board in the base of the distributor.

The ECCS crank-angle sensor is like a Hall effect sensor in that it needs both voltage and ground to create a signal. The loss of either will prevent the LEDs and photo-cells from doing their job, and no timing signal will be generated. This is the most common source of trouble, rather than failure of the crank-angle sensor itself.

The computer uses the two signal inputs from the crank-angle sensor along with other inputs such as air flow, cylinder head temperature, and throttle position to control: ignition timing (by grounding the power transistor in the coil power module), duration of the injector pulses in fuel-injected engines, fuel pump relay operation, the Auxiliary Air Control system (AAC), exhaust gas recirculation (EGR), and the Idle Control Valve (IAC).

Nissan chose the ECCS system because the LED-triggered ignition is extremely accurate. Magnetic pickups and Hall effect sensors can't measure 1 degree of crankshaft rotation. But the LED crank-angle sensor can. This allows the computer to vary ignition timing and injector duration very precisely, depending on engine load, speed, temperature, and so on.

General Motor's Opti-Spark ignition was introduced on the 1991 Corvette (Fig. 5-10). The Opti-Spark system uses a front-drive distributor driven off the cam gear. The Opti-Spark distributor is a maintenance-free unit with a "lifetime" rotor (or so GM says). The pickup uses a slotted shutter wheel and LEDs (like Nissan), rather than the more familiar magnetic or Hall effect pickup. This allows the computer to monitor crankshaft position within 1 degree of rotation, as compared to 90 degrees for a crank-mounted magnetic pickup. The benefit is more precise spark control for improved emissions, performance, and fuel economy.

DISTRIBUTORLESS IGNITION SYSTEMS

General Motors introduced a new kind of ignition system that uses no distributor back in 1984 on the Buick 3.8-liter turbocharged V6 engine (Fig. 5-11). Called "Computer Controlled Coil Ignition" or "C3I," the same system has gone through several generations, and is now found on numerous GM engines. Similar distributorless ignition system (DIS) have also been adopted by Ford, Chrysler, and most of the imports.

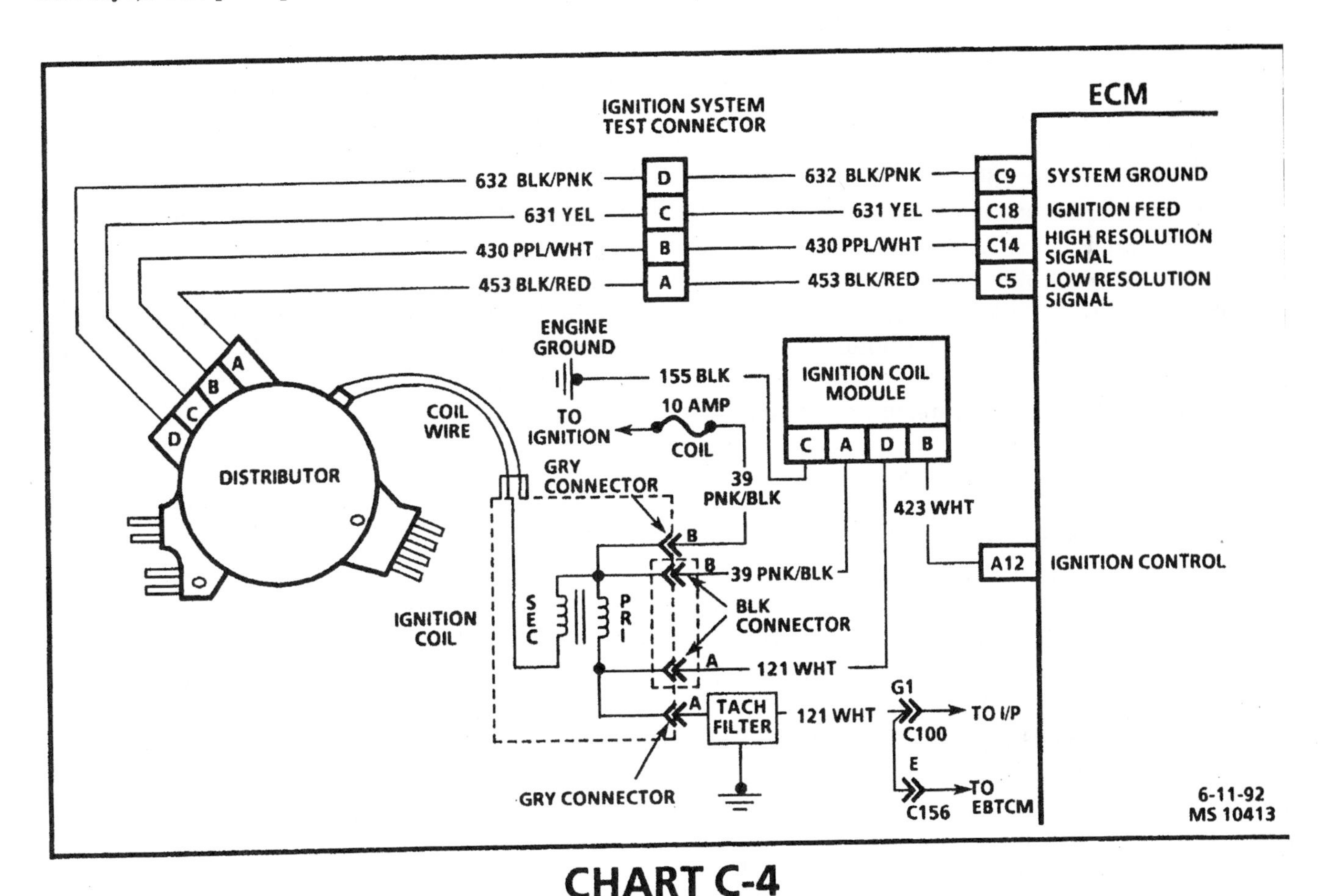

CHART C-4

Fig. 5-10 General Motor's Opti-Spark ignition schematic (courtesy GM).

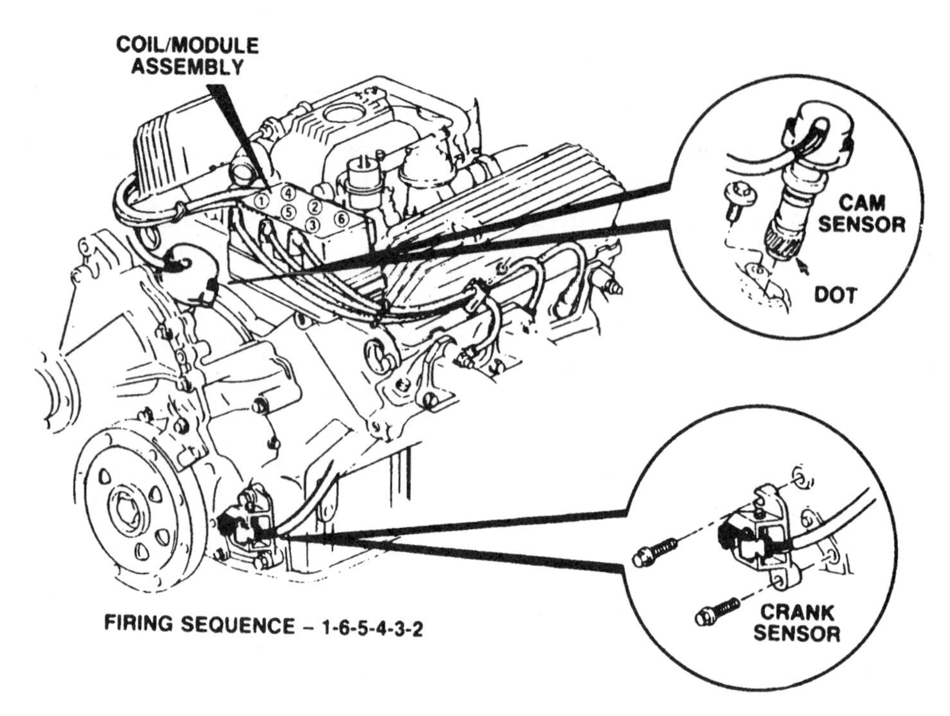

Fig. 5-11 Buick V6 distributorless ignition system (courtesy Delco Remy).

A distributorless ignition system is something like the ignition system on a motorcycle. The ignition coil is connected directly to the spark plug, and the plug fires every revolution of the engine. This eliminates the need for a distributor, distributor cap, rotor, and distributor pickup (and the potential problems these components can cause). No distributor means no manual timing adjustments (or misadjustments as the case may be), no cracked or burned caps, no eroded carbon buttons, no rotor arcing or wear, no moisture accumulation inside the cap to cause hard starting or misfiring, no distributor drive wear, and so on. Getting rid of the distributor also allows a lower engine profile, and the coils can be mounted almost anywhere on the engine providing increased design and packaging flexibility.

Because distributorless ignition systems have no distributor, the timing signal needed to trigger the ignition module is generated by a Hall effect or magnetic crankshaft position sensor mounted on the engine (Fig. 5-12). This sensor performs the same basic function as a pickup coil or Hall effect switch in a distributor. Most applications also have a separate camshaft position sensor (Fig. 5-13) so the computer can distinguish top dead center (TDC) of the compression stroke from TDC of the exhaust stroke. On some of these applications, the signal from the camshaft sensor is also generates the pulses that open and close the fuel injectors.

The signal from the crankshaft position sensor (and camshaft position sensor) are monitored by the engine computer and ignition module. The signal triggers the ignition module to fire the ignition coils, which in turn send sparks to the plugs.

Most distributorless ignition systems have a separate ignition coil for each pair of spark plugs (2 coils for a 4-cylinder, 3 coils for a V6, 4 coils for a V8, and in the case of a V10, 5 coils). Each coil fires its pair of spark plugs simultaneously. The paired spark plugs are located in opposite cylinders (such as 1-4

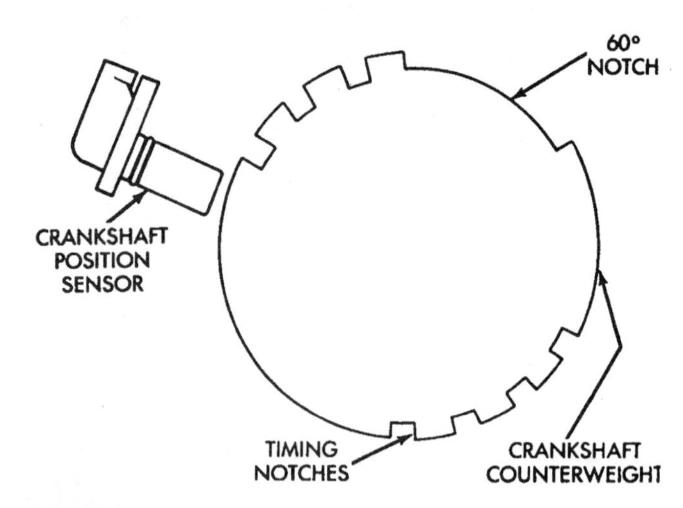

Fig. 5-12 The crankshaft position sensor generates a signal when notches in the crankshaft pass underneath it (courtesy Chrysler).

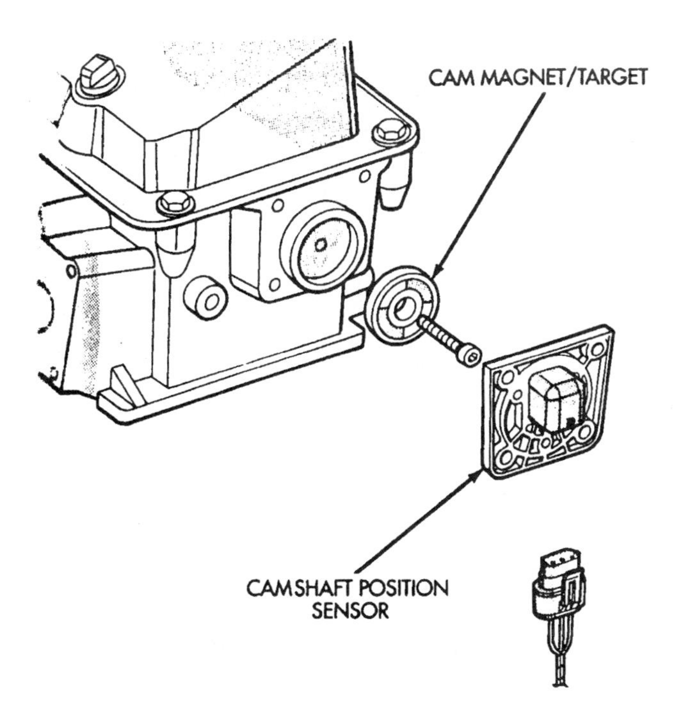

Fig. 5-13 The camshaft position sensor may be located on the cylinder head on some engines (courtesy Chrysler).

and 2-3 on a 4-cylinder, or 1-4, 2-5 and 3-6 on a V6). One cylinder fires on its power stroke, while the opposite cylinder fires on its exhaust stroke. The spark plug that fires in the cylinder during its exhaust stroke does nothing and is referred to as the "waste spark." There's little voltage resistance in the waste spark cylinder, so most of the firing voltage is available at the plug in the compression cylinder that needs it. When the engine turns another revolution and the cylinders reverse roles, the same process repeats with both plugs again firing simultaneously. One spark ignites the mixture, while the other does nothing.

A somewhat different setup is found on some engines such as GM's Quad Four (Fig. 5-14), 1990 and up Nissan 300ZX and 1992 and up Maxima with the DOHC 3.0L V6. On these engines, individual coils are located directly over each spark plug. On Saab's Direct Ignition (SDI) system, each spark plug has its own coil. But the coils run on a primary voltage of 400 volts, which is stepped up at the ignition module.

The firing of the coils in a distributorless ignition system is controlled by an ignition module located in the base of the coil pack, or elsewhere in the case of applications where individual coils for each cylinder are used. Spark timing is usually regulated by

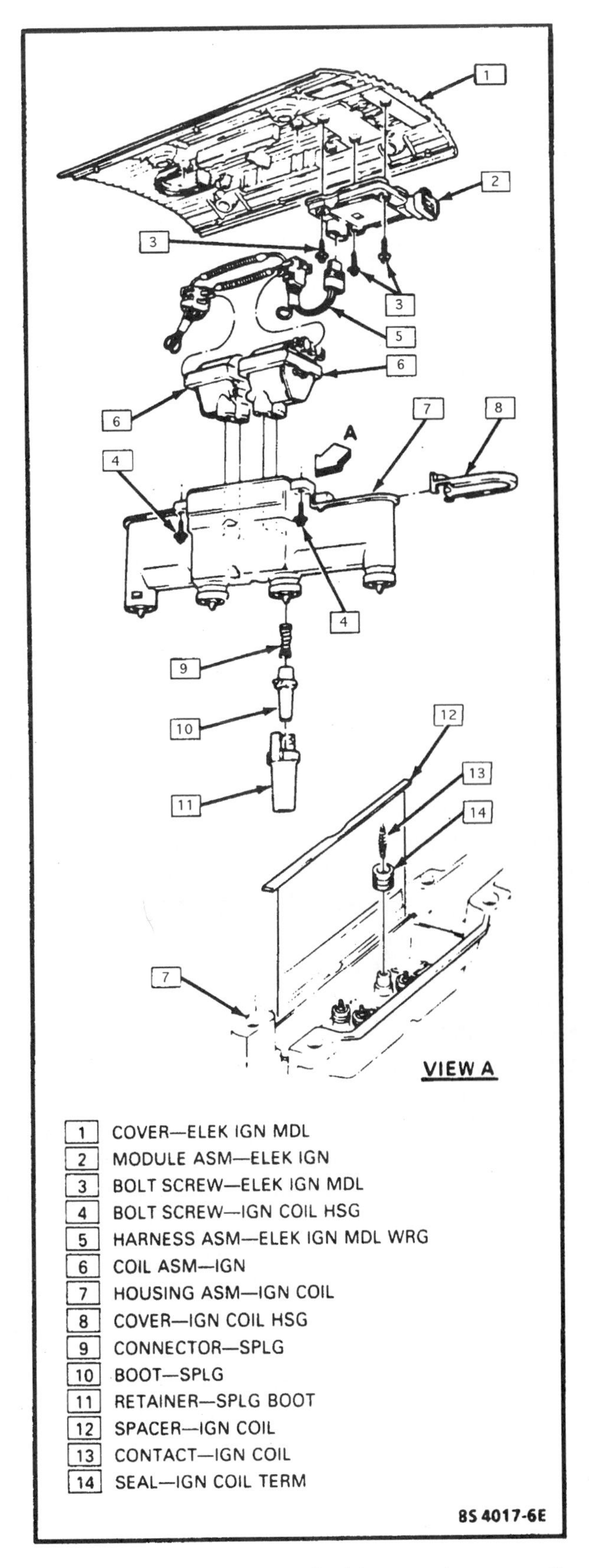

Fig. 5-14 Distributorless ignition system on the Oldsmobile Quad Four has individual coils that mount directly over each spark plug (courtesy Oldsmobile).

the engine computer. In some applications, the ignition module controls spark timing below 400 to 700 rpm while the engine is cranking. It then turns the job over to the computer once the engine starts. The ignition module may also have a built-in "fail-safe" capability that allows it to provide a fixed ignition timing signal (the "limp-in" mode) in case the spark control signal from the electronic contol module (ECM) is lost.

SECONDARY IGNITION COMPONENTS

The secondary ignition components include the secondary windings inside the coil, the distributor cap and rotor (on engines with distributors), ignition cables and spark plugs.

DISTRIBUTOR CAP AND ROTOR

The distributor cap is a redistribution terminal for high voltage. It takes in high-voltage current from the coil, passes the current along to the rotor, which then "distributes" the voltage to the appropriate plug wire as it turns inside the cap.

The cap is made from plastic or bakelite to contain the high voltage. The cap material must have a very high "dielectric" (insulating) value to prevent arcing and voltage leakage. In time, however, the dielectric strength of the cap can break down, allowing hairline cracks to form which create short circuits for the high voltage to follow. The result can be misfiring and erratic running. That's why the inside as well as the outside of the distributor cap should always be carefully inspected when performing a tune-up.

In some General Motors HEI ignition systems, the coil is located on top of the distributor cap. Deterioration of the carbon high-voltage center terminal is a common problem.

The plug wire terminals in the cap are usually aluminum or brass. The terminals may be female or male, depending on the application. The business end of the terminals inside the distributor cap are subjected to arcing every time they receive current from the rotor. This not only erodes the terminals but also forms an oxide layer that increases resistance. The center terminal is a carbon button against which the spring-loaded tang on the rotor rides. The constant turning combined with the high voltage that flows through it eventually wears the button down. So this area should be inspected for wear. Worn or cracked distributor caps should always be replaced.

The rotor sits atop the distributor shaft, which is driven by the camshaft and rotates at half engine speed (see Chapter 7). The rotor is made of plastic with a strip of metal (brass or steel) on top to carry current from the center terminal to the plug terminals.

The rotor does not touch the distributor plug terminals, however. There is a small air gap across which the voltage must jump in order to reach the plugs. The size of this gap is a critical dimension because it adds resistance and increases the required firing voltage approximately 3,000 volts. As the end of the rotor becomes worn, and the plug terminals in the cap erode, the distance of the air gap increases. This raises the voltage requirements of the system, and may push the coil beyond its ability to fire the plugs. The rotor should be replaced periodically to restore this critical dimension.

Like distributor caps, rotors are sometimes subject to cracking and arcing. But if the rotor arcs, the current goes to ground through the distributor shaft. No spark reaches the plugs and the engine ceases to run.

In some late-model, high-energy systems, there is so much voltage produced that it is hard to contain inside the distributor. Ford, for example, recommends coating the replacement rotor in certain applications with dielectric silicone grease to reduce the chance of arcing and misfire.

Ignition Cables

Ignition cables or spark plug wires are the conductors through which high-voltage current surges to fire the spark plugs. At one time, spark plug wires were copper, aluminum, or steel wire wrapped with insulation sufficient to contain about 12 KV. But when high-voltage current surges through a low resistance wire, the wire becomes a broadcast antenna and sends out radio waves. This causes radio frequency interference (RFI) which disrupts radio and television reception. So modern cars use resistor plugs and/or suppression wires to reduce RFI.

Metal blocks radio waves, so the outside world

is shielded somewhat from RFI by the metal hood and fenders surrounding the engine compartment. Metal shielding around the distributor and wires on some cars (notably older plastic-bodied Corvettes) also helps. But the best way to reduce RFI is to nip it at its source.

Adding resistance to the high-voltage secondary path greatly reduces the troublesome radio waves. Resistance is created by using "resistor" rather than solid-core spark plugs in conjunction with "suppression" cables.

Resistance in the spark plug cables can be created in one of two ways: by using graphite-impregnated fiberglass strands (commonly called "carbon-core" wires) to carry the high-voltage current, or by using a special nickel alloy resistance wire called "monel" wiring (Fig. 5-15). The monel wire is wound spring-like through the core of the cable. The spiral-wound suppression wire reduces RFI with less resistance and better conductivity than standard carbon-core cables. But monel wiring is expensive and is generally not used as original equipment.

Carbon-core cables have a poor reputation for longevity. The little particles of carbon impregnated in the fiberglass strands can sometimes migrate and separate over time. Heat and vibration contribute to the aging process, as does twisting or jerking on the cables themselves. Once the carbon begins to separate, resistance shoots up. This raises the voltage needed to pull current through the wires and to fire the plugs. And if the ignition system doesn't have the extra volts, the plugs will sputter.

The voltage demands are much higher in late-model engines for two reasons: wider spark plug gaps and leaner fuel mixtures. It takes more voltage to pull the spark across a wide gap than it does a narrow one, and it takes a stronger spark to fire a lean fuel mixture than a rich one. If the carbon inside the ignition cables has deteriorated to the point where

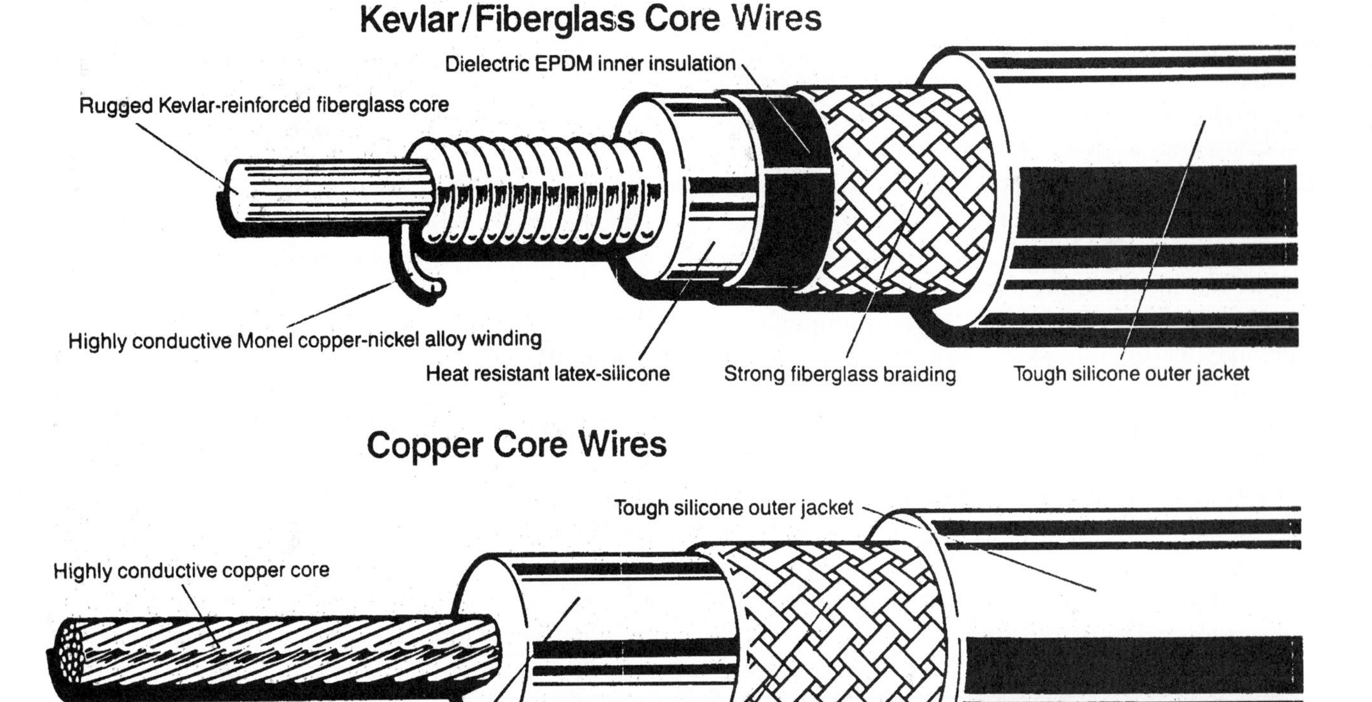

Fig. 5-15 Though most original equipment ignition wires on domestic cars are carbon core, many European wires have copper-core wires with resistor plugs. Magnetic suppression wires, with monel copper-nickel alloy windings wound around a fiberglass core, offer less resistance than carbon-core wires with performance similar to copper-core wires for domestic engine applications (courtesy Beck Arnley Worldparts).

the cumulative resistance between the spark plugs and coil exceeds the maximum voltage output of the system, the plugs will misfire.

A bad plug wire with excessive resistance usually causes an intermittent misfire, rather than a steady miss, because the voltage required to fire the plugs changes with engine speed and load. It is lowest at idle and under light load, but rises sharply as the throttle opens or the load increases. Thus the engine may run fine most of the time, but then misfire or stumble during acceleration. Unless the cables are replaced, the carbon-core will continue to deteriorate. The symptoms will worsen, and eventually the engine will refuse to start.

A bad plug wire with too much resistance will reveal itself on a scope by producing an unusually high firing voltage. Resistance can also be checked with an ohmmeter (see Chapter 6).

If an ignition cable has age cracks or chaffed or burned spots in its insulation, the secondary voltage will seek out the path of least resistance, possibly bypassing the spark plug. The typical symptom in this case will likely be a steady miss, often accompanied by a snapping noise as the spark arcs between the cable and engine. Arcing of this nature can usually be detected by watching the engine in the dark. Any visible fireworks around the plugs or cables means voltage is leaking. On a scope, the shorted cable will show an unusually low firing voltage on the affected cylinder.

Silicone insulation is superior to Hypalon insulation in terms of heat resistance and longevity. A Hypalon cable touching a hot exhaust manifold will likely melt through, while a silicone cable usually will not. The thickness of the insulation is also important. The thicker the cable, generally speaking, the more voltage it can safely handle. That's why many late-model, high-voltage electronic ignition systems have gone to 8 mm rather than 7 mm silicone ignition cables. Silicone cables are more expensive but also more durable.

A single bad plug wire can have a major impact on emissions and performance. One misfiring plug in a 4-cylinder engine can cause a 25% drop in power and fuel economy, not to mention an enormous increase in exhaust emissions. There is also the possibility that the misfiring may damage the catalytic converter. The converter's job is to reburn exhaust pollutants, not raw fuel. But that's what happens when a plug misfires. With every other revolution of the engine, a cylinder-full of unburned gasoline can be dumped into the exhaust. When the raw fuel hits the converter, temperatures soar, collapsing (and sometimes even melting) the innards of the converter.

REPLACING IGNITION CABLES

- Handle the cables with care during installation. Do not jerk, force, twist, or sharply bend.
- Change one wire at a time to avoid mixing up the firing order.
- If one or more wires are installed out of sequence, refer to the firing order cast on the intake manifold or printed on the underhood emissions decal.
- Don't run the cables of cylinders that fire consecutively parallel to one another. Separate them by several inches or crisscross them to avoid crossfire through induction.
- Make sure the terminals fit snugly into the distributor cap, and that the cap terminals are clean, dry and corrosion-free.
- The cables should be supported by wire looms, and not allowed to touch the exhaust manifold or rub against sharp objects.
- Cables should be routed several inches away from any wiring harness on late-model cars with computerized engine controls.

SPARK PLUGS

The spark plugs are the business end of the secondary side of the ignition system. Plugs are fairly simple in design, yet designed for a specific job. The steel shell that screws into the head grounds the plug. The ceramic shell insulates the center electrode, forcing the high-voltage current to take the only path provided, which is to jump across the air gap between the center and ground electrode. The ceramic insulator also acts as a heat barrier to keep the center electrode hot. This reduces the voltage required to fire the plug and helps burn off fouling deposits that otherwise would form on the nose of the plug. At the same time, however, the ceramic core provides a path through which some heat can be conducted away from the firing tip. The tip needs to be hot for easy firing and to prevent fouling, but not so hot that it becomes a source of pre-ignition.

Spark Plug Construction

There are four main differences between spark plug designs: reach, heat range, thread and seat, and gap (refer to Figs. 5-16 through 5-20).

"Reach" is the length of the threaded portion of the shell from the seat to the firing tip. It is determined by the design of the cylinder head. A replacement spark plug must be the correct each to fit properly in the head. If too short, the tip won't extend far enough into the combustion chamber to ignite the mixture properly. If too long, the plug may protrude too far into the combustion chamber and interfere with the valves or pistons.

"Heat range" refers to a plug's ability to dissipate heat from the firing tip. A plug's heat range is determined by the distance heat has to travel from the firing tip through the ceramic insulator to the shell and finally to the head (Figs. 5-17 and 5-18). A "cold" plug is one with a relatively short heat path. The shell will be shorter to provide a more direct route for heat to leave the plug. A "hot" plug will have a longer insulator and taller shell to create a longer path for the shell.

The reason for having spark plugs with different heat ranges is to prevent fouling under different engine applications. For normal driving, an engine usually needs a sort of a "middle-heat-range" plug. The plug must run hot enough at low speed to prevent fouling, but cool enough at high speed to prevent pre-ignition and detonation. An engine that runs continually at high speed (a high-performance engine, for example) needs a cooler plug (Fig. 5-19) to perform well. But if a cold plug is operated at low speed for extended periods of time, it can foul and misfire causing an increase in hydrocarbon emissions. An engine that always runs at low speed, is used for short trips, or is using oil typically runs best with a hotter plug to prevent fouling.

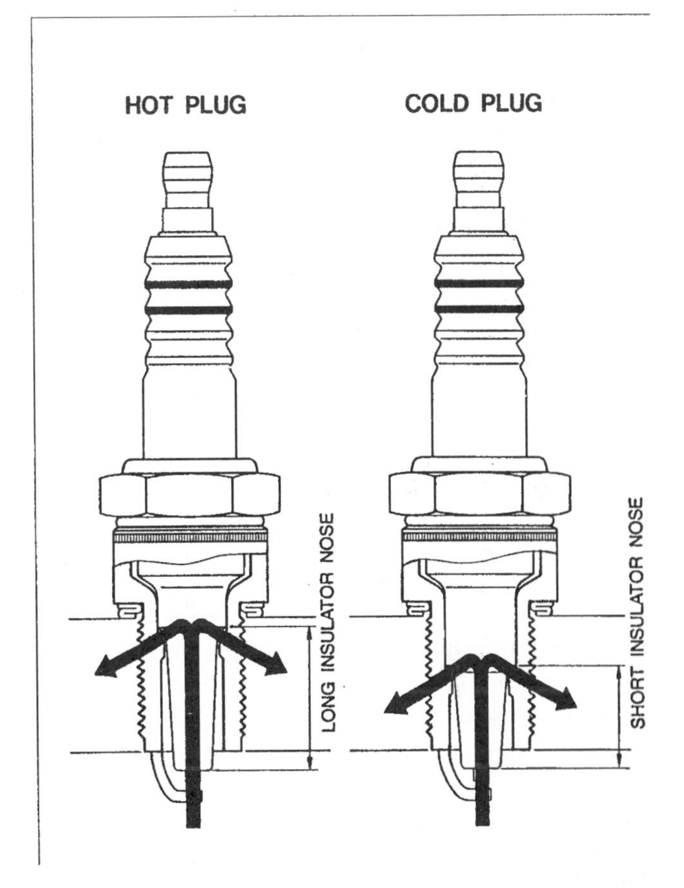

Fig. 5-17 The length and design of the insulator and shell determine a spark plug's heat range. The shorter the path, the colder the plug (courtesy Nippondenso).

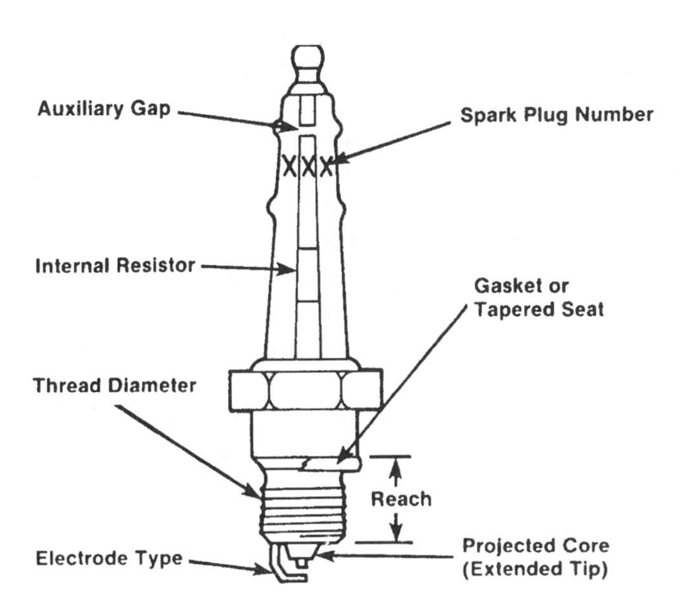

Fig. 5-16 Spark plug construction (reprinted, by permission, from Santini, *Automotive Electricity and Electronics, 2E.* Copyright 1992 by Delmar Publishers Inc.).

The need for various heat range plugs for different engine and driving applications can't be totally eliminated. But spark plug manufacturers have consolidated the need for so many different heat ranges with a plug design that provides a much broader heat range. By using a copper-core center electrode (Fig. 5-20), a single plug can now handle a much wider range of operating conditions and engine applications.

Another difference between spark plugs is different thread and seat sizes. There are two different thread diameters used: 18 mm (all of which have ta-

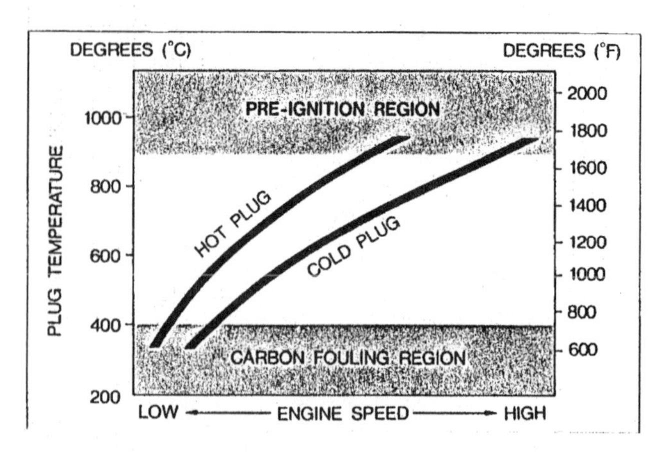

Fig. 5-18 Temperature performance comparisons between hot and cold spark plugs (courtesy Nippondenso).

pered seats), and 14 mm (which come with either flat or tapered seats). Tapered-seat spark plugs do not use a gasket, while flat seat plugs do. The exterior shell dimensions of 14 mm tapered-seat plugs are ⅝-inch hex, while 14 mm flat-seat and 18 mm tapered-seat plugs use the larger 13/16 inch hex.

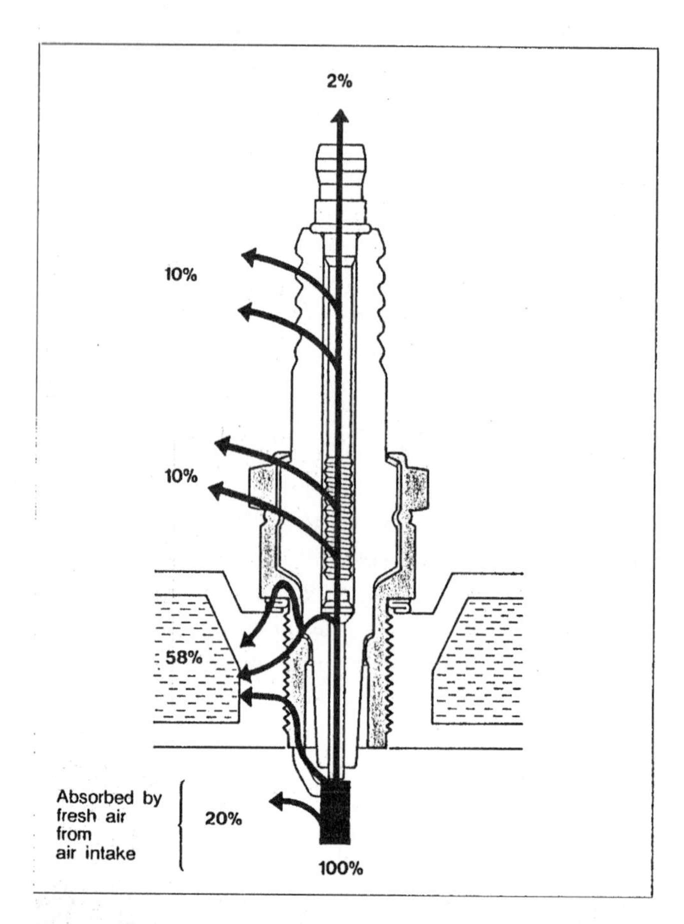

Fig. 5-19 Heat dissipation in a spark plug (courtesy Nippondenso).

Fig. 5-20 Copper-core spark plugs with platinum tips offer a broad heat range and significantly longer service life (courtesy Autolite).

A word about spark plug wrenches is in order here. When using a deep well socket to remove a spark plug, be sure that it is a "spark plug" socket, not an ordinary socket. A spark plug socket has a rubber insert that cushions and protects the plug against breakage. It also holds the plug while you remove it from the engine.

The spark plug electrode gap (also called the "air gap") determines the firing voltage requirements of the plug. The wider the gap, the more voltage it takes to create a spark (Fig. 5-21). At the same time, however, too narrow a gap may not make a strong enough spark to ignite the fuel mixture. In late-model engines with lean fuel mixtures, a wider gap is often necessary to prevent misfiring. For breaker point ignition systems, the spark plugs were usually gaped .030 inches. In late-model electronic ignition systems, .045- to .080-inch gaps are common. The electrode gap should be measured with either a wire or feeler gauge. If adjustment is necessary, carefully bend the outer electrode until the gauge just slips between the electrodes.

European ignition systems tend to use a higher firing amperage, rather than voltage, to accomplish

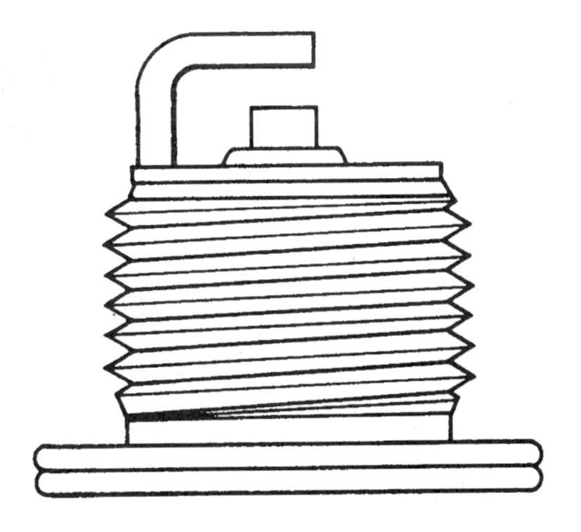

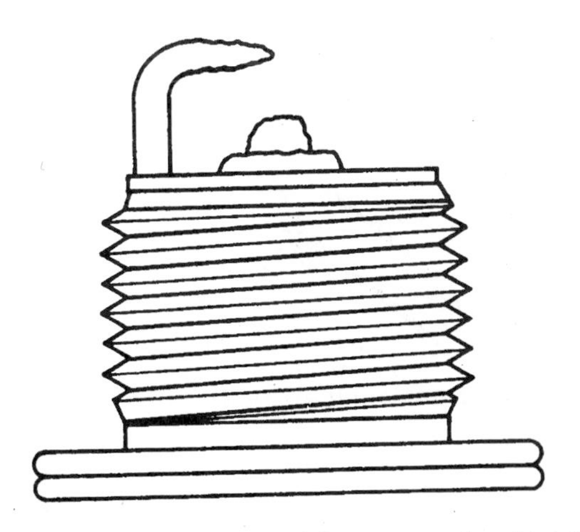

Fig. 5-21 A new spark plug (left) will have a flat, straight surface on both electrodes, while a worn plug (right) will show the effects of erosion (reprinted, by permission, from Santini, *Automotive Electricity and Electronics, 2E.* Copyright 1992 by Delmar Publishers Inc.).

this because it produces a hotter spark. But a hotter spark eats up the plugs more quickly. That's why most Japanese and American ignition systems use a lower-amperage, higher-voltage spark. To compensate for plug wear, the plugs for many European applications have somewhat larger diameter center electrodes (3.0 mm vs. 2.7 mm). This helps extend plug life. But even with platinum, some of the European auto makers still recommend 30,000 miles for plug-change intervals.

"Resistor"-type spark plugs have a resistor located in the center electrode to reduce radio interference (RFI). The resistor reduces the "inductive" portion of the spark discharge. This not only reduces radio interference but also extends the life of the electrodes.

"Extended-tip" spark plugs have longer tips that reach further into the combustion chamber than other spark plugs. Extending the tip generally provides better fouling resistance because the plug runs hotter. Repositioning the spark may also improve performance somewhat by moving the point of ignition further into the combustion chamber.

A "series-gap" or "booster" spark plug has an extra gap in the center electrode. This allows the plug to build up a slightly higher voltage before firing to overcome fouling. This type of plug may be used in an engine that has an oil-consumption problem.

Some plugs have two- or three-ground electrodes. Such spark plugs still only produce one spark. But by providing more than one path for the spark to take, it lessens the odds of a misfire. Such plugs are often used in high-revving or high-compression European engines.

There are also special "long-life" spark plugs that use platinum- or gold-palladium-tipped center electrodes (Fig. 5-22). Because these precious metals can withstand high voltages and operating temperatures without eroding, such plugs wear at a much slower rate. Some OEM long-life plugs now have a recommended replacement interval of 100,000 miles.

Reading The Plugs

Examining the tips of the spark plugs as they are removed can reveal a great deal about the health and performance of an engine (Fig. 5-22). The appearance of the plugs as well as the color and kind of deposits can alert you to various kinds of problems.

- *Normal worn plug*—the electrodes will be worn down and the insulator will have light brown or tan-colored deposits.
- *Fuel fouled plug*—black, fluffy carbon deposits indicate an overly rich fuel mixture and/or a weak spark. Check for such things as a stuck choke, a heavy carburetor float, a fuel leak in the carburetor, a leaky injector, low coil output, high resistance in plug wires, etc.
- *Wet plug*—a wet plug means the plug has not been firing. The problem is often due to a bad ignition cable (excessive resistance, shorted, or

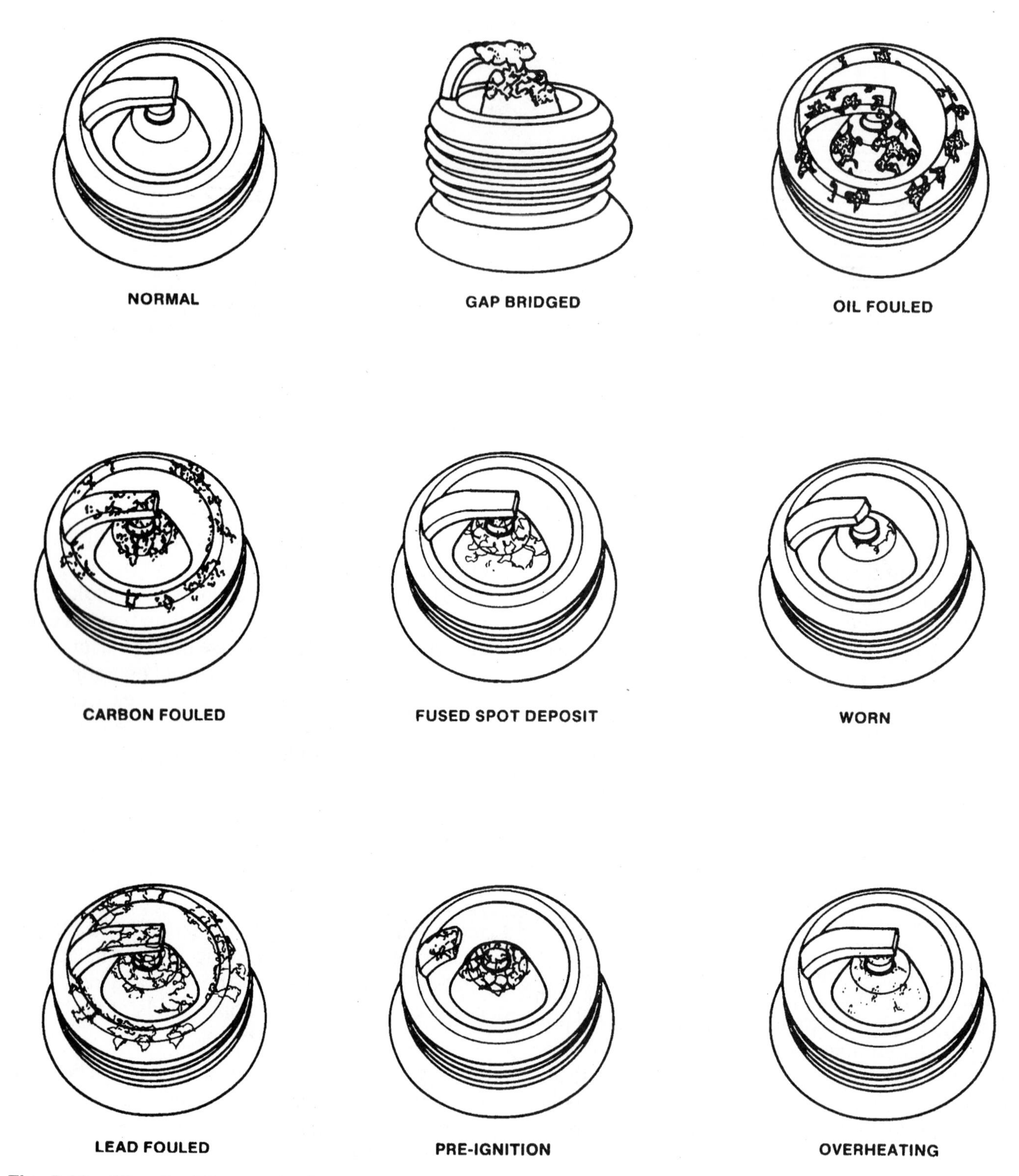

Fig. 5-22 "Reading" the spark plugs can reveal a great deal about what's going on inside the engine (courtesy Ford).

arcing). It can also be caused by dirt or moisture on the outside of the plug providing a conductive path for the voltage to bypass the electrodes, or by an internal crack in the ceramic insulator that shorts the plug to ground.

- *Oil-fouled plug*—heavy, black, often oily deposits indicate oil burning. Oil drawn past the valve guides and/or the rings is entering the combustion chamber and fouling the plugs. Switching to a hotter plug may help prolong the life of the replacement plug somewhat, but such an engine is in need of major repairs. No plug is going to survive long under such conditions.

- *Glazed plug*—yellowish, melted-appearing deposits on the insulator tip form when an engine that normally does stop and go driving is subjected to high-temperature operation. A plug with a broader heat range may be recommended.
- *Damaged plug*—if the electrodes have been smashed flat or broken, somebody put the wrong plug in the engine. If a plug protrudes too far into the combustion chamber, it may hit the piston or a valve. A foreign object in the combustion chamber may also cause such damage.
- *Overheating*—if the insulator is blistered, white, and free from deposits, something is making the plug run too hot. Check to see if the plug is the correct heat range. Also check for cooling problems, overadvanced ignition timing, lean fuel mixtures, or an air leak or other conditions that may be causing it to run hot.
- *Melted electrode*—a symptom of severe preignition. The plug has been running too hot for a long time (see overheating above).
- *Detonation*—if the insulator is split or chipped, detonation may be occurring in the engine. Check for overadvanced ignition timing, excessive compression due to accumulated deposits in the combustion chamber, or an inoperative EGR valve. Switching to a higher-octane fuel may be recommended.

IGNITION TIMING AND HOW IT AFFECTS EMISSIONS

Consistent ignition is essential for low emissions. When there's no ignition, there's no combustion (except in a diesel, of course). And when there is no combustion in a cylinder, raw unburned fuel passes through into the exhaust causing hydrocarbon emissions (HC) to soar.

For the fuel mixture to burn, therefore, the ignition system must be capable of producing a reliable hot spark. This means the ignition coil (or coils in the case of distributorless ignition systems) must be capable of delivering sufficient firing voltage to the plugs when the spark is needed. The secondary ignition components (distributor cap, rotor, and wires) must also be capable of transferring the voltage to each plug. Cracks or carbon tracks on the coil, distributor cap, or rotor can all short-circuit the voltage before it reaches its destination. So too can an excessive air gap between the rotor and cap and excessive resistance in the plug wires. And if the spark plugs are fouled, worn out, or incorrectly gaped or shorted, no spark will occur to ignite the fuel mixture. That's why the ignition system must be in perfect condition for the engine to fire reliably.

One of the most common causes of abnormally high HC emissions, therefore, is ignition misfiring. So unless the engine has a bad valve that's leaking compression (which a compression test will reveal), or a fuel problem such as a vacuum leak or bad injector that's leaning out the fuel mixture to the point where it won't fire reliably (a condition known as "lean misfire"), high hydrocarbon emissions are usually ignition related.

TIMING ADVANCE

Another way in which ignition affects emissions is through spark or timing advance. The instant at which a spark plug fires relative to the position of the piston during the piston's compression stroke determines the amount of timing advance. All engines require a certain amount of timing advance so the fuel can finish burning before the exhaust valve opens.

At an idle speed of 600 rpm, the crankshaft is spinning around 600 times a minute or 10 times a second. Each piston is also reciprocating up and down at the same rate, 10 times a second. Since combustion occurs only on the power stroke every other revolution, ignition takes place 5 times a second at 600 rpm.

The air/fuel mixture has plenty of time to burn at idle because the duration of the power stroke at 600 rpm is on the order of 1/20 (0.05) second. The actual combustion of the air/fuel mixture takes place in about 1/200 (0.005) second, so you can see that there is plenty of time for the fuel to burn and expand before it is pushed out of the cylinder on the exhaust stroke.

At highway speeds, a typical engine might be turning 3,000 rpm. At this speed, the pistons will be traveling up and down at the rate of 50 times a second. With ignition taking place every other revolution, combustion occurs 25 times a second.

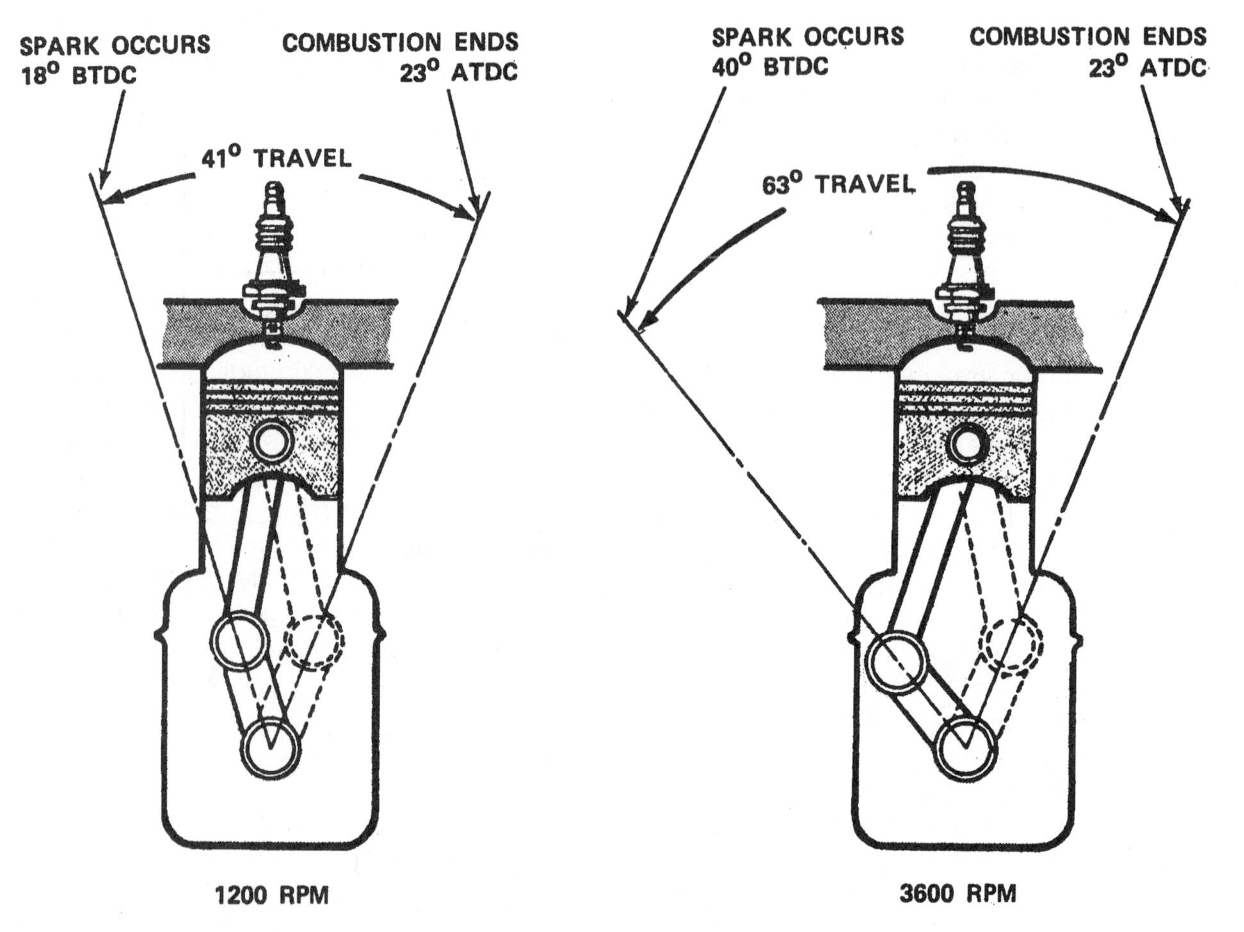

Fig. 5-23 As engine speed increases, there's less time between piston strokes for the mixture to burn, so ignition must be advanced (courtesy Sun Electric).

The time available during the power stroke at 3,000 rpm is now on the order of 1/100 (0.01) second, still enough time for the fuel to complete burning on the power stroke but not enough time for cylinder pressures to reach a maximum at the right point in the power stroke.

For an engine to deliver maximum fuel economy and power, cylinder pressure should reach a maximum early in the power stroke while the piston is accelerating downward. If the instant of maximum pressure occurs too late in the power stroke, it will not produce as much horsepower. And if maximum pressures are not achieved until very late in the power stroke, some of the "oomph" will be lost out the exhaust valve when the exhaust stroke begins. Timing, therefore, plays a very important role in both power output and economy.

To give the air/fuel mixture sufficient time to burn so that maximum cylinder pressures can be achieved at the best point in the power stroke, the ignition timing is advanced before top dead center (Fig. 5-23). Instead of happening exactly at TDC as one might assume, ignition takes place farther and farther before TDC as engine speed increases. In other words, ignition takes place toward the end of the compression stroke.

For example (Fig. 5-24), a typical engine at idle may be timed at anywhere from several degrees after top dead center (ATDC) to 10 or 12 degrees before top dead center (BTDC). Remember that at idle there is plenty of time for combustion, so little or no initial advance is needed. At 3,000 rpm, however, considerable advance is needed to achieve maximum cylinder pressure early in the power stroke. Such an engine might have 28 to 34 degrees of ignition advance, the amount of advance increasing in proportion to engine speed.

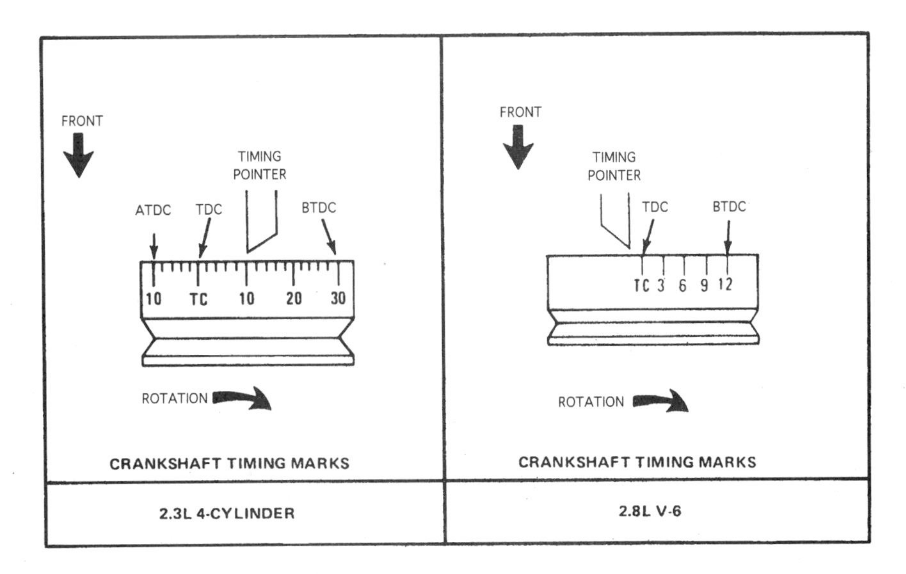

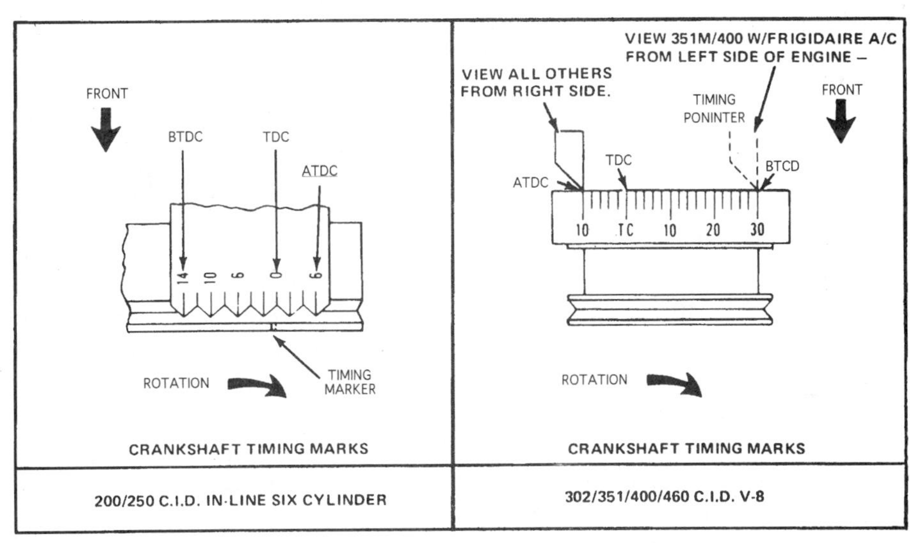

Fig. 5-24 Typical timing marks (courtesy of Ford Motor Corporation).

TIMING AND DETONATION

The higher the rpm, the more timing advance an engine needs to maximize power output and fuel economy. The timing must advance in proportion to the rpm rate. If the fuel is ignited too far in advance, however, the pressure of the expanding gases rises too quickly and peaks before the piston can respond. This causes detonation and increases HC emissions.

Think of combustion as an expanding balloon. Under normal circumstances, the flame kernel expands and fills the combustion chamber outward from the spark plug. When there is too much advance, though, the rapidly increasing pressure inside the combustion chamber causes fuel to ignite spontaneously in other areas of the chamber. This is akin to several balloons expanding all at once. And when the flames collide, they do so with great force. This produces the sharp knock or ping noise characteristic of detonation. Under severe circumstances, detonation can crack or punch holes in pistons, crack heads and piston rings, flatten connecting-rod bearings, and blow out head gaskets. Hydrocarbon emissions are also increased because erratic combustion leaves little pockets of unburned fuel.

For optimum performance and power, an engine should be timed so that it is just on the verge of detonation. The only drawback to this approach is that an engine's detonation resistance changes. It can vary with the quality of gasoline being used, the additives in the fuel, or be affected by humidity and temperature. High relative humidity increases the effective octane rating of gasoline, while dry weather does just the opposite. What this means to real-life driving is that an engine that's tuned for optimum performance on a rainy day may tend to knock and ping during dry weather. To play it safe, therefore, the manufacturer's recommended timing settings have a built-in safety margin to account for some variation in fuel quality and weather effects. Late-model cars with computerized engine controls can push these limits somewhat if they are equipped with a knock sensor. This device picks up engine vibrations produced when detonation occurs and signals the computer to momentarily retard timing until the detonation ceases.

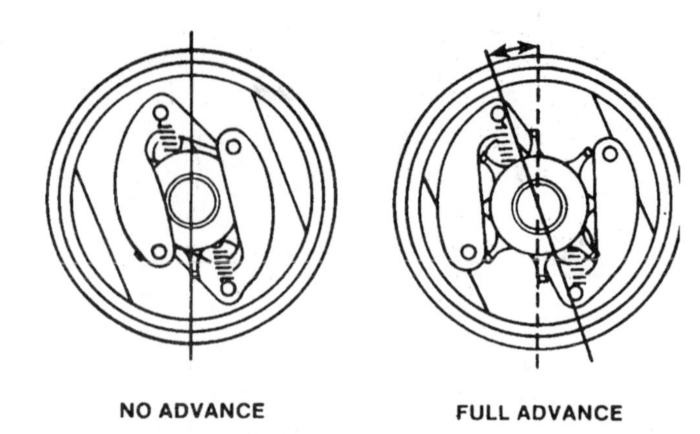

Fig. 5-25 Centrifugal advance weights in a distributor (reprinted, by permission, from Santini, *Fundamental Electronics Engine Training.* Copyright 1992 by Delmar Publishers Inc.).

TIMING BASICS

- If ignition occurs exactly at top dead center, there is zero degrees of timing advance.
- If ignition occurs after TDC, the timing is said to be "retarded."
- If ignition occurs before TDC, the timing is said to be "advanced."
- "Initial" timing (also called "basic" timing) is the amount of advance or retard the ignition has at idle as set in accordance with the manufacturer's instructions. Setting timing on vehicles with vacuum-advance distributors usually requires disconnecting the vacuum-advance hose from the distributor.
- Increasing the amount of ignition advance is called advancing the timing. Decreasing the amount of advance is called retarding the timing.

CENTRIFUGAL ADVANCE

As described earlier, the amount of timing advance must increase in proportion to engine speed. This is called "centrifugal" advance and it is accomplished in two ways: mechanically or electronically.

In distributors with centrifugal-advance mechanisms (Fig. 5-25), two small, spring-loaded flyweights control the rate of advance. As engine rpm increases, the weights are thrown outward against spring tension. The movement of the weights rotates the rotor and trigger wheel (on electronic ignitions) or breaker cam (on point ignitions) into an advanced position. This fires the coil and spark plugs sooner to advance timing. The size of the weights and the strength of the springs that resist the weights determine the rate of advance (Fig. 5-26). Changing the weights and/or springs will change the advance curve.

On cars equipped with computerized engine controls and electronic spark timing, no flyweights or advance mechanism are used in the distributor to advance timing. It's done electronically by the computer or ignition module. The computer or module calculates the equivalent amount of "centrifugal" advance the engine needs based on rpm. It then times the firing of the coil to create the required amount of advance. The amount of advance may be further modified by inputs from other engine sensors (such as the coolant sensor, manifold absolute pressure (MAP) sensor, barometric pressure sensor, or throttle position sensor), depending on the vehicle application and operating conditions.

VACUUM ADVANCE

To improve fuel economy, most engines also employ "vacuum" advance, which differs from centrifugal advance in that it is engine-load sensitive. Vacuum advance is added when there is a light load on the engine (as in cruising or decelerating), and subtracted when the engine is under load or accelerating at wide-open throttle. By comparison, centrifugal advance is speed-sensitive and changes with engine rpm, not load. So the two together allow

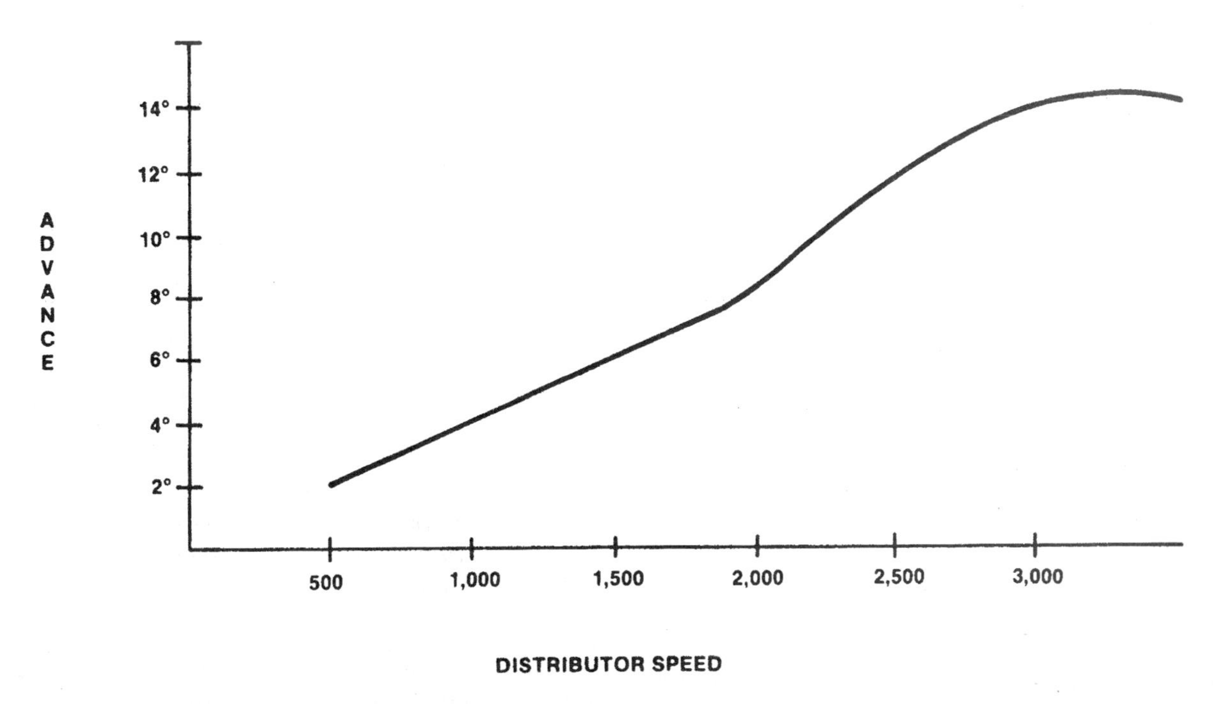

Fig. 5-26 The "spark curve" of a distributor depends upon how the centrifugal advance mechanism is calibrated. Heavier weights and/or lighter springs provide advance at a faster rate (reprinted, by permission, from Santini, *Automotive Electricity and Electronics, 2E.* Copyright 1992 by Delmar Publishers Inc.).

ignition timing to be varied according to changes both in speed and load. Centrifugal advance plus vacuum advance equals the "total" amount of timing advance the engine has at any given instant in time.

Since engine vacuum drops in proportion to the load applied, using or monitoring intake vacuum to control timing advance allows the timing to respond to the conditions (refer to Figs. 5-27 through 5-28). At idle, light load, and during deceleration, intake manifold vacuum is very high. As the throttle is opened wider, intake manifold vacuum drops. At full throttle there is very little manifold vacuum.

The vacuum signal for vacuum advance comes from one of three sources: intake-manifold, ported, or venturi vacuum. With intake-manifold vacuum, the vacuum hose is simply connected to a fitting on

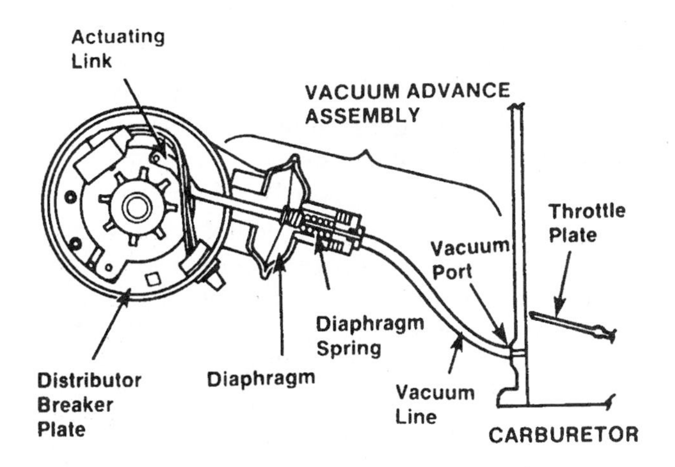

Fig. 5-27 Manifold vacuum is sometimes used for the distributor vacuum advance (reprinted, by permission, from Santini, *Automotive Electricity and Electronics, 2E.* Copyright 1992 by Delmar Publishers Inc.).

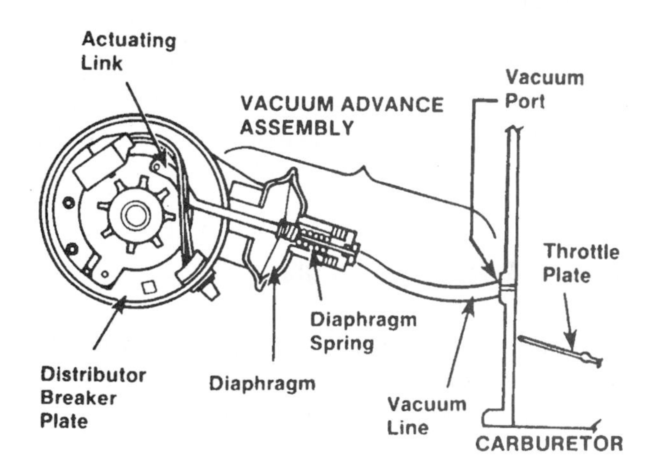

Fig. 5-28 Ported vacuum is also used for vacuum advance (reprinted, by permission, from Santini, *Automotive Electricity and Electronics, 2E.* Copyright 1992 by Delmar Publishers Inc.).

the manifold or the base of the carburetor or throttle body. With ported vacuum, the hose is connected to a fitting just above the throttle plates. At idle, there is no vacuum signal because the port is above the throttle plates (which are closed). As the throttle is opened, the port is exposed to intake vacuum. The vacuum signal passes through the hose and timing is advanced. On some vehicles, the vacuum hose is connected to a fitting on the carburetor that vents into the venturi. Reading engine vacuum at this point produces a faster response, but the venturi-vacuum signal is typically too weak to move the distributor diaphragm. The vacuum hose from the carburetor, therefore, is usually connected to a vacuum amplifier to boost the strength of the signal.

The vacuum-advance mechanism itself is fairly simple. A vacuum hose from the carburetor or intake manifold is connected to a vacuum diaphragm on the distributor. The diaphragm moves the breaker point or pickup plate to change the relative position of the plate and advance timing. Many older Fords have a dual-vacuum diaphragm that both advances and retards timing according to changing throttle position.

On Chryslers with Lean Burn computers and later models with a vacuum transducer on the computer, no vacuum-advance mechanism is used on the distributor. The same is true on other vehicles with computerized engine controls and electronic spark timing. The computer monitors engine vacuum through a transducer or MAP sensor to calculate the equivalent amount of vacuum advance that's needed. The computer or control module then alters the firing of the coil as needed to add or subtract advance according to engine load. The amount of advance that's added or subtracted may be further modified by other sensor inputs such as the throttle-position sensor and so on.

TIMING AND EMISSIONS

Ignition timing has a significant impact on tailpipe emissions. Generally speaking, retarded timing at idle and during deceleration reduce hydrocarbon emissions. When the timing is retarded, combustion occurs later in the power stroke. This increases exhaust-gas temperatures and promotes more complete burning of hydrocarbons in the exhaust. When the hot exhaust gases enter the exhaust manifold and meet fresh oxygen supplied by an air pump or aspirator valve, the unburned HC continues to burn. Add a catalytic converter to accelerate the process, and the result is almost complete combustion of any hydrocarbons that were not already burned inside the engine. But retarded ignition timing also requires a slightly wider throttle opening. This is necessary to increase the flow of air and fuel so that the idle speed can be maintained. The wider opening and increased flow promote better mixing of air and fuel. This aids combustion and reduces HC emissions.

Controlling Emissions

The easiest way to retard ignition timing for reduced emissions during idle and deceleration is to use ported-vacuum advance. At idle there is no vacuum advance because the port is located above the throttle plates. During deceleration there is no vacuum advance either because, again, the throttle is closed.

Another way to retard ignition timing during idle and closed-throttle deceleration is to use a combination of ported and intake-manifold vacuum. The two vacuum sources are balanced against one another through a spring-loaded valve. Manifold vacuum reaches the distributor only when ported vacuum is sufficient to open the valve. On the older Ford dual-vacuum diaphragm distributors, carburetor-ported vacuum is connected to one side of the diaphragm and intake-manifold vacuum to the other. At idle and during closed-throttle deceleration, ported vacuum is zero, and intake-manifold vacuum is high. The ported-vacuum side of the diaphragm advances timing, while the intake-manifold vacuum side retards it. Thus timing is retarded at idle and during deceleration when manifold vacuum is strongest, and advanced during other modes of operation when ported vacuum is strongest.

Many systems, including the Ford dual-vacuum diaphragm unit just described, use ported-vacuum switches (PVS) or temperature-vacuum switches (TVS) to reduce emissions (Fig. 5-29). A PVS or TVS installed in the vacuum line between the distributor and vacuum source prevents vacuum from reaching the distributor during certain modes of operation. For example, tailpipe emissions are lower on some cars if vacuum advance is restricted to cruising speeds only. A PVS is used to prevent the vacuum signal from reaching the distributor until the trans-

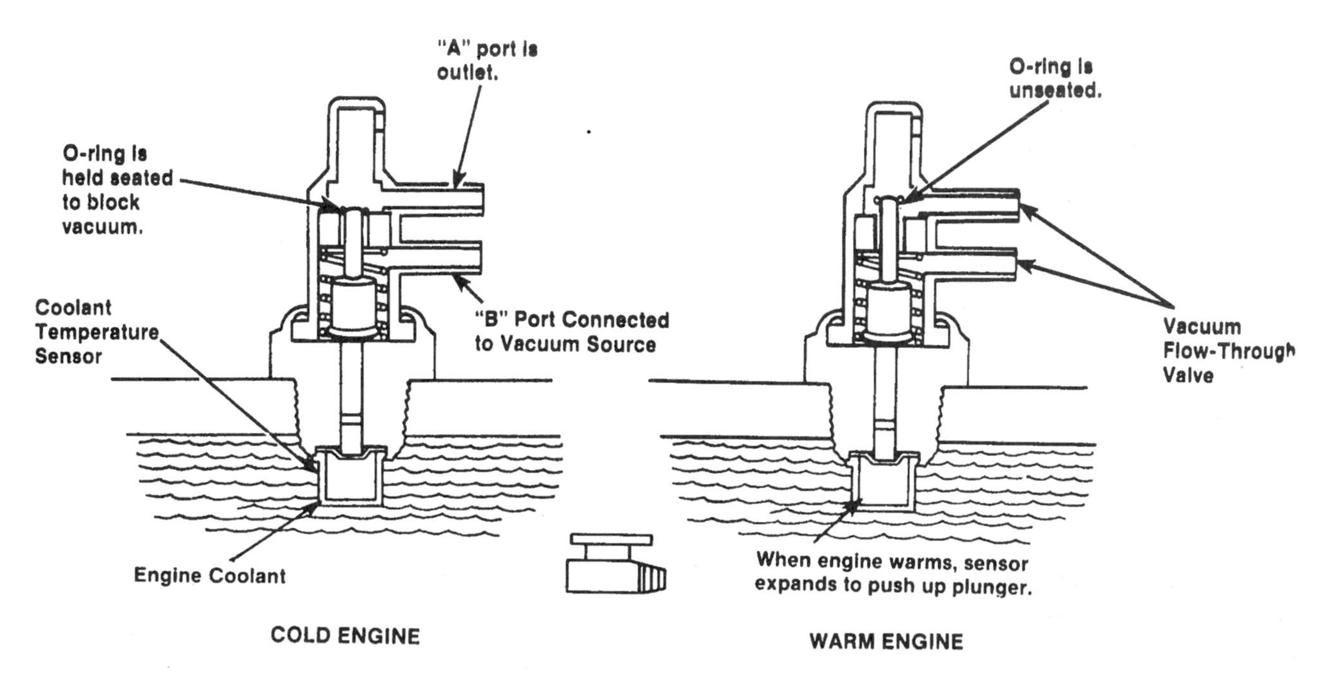

Fig. 5-29 A ported-vacuum switch (PVS) or thermal-vacuum switch (TVS) passes or blocks vacuum when the engine reaches a certain operating temperature (reprinted, by permission, from Scharff, *Complete Fuel Systems and Emissions Control.* Copyright 1989 by Delmar Publishers Inc.).

mission is shifted into high gear. This approach is known as "transmission-controlled spark." A small electrical switch located on the shift linkage controls a ported-vacuum solenoid. When the transmission is put into high gear, the switch grounds and energizes the solenoid, allowing vacuum to pass to the distributor.

On many engines, emissions are reduced by blocking vacuum advance to the distributor until the engine reaches operating temperature. This is because a relatively cold engine causes droplets of fuel to condense on the cylinder wall surfaces. This increases unburned hydrocarbon emissions so that higher exhaust temperatures are needed to burn the HC. A "thermal-vacuum switch" (TVS) is installed in the vacuum-advance line to block vacuum until the engine reaches operating temperature. The TVS is screwed into the engine block or intake manifold so that the tip of the TVS is in contact with the engine's coolant. When the coolant reaches normal operating temperature, the wax plug inside the TVS expands and opens the line between the vacuum source and the distributor.

Some thermal-vacuum switches work in just the opposite way. Some engines require a certain amount of vacuum advance when cold to improve idle quality and performance. On these applications, the TVS is designed to allow full vacuum advance until the engine is warmed up. The TVS then closes off the intake-manifold vacuum line and opens a ported-vacuum line.

On other applications, a TVS is used to perform yet another function. If the engine starts to overheat, the TVS temporarily opens up a line between intake-manifold vacuum and the distributor to advance the timing. This increases idle speed, which in turn circulates coolant through the engine more quickly. As the engine cools back down to normal temperature, the TVS closes the intake-manifold vacuum line and idle speed returns to normal. Such a switch is often called a coolant temperature override (CTO) switch.

Another device used to modify vacuum advance is the "spark-delay valve." This device works like a restriction in the vacuum line. Depending on the application, the valve can prevent full vacuum from reaching the distributor for a few seconds up to a half a minute or more. The delay valve is used to prevent sudden ignition advance that can cause combustion temperatures to soar, and thus increase NOX formation. By delaying the advance for a short period of time and allowing it to build up gradually,

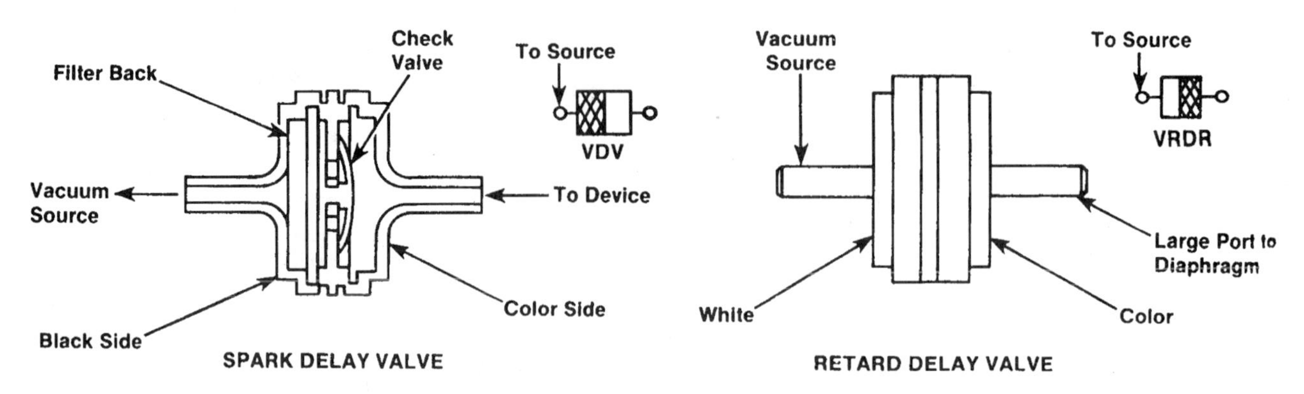

Fig. 5-30 Spark-delay valve and retard-delay valve (reprinted, by permission, from Scharff, *Complete Fuel Systems and Emissions Control.* Copyright 1989 by Delmar Publishers Inc.).

peak combustion temperatures can be avoided and NOX formation reduced.

Spark-delay valves (Fig. 5-30) go by a variety of names, but all do basically the same thing. Older Chryslers use an "orifice spark-advance control" (OSAC) valve. Ported vacuum from the carburetor is routed to the OSAC valve, then to the distributor. When a vacuum signal reaches the OSAC valve, it takes about 20 seconds to pass through the valve and reach the distributor. General Motors uses a "spark-delay valve" (SDV) or a "vacuum-delay valve" (VDV) for the same purpose. The SDV or VDV is located in the vacuum line between the ported-vacuum connection on the carburetor and a TVS. The valves have a 0.005-in. orifice restriction, which delays the vacuum signal from reaching the distributor by about 40 seconds.

Many spark-delay valves use a porous metal filter to restrict the flow of vacuum. Such delay valves are also fitted with a one-way rubber valve. The valve allows vacuum to escape from the distributor side of the line when the ported-vacuum signal drops to zero (as during deceleration or idle). In other words, the valves restrict vacuum one-way only. Therefore, it is very important that these valves be installed facing the right direction. Most are marked "DIST" on the distributor side or "CARB" on the carburetor side. They are also color coded according to the amount of delay they provide.

Spark-delay valves can become clogged with dirt. If this happens, the delay period may be longer than usual, or there may be complete blockage of the vacuum signal. Delay valves can be tested with a hand-held vacuum pump. Apply vacuum to the carb side and see how long it takes for the reading to drop. Compare this to the manufacturer's specs to see whether or not the valve is doing the job it is supposed to do. When vacuum is applied to the distributor side, there should be no restriction, and the reading should be zero.

ELECTRONIC-SPARK TIMING

Since the 1980s, most vehicles have gone to electronic-spark timing (EST) to control emissions (Fig. 5-31). Electronic-spark timing takes electronic ignition one step further by totally eliminating the centrifugal and vacuum-advance mechanisms. In many late-model cars with "direct ignition systems" or "distributorless ignition systems" (DIS), it has also eliminated the need for the distributor itself.

There are many different electronic-spark control systems on the road today, but all share essentially the same basic operating principle. Various sensors keep the computer informed about engine speed, coolant temperature, manifold vacuum, throttle position, ambient air temperature, transmission gear position, and anything else it needs to know to calculate the optimum spark timing based on the engine's current needs. This allows spark advance to be changed almost instantly and varied infinitely according to the preprogrammed spark curves in the computer.

Electronic-spark timing also increases the reliability of the ignition system by eliminating mechanical components that can wear out. And it eliminates much of the complex plumbing associated with older mechanical vacuum-spark control distributors.

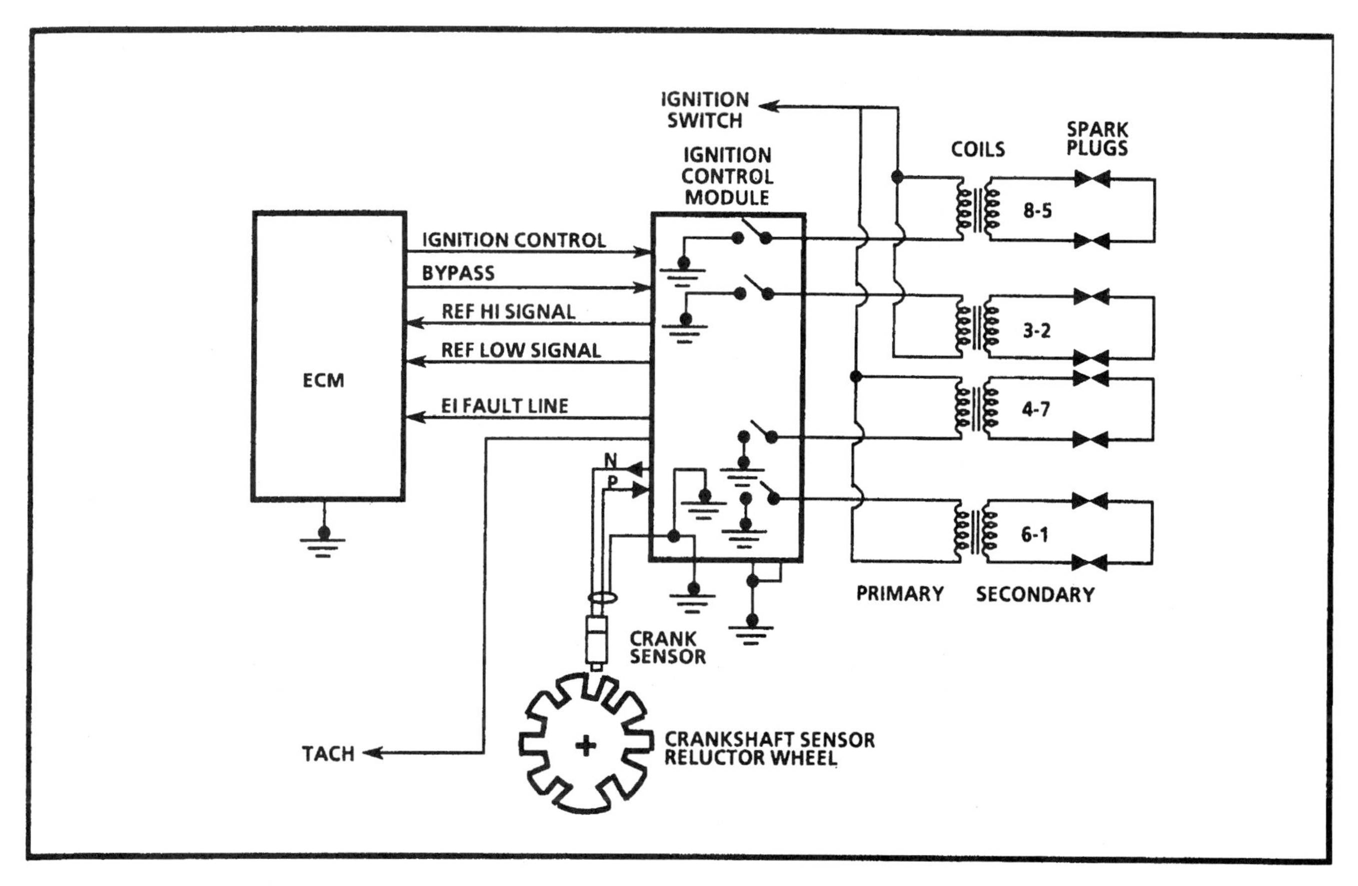

Fig. 5-31 Electronic-spark timing is regulated by the engine control module (ECM) in most late-model engines (courtesy GM).

Like electronic ignition, electronic-spark timing is a set-it-and-forget-it kind of system. Once set, electronic-spark timing will not change unless someone tampers with it, or there's a malfunction in the control module or sensors. On some vehicles with distributorless ignition systems, base timing isn't even adjustable, so all you can do it check it to see if its within specs.

Electronic-spark timing curves cannot be altered unless the computer itself is replaced, or the computer's "Program Read Only Memory" (PROM) chip is replaced (or reprogrammed in the case of "Electronically Erasable Program Read Only Memory" or EEPROM chips). Aftermarket performance PROMs have been a popular means of upgrading engine performance, but such products must be emissions-certified for street use.

CHAPTER 6

Ignition System Testing

Ignition performance checks such as timing and firing voltage were once part of the traditional tune-up. But since cars are no longer "tuned" but "maintained," ignition checks are now mostly made to confirm factory timing adjustments, to detect gross ignition problems, or when necessary to troubleshoot an emissions or driveability problem. The kind of checks we're talking about include:

- Checking base timing.
- Checking timing advance (mechanical, vacuum, or electronic).
- Checking firing voltage (peak Kv).
- Comparing primary and secondary voltage patterns (raster patterns) cylinder to cylinder.
- Observing firing duration (millisecond sweep) to detect abnormal combustion.

WHY TIMING CHECKS ARE SO IMPORTANT

Base timing should be checked when:

(1) Replacing spark plugs at the recommended replacement interval (usually every 30,000 miles),
(2) The engine has been experiencing a detonation (spark knock) problem,
(3) Idle speed is not within specs (too fast or too slow), and/or
(4) An engine is producing elevated emissions at idle.

The base timing should be within 1 or 2 degrees of the factory-specified setting (which can be found on the underhood emissions decal, a shop manual, or tune-up reference chart) for optimum performance and emissions. If the timing setting is off by more than 2 degrees, it should be reset as closely as possible back to the factory specs following the specified procedure for the vehicle.

On most vehicles with distributors, the timing is adjustable. But on many vehicles with distributorless ignition systems, timing is not adjustable. If the position of the crankshaft and camshaft position sensors are fixed, then no adjustments are possible. Timing advance is determined electronically so all you can do is check it to make sure it's where it is supposed to be. And if it isn't? Then you've found a sensor or computer problem that needs further investigation.

Engines with electronic ignition systems should not require periodic timing adjustments. There's nothing to wear that might affect the accuracy of the timing signal except the distributor drive gears and timing chain. (The timing chain can become worn and introduce slop into the otherwise accurate timing signal.) Even so, timing should always be checked to make sure someone hasn't changed it. It's not uncommon to find engines that are badly out of time because someone attempted to cure a detonation problem by retarding the ignition timing. If an engine won't run without detonating with base timing set to the stock specifications, something is amiss. The engine may have a cooling problem, an excessive buildup of carbon deposits in the combustion chambers, the wrong spark plugs (too hot), a lean fuel mixture or vacuum leak, or an inoperative EGR system. Retarding ignition timing to compensate for a problem elsewhere is treating the symptom, not the cause. The cure may be temporary at best, and only further hinder the engine's ability to run properly.

Retarded ignition timing can cause:

- Poor fuel economy
- Lack of power
- Hard starting
- Slow idle speed or stalling
- Burned exhaust valves (mixture still burning as it exits the cylinders)

Causes of retarded timing include incorrect base-timing adjustment; sticking centrifugal-advance weights; leaking vacuum-advance diaphragm or hose; wrong vacuum-hose connections; blocked vacuum-delay valve, switch, or solenoid; or the wrong computer calibration (wrong PROM or computer for the application).

Over-advanced timing can cause:

- Detonation (spark knock)
- Overheating and engine run-on
- Hard starting (forces the starter to work harder)
- Elevated idle emissions

Causes of over-advanced timing include incorrect base timing adjustment; broken or weak centrifugal-advance springs; wrong vacuum-advance connection (manifold vacuum instead of ported vacuum); defective vacuum-delay valve, switch, or solenoid; or the wrong computer calibration (wrong PROM or computer for the application).

It makes no difference whether you check timing with an ordinary timing light, a magnetic timing meter, or an engine analyzer or a scan tool to read timing digitally. *The point is, check it.*

Timing Advance

In addition to checking base timing, it's also important to check timing advance (Fig. 6-1). On older

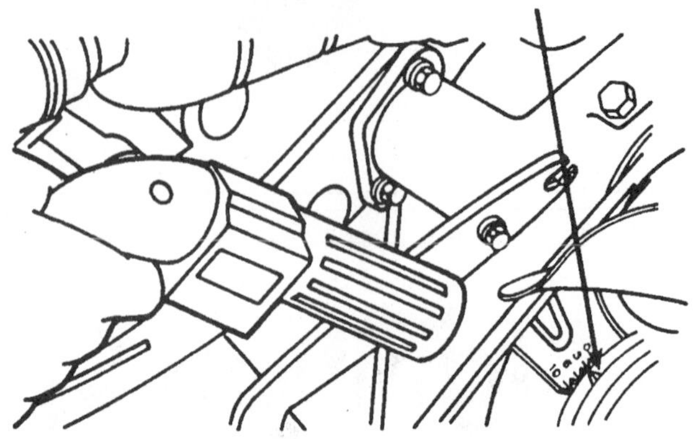

Fig. 6-1 Checking base ignition timing with a timing light (reprinted, by permission, from Santini, *Automotive Electricity and Electronics, 2E.* Copyright 1992 by Delmar Publishers Inc.).

vehicles with centrifugal- and vacuum-advance distributors, checking timing advance will tell you whether or not the mechanical components in the distributor are working properly.

To check centrifugal advance, disconnect and plug the hose to the vacuum-advance diaphragm. Then increase engine speed while observing the timing marks with a timing light. Timing should advance if the centrifugal-advance mechanism is working properly. The exact number of degrees of total advance can be determined by using an advance or adjustable timing light, or a digital mag timing meter and noting the total number of degrees of advance at a specific rpm. Centrifugal advance usually peaks out from 3,400 to 4,500 rpm depending on the application.

Vacuum advance can be checked by applying vacuum to the vacuum-advance diaphragm with a hand-held pump or attaching the hose to a source of manifold vacuum. As before, you should see timing advance if the vacuum diaphragm is working properly.

Timing Lights

Distributorless ignition systems and electronic timing have made a traditional timing light obsolete for many newer vehicle applications. But it's still a valid tool for checking timing on engines that have adjustable distributors and timing marks. A standard timing light simply flashes in response to ignition pulses it receives through its high-voltage pickup, which clamps around the number 1 spark-plug wire. For accurate timing checks, the timing light's pickup must be attached to the plug wire for cylinder number 1 (Fig. 6-2). You also have to make sure you're looking at the correct timing marks and know which marks mean what. If a pulley or flywheel has more than one timing mark, refer to a shop manual to determine the correct reference mark to use .

If you're using an "advance" timing light, the ad-

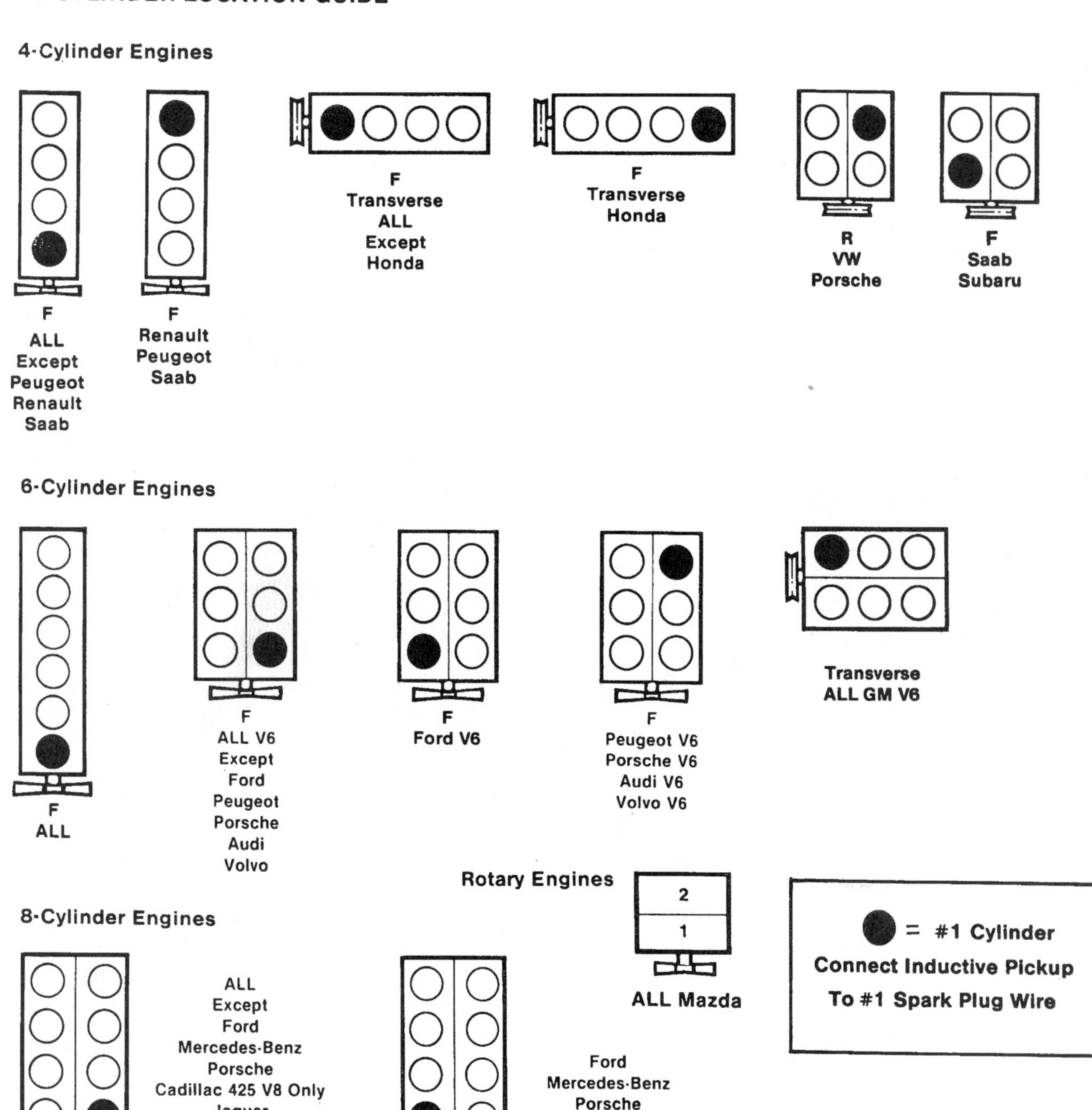

Fig. 6-2 Number 1 cylinder locations (courtesy All-Test).

TIMING MARK LOCATIONS

Crankshaft Pulley

AMC
Audi
Chrysler V8 & 6
Datsun
Fiat
Ford/Lincoln/Mercury
GM (all)
Jaguar
Mazda
Mercedes-Benz
MG
Peugeot
Toyota
Volvo
VW — air-cooled

Timing Mark

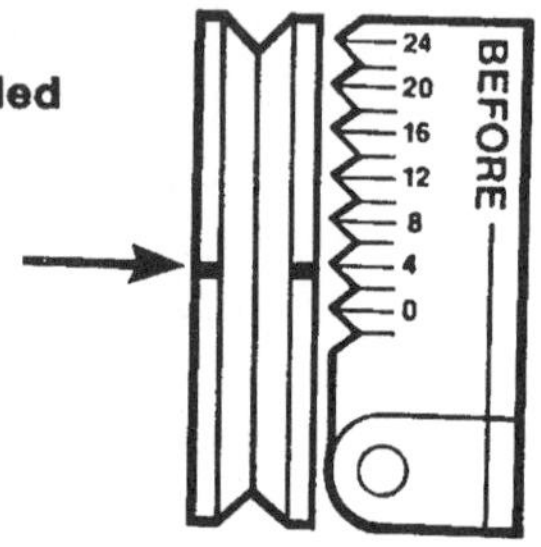

BEFORE

24 20 16 12 8 4 0

Flywheel

BMW
Chrysler/Dodge/Plymouth FWD 4-cyl.
Honda
Porsche 924 & 912E
Renault
Saab
Subaru
VW—water-cooled

FLYWHEEL ACCESS HOLE IN BELL HOUSING

POINTER

FLYWHEEL

Fig. 6-3 Timing mark locations (courtesy All-Test).

justment knob must be turned once the engine is running to change the apparent position of the timing mark on the engine to top dead center (TDC). Once the mark has been lined up (Fig. 6-3), the display on the timing light will show the number of degrees of timing advance. An advance timing light is considered to be better than a standard nonadjustable timing light (Fig. 6-4). The display on the timing light is usually easier to read than the timing marks on the engine. What's more, the advance/retard calibration marks on the engine may not be positioned accurately. But as long as TDC is accurate, an advance timing light should give you an accurate reading.

Digital Mag Timing

On some engines (late 1970s and up), timing can be checked and adjusted using a digital magnetic timing meter. Instead of using a flashing strobe light to illuminate timing marks on a pulley or flywheel, a digital mag timing meter uses an inductive pickup and magnetic probe to calculate ignition timing. The inductive pickup clamps on the number 1 spark-plug wire, and the magnetic probe inserts into a receptacle over the harmonic balancer or flywheel so it just touches the balancer or flywheel underneath. When the notches pass underneath the probe, an AC signal is generated that corresponds to TDC of cylinder number 1. The meter then calculates the base-timing setting by comparing the magnetic pulse against the actual firing of the spark plug.

To accurately calculate timing, however, there's one more step involved. The notch in the harmonic balancer or flywheel may be offset a certain number of degrees from actual TDC. GM typically uses a 135-degree offset, Ford a 52- or 135-degree offset, and Chrysler a 0-, 10- or 20-degree offset.

Mag timing came about as engine compartments became more and more cramped, making it difficult if not impossible to directly observe timing marks on the pulley or flywheel. Checking timing with a digital mag timing meter also allows a much higher degree of accuracy. Most units can display timing to within 1/10 of a degree! Such accuracy really isn't necessary for real-world applications, but it's better than trying to guess which timing mark means what, or whether a notch in a reference indicator stands for 1, 2 or 4 degrees of advance or retard (Fig. 6-5).

Electronic Checks

If you're using an engine analyzer that includes timing checks, then the results of those tests are typically displayed digitally on the analyzer's screen or

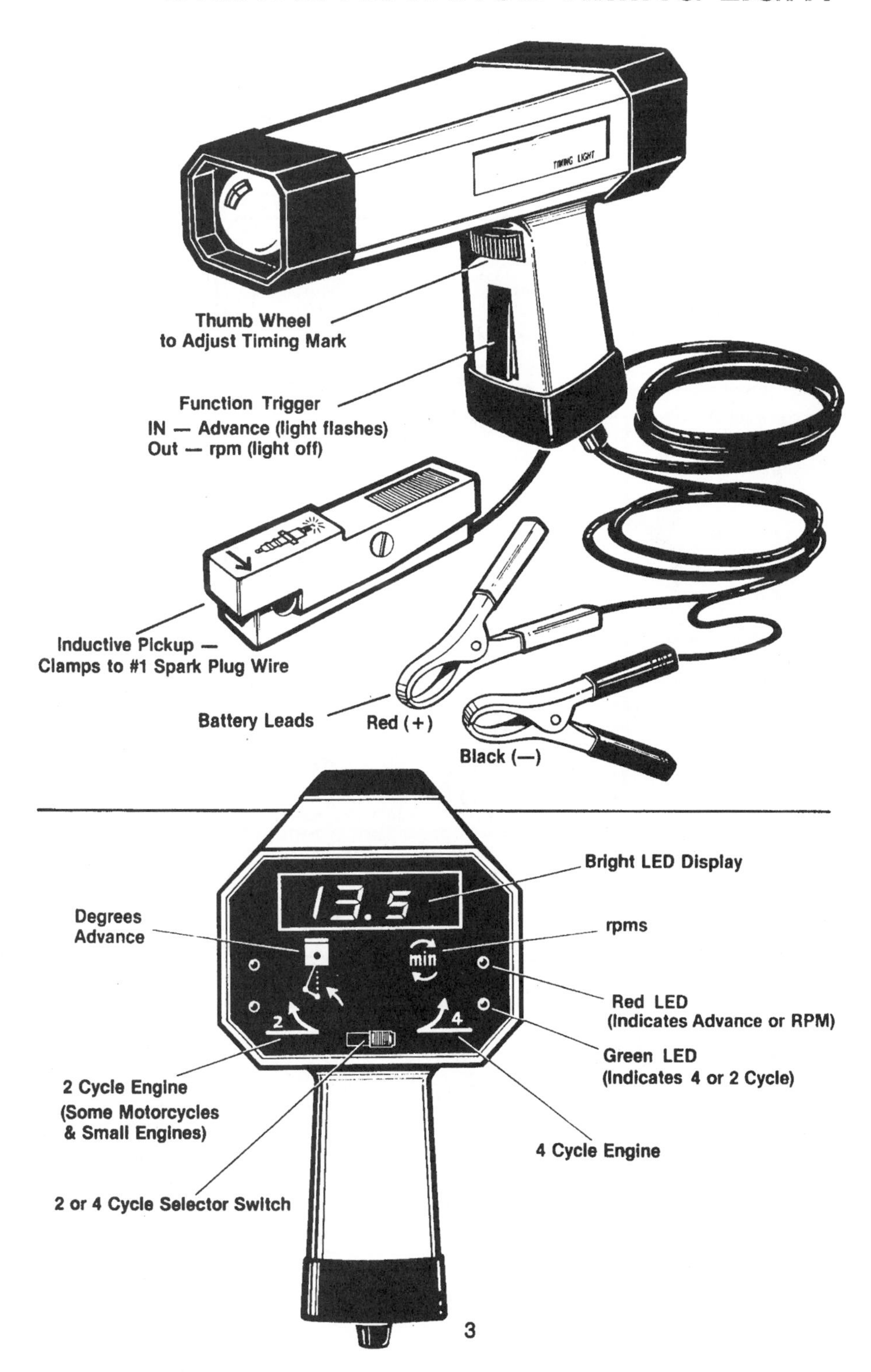

Fig. 6-4 Advance timing light (courtesy All-Test).

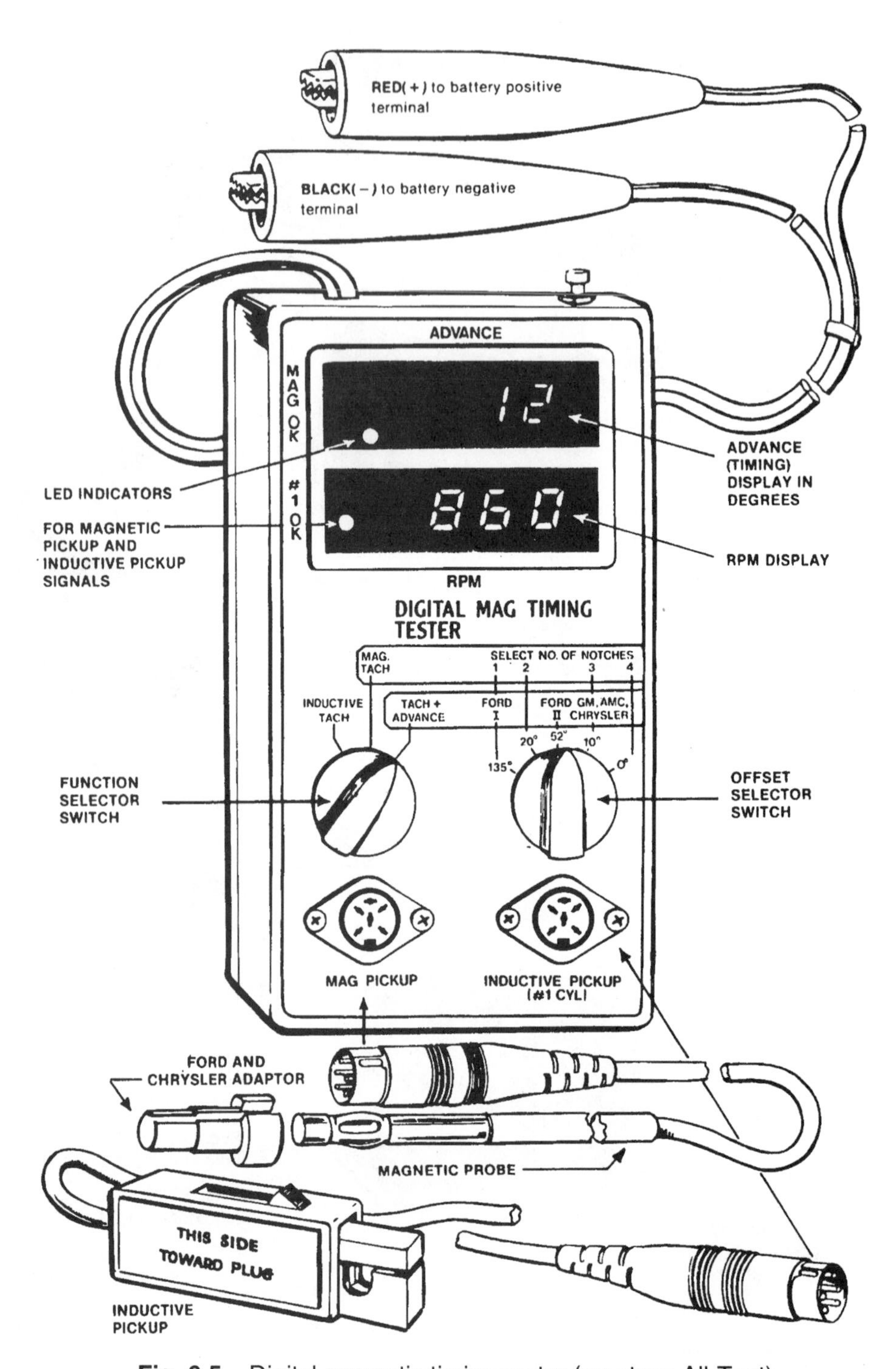

Fig. 6-5 Digital magnetic timing meter (courtesy All-Test).

printout. Newer computerized analyzers will usually indicate or red-flag a timing setting that is out of spec.

On an engine with a distributorless ignition system and nonadjustable timing (many of the newer GM engines, for example), base timing can be read with a scan tool. The information is for reference only, because there's nothing that can be adjusted. If timing is not within specifications, it indicates a sensor or computer problem that requires further diagnosis.

On some vehicles (Cadillac for example), you can also obtain a timing reading without a timing light by using the electronic climate-control system.

It's a complicated procedure that varies from year to year. By pressing the right combination of buttons, you can put the computer into a self-diagnostic mode and read the timing through the display panel.

TIMING CHECKS

Before you can accurately test base timing on an engine, certain procedures must usually be followed to assure an accurate reading. On most older vehicles with vacuum-advance distributors, the vacuum-advance hose must be disconnected and plugged. But this isn't true for all such applications, so be sure to refer to the vehicle manufacturer's procedure for checking base timing.

On vehicles with electronic-spark timing, various "tricks" may have to be performed to put the system into the proper mode for a base timing check. Again, unless you're familiar with the application, always refer to the vehicle manufacturer's procedure for making a base-timing check.

For example, the standard procedure for checking base ignition timing on GM engines with electronic-spark timing (EST) and feedback carburetion is to bypass the EST circuit by unplugging the four-wire EST connector from the distributor (Fig. 6-6). This trick doesn't work on engines with throttle body injection (TBI), tuned port injection (TPI), or port fuel injection (PFI) because the computer needs an ignition signal to pulse the injectors.

On Chevrolet 4-cylinder, V6, and V8 engines with TBI through 1985, the electronic-spark advance circuitry is bypassed by disconnecting only the connector on the tan wire of the four-wire EST harness. The tan wire connector may be at the distributor, or it may be hidden inside the black plastic wiring harness cover. Opening the EST connector will trigger the "Check Engine" light, just as it does when you pull the four-wire connector on an engine with a carburetor. To turn the "Check Engine" light off after you've finished with your timing checks, turn the ignition off and disconnect the electronic control module (ECM) fuse for ten seconds.

With Pontiac fuel-injected engines (including the Buick and Olds applications that use Pontiac engines) and 1986 Chevy multiport injected engines, yet another timing procedure is needed. The trick here is to connect a jumper wire between the A (ground) and B terminals of the underdash ALDL (assembly line data link) connector. This grounds the diagnostic connector and puts the computer into the self-diagnostic mode. While in this mode, the computer delivers a fixed timing signal to the distributor, which you can then check with your timing light against the specs listed on the underhood emissions decal.

On 1986 Chevy Cavalier 2.0-liter engines, GM recommends a rather unusual timing procedure. It is included here to illustrate the variety of timing techniques you're apt to encounter. Because the computer sets timing advance for each cylinder individually, the most accurate method of checking timing adjustment is obtained by reading the "average" timing of all the cylinders at once. This is done by connecting your timing light to the coil wire instead of the number 1 plug wire.

After disconnecting the tan connector on the four-wire EST distributor wiring harness and hooking up your timing light up to the coil, start the engine. When you point your timing light at the timing marks by the pulley, the notch in the pulley will appear blurred rather than the distinct line you're used to seeing. Don't worry, there's nothing wrong with your timing light. What you're seeing are the timing pulses of all four cylinders simultaneously. They won't align perfectly because of the slight variation in timing the computer is making between cylinder firings.

To adjust the timing, turn the distributor until the band of illuminated pulley notches are centered on the correct mark on the timing tab. For example, if the specified timing is supposed to be 6 degrees BTDC, and the pulley notches appear to cover a spread of 4 degrees, turn the distributor until the notches cover the span from 4 degrees to 8 degrees. Even though there is a 2-degree deviation either way, the "average" timing is 6-degrees, putting it within the correct range.

This technique is considered to be more accurate than the traditional method of reading timing off of the number 1 cylinder because it takes into account the range in timing variation between all the cylinders.

What about Ford applications? On microprocessor-controlled unit (MCU) systems, you'll get a retarded timing reading unless you first bypass the retard circuit. The MCU module will have 3 wiring connectors. One connector will have 2

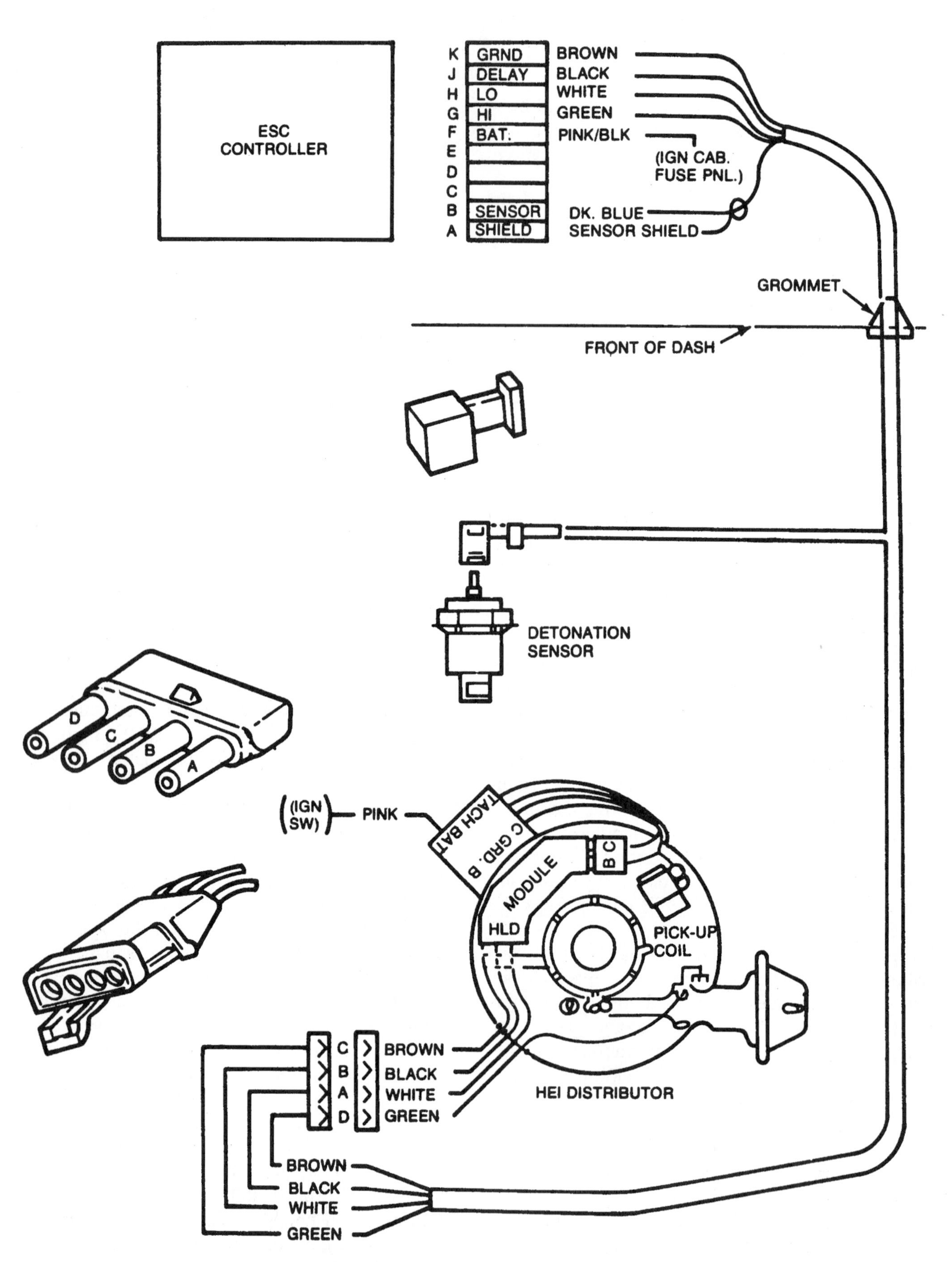

Fig. 6-6 Electronic-spark timing is disabled for a base timing check on this GM HEI system by unplugging the four-wire connector from the distributor (courtesy GM).

wires (one yellow, one black). Trace them back to the connector in the harness, unplug the connector, and install a jumper wire between the 2 terminals. Base timing should now match the specs on the emissions decal.

On Ford EEC-IV systems, the ignition module at the base of the distributor has a connector that links it to the computer. This connector can be unplugged (turn the ignition off first) to check base timing. The timing reading obtained with the connector unplugged should be within 2 degrees of the spec on the vehicle's emissions decal.

To check EEC-IV (electronic engine control) computer advance, the computer must be put into the self-diagnostic test routine. Reconnect the distributor-module connector, then jump the the single terminal self-test input connector to the #2 terminal in the four-terminal self-test connector, or connect a scan tool to the system. The system will now begin the self-diagnostic sequence when the ignition is turned on. Ignition timing during the self-test mode will remain fixed for 2 minutes and will be advanced 20 degrees over base timing. You can figure the base timing using this procedure alone by simply deducting 20 degrees from the reading.

On Chryslers with electronic fuel injection, spark timing can be checked by putting the computer into the "limp-in" mode. In this mode, the computer will hold the idle speed at 1,000 rpm and set the timing at a fixed specification. This can then be checked against the spec on the emissions decal. To put the computer into this mode, disconnect then reconnect the coolant sensor lead while the engine is idling at normal operating temperature. To check electronic-spark advance, connect a scan tool to the diagnostic connector and turn the ignition key on and off three times leaving it in the on position the third time. This will put the computer into a special self-test mode. It will display trouble codes, perform a sequence of actuator tests, and display the number of degrees of advance.

Knock Retard

Another aspect of ignition timing that may have to be checked is knock retard. On engines that have a knock sensor, the sensor signals the computer when it detects vibrations that indicate the engine is experiencing detonation (spark knock). The computer will then retard timing electronically until the knocking stops. Under normal operating conditions, the knock-sensor retard circuit plays no role in ignition timing or advance. But if an engine has a bad rod bearing or other problem that produces vibrations similar to those caused by detonation, it may trigger the knock sensor, causing the timing to retard.

The knock-sensor retard circuit can be checked by observing base timing, then rapping on the manifold or cylinder head with a wrench near the knock sensor (do not pound on the sensor itself!). The noise should cause the timing to retard momentarily if the circuit is functioning properly.

OSCILLOSCOPES

The traditional ignition oscilloscope was once essential for diagnosing primary and secondary ignition problems with breaker-point ignition systems. It almost faded into history with the advent of electronic ignition, but has now come full circle. Oscilloscopes have always been a good tool for identifying secondary ignition problems such as fouled spark plugs and bad plug wires as well as combustion and fuel problems. But now they're being used to analyze sensor signals and fuel injectors. Small hand-held "dual trace" scopes with liquid crystal displays (LCD) can be extremely helpful when it comes to finding intermittent sensor problems. So scopes are a very important diagnostic tool with which you should be familiar.

Most oscilloscopes have a cathode ray tube similar to a television minus the channel tuner (except for the new hand-held units with LCD displays). For ignition testing, the scope uses inductive pickups on the coil high-voltage wire and number 1 spark-plug cable, as well as primary leads that attach to the ignition coil-negative terminal.

An oscilloscope can be best described as a graphic voltmeter. The scope converts voltage signals into a visual wave pattern on its screen, thus painting a picture of what's happening inside the ignition system. It does this by using the ignition system's voltage to distort what would otherwise be a straight line sweeping across the screen. The wave pattern that emerges is "dynamic," showing the live action as it happens. The pattern is displayed rapidly from left to right, producing a continuous line that changes as the voltage changes.

The display is a graph that plots voltage versus time. Vertical (up and down) displacement on the

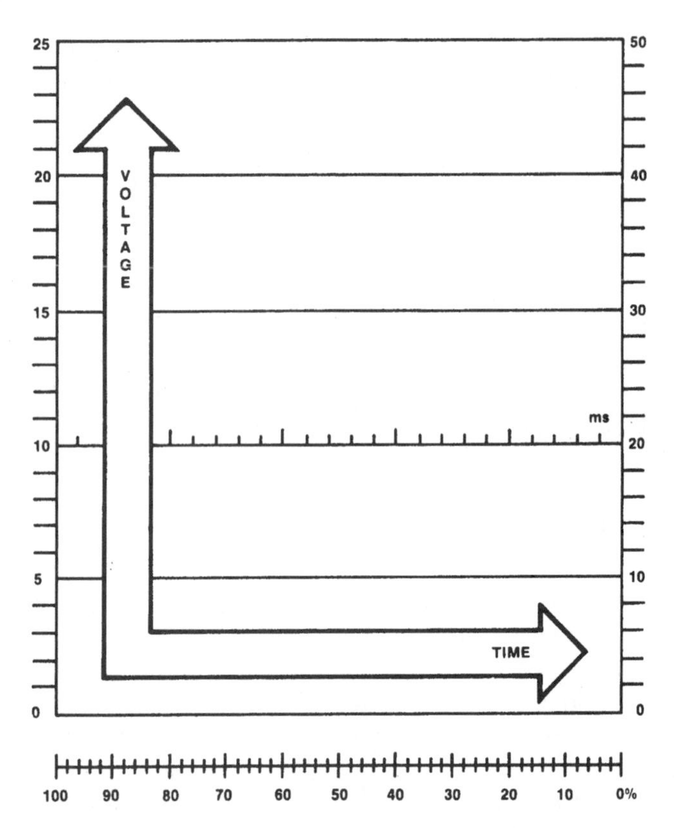

Fig. 6-7 An oscilloscope plots voltage versus time (reprinted, by permission, from Santini, *Automotive Electricity and Electronics, 2E.* Copyright 1992 by Delmar Publishers Inc.).

screen represents voltage. The taller the line, the higher the voltage (Fig. 6-7). The horizontal (sideways) sweep of the line represents time.

Scope Displays

The scope can display either primary or secondary ignition patterns one of three different ways. Each mode has its own advantages and uses:

- *Superimposed Pattern*—this pattern displays all engine cylinders directly on top of one another (Fig. 6-8). This mode is used to see problems that are common to one another.
- *Parade Pattern* (*Display*)—this pattern shows each cylinder in a sequence from left to right according to firing order (Fig. 6-9). The ignition pattern for each cylinder is compressed, so all will fit on the screen. This mode is used to compare firing-voltage variations cylinder to cylinder. It's a quick way to find a problem cylinder because the problem cylinder will have a taller or shorter firing line than the rest.

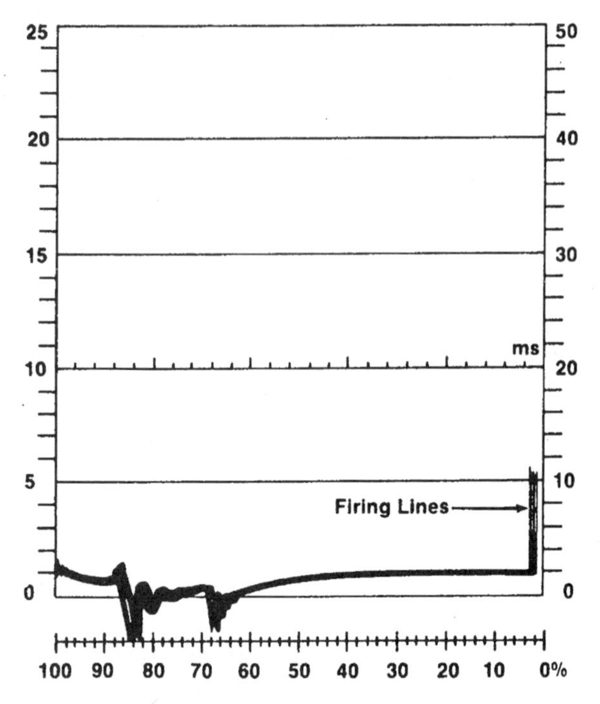

Fig. 6-8 Superimposed ignition pattern (courtesy Sun Electric).

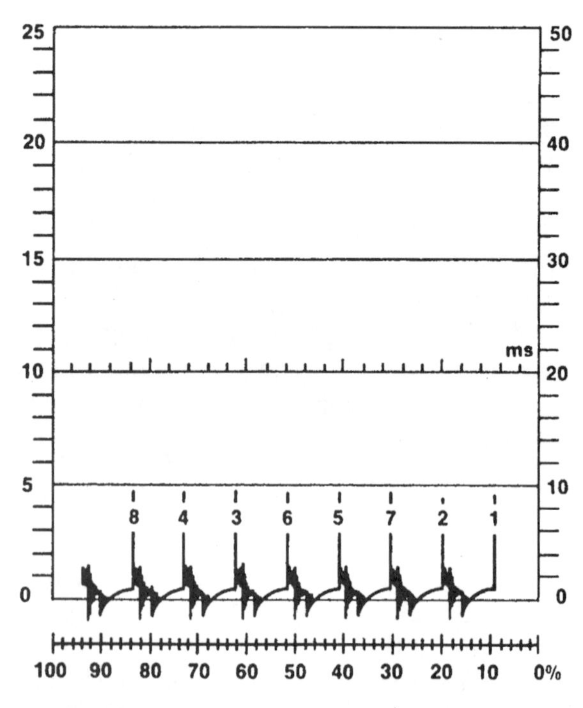

Fig. 6-9 Display or parade pattern (courtesy Sun Electric).

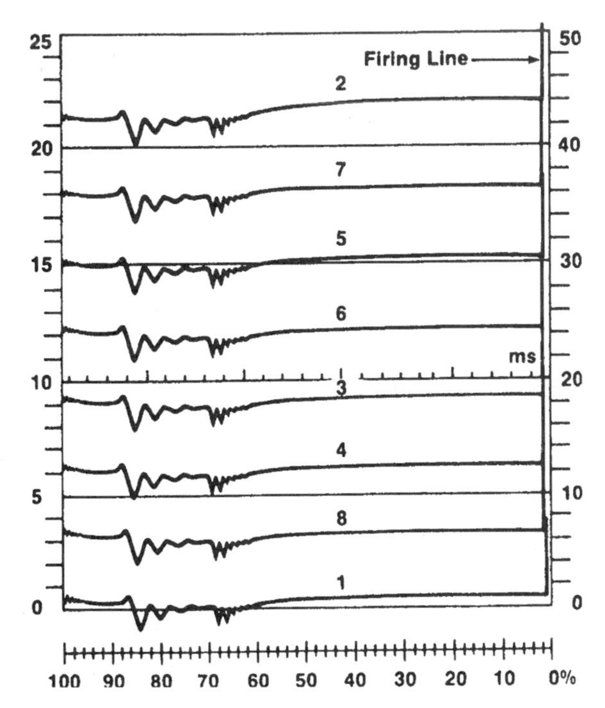

Fig. 6-10 Raster or stacked pattern (courtesy Sun Electric).

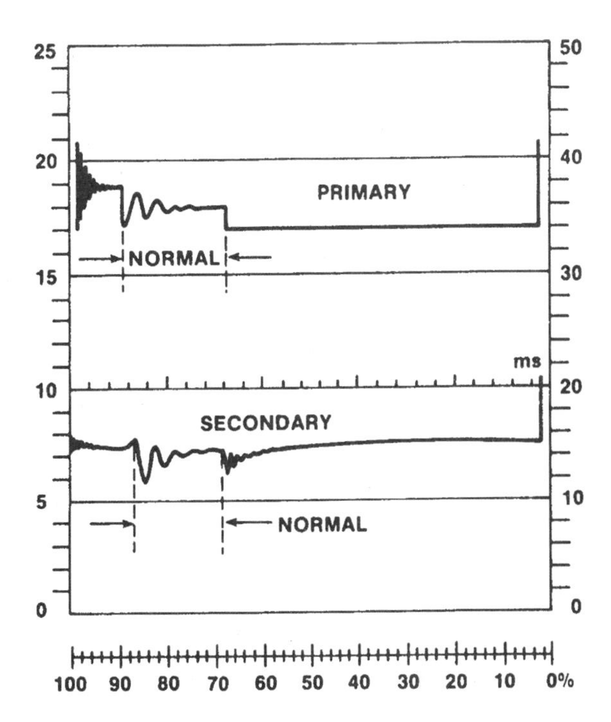

Fig. 6-11 Primary and secondary ignition patterns (courtesy Sun Electric).

- *Stack (or Raster) Pattern*—the stack or raster pattern displays each cylinder separately from top to bottom on the screen (Fig. 6-10). The cylinders may or may not be displayed in their firing order. This mode is used to compare spark duration and dwell times cylinder to cylinder.

PRIMARY PATTERNS

When the scope is set to read the primary pattern (Fig. 6-11), it shows what's happening in the primary side of the ignition circuit. When cars had breaker-point ignitions, the primary pattern showed: (1) point opening, (2) primary voltage, (3) primary coil oscillations, (4) coil/condenser oscillations, (5) point closing, and (6) dwell. The pattern will also show: (a) primary resistance, (b) coil energy, and (c) condenser energy.

Though primary patterns were very helpful when analyzing breaker-point ignitions, they are not as necessary with electronic-ignition systems because dwell is controlled electronically. The scope can still tell you, however, if dwell is increasing with changing engine rpm (most ignition modules will vary dwell with engine rpm).

Secondary Patterns

The secondary pattern shows what's happening in the secondary side of the ignition circuit (Fig. 6-12). This pattern will show: (1) plug firing voltage, (2) plug firing time, (3) coil/condenser oscillations, (4) coil buildup, and (5) dwell.

Plug Firing Voltage. The initial secondary spike shows you how much voltage it takes to start the spark. The spark must jump across not only the electrode gap at the end of the spark plug, but also the rotor air gap in the distributor cap. The height of this line is the firing voltage in thousands of volts (kV).

Although the coil is capable of producing more voltage than the initial secondary spike indicates, it usually takes only about 5,000 to 18,000 volts (5 to 18 kV) to fire the spark plugs.

The voltage required to fire a plug depends on a number of factors: the distance across the electrode gap, whether or not the electrodes are worn, the distances across the rotor air gap inside the distributor cap, the amount of resistance in the ignition cable and spark plug, the temperature of the spark plug, ignition timing, engine compression, fuel mixture, and load. Worn plugs, worn rotors, and high-resistance ignition cables can all increase the voltage

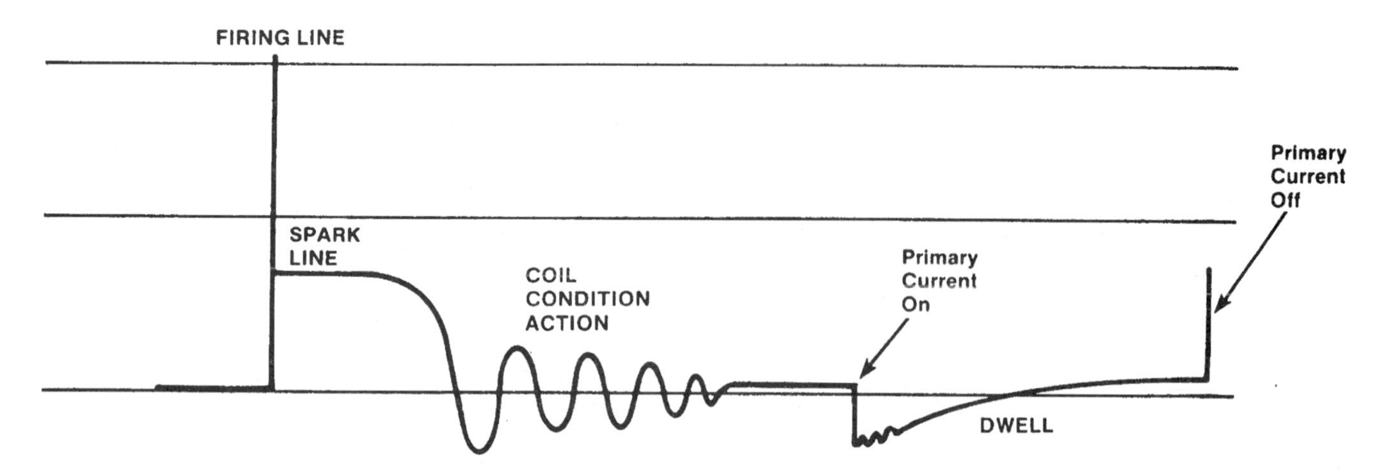

Fig. 6-12 Secondary pattern components (reprinted, by permission, from Santini, *Automotive Electricity and Electronics, 2E.* Copyright 1992 by Delmar Publishers Inc.).

requirements dramatically. And if the required voltage exceeds the maximum output capacity of the coil, the plug will misfire. If the initial secondary spike is approaching the upper limits of the system, it's a sure indication that repairs are needed.

It's not as important to know precisely how much voltage it takes to fires the plug in a particular cylinder as it is to compare the firing voltages of all the cylinders against one another. Firing voltages should not vary by more than 3 kV cylinder to cylinder. A greater variation indicates problems which will require further investigation.

A cylinder that shows an abnormally low firing voltage probably has a grounded spark plug (deposits bridging the electrode gap), or a shorted ignition cable. A cylinder that shows an abnormally high firing voltage likely suffers from an open ignition cable or a plug with a wide gap.

Plug Firing Time. This portion of the secondary display shows how long the spark lasts at the spark plug. This is the spark firing line. The duration of the spark is measured in milliseconds (thousandths of a second). The average spark duration with the engine running will be about 1.5 milliseconds. Duration is an important factor to note because it reveals the condition of the secondary components in the ignition system. A duration of less than 0.8 milliseconds, for example, would tell you that there either isn't enough voltage to keep the spark going, or that the voltage is having trouble getting through.

Low coil output, for example, will not produce enough energy to sustain a long spark. Consequently, the firing time will be shortened—possibly to the point that the spark may not burn long enough to keep the fuel mixture lit. This can be a real problem in late-model engines with lean mixtures. It takes a much longer spark to assure complete ignition with a lean mixture than it does a rich mixture.

High resistance in the plug or ignition cable may not increase the initial firing voltage significantly. But it will make it harder for the spark to sustain itself once it starts. The available voltage is quickly used up trying to overcome the resistance. This shortens the duration of the spark and possibly causes a misfire situation.

A longer-than-normal spark indicates that the firing voltage is experiencing little resistance. A spark duration of 1.8 milliseconds or more usually means a plug is grounded. Carbon deposits have bridged the electrode, fouling the plug, and providing an easy path to ground for the voltage to follow. It may also indicate a shorted ignition cable. A narrow plug gap, an overly rich fuel mixture, or low compression normally don't make enough of a difference in resistance to affect the burn time significantly.

As the duration of the spark nears its completion, the burning fuel mixture makes it harder for the remaining coil voltage to keep the spark going. This causes a momentary rise in the firing voltage level, just as the spark flickers out.

When one or more (but not all) of the spark firing lines slope up to the right and firing time shortens, it indicates a lean fuel condition in the affected cylinders. A rich mixture is more conductive than a lean one, and requires less voltage to sustain a spark. On a fuel-injected engine, this could be a symptom of a dirty, clogged, or inoperative fuel injector. The premature "snuffing out" of the spark can also be caused by poor cylinder breathing. A rounded cam lobe that prevents either an intake or exhaust valve from opening completely, or an exhaust valve that leaks, can interfere with proper breathing and leave exhaust in the cylinder.

Turbulence in the combustion chamber also shows up in the spark plug firing line. An erratic line that shows little spikes (voltage rises) could mean the spark is encountering little pockets of lean mixture (which is normal on some engines). This could be due to an air leak or an open EGR valve.

Coil Buildup. This part of the secondary pattern shows what happens when the electronic module has switched back on and current starts to flow back into the coil's primary windings.

The pattern shows a dip here because the coil tries to resist the sudden change in current. During the initial surge of current, the magnetic field starts to grow in strength. This self-induces a current in the opposite direction. The line takes a downward plunge momentarily because the polarity of the induced voltage is reversed. The line shows up in the secondary pattern because the induced voltage is in the secondary windings. Distortions in this area can help you evaluate the condition of the coil. For that, you should switch to the primary pattern.

As the current builds in the primary coil winding, the magnetic field begins to reach the saturation point. This can be seen by the gradual weakening of the voltage oscillations at this point.

Dwell. The time until the next cylinder firing will fill out the remainder of the secondary pattern line. This portion of the line corresponds to dwell time (on time) of the coil. This is the period during which voltage passes through the control-module power transistor to the primary windings in the coil.

ANALYZING SCOPE PATTERNS

Learning how to read and interpret scope patterns will enable you to quickly identify various kinds of ignition problems as well as other engine problems.

- *Too Much Primary Resistance*—When this condition exists, the first oscillation in primary coil voltage will be greatly reduced because less-than-normal current is reaching the windings. The line will also show fewer and small coil oscillations. Consequently, the coil's voltage output will be less than normal, likely resulting in misfiring.
- *Check the ballast resistance, coil resistance, and ignition switch for excessive resistance. Check the battery for low voltage.*
- *Open Plug or Wire*—In the case of an open plug or wire, there is no spark across the electrodes. Therefore, the firing voltage will shoot up to the maximum output of the coil (Fig. 6-13). The insulation on the plug cables can be checked by running a grounded probe along each cable, while watching the firing lines in the secondary pattern. If there is any weakness in the cable insulation, you'll see a slight dip in the height of the firing line as the probe provides an easier path to ground.
- *Fouled Plug*—An unusually low firing voltage indicates a fouled plug or shorted ignition cable. The voltage required will be no greater than what's required to jump the rotor air gap in the distributor. Replace the plug, or replace the defective cable.

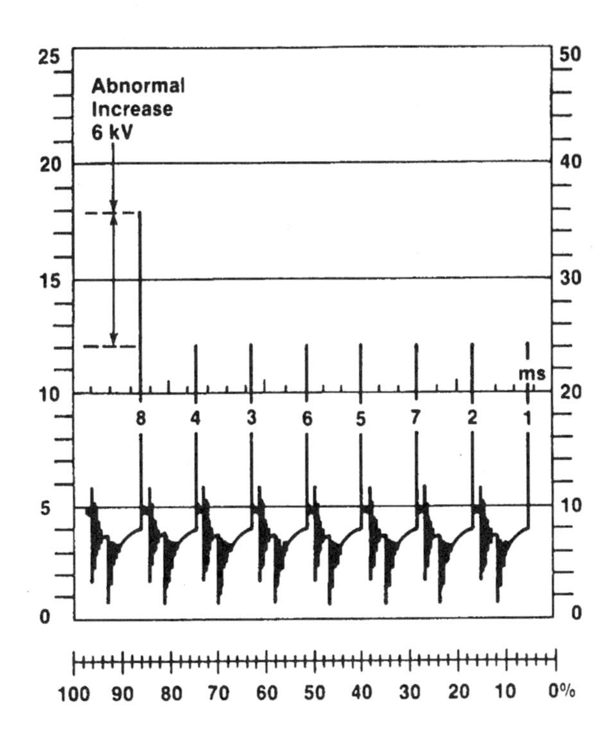

Fig. 6-13 Pattern for a misfiring plug or open plug wire (courtesy Sun Electric).

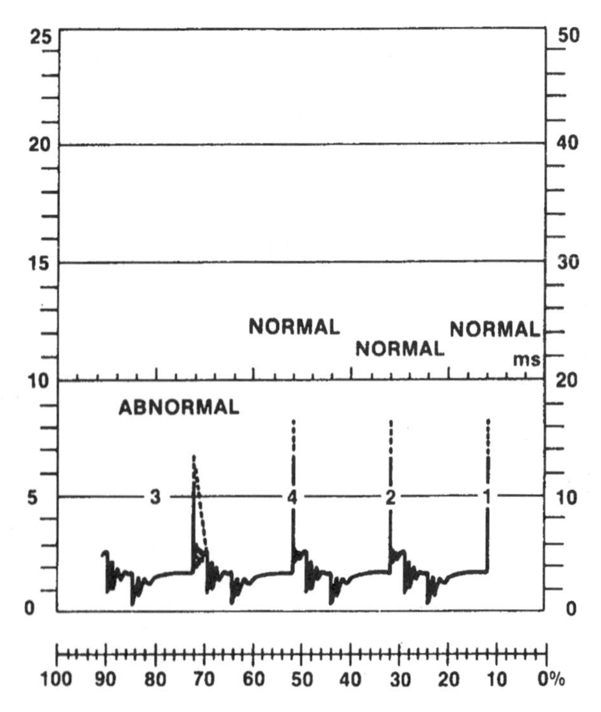

Fig. 6-14 A fouled plug produces an abnormally low firing voltage (courtesy Sun Electric).

- *Reversed Coil Polarity*—If the secondary pattern appears to be upside down, the leads to the coil primary terminals have been reversed. The engine will still run with this condition, but the voltage required to fire the plugs can be up to 40% greater than usual because the spark is forced to jump from the cooler ground electrode to the center electrode (Fig. 6-14).
- *High Secondary Resistance*—When one of the cylinders in the superimposed display shows the firing line higher than the rest of the cylinders and a shorter spark duration, high secondary resistance is the problem. High secondary resistance may be due to worn spark plugs, bad ignition cables, or a lean fuel condition.

  To further isolate the cause, the kV demand for the affected cylinder should be compared to the rest of the cylinders. If the required firing voltage is 20% or more higher than the other cylinders, then there is a strong possibility of a wide plug gap or an air leak into the combustion chamber. Wide plug gaps require more voltage to initiate a spark, and use up the coil's output much more quickly. Likewise, a lean mixture is less conductive, which also requires more voltage and uses up the spark more quickly.

  If the firing voltage is roughly equal to the other cylinders, then the likely cause is high resistance in the spark plug or ignition cable. The initial firing voltage required to start the spark is no greater, but the remaining voltage is quickly consumed trying to overcome the resistance.
- *Internal Resistance*—If the firing line rises at all engine speeds, then the problem is likely wide plug gaps. Once the spark begins, the wide gap will require more voltage to sustain the spark, and resistance will rise as the fuel is burned. The same kind of pattern can also be produced by the EGR valve. Check the operation of the EGR valve for sticking, and check the condition and gap of the spark plugs.
- *Rich Fuel Mixtures*—The firing-line portion of the secondary pattern will show hair-like extensions hanging from the spark line as a result of the high conductivity of the rich fuel mixture. As the fuel mixture moves in and out of the plug gap, the spark finds it less difficult to bridge the gap. This condition is usually characterized by a lower-than-normal plug firing voltage (kV) and a longer-than-normal spark duration. Be careful not to confuse this condition with a fouled plug.

A rich mixture can be caused by a misadjusted idle mixture, a "heavy" carburetor float or too-high a float level, or a defective oxygen sensor.

MILLISECOND SWEEP

One way to get more information out of the patterns displayed on the oscilloscope screen is to spread out a portion of the pattern so you only see what happens during a certain interval of time. This can be done by using the millisecond sweep (Figs. 6-15 & 6-16).

When the primary and secondary pattern is displayed on the scope, there is no time scale to judge the actual duration of the various wiggles in the line. What the screen shows is the complete ignition cycle from start to finish. But it doesn't tell you how long the ignition cycle actually takes. The reason why the scope doesn't plot the pattern against actual time increments is because the length of the ignition cycle changes with engine speed. The faster the engine runs, the shorter the time between cylinder firings and less time it takes to complete each ignition cycle. Double the rpm, and you halve the time

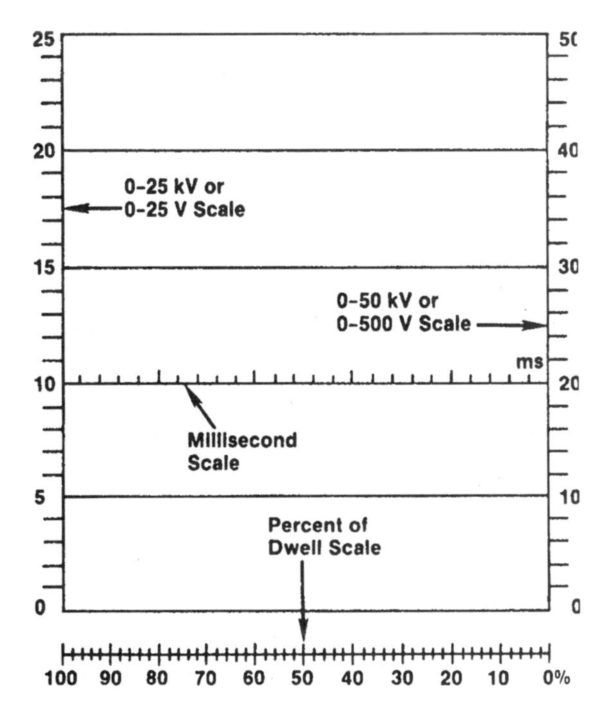

Fig. 6-15 Millisecond scale (courtesy Sun Electric).

of each ignition cycle. Yet the scope still displays the full pattern.

To "reset" the scope so it displays the pattern according to a fixed length of time requires using the millisecond sweep function. Not all scopes have this feature, but on those that do, you are often given a choice as to the time frame.

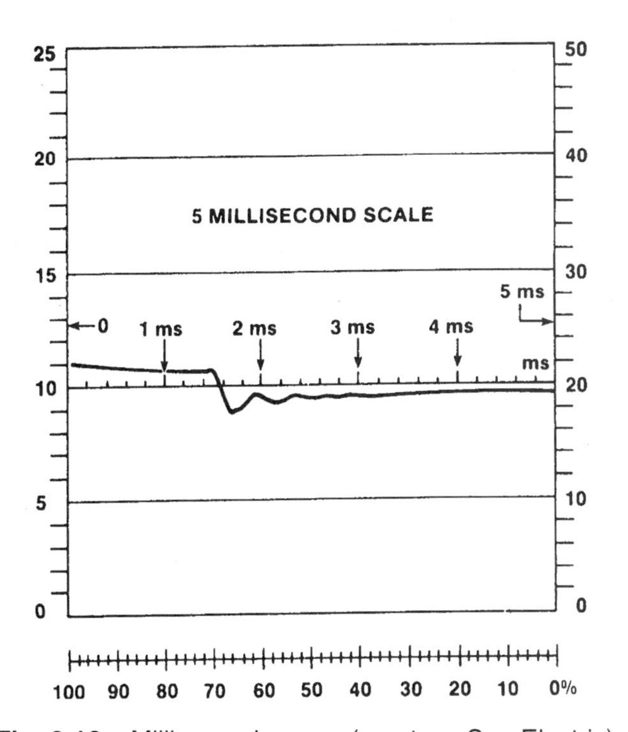

Fig. 6-16 Millisecond sweep (courtesy Sun Electric).

FLAME FRONT PROPAGATION

Flame front propagation is the burning of the fuel mixture starting at the spark plug. It proceeds outward like an expanding balloon, until all the fuel in the combustion chamber is consumed. The flame front should sustain itself and continue burning, once it has been ignited by the spark.

Flame front propagation was never a problem until lean fuel mixtures arrived on the scene. Because there is less fuel in a lean mixture, the mixture may not continue to burn once ignited, if the duration of the spark is too brief. It's become important, therefore, to know how long the spark burns to make sure it lasts long enough to ensure complete combustion.

The duration of the spark should last a minimum of 0.8 milliseconds with today's lean fuel mixtures. Otherwise, the mixture may not burn completely, increasing exhaust emissions and reducing power and fuel economy.

If the duration of the spark is shorter than approximately 0.7 ms, it indicates excessive secondary resistance. If the firing time is longer than 1.8 ms, it usually indicates a fouled spark plug (or a shorted plug wire or narrow plug gap).

INTERMITTENT MISFIRING

An intermittent misfire can be caused by a variety of ignition, fuel, or mechanical problems. A scope can help you "see" the occasional misfire, which should help you narrow down the list of possible causes.

Lean misfire caused by an air or vacuum leak, a badly clogged fuel injector, or a leaky EGR valve is a common condition you're apt to encounter. If the problem appears to be isolated to one cylinder on a fuel-injected engine, check the O-ring at the base of the injector for air leaks. Fuel injection is covered in Chapter 9, and fuel system service in Chapter 13. If the problem appears to "jump around" from cylinder to cylinder, then suspect a manifold vacuum leak or a leaky EGR valve.

Badly worn or fouled spark plugs may also fire intermittently. If only one plug is misfiring, the

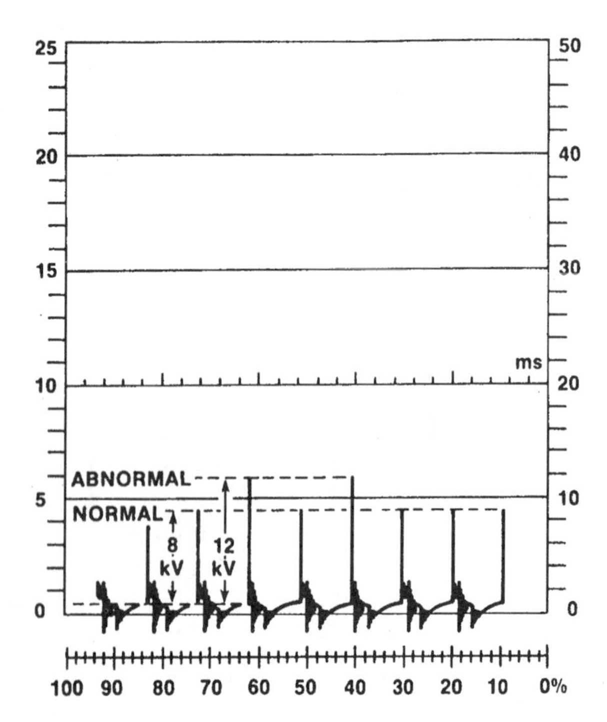

Fig. 6-17 Occasional misfires will cause the firing line to jump (courtesy Sun Electric).

scope pattern for that cylinder will show a variation in the firing voltage (Fig. 6-17).

Misfiring which occurs under load or acceleration may be due to cracking or arcing around the high-voltage center terminal of the ignition coil. Hairline cracks in the tower open a path for the coil's high-voltage output to ground. With GM HEI distributors where the coil is inside the cap, deterioration around the coil button can cause a similar condition. An arcing coil may prevent an engine from starting, or may cause erratic misfiring or misfiring only under load. Arcing can usually be observed by watching the coil in the dark. If the coil shows no signs of arcing at idle, revving the engine may increase the firing voltage to the threshold where it will arc.

Dirt and moisture, and/or a loose-fitting high-voltage cable can also cause arcing. If the high-voltage cable doesn't fit snugly in the coil tower, arcing can deteriorate the terminal, creating resistance that reduces available firing voltage. Cleaning out the tower terminal and replacing the high-voltage cable can solve the problem.

Misfiring can also be caused by bad plug wires which are shorting or arcing. Plug boots that fit loosely around the spark plugs, or plugs that are coated with grease, can also allow the high-voltage current to find a shortcut to ground.

DIAGNOSING A NO-SPARK CONDITION

If an engine won't start because it has no spark, any of the following may be causing the problem:

- *Defective coil*—The first thing to check is whether or not voltage is reaching the coil. This can be done by attaching a test light or voltmeter to the coil positive terminal. There should be voltage when the ignition is on. No voltage means an open somewhere in the primary ignition circuit.

  If voltage is reaching the coil, but the coil is not producing a spark, you need to determine if the ignition module is switching the coil on and off. Check for voltage at the coil negative terminal while cranking the engine. No voltage indicates a switching problem in the module, distributor pickup (or crank/cam position sensor), engine computer, or wiring. If the coil is being switched on and off, however, but is not sparking, measure the coil primary and secondary resistance with an ohmmeter (Fig. 6-18). Primary resistance is checked between the positive and negative coil terminals. As a rule, primary resistance should be 2 ohms or less. Secondary resistance is tested between the high-voltage terminal and the negative terminal. Secondary resistance should be high, ranging anywhere from 8,000 to 20,000 ohms. The exact specs will vary from one application to another, so refer to a manual for the exact specifications.
- *Defective distributor pickup or crank/cam position sensor*—On most late-model vehicles with computerized ignition systems, this condition will set a trouble code indicating the loss of an ignition pickup signal. The exact test procedures are so varied that we'll summarize by saying that most checks involve using an ohmmeter to measure resistance (Fig. 6-19). If a magnetic pickup or Hall effect sensor's resistance is out of specs (a bad one will usually read excessively high resistance or open), it needs to be replaced. With Hall effect sensors, it's also important to make sure the sensor is receiving

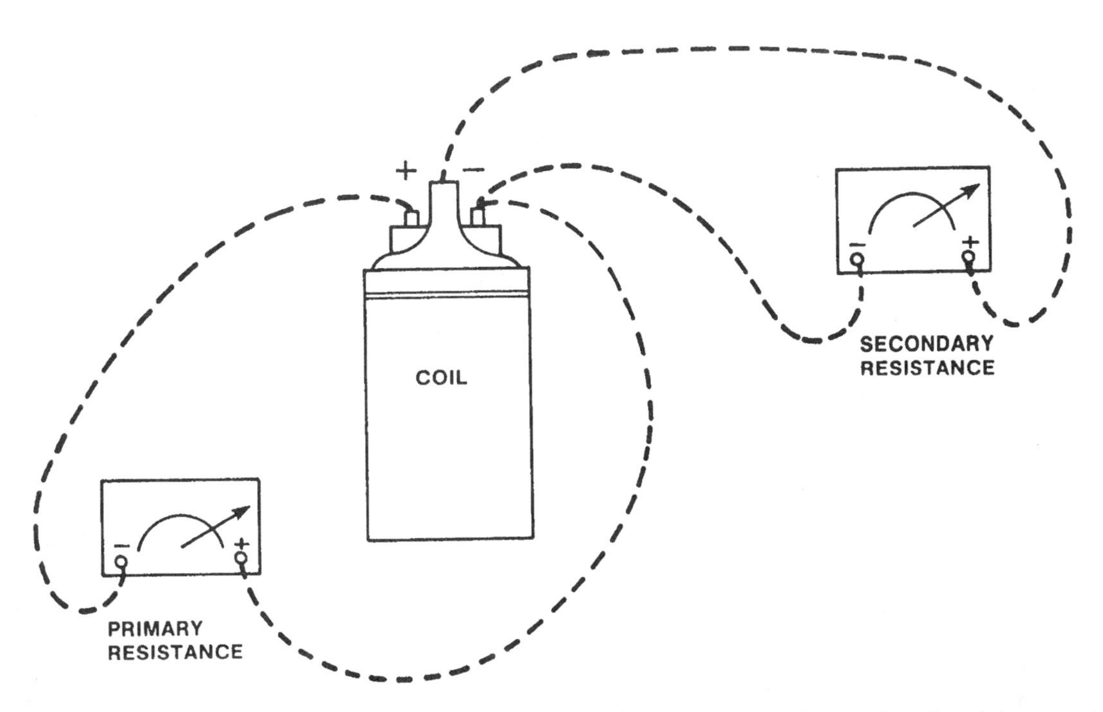

Fig. 6-18 Ohmmeter connections for coil primary and secondary resistance checks (reprinted, by permission, from Santini, *Automotive Electricity and Electronics, 2E.* Copyright 1992 by Delmar Publishers Inc.).

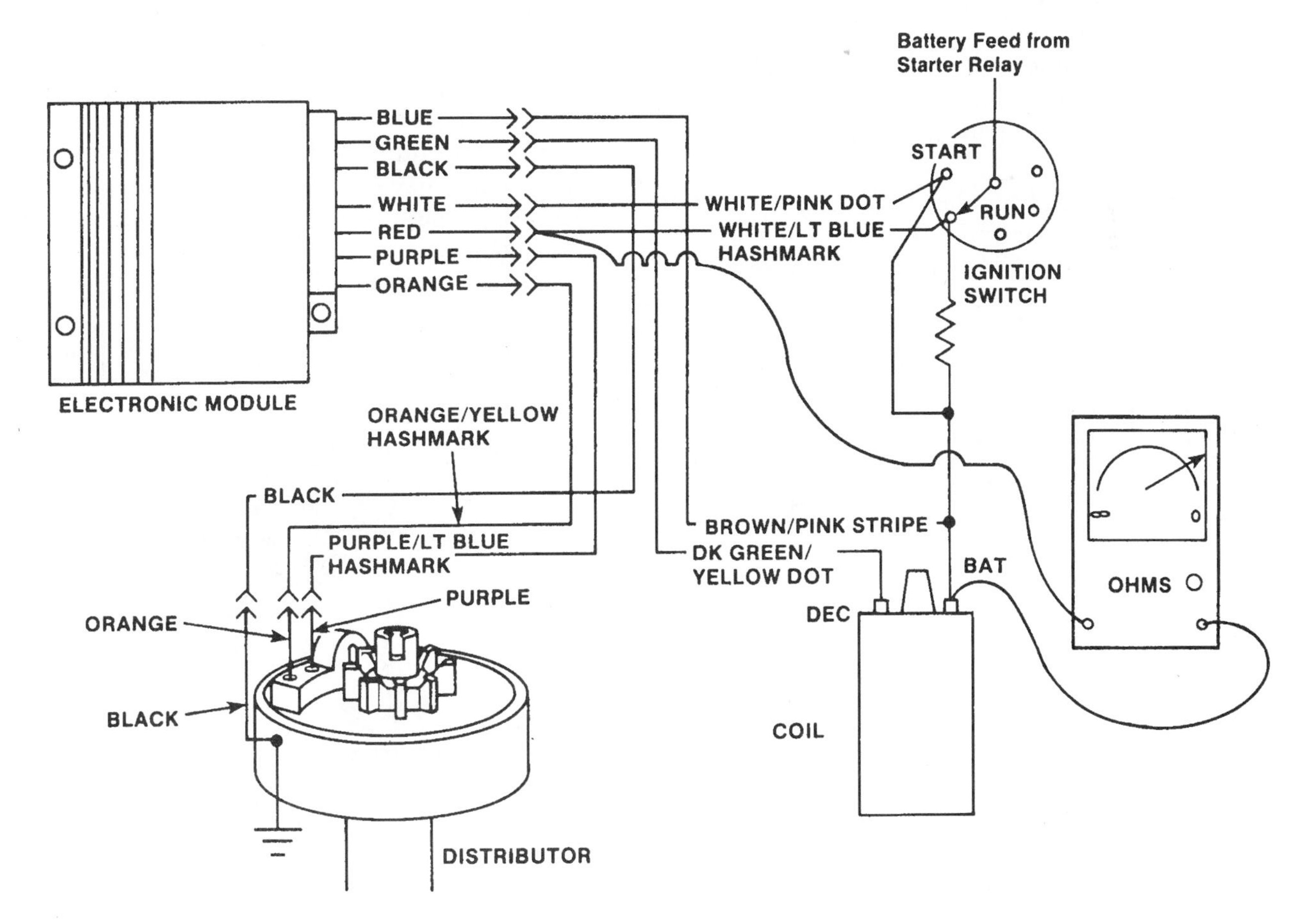

Fig. 6-19 Using an ohmmeter to measure primary resistance (reprinted, by permission, from Santini, *Automotive Electricity and Electronics, 2E.* Copyright 1992 by Delmar Publishers Inc.).

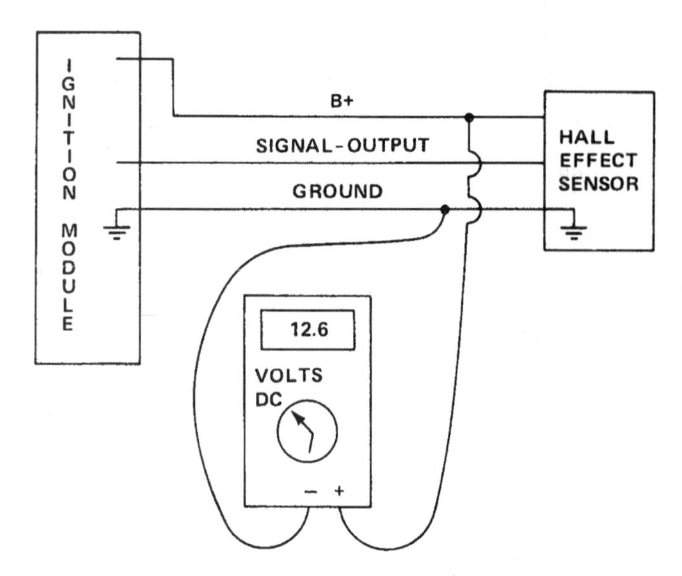

Fig. 6-20 Testing a Hall effect sensor with a voltmeter for power and ground (reprinted, by permission, from Santini, *Automotive Electricity and Electronics, 2E.* Copyright 1992 by Delmar Publishers Inc.).

voltage, because it can't generate a signal without voltage (Figs. 6-20 and 6-21).

- *Defective ignition module*—Modules frequently fail for three reasons: heat, vibration or voltage overload. Excessive heat can damage sensitive electronic chips. GM ignition modules and Ford TFI modules, in particular, rely on a layer of grease under the module to help carry heat away from the electronics. If someone replaced the module and forgot to apply the grease, the module may have failed due to overheating. The leads that connect the module to the wiring are also vulnerable to breakage, often as a result of vibration. Finally, a module can be destroyed if the high secondary voltage in the distributor somehow finds a path to the module.

  In most cases modules can be checked by using a high-impedance (10 megohms) digital ohmmeter to measure resistance between various terminals (Fig. 6-22). If any of the measurements are out of spec, the module is defective and needs to be replaced. Substituting a "known good module" for the old one is another technique that's often used to troubleshoot a no-start condition. If the engine starts, then the old module is assumed to be defective. But if the engine still refuses to start, then it's something else. The problem with this technique is that the new module may be damaged if there's a short in the wiring that caused the old module to fail.
- *Defective engine control module*—Computerized spark management is covered in Chapter 18, so we'll only say that computer failures are not that common. Most problems can be traced to opens in the wiring harness, or loose or corroded connectors. If the circuitry that regulates electronic-spark timing function in a computer fails, the engine may still run, but the computer will generate a fault code that corresponds to an internal fault.

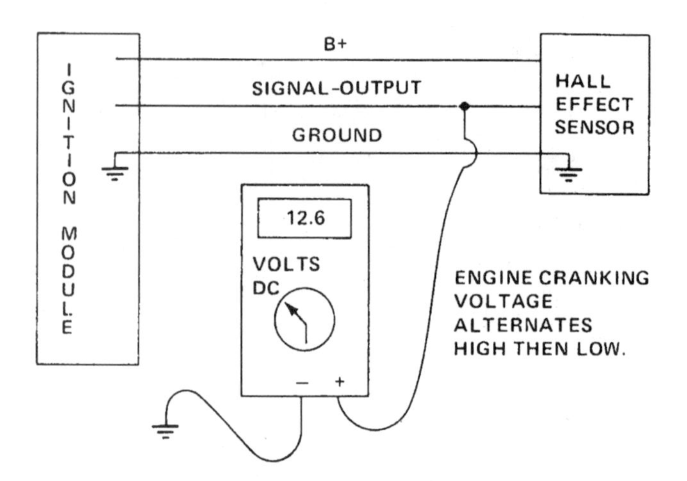

Fig. 6-21 Testing a Hall effect sensor for an output signal (reprinted, by permission, from Santini, *Automotive Electricity and Electronics, 2E.* Copyright 1992 by Delmar Publishers Inc.).

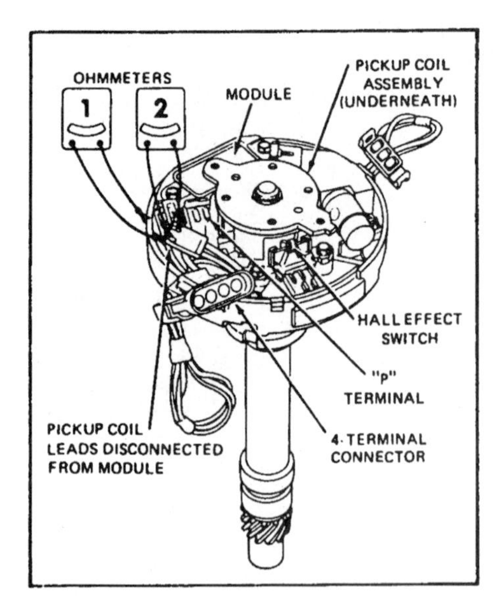

Fig. 6-22 Making resistance checks on a GM HEI ignition module (courtesy GM).

CHAPTER 7

Engine Fundamentals

First things first. An engine is not a "motor," although most of us use the two words interchangeably. Technically speaking, a "motor" is a device that converts electrical energy into mechanical work. An example of a motor would be the starter motor that cranks the engine. An "engine" is a machine that converts heat energy into mechanical work. An automobile engine does this by burning a mixture of gasoline vapor and air inside its cylinders.

To better understand the mechanical components of the engine itself, let's look at what happens inside the engine when heat is converted into mechanical motion. Air and fuel are mixed together in the right proportions. (14.7 times as much air as fuel is considered the ideal ratio.) This creates a vaporous fuel mixture that can be easily ignited. The source of ignition is a spark provided by the spark plug.

When the spark sets off the mixture, a tremendous amount of heat is suddenly released by the exploding mixture. A "flame front" spreads out from the point of ignition like an expanding balloon until the remaining air and fuel mixture in the combustion chamber is consumed. All this takes but a split second. Yet in that instant, temperatures soar several thousand degrees as the stored energy contained in the fuel is unleashed by the sudden chemical reaction. The heat created by the exploding fuel causes a sharp rise in pressure inside the cylinder, as the rapidly expanding gases push out in all directions. Thus heat creates pressure, and pressure creates force, which shoves the piston down and turns the engine's crankshaft. And that's how an engine transforms heat into useful mechanical work.

Think of the transformation of heat energy into work that occurs in an engine as the equivalent of setting off a stick of dynamite inside a sealed container. When the fuel explodes, the heat released has no place to go. So pressure builds until something gives. The "something" in this case is the piston, which is designed to slide down the cylinder and provide the necessary room for expansion. As the hot gases shove the piston down, the piston becomes the means by which the transfer of energy takes place. By attaching the piston through a connecting rod to a crankshaft, the piston's downward travel is redirected to a rotary motion. This can then be used to drive a vehicle or whatever. In essence, then, an internal combustion engine temporarily bottles up the energy released by exploding fuel so the pressure created can be captured and put to good use. That's all there is to it. But as you're about to learn, figuring out the means by which this could be accomplished took many years and the combined efforts of many great minds.

THE RECIPROCATING ENGINE AND FOUR-STROKE COMBUSTION

An internal combustion engine is so named because the fuel is burned inside the engine's cylinders. All automotive engines today are four-stroke, reciprocating piston designs. The only exceptions are some two-stroke engines that are now obsolete (but may make a comeback), and the Mazda rotary engine. The rotary is unique in that a triangular-shaped rotor orbits the crankshaft inside a figure-eight-shaped housing. This engine is not a reciprocating engine because the rotor always travels in the same direction and never reverses itself. The advantage of this design is less vibration, fewer moving parts, and higher potential engine speeds. But rotary engines are difficult to seal, and have not been able to achieve as good a fuel economy as their piston-powered counterparts.

The basic combustion cycle of a reciprocating piston four-stroke engine (Fig. 7-1) is as follows:

(1) *Intake stroke*—The intake valve opens as the piston travels down the cylinder. This creates a partial vacuum (lower pressure) in the cylinder. Atmospheric pressure (about 14 lbs. per square inch at sea level) pushes air past the throttle, through the intake manifold, the intake port in the cylinder head, and past the intake valve to fill the void. So although it seems that an engine "sucks" air into its cylinders, it's actually atmospheric pressure that's pushing air into the cylinders. With a gasoline-powered engine, fuel is mixed with the air by the carburetor, or is sprayed into the intake manifold by fuel injectors. With a diesel engine, only air is drawn into the cylinder. There's also no throttle with a diesel because engine speed is controlled by fuel delivery.

(2) *Compression stroke*—As the piston reaches the bottom of the cylinder (called "bottom dead center" or BDC) and reverses direction, the intake valve closes. The piston travels up, compressing the air and fuel. (In the case of a diesel engine, there's only air in the cylinder.) The compression stroke squeezes the air/fuel mixture to roughly one-eighth to one-ninth of its original volume, depending on the engine's compression ratio (more on this later). This heats and blends the air and fuel in preparation for ignition, as well as causes the tiny droplets of fuel to vaporize. During the intake and compression strokes, the piston acts like an air pump. It produces no power, and in fact takes power from the engine to complete these two steps.

(3) *Power stroke*—As the piston approaches the uppermost point in the cylinder (called "top dead

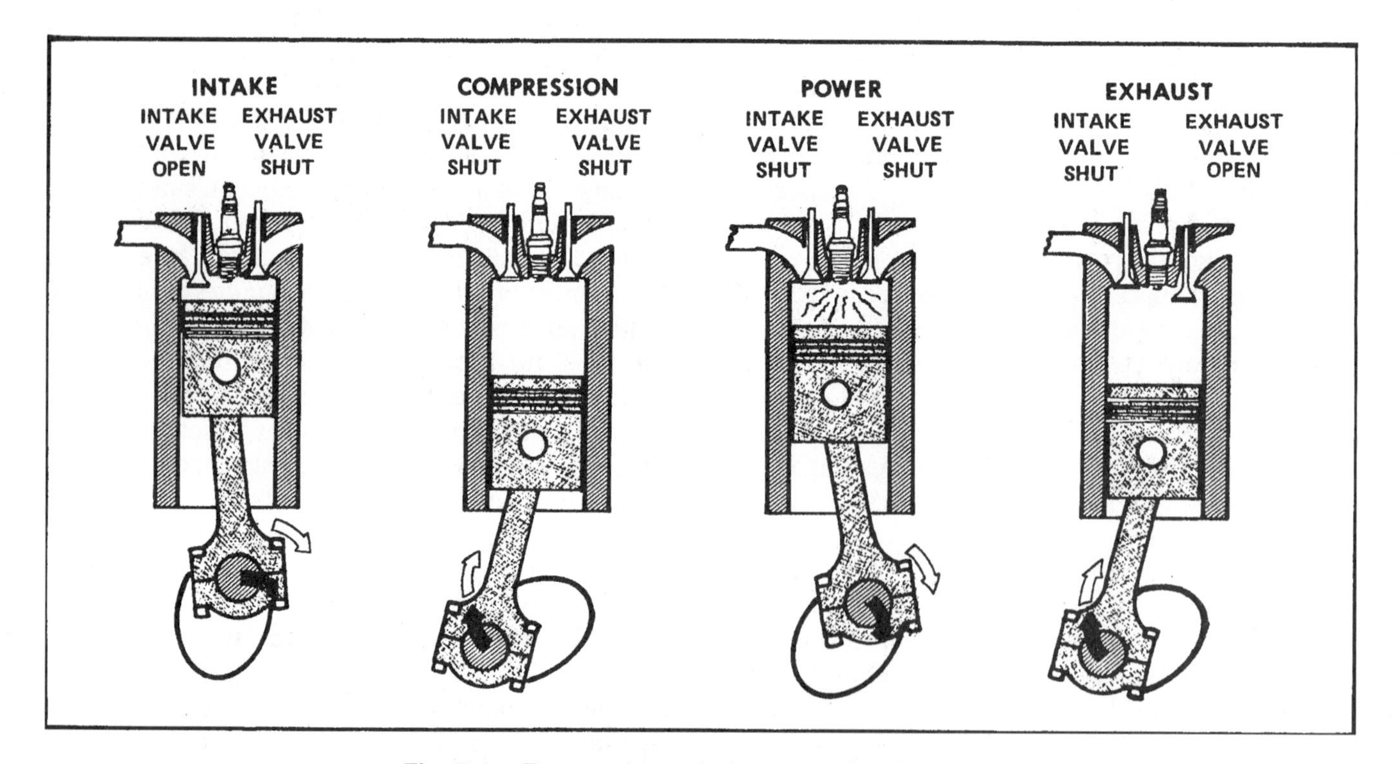

Fig. 7-1 Four-stroke cycle (courtesy Sun Electric).

center" or TDC) and the air/fuel mixture has been compressed to its maximum concentration, the spark plug fires igniting the mixture. The air/fuel mixture explodes and forces the piston down. This is the power stroke. It lasts from approximately TDC through BDC. This is the period during which the piston produces power to drive the engine and the other pistons. With a diesel engine, combustion occurs when fuel is injected directly into the cylinder at or near TDC. Heat and pressure cause the fuel to ignite spontaneously, so no spark plug is necessary.

(4) *Exhaust stroke*—As the piston passes bottom dead center and reverses travel, the exhaust valve opens allowing the piston to push the burnt gases from the cylinder. The exhaust gases must be pumped out of the cylinder so the piston can draw in a fresh mixture of air and fuel with the next intake stroke. As the piston approaches top dead center again, the exhaust valve closes, and the intake valve opens to repeat the four-stroke cycle all over again.

Another way to look at the four-stroke process is to break it down by degrees of crankshaft rotation. During the first 180 degrees of crankshaft rotation, the piston goes down (intake stroke). Then it reverses direction and goes back up (compression stroke) for the next 180 degrees of rotation. Combustion occurs near top dead center on the compression stroke. The piston then travels back down (power stroke) for the third 180 degrees of rotation, and then back up (exhaust stroke) during the final 180 degrees of crankshaft rotation.

Each piston, therefore, produces a power stroke *only once every other revolution* in a four-stroke engine (once every 720 degrees of crankshaft rotation). The other three strokes are parasitic in that they take power from the crankshaft to drive the piston.

By comparison, a two-stroke engine has a combustion cycle every stroke. Theoretically a two-stroke engine is capable of producing more power than a four-stroke because it has twice as many power strokes (one every 360 degrees). Two-stroke engines are also very compact and do not require a complex valvetrain. But two-stroke engines do not breathe as efficiently as four-stroke engines, so the actual power advantage isn't that great. What's more, two-strokes are notoriously dirty engines from an emissions standpoint. However, breakthroughs in direct fuel-injection techniques may allow the two-stroke to once again be accepted as an automotive powerplant.

ENGINE DISPLACEMENT

An engine's displacement in cubic inches (CID) or liters (L) is determined by its "bore" (the diameter of each cylinder), "stroke" (the distance which the piston travels up and down in the cylinder which equals the length of the "throw" on the crankshaft), and number of cylinders. It is figured by multiplying the bore area (the diameter of the cylinder times the constant *pi* which is 3.14) times stroke (length of the crankshaft throw) times the number of cylinders.

For example, a Chevrolet 350 V8 has a bore of 4 inches, and a stroke of 3.5 inches. The bore area of each cylinder is 4 x 3.14 (*pi*) = 12.56 square inches. Multiplying by a stroke of 3.5 inches gives 43.96 square inches per cylinder. Multiply the cylinder volume of 43.96 sq. in. times 8 cylinders, and you get 351.7 cubic inches. Round it off to 350 for the marketing boys, and you have the correct displacement.

Figuring metric displacement is the same, only different units of measurement are used. Instead of cubic inches, cubic centimeters are used. One thousand cubic centimeters equal one liter. Divide the number of cubic centimeters by 1,000 to get the number of liters. Example: 5,736 cc divided by 1,000 = 5.7 liters

For converting English units to metric and vice versa:

- Cubic inches to cubic centimeters: Multiply CID by 16.39.
- Cubic inches to liters: Multiply CID by 0.01639
- CCs to inches: Multiply cubic centimeters times 0.061.
- Liters to inches: Multiply liters by 61.02

Displacement is actually a fictitious number because it assumes each cylinder always has a full cylinder of air—which it doesn't. In a four-stroke piston engine, each piston is at a different point in its cycle. As one piston is going up, another is going down. In a 4-cylinder engine, for example, only one piston will be at bottom dead center and have a full cylinder's worth of air (the one just completing its intake stroke). One piston will be at the top of its compression stroke. Another will be halfway through its power stroke. And the other will be halfway through its exhaust stroke. The net displacement, therefore, is really only *half* the calculated displacement. A 350-cubic-inch V8 only has 175 cubic inches of air and exhaust in it at any given instant in time.

CYLINDERS AND RINGS

The condition of an engine's cylinders from an emissions standpoint is extremely important because they help seal combustion (Fig. 7-2). If the walls of the cylinders are worn, scratched, or cracked, combustion gases can blow past the rings and enter the crankcase. The "blowby" gases can contain the nor-

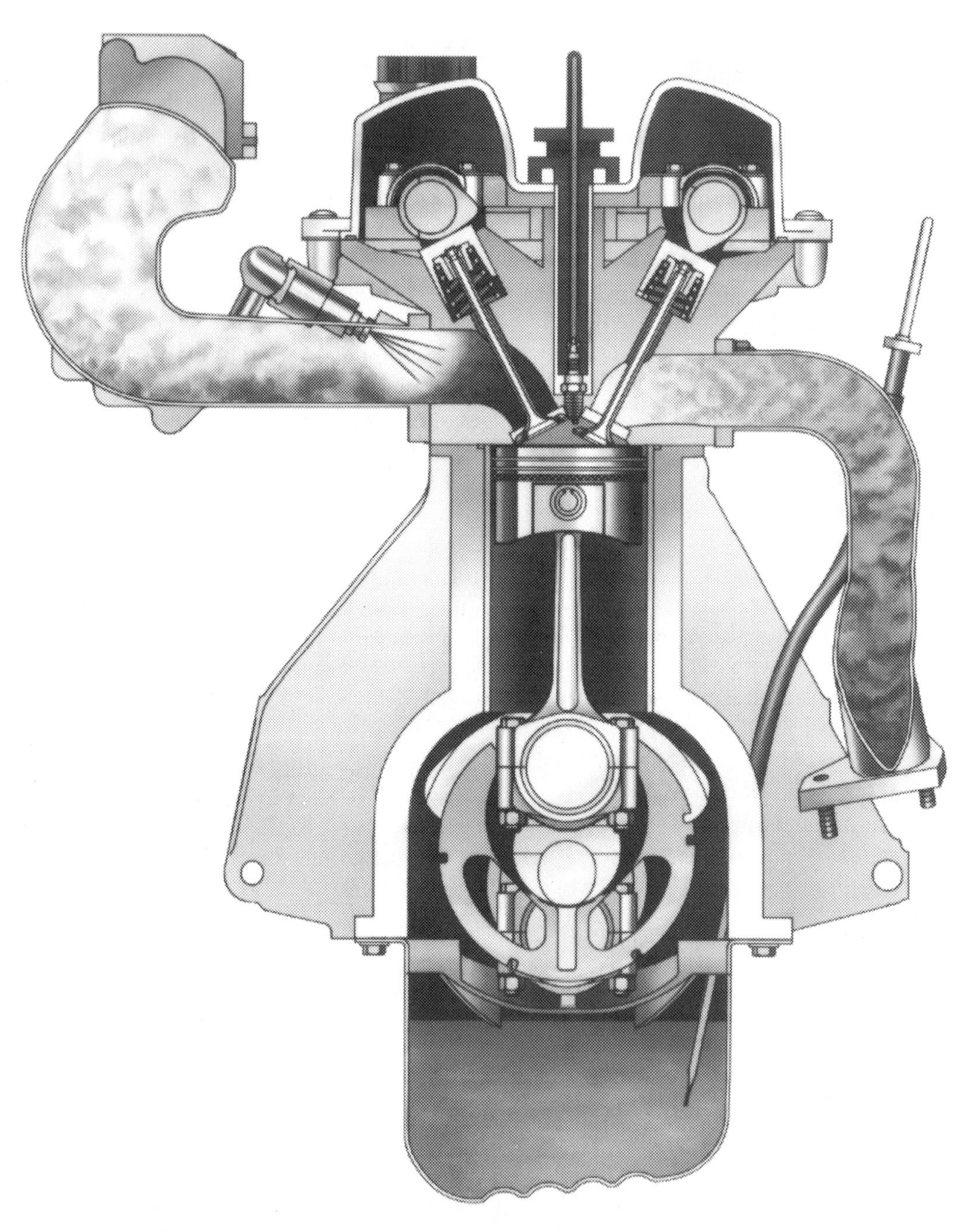

Fig. 7-2 Proper ring sealing is required to minimize blowby into the crankcase. At the same time, the valves must seal tightly to prevent leakage of unburned fuel into the exhaust (courtesy Saturn).

mal byproducts of combustion as well as droplets of unburned fuel. At the same time, oil can get past the rings and enter the combustion chamber. This creates an oil-consumption problem, possibly fouling the spark plug and causing an increase in hydrocarbon emissions.

As a cylinder wears, most of the wear occurs at the top. That is where the forces of combustion are the greatest. When fuel explodes, the piston rings are slammed against the walls of the cylinder. This scrapes away tiny flakes of metal. In time the cylinder develops a "tapered" wear pattern where the top is worn more than the bottom. This, in turn, allows increased blowby. It also increases ring flexing as the rings expand and contract with every stroke in an attempt to follow the contours of the cylinder walls. This can lead to ring breakage and a dramatic increase in oil consumption, blowby, and emissions. The only cure for this type of condition is to overbore the cylinders and install oversized pistons with new rings.

As for oil control, the rings scrape the cylinder walls with every stroke of the piston to keep oil out of the combustion chamber. If the rings are worn, improperly installed (wrong side up), have too much end-gap clearance, or are damaged or broken (often due to a detonation problem), they won't seal properly. This will allow oil to enter the combustion chamber and produce excessive blowby into the crankcase. Oil in the combustion chamber will increase hydrocarbon emissions, and excessive blowby will elevate carbon monoxide emissions. If the rings are in really bad shape, blue smoke will be visible in the exhaust accompanied by a strong odor.

Ring problems can be diagnosed with a compression check. If the compression readings are low or vary by more than 10% from one cylinder to another, squirt a few drops of 30-weight oil through the spark plug hole and crank the engine over a couple of times. The oil will seal the rings temporarily. Repeat the compression test. If the readings are now noticeably higher, the rings are worn and should be replaced. If there is no change in the compression readings after squirting oil in the cylinders, the valves (or head gasket) are leaking compression. A cylinder leakdown test can be used to achieve the same results.

The only cure for ring-sealing problems is a ring job. The engine must be disassembled, and the rings replaced. If the cylinders are worn, out of round, or have excessive taper, the block will have to be rebored and honed so oversized pistons and rings can be installed.

Newer engines are being manufactured with closer piston-to-cylinder tolerances than ever before to minimize blowby and emissions. This has become necessary to comply with tougher emission laws. At the same time, though, the size of piston rings and the tension they exert against the cylinders has also been reduced. This lowers internal friction for improved fuel economy and performance. Special cylinder bore-honing techniques (torque plate honing) are used to minimize cylinder-bore distortion for better sealing. New head gasket materials and designs, head bolts, and bolt torquing techniques have also become necessary to seal these engines. If internal engine repairs are performed, or an engine is overhauled and the work is not done correctly, the result can be cylinder-bore distortion, excessive blowby, increased emissions, and possible piston scuffing.

CONTROLLING BLOWBY

The engine's positive crankcase ventilation (PCV) system (see Chapter 10) helps control crankcase blowby emissions by siphoning the vapors out of the crankcase. Intake vacuum pulls the vapors back into the intake manifold where they are reburned in the engine. As long as blowby isn't excessive, the PCV system can handle the volume of blowby gases that normally end up in the crankcase. But in a high-mileage engine with poor compression, excessive blowby can overwhelm the PCV system.

Blowby gases contain water vapor, soot particles, and unburned fuel. When blowby is excessive, it dilutes and contaminates the oil in the crankcase. This shortens the life of the oil and promotes the formation of sludge. Unless the oil is changed regularly, accelerated engine wear will occur. At the same time, the increased quantity of blowby gases being siphoned back into the intake manifold will have a richening effect on the air/fuel mixture. This will cause elevated hydrocarbon emissions in the exhaust.

Blowby problems can usually be diagnosed by performing a compression test and/or leakdown test.

HOW OPERATING TEMPERATURE AFFECTS ENGINE SEALING

In spite of the fact that a piston looks round, it really isn't. It's actually slightly oval in shape, the reason being it needs room to expand once it gets hot. Most automotive pistons are made from cast aluminum because aluminum weighs less than half as much as steel. That reduces the reciprocating weight of the piston and allows for higher engine speeds. But aluminum expands much faster than steel when heated. So a certain amount of clearance is needed to allow for thermal expansion as the piston heats up. If there isn't enough clearance, the piston may scuff against the cylinder wall or even seize.

When there's too much clearance between the piston and cylinder, the piston rocks back and forth. This can produce a steady rapping noise called "piston slap." The rocking motion also increases the stress on the rings, as well as ring and cylinder bore wear. So, as an engine accumulates miles and the pistons become worn, the tendency to slap when cold will become more pronounced.

To maintain the correct piston-to-wall clearance, the engine needs to run at a fairly consistent operating temperature. If the operating temperature is too low, excessive piston-to-wall clearances can result in increased blowby and cylinder wear. (Blowby washes oil off the cylinder walls and accelerates wear.) If the engine runs too hot, the pistons may scuff or seize. So it's very important that the engine's cooling system be in good working condition. The engine should have a thermostat with the correct temperature rating for the application. Almost all engines today require 195-degree thermostats. Substituting a thermostat with a lower temperature rating is *not* recommended because it can prevent an engine with computerized fuel control from going into closed-loop operation (see Chapter 14).

COMPRESSION RATIOS

An engine's compression ratio is the amount of squeeze given the air/fuel mixture in each cylinder as the piston travels from bottom to top dead center. It is expressed as the numerical ratio between the total volume of air in the cylinder and combustion chamber when the piston is at the bottom (BDC), compared to that when the piston reaches the top (TDC). The volume of air left at the top of the cylinder after the piston reaches top dead center is called the "clearance volume." Compression, therefore, can be figured as total cylinder volume divided by the remaining clearance volume.

$$\text{COMPRESSION RATIO} = \frac{\text{TOTAL}}{\text{CLEARANCE VOLUME}}$$

In modern engines, compression ratios are generally between 8:1 and 9.5: 1. As a rule of thumb, the higher the compression ratio, the greater the thermal efficiency of combustion. Squeezing the air/fuel mixture into a tiny space before igniting it helps get more bang out of the resulting explosion. Raising an engine's compression ratio, therefore, is a way to improve its performance. But compression also heats the mixture, which increases the tendency for the fuel to ignite spontaneously without the benefit of a spark. That's basically how a diesel engine works. It uses the heat of compression to ignite the fuel. By using compression ratios in the 16:1 to 22:1 area, the air is heated to the point where diesel fuel will ignite spontaneously when it is injected directly into the combustion chamber. But diesel engines and gasoline engines operate on different concepts, and using too much compression with gasoline leads to detonation. To prevent detonation, therefore, a higher-octane gasoline must be used when the compression ratio is raised. Higher-octane gasoline is usually required when the compression ratio is above 9:1.

CYLINDER HEADS

The type of cylinder heads on an engine can affect the engine's emissions. The volume of the combustion chambers helps determine the engine's overall compression ratio. So it's important that a replacement head be the same as the original. Too much compression can result in detonation and pre-ignition.

The design or configuration of the combustion chamber can also influence emissions as well as performance. When the air/fuel mixture is drawn into the cylinder during the intake stroke, the tiny droplets of gasoline tend to condense on the surface of the combustion chamber. This interferes with complete combustion and leaves traces of unburned

HC in the exhaust. Minimizing the "quench area," therefore, is one way engineers have made today's engines cleaner.

The cleanest cylinder heads are usually those with "open"-type hemispherical or pent roof-shaped combustion chambers with centrally located spark plugs. Heads with wedge-shaped chambers, on the other hand, have a large quench area that also creates a dead space with very little clearance between the top of the piston and the head. Such a design can interfere with the propagation of the flame front and leave small pockets of unburned fuel.

Another way in which engineers have redesigned heads to lower emissions is to promote swirling of the air/fuel charge. This allows the flame kernel to grow in a more controlled manner for more complete combustion. Such heads are often dubbed "high-swirl combustion" (HSC) heads.

Another approach is that used in Honda's CVCC (controlled vortex combustion chamber) and Mitsubishi's Jet Valve heads. The Honda system uses a small auxiliary combustion chamber that receives a rich mixture (the spark plug is located here). This allows a small pocket of rich mixture to ignite the rest of the fuel mix which is very lean. Mitsubishi uses a tiny third valve that admits air into the combustion chamber to churn up the air/fuel charge so it will burn cleaner.

The seal between the cylinder head and block (and cylinders) is also extremely important. A leaky head gasket can allow gases from one cylinder to leak into another. Compression loss results causing misfiring and greatly elevated hydrocarbon emissions. It also results in coolant seepage into the cylinders which can cause piston and ring scuffing.

Most head-gasket leaks and failures are the result of head warpage (usually from overheating). Bimetal engines with aluminum heads are especially vulnerable to this type of problem because of the different rates of thermal expansion between the aluminum head and cast-iron block. If a head gasket is leaking, therefore, the surfaces of both the head and deck should be carefully inspected with a straight edge and feeler gauge for warpage. If out-of-flat exceeds the specs for the application, the head and/or block should be resurfaced. Aluminum heads can often be straightened minimizing the need for subsequent resurfacing.

Cracks are another problem that plague many aluminum heads. Cracks can allow coolant to enter the combustion chamber, and/or blowby gases to enter the cooling system, causing a loss of coolant and overheating. Detecting cracks can be tricky, and may require removal of the head for an accurate diagnosis. A cooling system pressure gauge attached to the radiator can reveal the presence of cracks or head gasket leaks that are allowing combustion gases to enter the cooling system. But the gauge can't pinpoint their location. For that, the head must be removed for inspection using pressure testing, penetrating dye, visual inspection and/or Magnaflux testing (cast iron only). Cracks can be repaired by welding (aluminum primarily) or pinning (cast iron).

VALVES AND COMPRESSION

A close fit between valve and seat is absolutely essential to prevent compression losses. If a valve doesn't seal properly, air and fuel will be pushed out of the cylinder during the compression stroke, resulting in a loss in power. Likewise, the valve won't hold pressure during the power stroke either, which only further compounds the loss of power.

The ability of a valve to seal tightly depends not only on perfect concentricity between its face and seat, but also on spring pressure that holds it shut. A weak or broken spring can cause an otherwise good valve to leak. Insufficient clearances in the valve train can prevent a valve from closing completely, as can binding that causes the valve to stick open.

The most common cause of valve leakage, however, is the normal wear and tear that eventually eats away the close relationship between valve face and seat. Each time the valve opens, microscopic particles of metal are ripped away from both valve and seat. Some of these particles are embedded into the valve and seat when the valve slams shut. There they act like abrasives to further wear away at both surfaces. Eventually the close fit is destroyed, and the valve begins to leak. Once this happens, the engine develops a steady miss and a significant loss of power.

Exhaust valves are more prone to accelerated wear than intake valves because they run considerably hotter. They do not have the benefit of incoming air and fuel to cool them. Consequently, the metal is subjected to more thermal stress which sometimes results in cracking and chipping. Leaking exhaust valves are sometimes called "burnt valves" because heat has destroyed their ability to seal.

When regular leaded gas was still used, tetraethyl lead (an octane-boosting additive) helped "lubricate" the valves and slow down wear. When unleaded gas came along, engine manufacturers had to change to hardened valve seats and higher-grade valve materials to prevent wear. Because of this difference, older engines (particularly trucks) originally designed to run on leaded gasoline often have worn valves because of using unleaded fuel.

Leaking valves can be detected by performing a compression test, a leakdown test, or checking intake vacuum with a vacuum gauge.

VALVE GUIDES AND SEALS

Most oil-consumption problems are due to worn valve guides and seals rather than worn or damaged piston rings. The guides support and position the valve stems in the cylinder head as the valves open and close. The guides help cool the valves by drawing heat away from the stems. And they keep the valve stems lubricated by allowing a small amount of oil into the gap between the stem and guide. The guides are subject to wear over time because of the constant opening and closing of the valves, and side forces created by the rocker arms or overhead cam. As clearances begin to loosen up, the guides allow more and more oil to leak down past the stems and be drawn past the valves into the engine (Fig. 7-3). The result is increased oil consumption and higher HC emissions.

If the guides are badly worn and passing a lot of oil, it can cause a buildup of heavy carbon deposits on the backside of the intake valves. That interferes with proper breathing. On some fuel-injected engines, the accumulation of carbon deposits can cause rough idle and hesitation problems. Worn guides can also accelerate valve wear and sometimes lead to valve failure because of excessive valve wobble and flexing.

Oil flow into the guides is controlled by the seals at the top of the guides. There are two basic types of seals: "umbrella" or "deflector" seals are little circular seals mounted on the valve stem that keep oil from splashing or running directly down and into the guides. "Positive" seals fit snugly around the valve guide boss and scrape excess oil off the valve stems. The latter are used on most late-model engines to reduce emissions.

One of the myths about controlling oil leakage past the valve guides is that it is a problem only on the intakes (Fig. 7-3). The vacuum in the intake port area during the intake stroke will suck oil through the guides like a straw if the seals are bad. But the same thing can also happen with the exhaust valves. Even though the exhaust gases are pushed out of the cylinders under pressure, the flow of gases past the exhaust-valve guide creates a partial vacuum, which soars right after each exhaust pulse. This can suck oil down the guides and into the exhaust system just as effectively as on the intake side. To make matters worse, exhaust valves usually have larger valve stem-to-guide clearances because they run hotter and need more room for thermal expansion than do intake valves. This can accelerate an oil-consumption problem if the exhaust-valve seals go bad.

When an engine has worn valve guides and/or seals, the problem can be cured one of two ways. The least expensive fix is to install new valve guide seals. This can sometimes be done without removing the cylinder head. By connecting an air hose to the spark-plug hole and filling the cylinder with about 100 psi, the valve will be held shut so that the spring and retainer can be removed. The seal can then be easily replaced. The other alternative is to pull the head and do a complete valve job. Chances

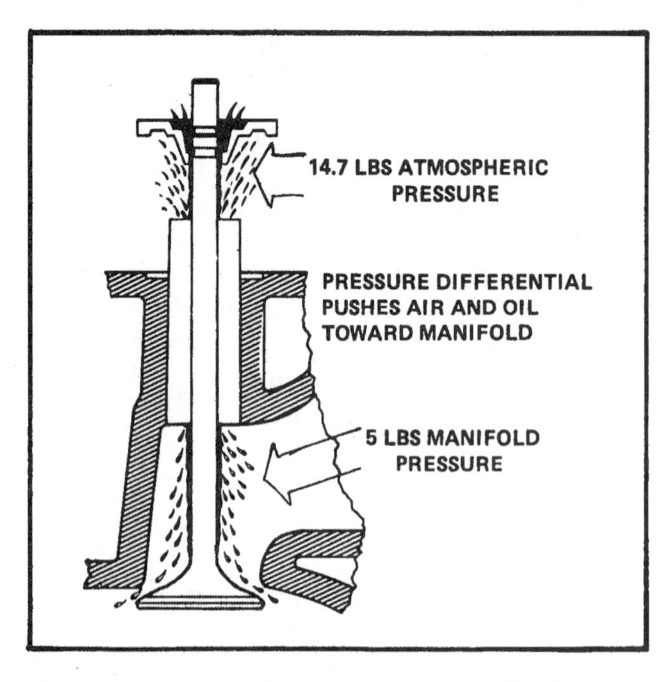

Fig. 7-3 Value guide leakage (courtesy Sun Electric).

are that if the seals are bad, the guides, valves, and seats will also need attention. When the head is disassembled, the valve stem-to-guide clearances should be carefully checked. If the guides are worn, they will have to be replaced, relined, knurled, or reamed out so valves with oversize stems can be installed.

HOW THE CAMSHAFT CAN AFFECT EMISSIONS

Another factor that influences the emissions produced by an engine is valve timing. The camshaft that controls the opening and closing of the valves influences not only the rpm range where the engine develops the most horsepower and torque, but also such things as: idle vacuum, which affects the operation of other emission control devices such as PCV and exhaust gas recirculation (EGR), as well as manifold absolute pressure (MAP) sensors on late-model cars with computerized engine controls; idle quality; idle emissions; and emissions throughout the rpm range.

Cars and light trucks have to meet strict emission and fuel-economy requirements. So most stock cams are a three-way compromise that attempt to achieve as broad a usable power curve as possible, while providing good fuel economy and low emissions. The requirements often conflict, so the car makers are forced to give emissions top priority, followed by fuel economy, then performance.

The emissions performance of a camshaft depends on its duration and timing. Duration is the amount of time, specified in degrees of crankshaft rotation, that a valve is open. Lift, on the other hand, is how far the valve is pushed open. This depends on the height of the cam lobe and the geometric ratio of the rocker arms. Short-duration cams (less than about 214 degrees of duration) are good for high-volumetric efficiency at low rpm because the intake valves close before the piston bottoms out and starts back up on the compression stroke. A short-duration cam also has minimal valve overlap. Overlap between the closing of the exhaust valves and opening of the intake valves between the exhaust and intake strokes is bad for emissions because unburned fuel can pass through into the exhaust (Figs. 7-4 and 7-5).

By comparison, high-rpm performance requires more valve lift, duration, and overlap to make more horsepower. A cam designed for high-rpm power usually sacrifices low-end torque by moving the engine's power range up the rpm curve. At the same time, it also increases hydrocarbon emissions because excessive overlap allows unburned fuel to enter the exhaust. Increased duration and overlap also have an adverse effect on idle quality and intake

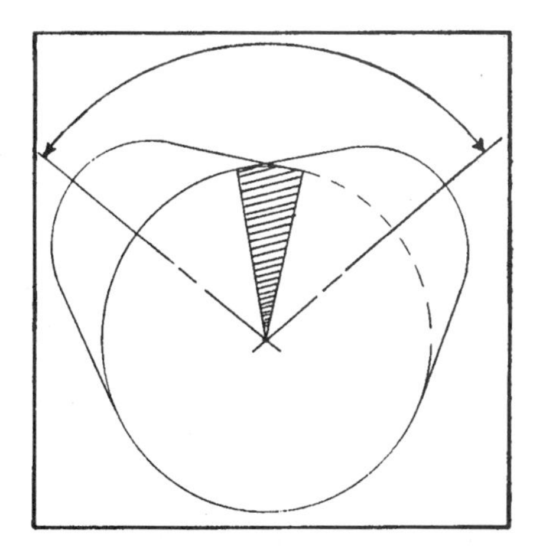

Fig. 7-4 The "overlap" between the intake and exhaust lobes on a camshaft is the shaded area. During the overlap period, the intake valve is beginning to open while the exhaust valve is still closing. Overlap improves high-rpm performance but can create emission problems by allowing unburned fuel to enter the exhaust (courtesy TRW).

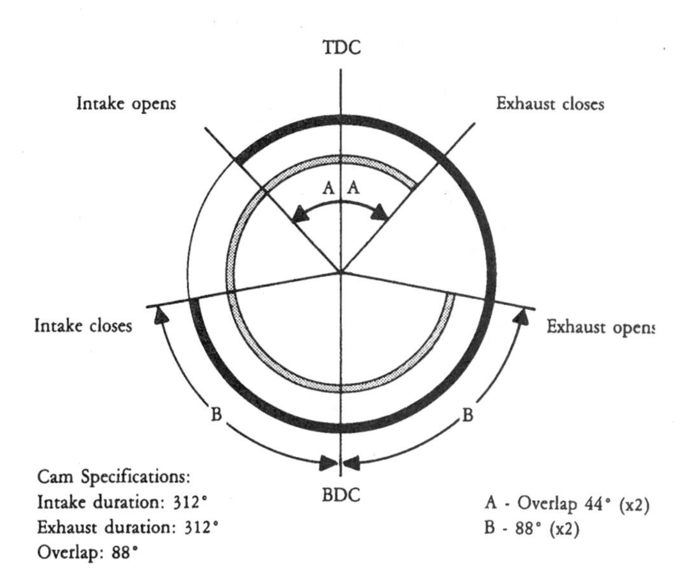

Fig. 7-5 A representative valve timing diagram for a camshaft, showing overlap between the intake and exhaust valves (courtesy TRW).

vacuum. This in turn affects the fuel mixture, vacuum-operated accessories, and the calibration of the electronic engine control system. Other computer-controlled functions such as idle speed, torque converter lockup, and EGR, can also be upset.

Aftermarket camshafts that alter stock cam timing for improved performance may or may not be "emissions legal." If a replacement cam is a non-stock grind, it can only be used on a street-driven vehicle if the camshaft manufacturer has certified that it does not cause an increase in emissions. The manufacturer must do this by submitting test results that prove this. Obtaining an "exemption order" number from the California Air Resources Board (CARB) meets the Environmental Protection Agency's criteria for being emissions legal. So if the cam doesn't have this number, it is not an emissions legal. Use of a noncertified cam (offroad or racing cam) on a street-driven vehicle is a violation of the federal clean air laws. This may make the vehicle's owner, the person who installed the cam, and the dealer that sold the cam all eligible for up to a $2,500 fine!

The only emission problems a stock camshaft can create is if it develops a flat lobe. A lobe can be flattened (rounded off) or worn down by improper cam break-in (new engines only), lack of adequate lubrication (an oil starvation or breakdown problem), or reusing worn lifters with a new cam. Sometimes a lobe will go flat because the cam was not properly heat treated (hardened) when manufactured. If a cam lobe goes flat, it won't open its valve as far, resulting in a loss of power in the affected cylinder. If the lobe that's gone flat prevents an intake valve from opening, it can cause the cylinder to suck oil past the rings and valve guides, increasing HC emissions. If the bad lobe affects an exhaust valve, the combustion gases will be forced past the rings into the crankcase. This will increase crankcase emissions and possibly overload the PCV system, causing an increase in both HC and CO emissions.

Worn timing gears, a stretched timing chain, or a loose timing belt can have an adverse effect on valve timing as well as ignition timing. Worn gears or a loose chain typically cause retarded valve and ignition timing, which will affect HC emissions. The cure is to replace the gears and/or timing chain.

You can check for a loose timing chain or excessive play in a set of cam drive gears. Remove the distributor cap, then watch the motion of the rotor as a wrench is used to turn the crankshaft back and forth. If the movement of the rotor lags behind that of the crank, it indicates excessive play. To determine if the play is in the timing chain or gears (and not the distributor drive gear), the timing cover needs to be removed from the engine. Then, the timing chain or gears can be inspected directly.

When inspecting timing gears, use a feeler gauge and refer to a shop manual for the maximum allowable backlash. With timing chains, up to about half an inch of sideways movement or slack is usually allowed between the crankshaft and camshaft sprockets on pushrod 4-cylinder, V6, and V8 engines. But always refer to a manual for the service specifications. On overhead cam engines, chain slack is controlled by an automatic or adjustable chain tensioner. If the adjuster is at its maximum limit, and/or the chain appears to be worn, it needs to be replaced.

DISTRIBUTOR DRIVES

The distributor, like the cam, only turns at half engine speed because each cylinder in a four-stroke engine only needs ignition every other revolution. The distributor is usually driven off the camshaft, but may be driven directly by the crankshaft on some engines.

To maintain precise ignition timing, the relative positions of the distributor gear and cam are critical. Pulling the distributor out of the engine and dropping it back in with the gear off by one or more teeth from its original position has the same effect as turning the distributor to advance or retard timing. Proper "indexing" of the distributor, like that of the cam, is essential.

Anything that affects the distributor drive will affect ignition timing. If the distributor drive gears become worn, timing can become erratic. Likewise, if the engine has a stretched timing chain or worn timing gears, ignition timing will also be affected.

FIRING ORDERS

The relative positions of the pistons inside a four-stroke engine vary because each of the pistons operates at a different point in its four-stroke cycle.

The reason for doing this is to space out the power strokes. By staggering the relative positions of the pistons in a multi-cylinder engine, the power strokes come at spaced intervals. This smooths out the flow of power and makes for a much smoother running engine. If all the pistons hit their power stroke at the same instant, the engine would produce a sudden surge of power, followed by a lull, followed by another surge of power, followed by another lull, and so on. Spacing out the firing cycles into a balanced sequence means at least one piston will be on its power stroke at all times to help drive the other pistons.

In 4-cylinder engines, the firing intervals are spaced 180 degrees apart. There are four cylinders, each one of which produces a power stroke every 720 degrees of rotation. Divide 720 degrees by 4 and get 180. In an even-firing V6, the firing pulses are spaced every 120 degrees (720 divided by 6), and in a V8 every 90 degrees (720 divided by 8).

Which cylinder fires first and the ignition sequence that follows is determined to balance loads on the crankshaft and to reduce "scavenging" losses between adjacent cylinders. If two adjacent cylinders fire in succession, one can steal air and fuel from the next, unless the intake manifold has long runners to keep the two separated. This can lead to uneven fuel distribution cylinder-to-cylinder in an engine with a carburetor or throttle body injector.

Typical firing orders are as follows:

- Most V8s: 1-8-4-3-6-5-7-2 or 1-5-4-2-6-3-7-8
- Most straight 6-cylinders: 1-5-3-6-2-4
- Most V6s: 1-6-5-4-3-2
- Most 4-cylinders: 1-3-4-2

CHAPTER 8

Basics of Carburetion

To understand the basics of carburetion, you first have to understand the combustion process itself. Gasoline fuels most internal combustion engines, with the exception of diesels (which run on diesel oil), or vehicles powered by alternative fuels such as propane or methane. Gasoline is a "hydrocarbon" fuel made up of hydrogen and carbon atoms. It is also a blend of different hydrocarbons, including lighter and heavier hydrocarbons that give various grades of gasoline specific fuel properties.

When gasoline is burned inside an engine, the combustion process splits apart the hydrogen and carbon atoms, which then release heat energy as they combine with oxygen. The air we breathe contains only about 21% oxygen (the rest is mostly nitrogen, with small amounts of carbon dioxide, water vapor, and other trace gases). As long as there's sufficient oxygen present to combine with all the hydrogen and carbon atoms in the fuel, we have "complete" combustion. There's no unburned fuel left over, and all that comes out the tailpipe is "carbon dioxide" (CO_2), which contains one atom of carbon and two of oxygen, and water vapor (H_2O) which contains two atoms of hydrogen and one atom of oxygen. The whole process can be summed up with the following equation:

$$O_2 + HC = CO_2 + H_2O$$

Oxygen combines with the hydrogen and carbon atoms in gasoline to create carbon dioxide and water vapor. The formula as written isn't "balanced" because the number of atoms on each side is not equal. If the equation is rewritten using a molecule of iso-octane as one of the gasoline hydrocarbons, the equation would look like this:

$$7O_2 + C_8H_{18} = 4CO_2 + 6H_2O$$

The equation tells us that it takes seven molecules of oxygen (which likes to float around in pairs of two atoms) to burn a molecule of gasoline (C_8H_{18}). The result is four molecules of carbon dioxide and six of water vapor.

When translated into more familiar air/fuel ratio terms, a balanced mixture of oxygen and fuel, called a "stoichiometric" mixture, would be 14.7 lbs. of air to 1 lb. of fuel. Since we're talking weight here instead of molecules, the numerical ratio is different.

STOICHIOMETRIC COMBUSTION

Gasoline weighs 6 1/2 lbs. per gallon. The hydrocarbons in each gallon of fuel contain about 5 lbs. of carbon and 1 1/2 lbs. of hydrogen. If it takes 14.7 times as much air as fuel to completely burn the fuel, the amount of air required is 14.7 x 6.5 = 95.55 lbs. Why so much air? Because as we said earlier, only 21% of the air we breathe is oxygen (21% of 95.55 lbs. is 20 lbs. of oxygen). So in reality, an engine has to inhale 20 lbs. of oxygen to convert one gallon (6 1/2 lbs.) of fuel to heat, carbon dioxide, and water vapor.

$$\text{IDEAL COMBUSTION} = \text{Fuel} + O_2 \rightleftharpoons \text{Heat} + CO_2 + H_2O$$

The 20 lbs. of oxygen and 6 1/2 lbs. of fuel consumed during the combustion process combine as follows: 15.4 lbs. of oxygen combines with 5 lbs. of carbon to make 20.4 lbs. of carbon dioxide. The remaining 4.6 lbs. of oxygen combines with 1 1/2 lbs. of hydrogen to make 6.1 lbs. of water vapor (nearly a gallon). The total weight of the oxygen and fuel before burning (which is 26.5 lbs.) equals the total weight of the combustion byproducts (carbon dioxide and water vapor) after burning. Thus oxygen and fuel are not really consumed during combustion, but merely transformed into new compounds.

The amount of heat released by burning one gallon of gasoline is about 116,000 BTUs. The actual amount of heat produced will vary somewhat depending on the fuel blend (gasoline containing a higher concentration of heavier hydrocarbons makes more heat). The heat that's produced by the combustion process plus the byproduct gases (CO_2 and H_2O) create tremendous pressure (up to 600 psi or more) inside the engine's cylinders. This pressure drives the pistons to make power and torque.

COMPLETE AND INCOMPLETE COMBUSTION

When there is just the right amount of oxygen to convert all the carbon to carbon dioxide and all the hydrogen to water vapor, the burning of the fuel is called "complete" combustion. It is complete because all the hydrocarbon fuel has been completely converted to its ultimate end products of CO_2 and H_2O.

A balanced, (ideal or stoichiometric) air/fuel ratio results in almost complete combustion (Fig. 8-1). We say "almost complete" because 100%-complete combustion is nearly impossible to achieve in a piston engine. A small fraction of the fuel will stick to the cylinder walls or become trapped in the crevice between the piston crown and rings where it won't burn. It is also very difficult to maintain a perfectly even blending of air and fuel throughout the combustion chamber during compression and ignition. Swirling and turbulence help to blend the mixture, but there are always small regions where the mixture is slightly "richer" (too much fuel, not enough air) or "leaner" (too much air, not enough fuel) than the ideal blend. The result is then "incomplete" combustion for a fraction of the mixture.

Incomplete combustion also occurs when the relative proportions of air and fuel are out of balance with one another. When there is not enough oxygen to convert all the carbon to carbon dioxide, the resulting byproduct can be carbon monoxide (CO), or even straight carbon particles (soot). Hydrogen combines more readily with oxygen than carbon, so water vapor will always be produced no matter what the imbalance during combustion. If the fuel itself fails to burn, the result is unburned hydrocarbons (HC) in the exhaust.

Reading the amount of carbon dioxide (CO_2) in the exhaust with an exhaust analyzer (see Chapter 11) will reveal combustion efficiency. CO_2 readings are highest when the air/fuel mixture is at the ideal ratio, and complete combustion is occurring. As the mixture becomes richer or leaner, combustion efficiency falls off and the amount of CO_2 in the exhaust drops.

Why not read the level of carbon monoxide in the exhaust to get an indication of combustion efficiency? You can, but it isn't as accurate an indicator as CO_2 because the level of carbon monoxide rises sharply when the fuel mixture goes rich, but remains relatively flat if the mixture is lean. So reading CO by itself would only tells you when the mixture is too rich, not when it is too lean. What's more, the amount of CO in the exhaust is reduced by the catalytic converter, while CO_2 is not.

AIR/FUEL RATIOS: RICH AND LEAN

The reason for having the right amount of air to balance the amount of fuel being burned should now be obvious. Without enough air, the fuel mixture won't burn completely, reducing power and creating exhaust pollutants.

When the air/fuel mixture contains too much fuel and not enough air, the mixture is "rich" (Fig. 8-1). A rich mixture would be one with a numerically lower ratio than 14.7: 1. Rich fuel mixtures do not have enough oxygen to convert all the carbon in gasoline into carbon dioxide, so the combustion

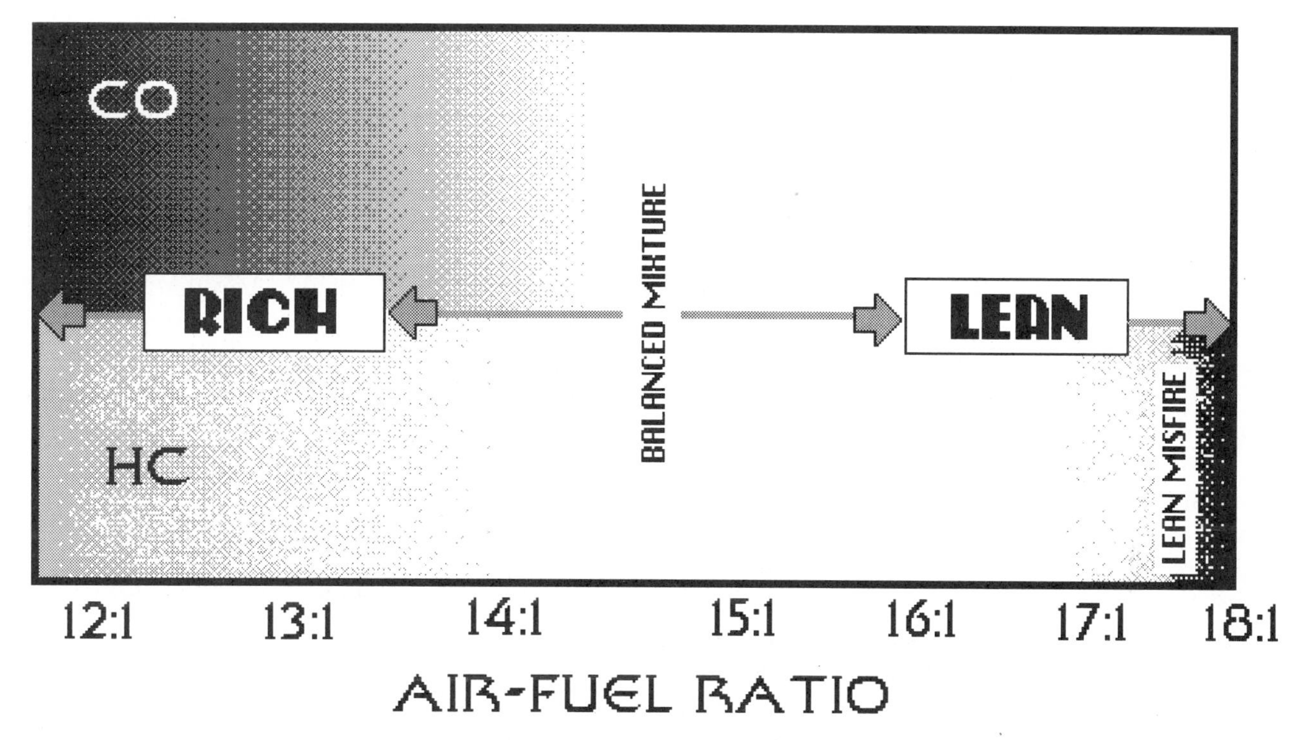

Fig. 8-1 How changes in the air-fuel ratio affect emissions.

gases contain a higher proportion or carbon monoxide. The richer the mixture, the greater the concentration of carbon monoxide in the exhaust. Reading the level of CO in the exhaust, therefore, becomes a good way to identify a rich fuel mixture.

A "lean" fuel mixture is one that contains a greater percentage of air than fuel compared to the ideal ratio of 14.7: 1. Thus, a lean mixture would be one with a higher numerical ratio than 14.7: 1. The combustion byproducts for a lean mixture are identical to those of a perfectly balanced mixture up to a point—the point at which lean misfire (see below) occurs. Up to this point, there is more than enough oxygen to convert all the carbon to carbon dioxide. But once the mixture becomes so lean that it fails to ignite, the amount of oxygen that's available becomes irrelevant, because the fuel is passing unburned through the engine and into the exhaust, creating high unburned hydrocarbon emissions.

Reading the level of oxygen in the exhaust with an exhaust analyzer is the best way to spot a lean fuel mixture problem. Below the ideal ratio, very little oxygen is left in the exhaust after combustion (only 1-2%). But as the mixture goes lean, the amount that's left over begins to rise.

Lean fuel mixtures are clean-burning mixtures, but they produce less power than rich or balanced mixtures. The extra oxygen takes up space in the combustion chamber, which reduces the total volume of fuel that can be burned in each cylinder. Lean fuel mixtures are also more prone to detonation. On the other hand, a lean mixture improves fuel economy because less fuel is consumed. Under certain driving conditions (cruising and deceleration, for example), a lean fuel mixture is desirable because it can save gas. A lean mixture would not be desirable, however, when maximum power is needed (during acceleration, for example) or during engine warm-up when the fuel is slow to vaporize.

When a fuel mixture is excessively lean (leaner than about 22:1), it can fail to ignite, a condition known as "lean misfire." The result is a loss of power, stalling, hesitation, and increased pollution. The unburned fuel passes through the engine and sharply raises the level of unburned hydrocarbons

(HC) in the exhaust. A overly lean fuel mixture can be caused by a variety of ailments, but often it is due to a vacuum leak that allows extra air to enter the intake manifold.

For combustion to occur, the air/fuel mixture must be between certain minimum or maximum limits. This will vary somewhat depending on the engine design, but for most engines the lean limit is around 22: 1. The rich limit is about 8: 1. If the air/fuel mixture is leaner or richer than these limits, the mixture won't ignite, and the cylinder will misfire.

Rich fuel mixtures require less firing voltage than lean mixtures because the denser concentration of fuel molecules between the spark gap provides an more conductive path for the voltage to follow. Thus, the "ionization" voltage is lower, and the spark jumps more easily across the gap to ignite the mixture. With a lean mixture, there is less fuel, and the fuel molecules are spaced more widely apart. It therefore takes more voltage to ionize the molecules to create a conductive path for the spark.

FUEL CALIBRATION

The calibration of the air/fuel ratio is determined by the metering circuits in the carburetor, or in a fuel-injected engine by the duration or "on time" of the injectors (see Chapter 9).

One thing that's very important to note about air/fuel ratios is that the engine can benefit from a slightly richer mixture at some times, and a leaner mixture at others. When the engine is first started and is cold, the fuel vaporizes more slowly because the intake manifold and cylinders are not that hot yet. To keep the engine running, a richer fuel mixture is required. As the engine warms up, the mixture can be gradually leaned out until it will idle smoothly on its normal mixture. During acceleration or periods of increased load, a richer mixture can boost power and help the engine resist detonation. During deceleration, the fuel supply can be momentarily cutoff or leaned out to save fuel.

On 1981 and later model engines that have "oxygen sensors" and either "feedback" carburetion or electronic fuel injection, the air/fuel mixture can be constantly adjusted by the engine computer based on oxygen-sensor input (see Chapter 17).

Another thing you should know about the air/fuel ratio is that it is affected by altitude. Atmospheric pressure at sea level is about 14 pounds per square inch. But as you go up in elevation, there's less and less air overhead to press down. Consequently, atmospheric pressure drops about one psi for every 1,000 feet in altitude. Since the air/fuel ratio depend on the relative proportions of air and fuel entering the engine, the fuel mixture can become progressively richer as air density decreases with altitude. On later-model engines with computerized fuel control, a barometric pressure sensor allows the computer to compensate for changes in altitude and atmospheric pressure. On older vehicles, though, carburetor jetting and ignition timing had to be recalibrated to compensate for thinner air.

FUNCTIONS OF A CARBURETOR

The basic function of a carburetor is to deliver a combustible air/fuel mixture to the engine over a wide range of operating speeds and loads. It does this by metering the flow of both air and fuel so that the engine will receive the right mixture of air and gasoline.

The throttle is the master control valve in the carburetor because it restricts the amount of air that can enter the engine. The rate of airflow, in turn, determines the amount of fuel that is siphoned through the carburetor's idle and main metering circuits. The relative position of the throttle also determines which carburetor metering circuit is used to regulate the fuel mixture (idle, transition, cruise, or acceleration).

The importance of the carburetor's role in engine performance and emissions can't be overemphasized because almost everything the engine does depends upon it. The carburetor controls engine speed and power, aids starting, regulates idle speed (which affects driveability and idle emissions). It also provides ported vacuum to help regulate spark advance, various emissions functions (such as EGR), and even the speed at which the automatic transmission shifts gears. All things considered, the carburetor is probably the single most important component in terms of emission, driveability, and performance.

Supplying the engine with the proper air/fuel mixture may not seem like such a difficult task—until you consider all the conditions under which the air/fuel ratio needs to vary to meet the engine's needs:

- *For starting,* the engine needs a super rich mixture (which the choke provides).
- *For warm-up,* the engine needs a rich mixture that becomes gradually leaner (which again the choke regulates).
- *For good idle,* the engine needs a slightly rich mixture (provided by the idle circuit).
- *For good fuel economy,* the engine needs a lean mixture at cruise (provided by the main metering circuit).
- *To accelerate without stumbling or hesitating,* the engine needs a momentarily richer mixture (which the accelerator pump supplies).
- *To minimize the risk of detonation and to increase power,* the engine needs a richer mixture under load (which the power circuit provides).
- *To keep from running on after the ignition is turned off,* the engine needs to have its fuel supply cut off (which the idle-stop solenoid or fuel cut-off solenoid does).

FUEL VAPORIZATION

For gasoline to burn properly, it must first be "atomized" into small droplets and mixed with air. The atomization process is accomplished by a variety of means. The vacuum that exists beneath the throttle plate inside the intake manifold lowers the boiling point of the fuel. This makes it easier for the fuel to vaporize.

There is a direct relationship between pressure and temperature: lowering pressure reduces the temperature at which a liquid boils (evaporates). The drop in pressure from atmospheric to 18 inches or more of vacuum lowers the boiling point of the fuel drastically. (The actual boiling points of the various hydrocarbons in gasoline range from 100 to 400 degrees F.) The stronger the vacuum, the more readily the fuel evaporates. Thus the fuel evaporates more quickly at idle and light throttle than it does at wide-open throttle when there is very little vacuum in the intake manifold.

Heat is also a necessary ingredient. A hot intake manifold speeds the vaporization process. An exhaust crossover passageway through the intake manifold on a V8 or V6 engine is often used to heat

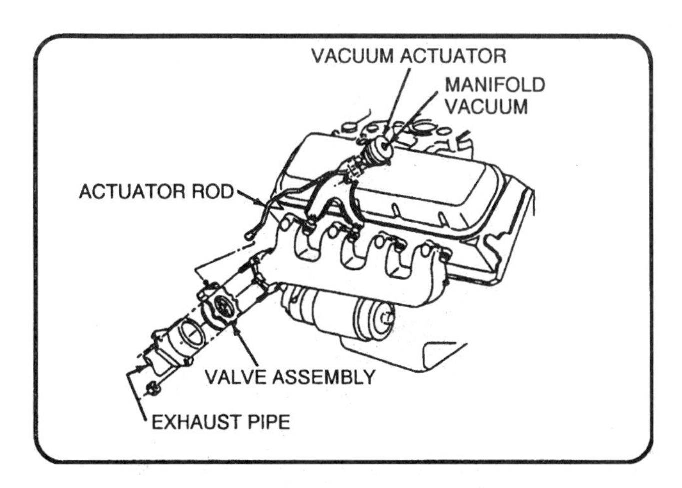

Fig. 8-2 The heat riser valve on a V6 or V8 engine temporarily blocks one manifold, forcing hot exhaust to flow through a crossover passage under the intake manifold. This speeds engine warm-up and promotes better fuel vaporization.

the manifold when the engine is cold. A "heat riser valve" (Fig. 8-2) on one of the exhaust manifolds partially blocks the flow of exhaust, forcing it to back up and flow through the passageway. As the engine warms up and additional heating is no longer required, the heat riser valve opens allowing the exhaust to exit its normal route. If the intake manifold gets too hot, however, it can cause problems. Too much heat causes the incoming air to expand and become less dense. This reduces power by decreasing the density of the air/fuel mixture. Overheated air also increases the likelihood of detonation.

On many inline engines, the intake manifold is located on the same side of the head as the exhaust manifold, so heat from the exhaust can provide heat. Coolant from the engine block may be circulated through a passage in the intake manifold to even out the distribution of heat, and to prevent the intake manifold from getting too hot.

On some engines to speed fuel vaporization immediately after a cold start, an electrically heated ceramic grid under the carburetor (Fig. 8-3) may provide extra heat for up to a minute or so. This is part of the "early fuel evaporation system." The atomized fuel droplets passing through the grid are heated and turned to vapor. This speeds engine warm-up and cold idleability.

Another aid to vaporization is the "thermostatically controlled air cleaner." This picks up heated air from a stove around the exhaust manifold when the

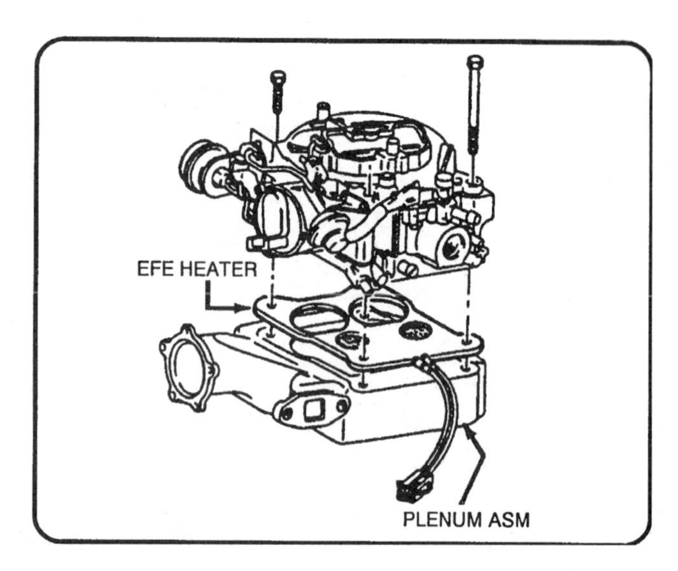

Fig. 8-3 Heat provided by the early fuel evaporation grid under the carburetor promotes better fuel vaporization when the engine is cold.

engine is cold. By preheating the incoming air, the carburetor is better able to maintain a more closely controlled fuel mixture. The added heat also helps the fuel evaporate.

The final point of fuel vaporization occurs within the combustion chamber itself. The heat of compression combines with the turbulence and swirling that's created as the piston squeezes the fuel mixture into the combustion chamber. This helps to vaporize the remaining droplets of fuel.

Compressing the fuel helps to remix the air and fuel that becomes separated during the journey from carburetor (or throttle body injector) to the cylinder. (This isn't a problem with multipoint fuel injection, because only air flows through the intake manifold.) Some separation between air and fuel is unavoidable in the intake manifold because fuel is heavier than air and doesn't turn corners as quickly. This is also why the end cylinders in an inline engine tend to run leaner than those towards the middle. The end cylinders often have the longest intake runners, and thus a greater change for fuel separation to occur. Those nearest the center usually receive the richest mixture because the fuel has less distance to travel.

ROLE OF VACUUM IN CARBURETION

One of the fundamental differences between carburetion and fuel injection is that carburetion is totally dependent on intake vacuum for fuel metering. Fuel injection is not. A carburetor needs vacuum to pull fuel through its metering circuits. A carburetor also needs vacuum to help atomize the fuel. A fuel injector, on the other hand, simply sprays a fine mist of atomized fuel directly into the intake manifold (throttle body injection) or intake port (multiport injection). Vacuum is only needed to "fine-tune" the volume of fuel delivered.

Vacuum exists in the intake manifold as a result of the pumping action of the engine's pistons and the restriction created by the throttle valve in the carburetor. Were it not for the throttle choking off the flow of air into the engine, there would be little if any vacuum in the intake manifold (as in a diesel).

A carburetor must have vacuum, because without it the carburetor has no way of delivering fuel to the engine (except for the accelerator pump, which by itself cannot supply enough fuel to keep an engine running). Vacuum is what siphons fuel through the idle, main metering, and power circuits. An engine that has a vacuum leak, therefore, will likely be an engine that suffers from a carburetion problem. The same holds true for any engine with severe backpressure (caused by an exhaust restriction), or one with very low compression or valve timing problems that reduces intake vacuum.

If intake vacuum is unusually low, the carburetor can't do its job properly, and the result is carburetion-related driveability symptoms. On most engines, the idle vacuum reading should be steady between 16 and 22 inches. A low reading usually indicates a vacuum leak, or in some cases an exhaust blockage. An oscillating reading usually indicates a leaky valve or badly worn valve guides.

THREE TYPES OF VACUUM

The carburetor makes use of three kinds of vacuum (Fig. 8-4)

- *Manifold vacuum,* which is the vacuum that exists under the throttle plates in the intake manifold. Manifold vacuum is used to: pull fuel through the idle circuit at idle; to open the power valve in the power circuit when the throttle opens and vacuum drops; to pull hot air through the choke housing on some carburetors; to pull the choke open when the engine starts, to supply vacuum to the heated air-intake control mo-

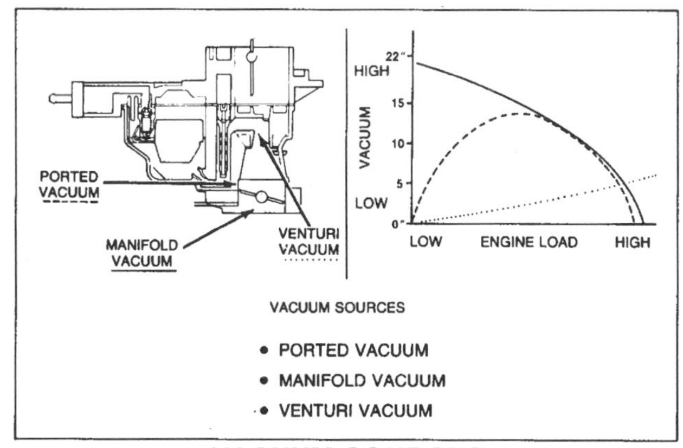

Fig. 8-4 Vacuum sources (courtesy Chrysler).

tor; and to pull crankcase blowby vapors through the PCV system.

Manifold vacuum is affected by the position of the throttle. At idle or closed-throttle deceleration, manifold vacuum is highest. At wide-open throttle, it is lowest. Because of this relationship, manifold vacuum is an excellent indicator of engine load. When vacuum is high, the load on the engine is low. When vacuum is low, engine load is high.

One point worth noting here is that while the engine is being cranked during starting, manifold vacuum is very low. Vacuum is low because the engine is turning over at less than half its normal idle speed so the volume of air being drawn into the engine is not great. To help create a stronger vacuum in the engine, the choke at the top of the carburetor is held shut. This cuts off the engine's air supply, creating a temporary area of low pressure inside the throat of the carburetor. This pulls fuel through the idle and main metering circuits to get the engine started.

- *Ported Vacuum,* which is produced when an opening or "port" in the carburetor bore above the throttle plate is exposed to manifold vacuum when the throttle opens. Ported vacuum is used to continue fuel delivery during the transition from idle to off-idle operation. It also supplies vacuum to other components such as the distributor vacuum spark advance diaphragm and EGR valve.

  Ported vacuum is totally dependent on throttle position. At idle, the opening is above the throttle plates so no vacuum exists in the circuit or line. Only when the throttle opens far enough to expose the port to intake vacuum does vacuum exist in the circuit or line. The strength of the vacuum, however, is moderate because the throttle is partially open and manifold vacuum is reduced.
- *Venturi Vacuum,* which is created by a low-pressure area in the narrowest part of the carburetor throat (the venturi). Venturi vacuum is used to pull fuel from the fuel bowl through the main metering circuit and discharge tube during part-throttle cruise operation. Venturi vacuum is not as strong as ported or manifold vacuum. So, if used for other control functions (EGR, for example), it is usually connected with a vacuum booster or amplifier that uses the venturi vacuum signal as a control signal.

  As with manifold and ported vacuum, venturi vacuum too is dependent on throttle position. At idle there is not enough air flowing down through the throat of the carburetor to create a pressure drop in the venturi. So no venturi vacuum exists at idle. At part throttle, there is enough air flowing down through the carburetor to create moderate vacuum. At wide-open throttle, venturi vacuum reaches a maximum because the volume of air flow is greatest. Manifold and ported vacuum at wide-open throttle, however, become weak.

BASIC CARBURETOR COMPONENTS AND CIRCUITS

Fuel Bowl, Needle Valve and Float

Liquid gasoline enters the carburetor "bowl" through a "needle valve," a simple open-shut valve that uses a tapered needle to close off the fuel-inlet orifice (Fig. 8-5). An arm that extends from the hinged "float" inside the bowl pushes the needle valve shut as the fuel level rises. This prevents the bowl from overflowing because fuel is being fed to the carburetor all the while under pressure from the fuel pump.

The level of the fuel within the bowl is critical because it affects fuel metering and the ratio of the air/fuel mixture. The height of the float, therefore, must be carefully adjusted to exact specifications

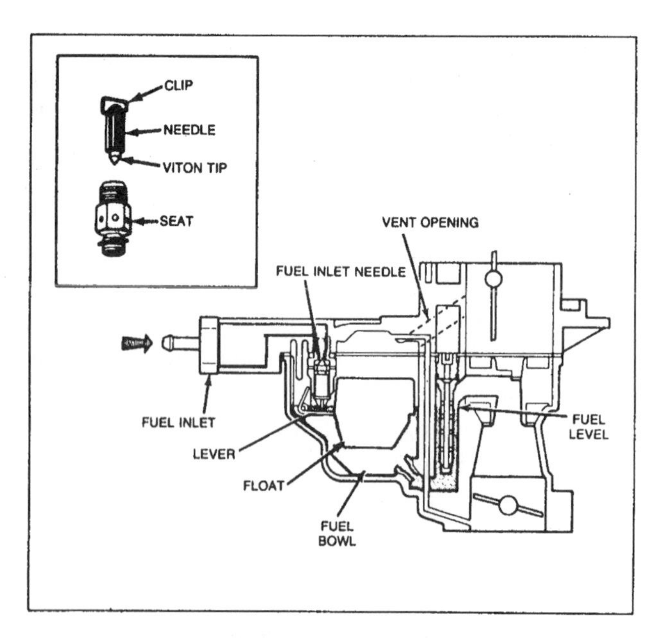

Fig. 8-5 Float circuit with needle valve and seat (courtesy Chrysler).

following the vehicle manufacturer's procedure (Fig. 8-6). The wrong height adjustment may set the fuel level too high or too low. Too much fuel in the bowl can make the air/fuel mixture run rich. Too little and it can run lean.

As fuel is drawn through the carburetor's metering circuits into the engine, the fuel level in the bowl drops. This lowers the float, allowing fuel pressure to push the needle valve open. As more fuel enters the bowl, the float rises and closes the needle valve to maintain a consistent fuel level in the bowl.

As long as the needle valve and float do their job and are adjusted correctly, the fuel level in the bowl will stay at the desired level. But several things can upset this delicate balance. If a piece of dirt or rust gets past the fuel filter and enters the fuel inlet, it can lodge between the needle and seat, creating a leak that can cause the bowl to flood. Leaks can also occur when the needle and seat become worn and don't seal well against one another. Fuel leaking into the bowl will raise the fuel level and cause a rich fuel mixture. Symptoms of a rich mixture include a rough idle, carbon-fouled spark plugs, increased CO emissions, soot in the exhaust, and stalling. If the bowl floods, fuel may gush out of the bowl vents in the carburetor and enter the engine.

Sometimes the needle valve can stick. The needle valve in some carburetors is rubber tipped because rubber is pliant and seals more easily against an irregular surface. But rubber can deteriorate with age, and can swell if exposed to excessive concentrations of alcohol in gasoline. If the valve sticks, it prevents fuel from entering the bowl, starving the carburetor, and causing the engine to stall. A leaking or sticking needle valve should be replaced.

Another problem that can affect the fuel level in the bowl (which in turn affects the air/fuel mixture and emissions) is a sunken or heavy (fuel-saturated) float. Floats are typically made of one of three different materials: brass, plastic, or foam. Hollow brass floats sometimes develop leaks in the soldered joint that joins the two halves together. A leak usually fills the float with fuel, causing it to sink. This causes the needle valve to remain open, flooding the bowl and carburetor with fuel. In some cases a leak may not be severe enough to entirely fill the float, but enough to reduce the float's buoyancy. This kind of problem causes the fuel level to run higher, resulting in a richer mixture. The same thing can happen to a hollow plastic float if a leak develops.

If a float appears sunken or low, it should be visually inspected for damage or leaks, and shaken to determine if there's any fuel sloshing around inside it. A leaky float should be replaced.

Many carburetors have rigid foam floats, made of a material called "Nitrophyl." The closed-cell-foam plastic is not supposed to absorb fuel. But sometimes the plastic is not cured properly, allowing it to gradually absorb gasoline over a period of many months, even years. This causes the float to become progressively heavier and to slowly lose buoyancy. As the float sinks lower and lower in the

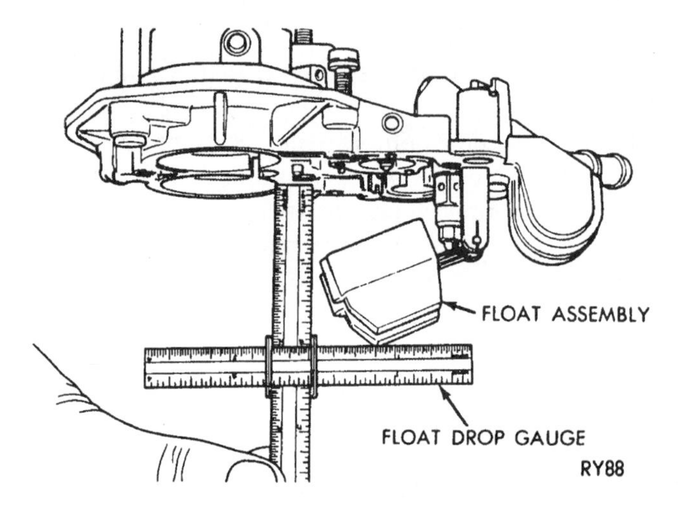

Fig. 8-6 Measuring float level (courtesy Chrysler).

fuel bowl, the fuel level rises and the carburetor runs progressively richer. Fuel economy declines and emissions creep up. The only way to accurately diagnose this type of problem is to remove the float and weigh it. Special scales are available for this purpose, along with charts that list how much a given float should weigh for a specific carburetor application. If a float is found to be heavy, replacement is the only repair option because there's no way to "dry out" the float or remove the fuel that's been absorbed. A heavy float will often ooze gasoline when squeezed with a fingernail or sharp object months after its removal. If a hollow brass float is substituted for a foam plastic float, a somewhat different level adjustment will probably be necessary because of the brass float's heavier weight.

Bowl Vent

To prevent fuel vapors from building up pressure inside the fuel bowl due to engine heat, the bowl is vented. Before the days of sealed fuel systems, the carburetor was simply vented to the atmosphere. Some used a device called an "anti-percolation vent" to keep fuel vapors from bubbling through the main metering circuit into the engine. But since the mid-1970s, the fuel system has been sealed to eliminate evaporative emissions (see Chapter 10). The bowl vents are connected to a charcoal-filled canister so fuel vapors cannot escape into the atmosphere (Fig. 8-7). The vapors are stored in the canister until they

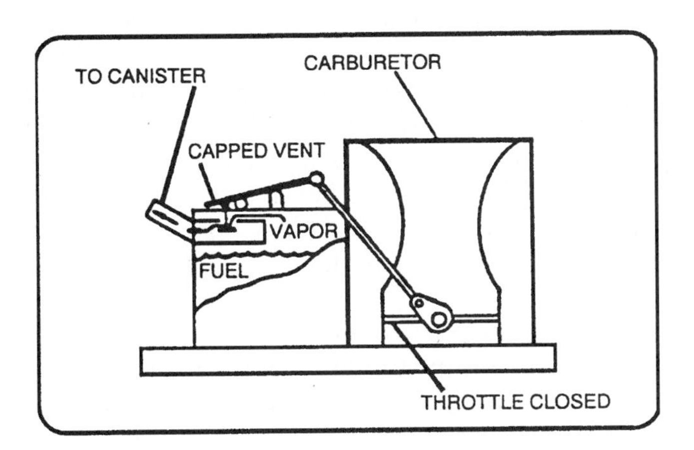

Fig. 8-7 The carburetor bowl is vented to a vapor storage canister to prevent the escape of fuel vapors into the atmosphere (reprinted, by permission, from Scharff, *Complete Fuel Systems and Emissions Control.* Copyright 1989 by Delmar Publishers Inc.).

can be purged back into the intake manifold when the engine starts. Some carburetors are designed with such compact fuel bowls that little if any venting is necessary.

The bowl may also be vented internally through a "balance tube" that extends into the upper portion of the carburetor (called the "air horn"). On some carburetors, the balance tube is part of the air horn itself. As air enters the carburetor, the downward flow pushes air through the balance tube and into the bowl. This equalizes the pressure in the bowl with that in the air horn to compensate for any restriction caused by the air cleaner. Were it not for the balance tube, pressure could be greater inside the bowl. This could push more fuel through the main metering circuit, making the air/fuel ratio run rich.

Throttle

The throttle is the main control valve in a carburetor or fuel-injection throttle body for regulating air flow into the engine. Just below the throttle plate on a carburetor are one or more idle ports though which fuel flows at idle. (There are no such ports in a fuel-injection application, but there is a "throttle bypass" circuit that controls idle speed.) Additional ports may be located just above the throttle in a carburetor to supply additional fuel as the throttle opens or to pass vacuum to ported-vacuum circuits. Everything below the throttle plate is exposed to engine vacuum, and everything above it is not until the throttle opens.

When the engine is idling, cruising under light load, or decelerating, the throttle is nearly closed and there isn't enough air flowing past the throttle to completely satisfy the engine's breathing requirements. As a result, intake vacuum is high. But when the throttle is open wide, as when accelerating or pulling a heavy load, there's plenty of airflow and intake vacuum is low.

Idle Circuit

One of the ports just below the throttle is for the carburetor's "idle circuit." Air doesn't flow fast enough down through the throat of the carburetor at idle to draw fuel through the main fuel-metering circuit. So a special idle circuit is needed to keep the engine running. The idle circuit relies on the strong vac-

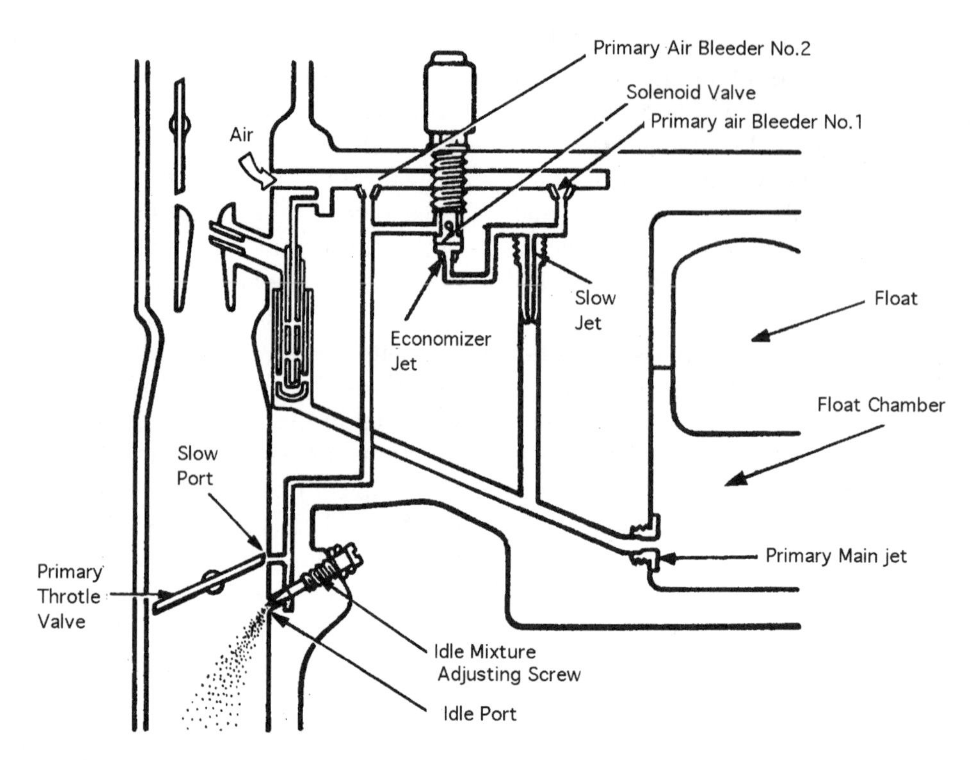

Fig. 8-8 Idle circuit and low-speed transfer (slow) port (courtesy Toyota).

uum that exists in the intake manifold to pull fuel through the circuit and out the "idle port" just below the throttle plate.

The fuel that's drawn through the idle circuit first passes through the main metering jet in the bottom of the carburetor bowl (Fig. 8-8). It then enters the "idle tube" where it passes through a second restriction that limits the maximum amount of fuel that can flow through the idle circuit. At the top of the idle tube, the fuel makes a pair of 90-degree turns before heading down to the idle port.

Because the idle port is located below the fuel bowl, "air bleeds" at the top of the idle tube are needed to prevent fuel from siphoning out of the bowl when the engine is turned off. Were it not for the anti-siphon function of the air bleeds, fuel would continue to dribble through the idle circuit until the fuel bowl ran dry.

The air bleeds also perform another important function. By introducing air bubbles into the fuel passing through the idle circuit, the fuel is "aerated." This helps to break it up so it will atomize and mix more easily with the air flowing by the throttle plate.

On many carburetors, a port just above the throttle plate serves as an additional air bleed. This is the "transfer port." When the throttle is nearly closed at idle, vacuum below the throttle plate pulls air in through the transfer port to further aerate the fuel as it flows down and out the idle port. The transfer port's primary function, however, is to keep the fuel mixture from leaning out as the throttle comes off idle. When the throttle begins to open, the plate passes by the transfer port. As soon as the transfer port is exposed to vacuum, it stops sucking air and begins to draw additional fuel from the idle circuit. The effect is like doubling the discharge capacity of the idle port. The extra fuel from the transfer port is enough to keep the air/fuel mixture from leaning out—which otherwise would result in a flat spot off idle. Extra fuel is needed because the volume of air flowing past the throttle plate increases faster than the idle port's ability to deliver fuel as the throttle opens up.

The relative proportions of air and fuel at idle (called the "idle mixture") are adjusted by a tapered "idle-mixture adjustment screw". On older carburetors, the idle-mixture screw extends through the cavity behind the idle port. Turning the screw in decreases the size of the opening to reduce the flow of fuel which leans out the idle mixture. Backing the screw out allows more fuel to flow through the idle port, which enrichens the mixture.

On later-model carburetors with "sealed" idle-mixture adjustment screws (to discourage tampering), a second adjustment screw is provided to control the amount of air bleeding into the idle circuit. Thus the idle-mixture screw is really an air-bleed screw. The effects of turning the air-bleed screw in and out are reversed: turning the screw in enrichens the mixture, while turning it out leans the mixture. The screw has a left-hand thread, however, which means you still turn it in the same direction of rotation to adjust the mixture: counterclockwise to enrich the mixture, or clockwise to make it leaner.

Idle-limiter caps are used on many carburetors to limit the range of adjustment in the idle mixture. The reason for limiting the adjustment is to prevent overly rich or lean idle mixtures that could adversely affect exhaust emissions. Tabs that project from the plastic caps restrict the amount of adjustment to about half a turn. The caps can be pried or pulled off, but even with the caps removed the range of adjustment is limited by the internal restriction in the idle tube.

Idle-mixture screws are often concealed beneath a metal plug to discourage tampering. If a satisfactory idle cannot be obtained by adjusting the idle air-bleed screw, it may be necessary to remove the plug and readjust the factory-set mixture screw. The plug can be removed by drilling, or by punching a small hole in it and then using a screw-in slide puller to pull it out. On some applications, the carburetor must be removed from the manifold to knock out the plug. Care must be taken not to damage the adjustment screw. Once the adjustment has been made, the screw should be resealed by inserting a new plug or covering the hole with RTV silicone sealer.

Idle-mixture problems can be caused by dirt or varnish plugging the air bleeds, idle-discharge port, or transfer port; by a damaged idle-mixture screw; or by misadjustment. When turning a mixture screw in, go easy on the screwdriver. The screwdriver must be seated gently to prevent damaging the end of the screw. Light pressure should be used; otherwise the tapered tip of the screw can be bent, broken off, or distorted.

COLD-IDLE ENRICHMENT

On some carburetors, a special "cold-idle enrichment" circuit is included to provide a richer fuel idle mixture while the engine is warming up. The reason for doing this is because the carburetor is calibrated so lean for emission purposes that it may not idle well when cold. (A cold engine needs a richer mixture to idle well.)

The cold-idle enrichment circuit contains an extra air bleed that is blocked off by the "idle-enrichment valve" when the engine is cold. This increases the relative proportion of fuel to air flowing through the idle circuit, causing the idle mixture to run richer. The idle-enrichment valve is vacuum-actuated and is connected to a thermal vacuum switch (TVS). The thermal vacuum switch passes vacuum to the idle-enrichment valve when the engine is cold. This pulls the valve shut, blocking off the idle-enrichment air bleed. As the engine warms up, the thermal vacuum switch blocks vacuum, which allows the enrichment valve to open the air-bleed passage. The extra air leans out the idle mixture to its normal ratio.

If the cold enrichment valve sticks shut, the idle mixture will remain rich causing increased idle emissions. If the cold enrichment valve fails to open, the idle mixture will run too lean possibly causing poor idle (lean misfire), stalling, or hesitation. Keep in mind that the function of the thermal vacuum switch is just as important as the idle-enrichment valve. An inoperative TVS, one that is incorrectly connected (wrong vacuum connections) or has leaky or plugged vacuum hoses will cause similar problems.

TRANSFER CIRCUIT

As the throttle comes off idle, a carburetor must make a transition from the idle circuit to running on the main fuel-metering circuit. The "transfer circuit" is the means by which this transition is accomplished.

Just above the throttle plate is an additional fuel port (or ports), called the "transfer port" or "off-idle port." As the throttle opens, vacuum below the throttle begins to drop, reducing the amount of fuel being siphoned through the idle port. As the throttle opens and moves past the transfer port (located just above the idle port), the transfer port is exposed to manifold vacuum. As soon as the transfer port begins to see vacuum, fuel is pulled through it. The extra fuel compensates for the reduced fuel flow through the idle circuit. It also prevents the air/fuel mixture from leaning out, until there is sufficient air-flow through the venturi to start pulling fuel through the main circuit discharge tube.

At just above idle, therefore, the carburetor supplies the engine with fuel through both the idle and transfer ports. As the throttle continues to open, vacuum continues to drop. The flow of fuel through the idle and transfer ports begins to taper off, as air velocity through the venturi picks up. Thus, the transition from idle to main circuit is gradually made until the carburetor is supplying all its fuel through the main metering circuit.

On some carburetors, the idle system quits completely once the main circuit takes over (Fig. 8-9). But on most carburetors, the idle system never stops completely. A small amount of fuel still passes through the idle circuit even when the engine is cruising at high speed. The amount of fuel is reduced, however, in proportion to the level of manifold vacuum and air flowing through the venturi.

MAIN METERING CIRCUIT

Once the throttle opens up, the volume of air flowing down through the carburetor begins to increase. As it does so, it speeds up as it passes through the narrowest part of the throat, called the "venturi."

A very important change takes place at this point called the "venturi effect," which is based on Bernoulli's Principle. The restriction created by the venturi causes the velocity of the air to speed up as it squeezes through the narrow opening. The increase in velocity creates a zone of low pressure (vacuum) in the venturi. This pulls fuel from the fuel bowl through the main jet, into the main fuel well, and out the venturi discharge nozzle (Fig. 8-10).

In many carburetors a "secondary venturi" or "booster venturi" is located in the main venturi to further increase the drop in pressure. The main metering circuit fuel-discharge nozzle will be located at this point. Besides pulling fuel through the main metering circuit, the pressure drop also helps to atomize the fuel droplets as they emerge from the nozzle. This promotes better fuel atomization and mixing.

The main metering circuit may also use air bleeds to aerate the fuel for better atomization (just as in the idle circuit). The air bleeds are located at the top of the main fuel well, and may be part of, or screwed into, "emulsifier tubes" in the well. Small holes in the emulsifier tubes extend down from the air bleeds into the main fuel well. These help to mix air bubbles into the fuel so the fuel will atomize more easily.

The air/fuel ratio in the main metering circuit is the leanest of all the circuits, and is controlled pri-

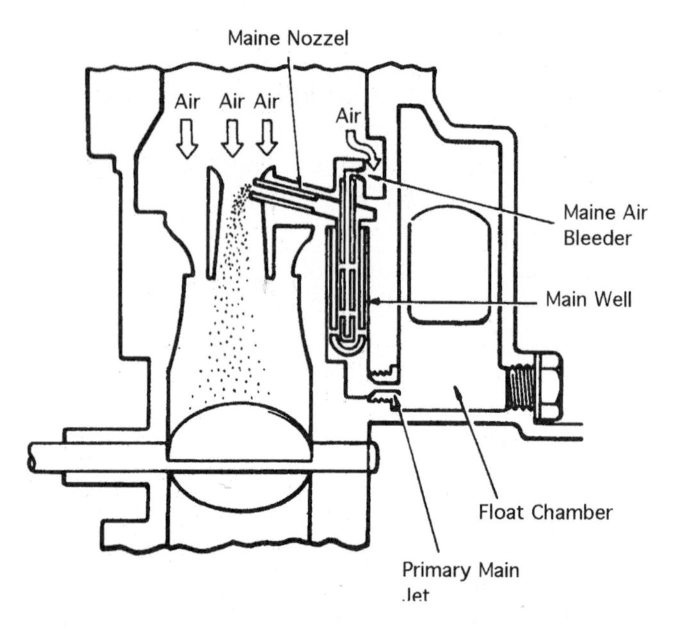

Fig. 8-9 Main metering circuit (courtesy Toyota).

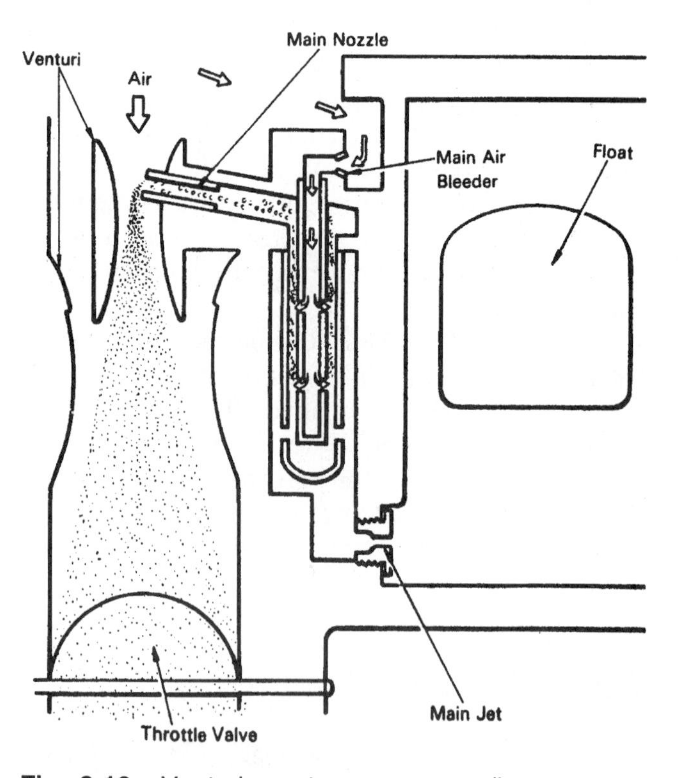

Fig. 8-10 Venturi creates a vacuum (low-pressure area (courtesy Toyota).

marily by the size of the main jet orifice. On some carburetors, a metering rod that extends into the orifice is also used. The smaller the opening in the jet, the less the volume of fuel that flows through the circuit, and the leaner the mixture. Conversely, increasing the size of the opening allows more fuel to be drawn into the venturi, enrichening the air/fuel mixture.

Main jets usually have a number stamped on the side that indicates the jet's flow capacity. The higher the number, the greater the jet's flow capacity. The amount of fuel that flows through the jet also depends on the float level inside the bowl. The higher the fuel level in the bowl, the less vacuum is needed at the venturi to pull fuel through the metering circuit. The lower the float level, the greater the amount of venturi vacuum needed to draw fuel through the circuit.

When an engine manufacturer specifies a certain float level, therefore, it is based on the flow characteristics of the carburetor and the size of the main jet. If someone sets the float level too high (or if the float is heavy and loses buoyancy), the air/fuel mixture goes rich. If the float level is set too low, the air/fuel mixture goes lean. So the only way to assure the air/fuel ratio will be correct through the main metering circuit is to make sure the float level is set accurately to specifications.

Power Circuit

Because the main fuel metering circuit is calibrated to provide a lean fuel mixture for good fuel economy, the engine may run too lean when under heavy load. A richer mixture under such circumstances helps lessen the chance of detonation, while at the same time increases power. The power circuit does just that.

The amount of vacuum in the intake manifold changes in direct proportion to engine load: the wider the throttle opening, the lower the vacuum. So manifold vacuum is used to gauge engine load for the purpose of making the desired air/fuel ratio changes. The power circuit provides additional fuel enrichment when the engine is under load by means of a "power valve" (also called a "power jet" or "enrichment valve").

The power valve (Fig. 8-11) has a spring-loaded diaphragm or piston that opens a fuel valve. Vacuum holds the diaphragm shut and prevents the valve from opening until vacuum drops. As engine load increases and vacuum drops, the diaphragm is pushed open by the spring, allowing additional fuel to flow into the main fuel well and to the venturi to enrich the fuel mixture. The point at which the power valve opens is determined by the spring tension against the diaphragm. As engine load eases and the need for additional fuel is no longer required, the increase in vacuum pulls the diaphragm shut, closing the valve, and cutting off the extra fuel. The air/fuel ratio then returns to its former lean setting.

A defective power valve that sticks open will cause the carburetor to run too rich during part-throttle, light-load operation, resulting in a loss of fuel economy. A power valve that remains shut and fails to open when manifold vacuum drops will prevent the fuel mixture from enrichening under load.

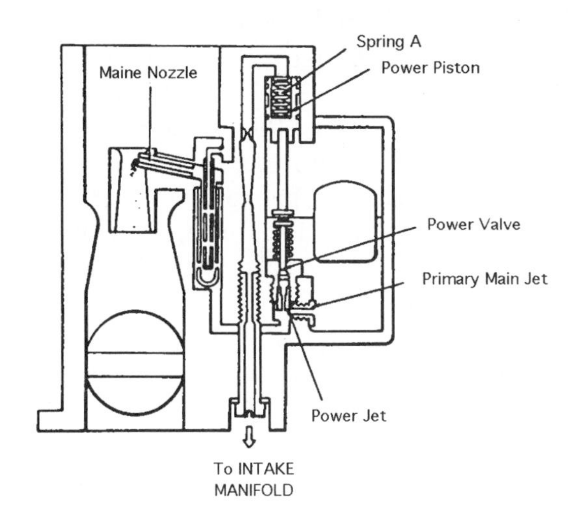

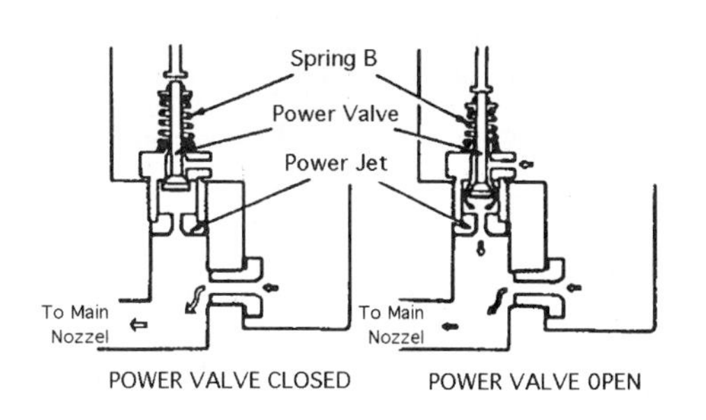

Fig. 8-11 Power valve circuit (courtesy Toyota).

The result, in this case, can be hesitation and detonation. A leaky or ruptured diaphragm, or sticking piston are the main causes of failure for most power valves. Varnish or dirt buildup in the valve itself can also cause binding or clogging.

Metering Rod Power Enrichment

Another means by which temporary fuel enrichment can be provided under load is to use a vacuum-actuated "metering rod" in the main metering jet (Fig. 8-12). Metering rods are typically used in Carter and Rochester carburetors. By inserting the tapered tip of a metering rod into the main jet, the size of the orifice can be reduced to restrict the flow of fuel and making the mixture leaner. By pulling the rod out, more fuel can pass through the jet making the mixture richer.

A small spring-loaded "power" piston attached to the metering rod allows vacuum to raise or lower the rod in response to changing engine loads. As the throttle opens and vacuum drops, the metering rod pulls out of the jet to enrichen the mixture. Then as the throttle closes and high vacuum returns, the rod plunges back into the jet and leans the mixture. Thus the vacuum-actuated metering rod enables a carburetor to alter the air/fuel ratio in response to changing engine loads.

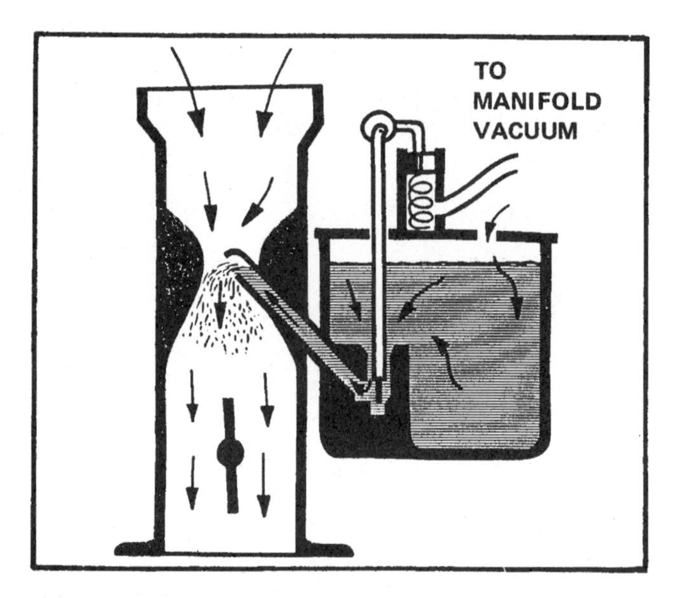

Fig. 8-12 Some power enrichment systems consist of metering rods placed in the main jets (courtesy Sun Electric).

Some carburetors use a metering rod with a two-step tip. When the largest portion of the rod extends into the jet, the main metering circuit delivers a lean mixture. When vacuum drops, the piston lifts the rod so that only the smaller diameter of the rod remains in the jet providing a richer mixture.

Three-step metering rods are used in other carburetors that have a tapered region between the largest and smallest part of the rod. The thickest part of the rod provides the leanest mixture while cruising. The tapered portion provides a gradual transition from lean to rich as vacuum drops.

Accelerator Pump Circuit

When the throttle snaps open, the engine ingests a big gulp of air. Vacuum drops as the incoming air satisfies the engine's desire for air. This causes a momentarily lag in fuel delivery through the main metering circuit because the flow of air into the engine accelerates more quickly than the flow of fuel through the main metering circuit. Were it not for the accelerator pump circuit (Fig. 8-13), the fuel mixture would lean out to the point of misfire causing the engine to hesitate or stumble every time the throttle opened. In fact, the classic symptom of a defective accelerator pump circuit is a severe "bog" or stumble during acceleration.

Unlike the power valve which relies on dropping engine vacuum to provide additional fuel, the accelerator pump is strictly mechanical. A spring-loaded pump plunger (Carter and Rochester carburetors) or diaphragm (Holley and Motorcraft) connected to the throttle linkage is depressed when the throttle is opened. This forces fuel past a check ball and out through a nozzle in the venturi.

The amount of extra fuel delivered by the accelerator pump is determined by the stroke of the plunger and the size of the orifice in the discharge nozzle. The accelerator pump is calibrated to furnish just enough extra fuel to keep the mixture from going lean, until fuel flow through the main metering circuit and power valve catch up with the engine's demands.

If the accelerator pump plunger becomes worn or the diaphragm develops a leak, insufficient fuel or no fuel at all will be pumped through the accelerator circuit when the throttle opens. Dirt in the discharge nozzle or an obstruction which prevents fuel from refilling the pump can also reduce fuel

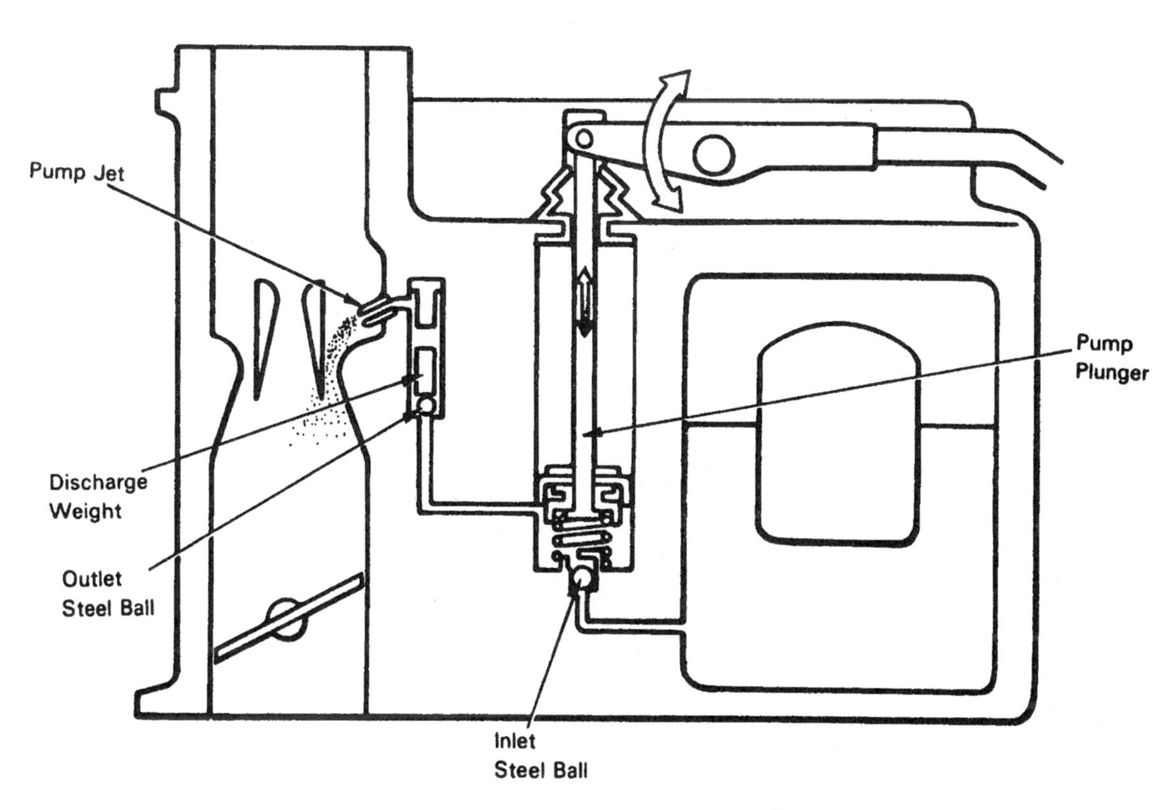

Fig. 8-13 Acceleration circuit (courtesy Toyota).

delivery. This results in the characteristic flat spot or bog during acceleration.

Working the throttle linkage to see if fuel squirts into the carburetor is a good way to visually check the gross operation of the accelerator pump. But it is very difficult to judge by eye alone whether or not the accelerator pump is delivering the right amount of fuel. A weak discharge, dribble, or no discharge at all would be an obvious indication of trouble in the accelerator pump circuit. The problem may be due to an obstruction, or a leaky pump diaphragm or pump plunger. But if the pump appears to squirt normally, you still don't know if there's the right amount of fuel or not. Sometimes a flat spot during acceleration may be due to too much fuel. If the nozzle has been drilled out by some misinformed technician or do-it-yourselfer in an attempt to "cure" a flat spot, the problem is usually made worse, because the pump will flood the engine with too much fuel.

Choke

When the engine is first started, a very rich air/fuel mixture is required because cold fuel vaporizes slowly. The "choke" at the top of the carburetor provides the richer mixture by closing and "choking off" the carburetor's air supply (Fig. 8-14). The choking effect also creates an area of low pressure inside the throat of the carburetor that helps to pull additional fuel through the main metering circuit.

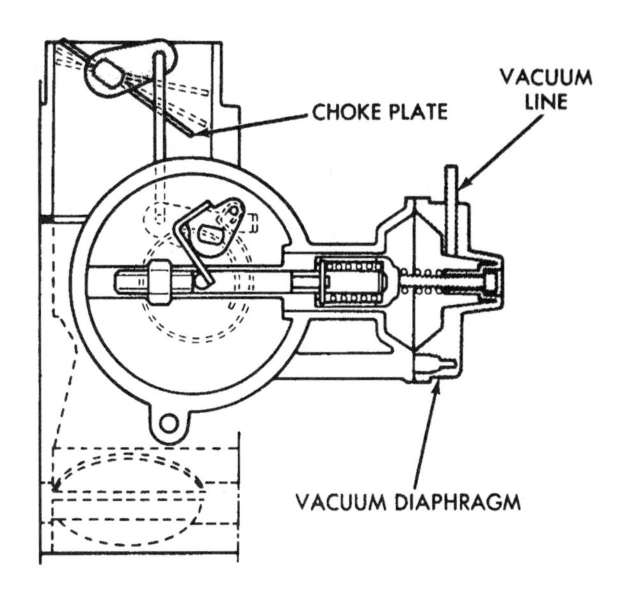

Fig. 8-14 Choke circuit (courtesy Chrysler).

At idle there isn't enough air flowing through the venturi to pull fuel through the venturi discharge nozzle. By temporarily choking off the air supply, however, manifold vacuum rather than venturi vacuum helps to draw the extra fuel through the main metering circuit.

When the choked engine is cranked, fuel is pulled into the engine through the idle port, the transfer port, and the main metering circuit. Combined with the reduced air supply, it creates the extremely rich mixture that's needed to start a cold engine.

As soon as the engine starts, it needs air to keep running and to offset the super-rich fuel mixture. The choke plate shaft is offset slightly to one side so incoming air will tend to push it open. On many older carburetors, a vacuum-operated piston in the choke housing or carburetor casting was used to pull the choke plate partially open so the engine could receive sufficient air to keep running. These pistons were prone to gum up and stick, causing hard starting and stalling problems.

In most later-model carburetors (but not all), the vacuum piston has been replaced with an external "choke pull-off" vacuum diaphragm. The choke pull-off is also attached to the choke linkage where it pulls the choke plate open slightly as soon as the engine starts. On some carburetors, a pair of choke pull-offs are used to give a more progressive opening. If the choke pull-off fails to work (because of a vacuum leak or ruptured diaphragm), the engine may be hard to start, or stall. The amount by which the choke is pulled open can be adjusted by bending the U-shaped pull-off linkage. If the linkage is not properly adjusted, however, it may open the choke too far, admitting too much air causing the engine to stall.

Once the engine starts and begins to warm up, the fuel mixture is gradually leaned out until the choke is fully open. This job is performed by a temperature-sensitive bimetal spring attached to the choke plate. The bimetal spring may be located in a well in the intake manifold where it is heated by exhaust gases flowing thorough the crossover passage. Or it may be located inside a plastic housing on the carburetor itself.

The choke housing may be heated one of three ways: by hot air siphoned up through a pipe in the exhaust manifold that then flows through the housing and enters the carburetor; by an electrically heated element inside the housing (Fig. 8-15); or by engine coolant circulating through a hose attached to the housing. Whatever the source of heat, the bimetal spring reacts by unwinding to open the choke (which may occur in as little as 45 seconds or less on an emissions-controlled engine). When the engine is shut off and allowed to cool, the bimetal spring contracts, winding up tighter to pull the choke shut. But since the choke linkage is also attached to the "fast-idle cam," the choke won't close

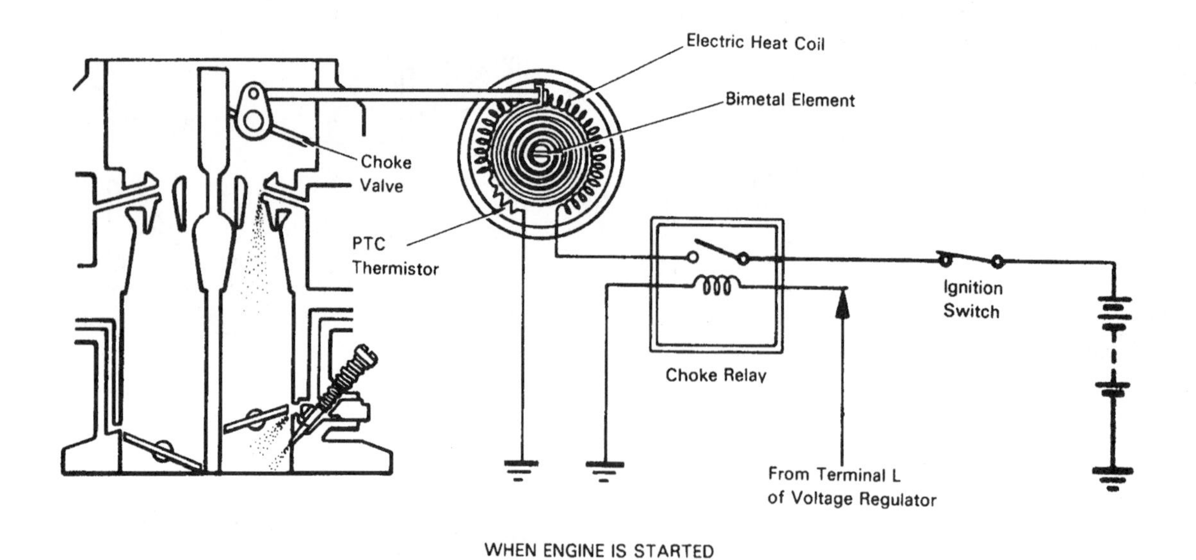

Fig. 8-15 An electrically heated automatic choke (reprinted, by permission, from Scharff, *Complete Fuel Systems and Emissions Control.* Copyright 1989 by Delmar Publishers Inc.).

until the linkage is pumped at least once to reset the cam. On automatic chokes that use an electric heating element, a temperature-sensitive contact switch may be used to break the flow of current to the heating coil when the housing is hot.

Problems with the operation of the choke will result whenever heating is lost at the choke housing. On carburetors that draw heated air through a pipe in the exhaust manifold, rust can corrode the inside of the pipe, blocking air flow. Or the pipe can simply rust off. When this happens, the bimetal spring doesn't heat up quickly enough, causing the choke to open too slowly. The result is a rich fuel condition resulting in a rough idle, excessive fuel consumption, and high emissions. On carburetors that use an electric heating element, a loose wire or poor ground connection can prevent the choke from opening. The choke housing should feel hot to the touch within a minute or so after the engine is started if its heat source is working correctly.

The choke can be adjusted to alter the temperature at which it closes and opens, which also enrichens or leans the starting fuel mixture. The choke is adjusted by loosening the screws that hold the choke housing and then rotating the housing. This changes the relative position of the bimetal spring inside, which puts either more or less tension on the choke. Rotating the housing to increase tension on the choke (which you can detect by holding the choke plate or watching it move) will make it close at a higher temperature and enrichen the mixture. Rotating the choke housing to decrease tension will make it close fully at a lower temperature and produce a somewhat leaner mixture. Notches are provided on some housings for reference. The vehicle manufacturer will often specify how many notches rich or lean the choke is to be adjusted. On many late-model carburetors, however, rivets are used instead of screws to discourage tampering with the choke housing adjustment. But this doesn't prevent adjustments, because the rivets can be drilled out. Once the adjustment is made, new rivets or screws can be installed.

The rate at which the choke opens is critical. If the choke doesn't open quickly enough, especially during warm weather, the mixture becomes too rich and increases carbon monoxide emissions. On the other hand, if the choke comes off too quickly, especially during cold weather, the mixture can lean out causing the engine to stall or to stumble when the throttle is opened quickly.

The choke pull-off plays an important role here by modifying the rate at which the choke opens during warm weather. Some carburetors are equipped with two choke pull-offs. The combination of two choke pull-offs provides a progressive rate of opening that changes according to temperature. The primary pull-off opens the choke a bit when the engine starts to keep it running. The second choke pull-off operates through a thermal vacuum switch that senses the temperature of the heated air entering the carburetor. During cold weather, the bimetal spring in the choke housing will open the choke fully before the second choke pull-off has any effect, because the choke housing warms up faster than the air entering the carburetor. So during cold weather, the second choke pull-off has no effect. But during warm weather, the air entering the carburetor is already warm. This causes the thermal vacuum valve to open, passing vacuum to the second choke pull-off, which pulls the choke open sooner than it would open otherwise. A vacuum-delay valve is often used in the secondary choke pull-off vacuum line to delay full choke opening for 8 to 20 seconds (to give the engine a little more time to warm up).

When the choke is defective or not adjusted properly, it can make an engine hard to start.

FAST-IDLE CAM

The fast-idle cam's purpose is to provide more throttle opening for starting, and to increase idle speed to prevent stalling while the engine is warming up. A series of progressively shorter steps on the cam are used to hold the throttle open and thus vary idle speed as the engine warms. The tallest step provides the fastest idle speed and should be in position when the choke is fully closed. As the engine warms up and the choke gradually opens, the cam rotates allowing the throttle to progressively close as idle speed slows from fast to normal.

A separate "fast-idle screw" is often used on the throttle linkage for this adjustment. When the engine is fully warmed and at normal curb idle, the adjustment screw should be on the lowest step of the cam (if only one idle-speed screw is used), or completely off the cam (if a separate curb idle-speed screw is used).

Choke and idle problems can result if corrosion or binding in the choke linkage occurs. The external linkage can often be freed by cleaning with solvent (carburetor cleaner) and/or lubricating with a greasless lubricant. Oil should never be used because it will attract dirt which will only further aggravate a binding problem.

IDLE SPEED

The idle speed on a carbureted engine is set by an adjustment screw on the throttle linkage (Fig. 8-16). The degree of throttle opening will determine the idle speed by regulating how much air enters the engine. The idle speed can be affected by the idle mixture or vacuum leaks. Changing the idle mixture can increase or decrease the idle speed depending on the relative richness or leanness of the idle mixture prior to adjustment. A small vacuum leak will often make an engine idle faster than usual. A large leak will lean the mixture out to the point of misfire, causing it to stall.

Various devices may be used in conjunction with the throttle linkage to alter idle speed:

- A "dashpot" to cushion throttle return, preventing the throttle from closing too rapidly.
- An "idle stop solenoid" to completely close the throttle when the ignition is turned off to prevent run-on. The idle-speed adjustment may be located on the idle stop solenoid plunger. If the solenoid fails, the engine may start but then stall after the engine warms up and the fast-idle screw comes off the choke cam. If the solenoid has been disconnected or improperly adjusted, it may fail to close the throttle when the ignition is turned off, allowing the engine to diesel and run-on.
- An "idle kicker solenoid" or "throttle solenoid positioner" (TSP) to increase idle speed under certain conditions (Fig. 8-17). When the air conditioner compressor is on, for example, the added burden on a small-displacement engine could cause it to stall. By increasing the idle speed, stalling is prevented.

  On some vehicles, an vacuum-actuated idle kicker or "decelerator valve" is used as an emissions-control device to hold the throttle open slightly during deceleration. This makes the engine vacuum draw more air into the intake manifold to prevent the fuel mixture from going too rich.
- An "idle-speed control" (ISC) motor to regulate idle speed. On later-model vehicles with com-

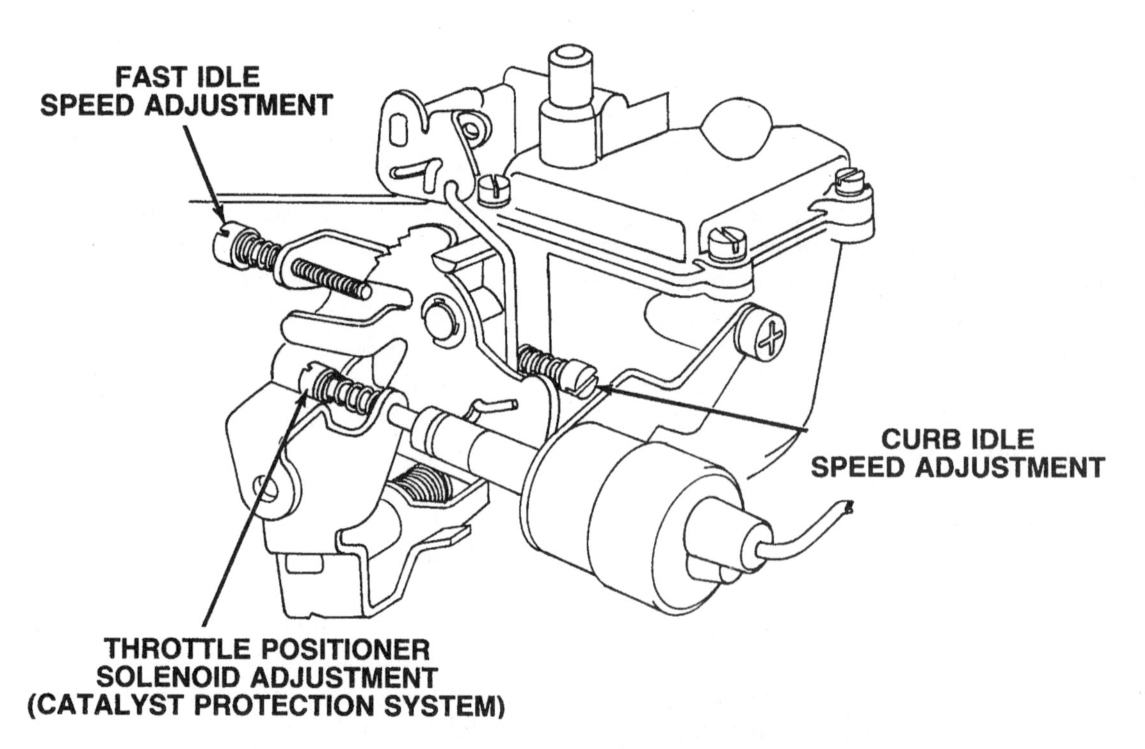

Fig. 8-16 Fast idle and curb idle adjustments (courtesy Chrysler).

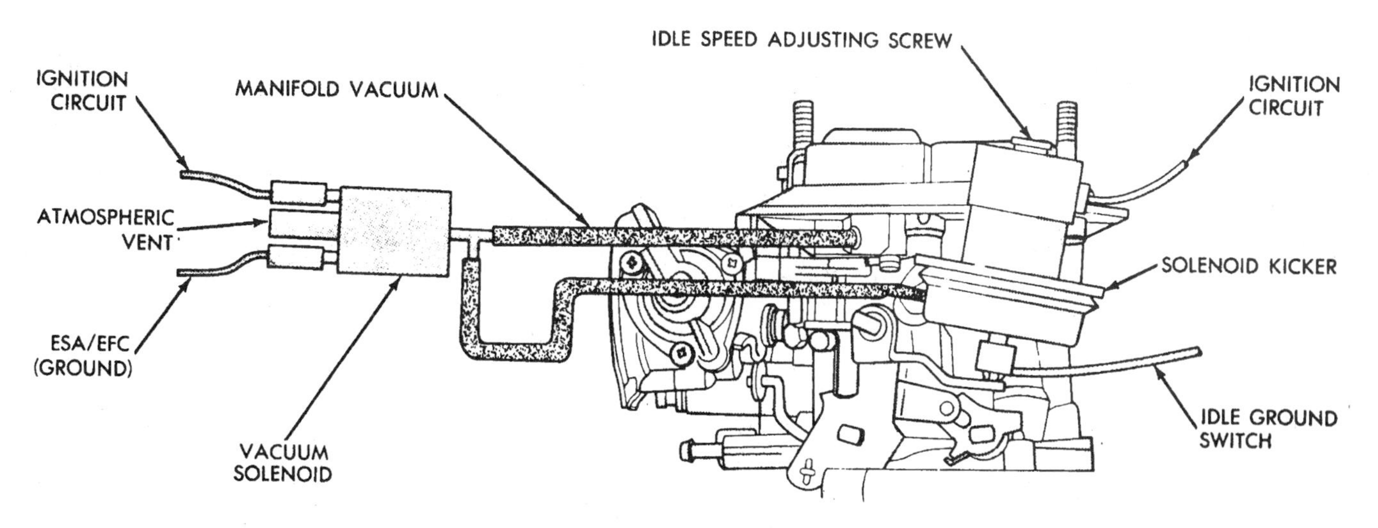

Fig. 8-17 Solenoid kicker system (courtesy Chrysler).

puterized engine controls, the computer monitors engine rpms by counting ignition pulses through the distributor. If the idle rpm is not within the programmed range, it runs a little plunger in or out to change the idle speed. Some computerized engine-control systems also use an "idle switch" and/or a "wide-open throttle switch" to tell the computer when the throttle is at idle or wide open. A "throttle-position sensor" (TPS) may also be used to signal the computer about the rate of throttle opening.

HOT-IDLE COMPENSATORS

When an engine is idling in traffic, particularly during hot weather, underhood temperatures can soar. Since hot air is less dense than cold air, extra air is needed to keep the air/fuel mixture from becoming too rich. A rich idle mixture causes a rough, lopping idle which not only can lead to stalling but also increases emissions. Some carburetors, therefore, use a "hot-idle compensator" (Fig. 8-18), a little valve that admits extra air into the intake manifold when

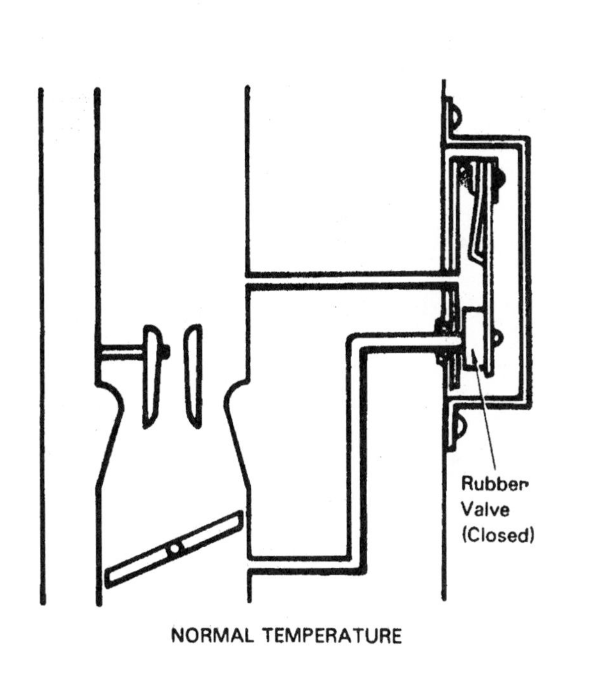

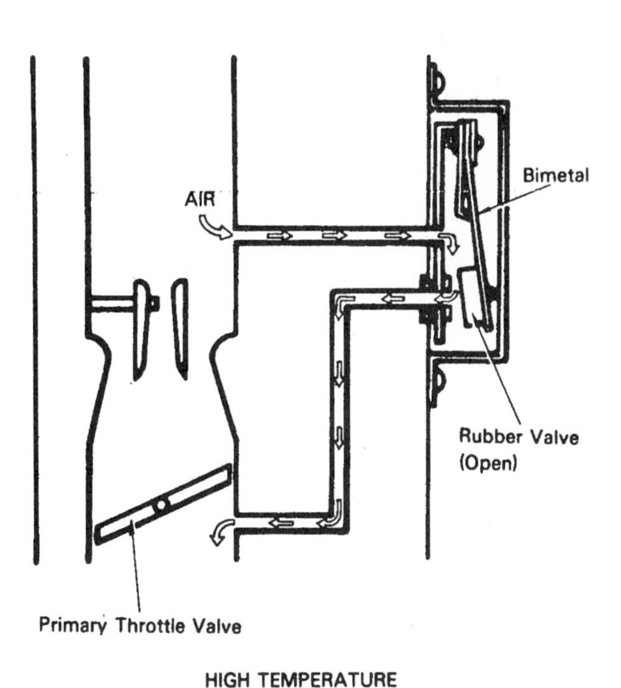

Fig. 8-18 Hot idle compensator. (courtesy Toyota).

the engine is hot. Some compensators are mounted inside the carburetor air horn, some are on the outside of the carburetor, and some are located in a vacuum line somewhere in the engine compartment.

The compensator on an older Carter AFB four-barrel carburetor will open and admit air to the intake manifold any time the air temperature exceeds 120 degrees F. The compensator could open at any throttle position, but it usually only does so at idle because at speeds above idle there is usually enough airflow past the compensator to cool it off.

Hot-idle compensators are not adjustable. If one sticks open, causing a lean mixture (stalling, hesitation, etc.), it must be replaced. Care must be used when handling a compensator so that the bimetal spring is not bent or distorted. If bent, it could change the calibration of the compensator.

CHAPTER 9

Fuel Injection

There are essentially two basic types of fuel injection: throttle body and multiport. Throttle body injection appeared on many domestic vehicles in the mid-1980s as the successor to electronic feedback carburetion. Electronic carburetors that could provide variable fuel mixture control became necessary in 1981 to meet federal emission standards, but were very complex and expensive (remanufactured feedback carburetors can cost hundreds of dollars!). So throttle body injection became the next step in reducing emissions.

THROTTLE BODY INJECTION (TBI)

Throttle body injection is much like a carburetor except that there's no fuel bowl, float, needle valve, venturi, fuel jets, accelerator pump, or choke. That's because throttle body injection does not depend on engine vacuum or venturi vacuum for fuel metering. A TBI fuel-delivery system consists of a throttle body with one or two injectors and a pressure regulator (Fig. 9-1). Fuel pressure is provided by an electric pump. It's a relatively simple setup and causes few problems.

The fuel mixture is managed electronically by the engine control computer. The computer uses inputs from its various sensors to determine how much fuel the engine needs under various operating conditions. The fuel injectors do not spray continuously, but rather in short pulses (which produce a buzzing sound). By varying the "on time" or duration of the one or two injectors, the fuel mixture can be varied to deliver more or less fuel as needed.

One of the main differences between throttle body injection and a carburetor is that fuel is sprayed under pressure into the engine, rather than siphoned through jets by engine vacuum. Consequently, cold starting is much better with TBI because there's no troublesome choke. Fuel economy

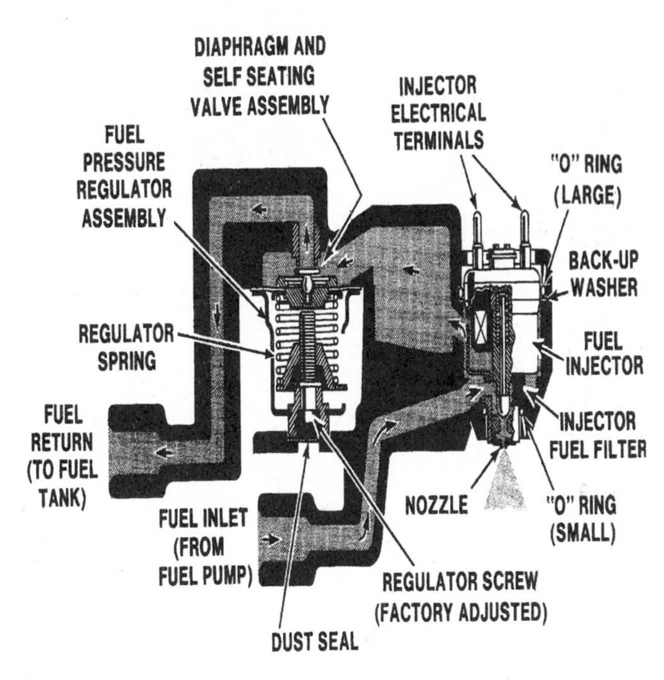

Fig. 9-1 TBI assembly (courtesy Rochester Products).

and emissions are also lower because the fuel mixture can be controlled much more precisely.

MULTIPORT INJECTION

The next step up from TBI is multiport injection. Engines with multiport injection have a separate fuel injector for each cylinder, mounted in the intake manifold or head just above the intake port. Thus, a 4-cylinder engine would have four injectors, a V6 six injectors, and a V8 eight injectors.

Multiport-injection systems are more expensive because of the added number of injectors. But having a separate injector for each cylinder makes a big difference in performance (Fig. 9-2). The same engine with multiport injection will typically produce 10 to 40 more horsepower than one with TBI because of better cylinder-to-cylinder fuel distribution. Injecting fuel directly into the intake ports also eliminates the need to preheat the intake manifold, since only air flows through the manifold. This, in turn, provides more freedom for tuning the intake plumbing to produce maximum torque. It also eliminates the need to preheat the incoming air by forcing it to pass through a stove around the exhaust manifold.

There are other differences between multiport-injection systems. One is the way in which the injectors are pulsed (Fig. 9-3). On some systems, all the injectors are wired together and pulse simultaneously (once every revolution of the crankshaft). On others, the injectors are wired separately and are

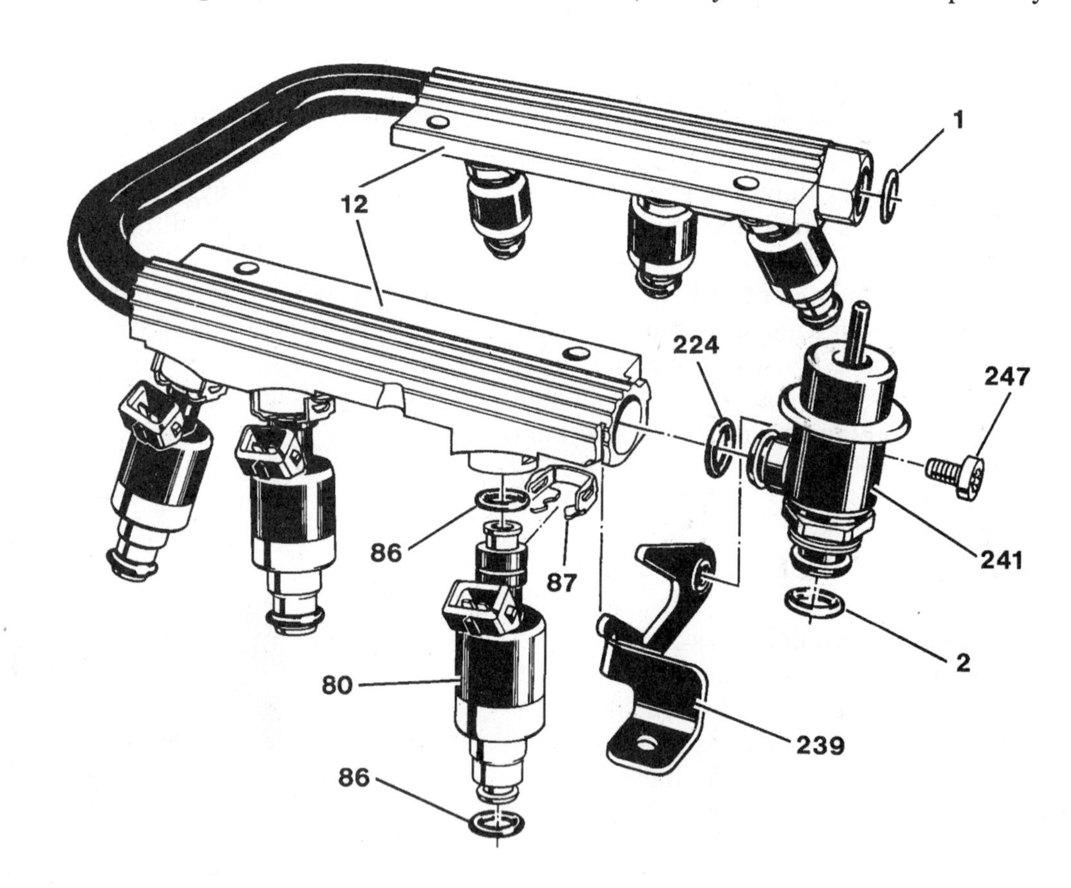

Part Names

1	O-ring - Fuel Inlet Line	87	Clip - Injector Retainer
2	O-ring - Fuel Return Line	224	Seal - O-ring, Fuel Outlet Tube
12	Front and Rear Fuel Rail Assembly	239	Retainer and Spacer Assembly
80	MPFI Injector Assembly	241	Pressure Regulator Assembly
86	Seal - O-ring - Injector	247	Screw - Regulator Attaching

Fig. 9-2 MPI fuel rail assembly (courtesy AC Rochester).

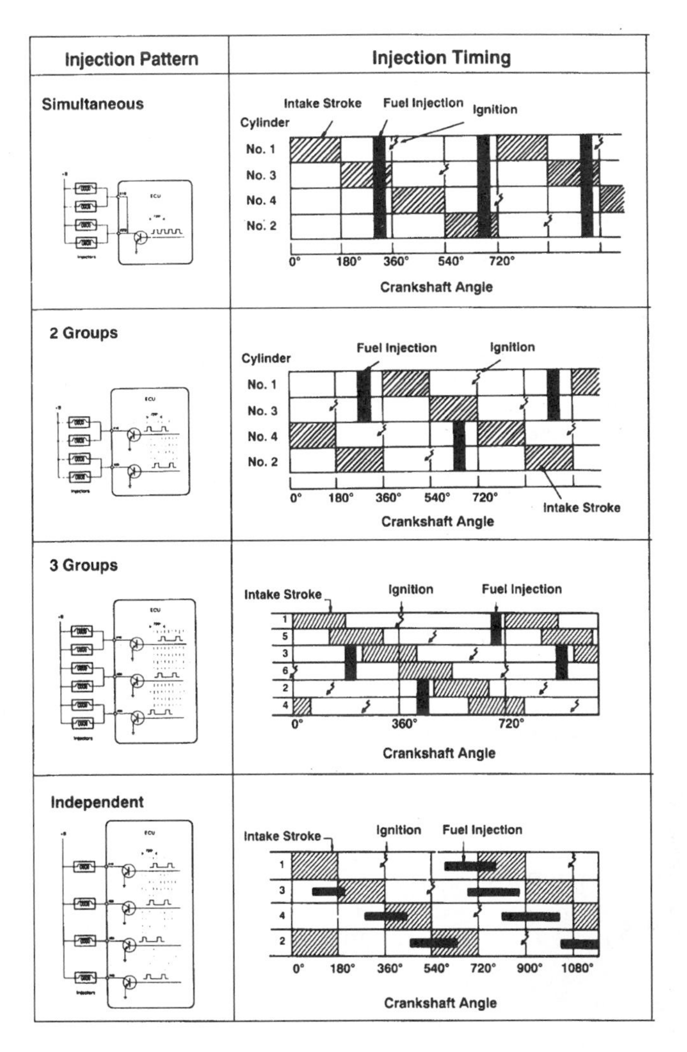

Fig. 9-3 Pulsing of injectors (courtesy Toyota).

pulsed sequentially (one after the other in their respective firing order). The latter approach is more complicated and requires more expensive electronic controls. But it provides better performance and throttle response by allowing more rapid changes in the fuel mixture.

MEASURING AIRFLOW

Another difference between multiport fuel-injection systems is the way in which they calculate airflow for fuel metering purposes. The engine computer has to estimate how much air is entering the engine so it can provide the appropriate amount of fuel. Too much fuel will make the mixture go rich, waste gas, and increase emissions. Not enough fuel will make the mixture go lean, increasing the danger of detonation, and possibly causing the engine to misfire. So the mixture must be carefully balanced according to airflow.

Some multiport-injection systems, such as the Bosch L and LH-Jetronic systems used on many import and domestic cars, use a "flap-style" airflow sensor to monitor airflow (Fig. 9-4). The airflow sensor is mounted between the air filter and throttle body (see Chapter 16 for additional information on airflow sensors).

On other multiport systems (General Motors primarily), the flap-style sensor is replaced by a solid-state electronic "mass-airflow sensor" (MAF). This type of airflow sensor contains a hot wire (Fig. 9-5) or heated filament that is cooled by the incoming air. The amount of current needed to keep the wire or filament at a constant temperature tells the computer how much air is entering the engine. Mass-airflow sensors are very sensitive to dirt contamination, and failures are common. Replacement costs are high, up to $750!

Another type of airflow sensor is the mechanical one that's used on Bosch CIS, K, and KE-Jetronic multipoint fuel-injection systems. Found on many import car applications, these Bosch systems use a mechanical "fuel distributor" and mechanical injectors rather than electronic injectors to deliver fuel. The later-model applications have electronic inputs for emission purposes, but are still essentially mechanical fuel-injection systems. Remanufactured fuel distributors and mechanical airflow sensors are available from various sources.

There are also fuel-injection systems such as those on many Chrysler engines, and 1990 and newer Chevrolets with tuned-port injection (TPI) that do not have an airflow sensor. These are called "speed-density" systems. The amount of fuel delivered by the injectors depends on engine rpm (speed), throttle position, engine load, and inlet air temperature. The engine computer estimates airflow at various speeds based on inputs from the throttle-position sensor (TPS), manifold absolute pressure (MAP) sensor (Fig. 9-6), and inlet air temperature (IAT). The speed-density approach isn't as sophisticated as actually measuring airflow with a mass airflow- or vane-style meter, but it gets the job done just the same. A speed-density system is also somewhat more reliable and less likely to experi-

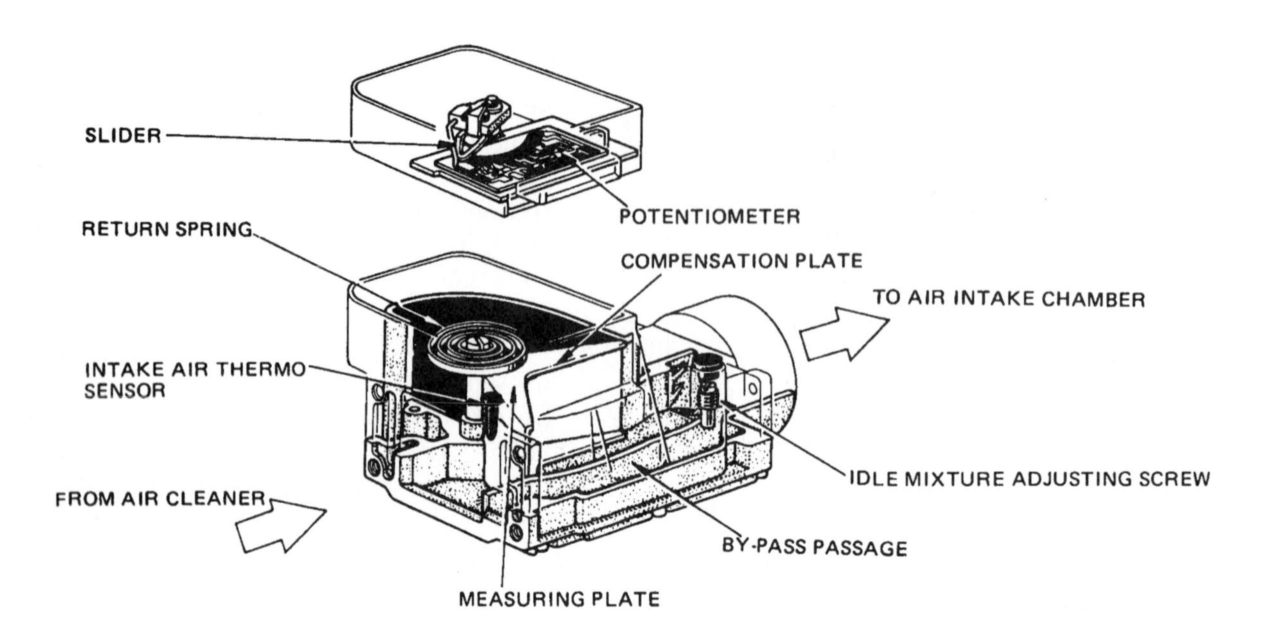

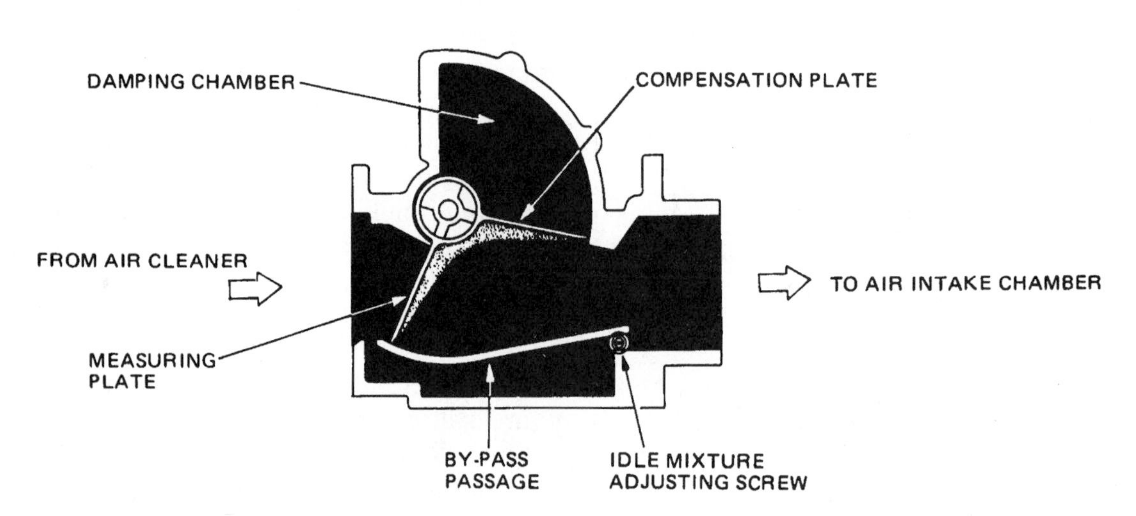

Fig. 9-4 Vane air flow meter (courtesy Toyota).

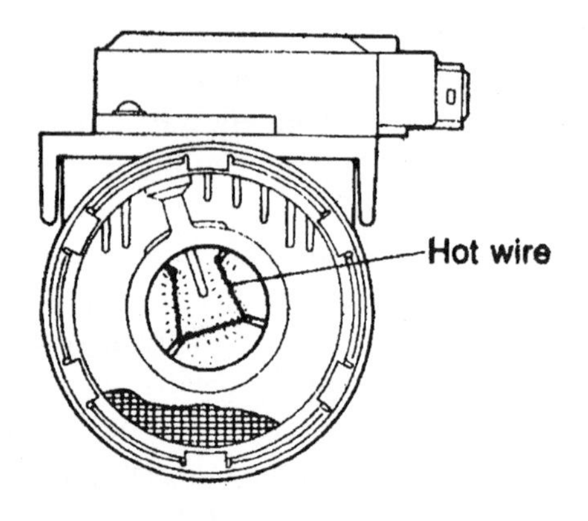

Fig. 9-5 Hot wire MAF (courtesy Mazda).

ence driveability problems because there's no troublesome airflow sensor in the system.

MAJOR COMPONENTS

Most of the major components in fuel-injection systems are considered emission-control devices and are covered by the federal (or California) emissions warranty. Covered parts include the injectors, throttle body, fuel-pressure regulator, idle air-speed-control motor, airflow sensor, other engine sensors,

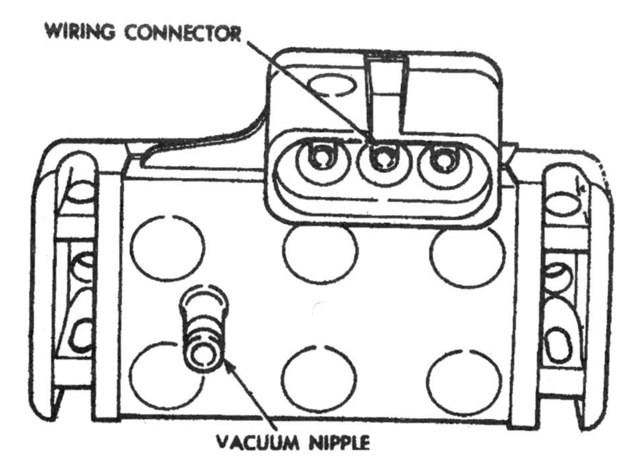

Fig. 9-6 MAP sensor (courtesy Chrysler).

electronic control module, fuel rail, hoses, and even clamps. The electric fuel pump is not covered under the emissions warranty, but may be included under an extended service contract or basic warranty.

Injectors

The injector is nothing more than fuel nozzle. With mechanical injectors, a spring-loaded valve allows fuel to squirt out of the nozzle when line pressure overcomes spring tension that holds the valve shut. With electronic injectors, a spring-loaded solenoid pulls open a pintle valve or ball-type valve when the injector is energized by the computer. This allows the pressurized fuel in the fuel rail to flow through the injector and squirt out the nozzle.

Injectors come in a variety of styles. Early Bosch injectors have a pintle valve and feed from the top. In 1989, General Motors introduced its new "Multec"-style injectors which have a ball-valve design. A third type are "Lucas" injectors that have a disc-valve design.

Most injectors are a "top-feed" design (Fig. 9-7), meaning fuel enters the top of the injector and flows down through the injector to the nozzle. One disadvantage of this type of injector is that fuel vapors created by engine heat tend to rise to the top of the injector and create vapor pockets in the fuel supply rail. This, in turn, can sometimes cause hot-starting problems after a brief period of heat soak. Some newer engines are now equipped with a new style of "bottom-feed" injector. With this design, fuel enters near the injector nozzle eliminating the problem of vapor backup. Bottom-feed injectors (Fig. 9-8) are

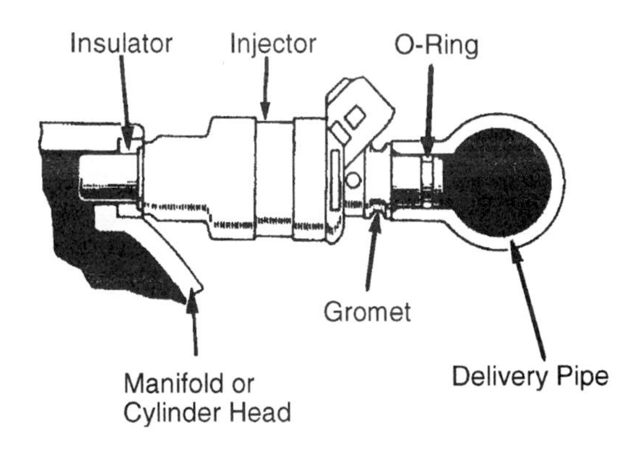

Fig. 9-7 Top feed injector (courtesy Toyota).

also designed to operate at much lower pressures (as low as 10 psi) which requires a less expensive fuel pump.

Two things can go wrong with injectors. The solenoid on electronic injectors can fail (which doesn't happen very often), resulting in a "dead" injector. More often than not, the real problem is a loose or corroded wiring connector or a bad injector-driver circuit in the engine computer. The other problem, which is much more likely to occur, is injector clogging. "Dirty" injectors, which aren't really clogged with dirt but rather a buildup of fuel-varnish deposits, can cause hesitation, and emissions and performance problems. Even minor deposits in the injector nozzle can cause "streamers" (Fig. 9-9) in the normal cone-shaped spray pattern that inhibit proper fuel atomization.

Injectors are expensive to replace. New domestic injectors sell for $60 to $100 each, with new import injectors fetching $125 to $175 each. Typical prices for rebuilt injectors range from as low as $20 to $50 or more.

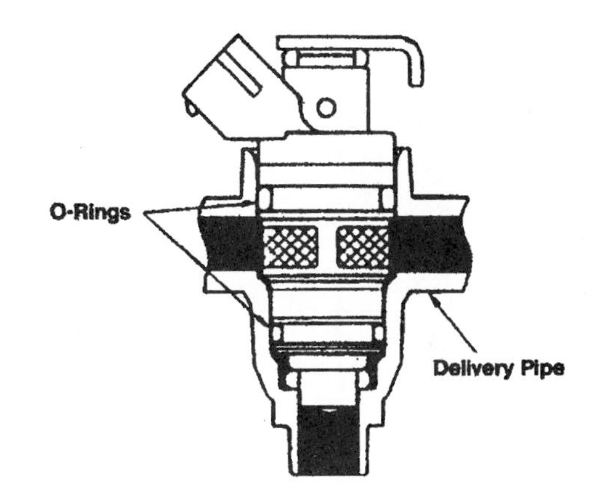

Fig. 9-8 Bottom feed injector (courtesy Toyota).

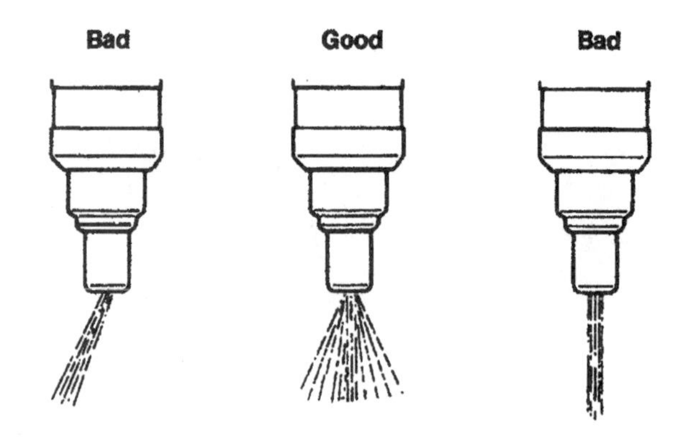

Fig. 9-9 Injector spray patterns (courtesy Toyota).

Injector clogging can be prevented by using quality gasoline that contains sufficient additives and detergents to keep the injectors clean. But fuel additives can only do so much. If the injectors are clogged with deposits, they have to be cleaned with pressurized solvent, or removed for off-car cleaning. On- car cleaning will often restore injectors to like-new performance and make a noticeable difference in idle smoothness, emissions, and fuel economy. If on-car cleaning doesn't do the trick, off-car cleaning using special fuel-injection cleaning equipment may be a more economical alternative to replacement. Injector cleaning is covered in Chapter 13.

The flow characteristics, spray pattern, and internal calibration of a particular set of injectors depends on the application. So it's important to make sure replacement injectors are the correct ones for the engine. The wrong injectors can cause major problems. Some aftermarket replacement injectors use a different type of nozzle design that is more resistant to clogging than the original equipment injectors. These might be a better alternative (where available) than remanufactured original equipment injectors.

Throttle Body

Throttle bodies are fairly reliable. Even so, they do wear around the throttle shafts. A worn throttle shaft will allow unmetered air to enter the engine, causing a lean condition. Defective O-rings can also create vacuum leaks. Leaky O-rings can be replaced, but worn throttle shafts call for replacement of the unit with either a new throttle body or a rebuilt throttle body that has had the shaft bores sleeved.

Idle speed on a fuel-injected engine is controlled by regulating the amount of air that's allowed to bypass the throttle plate. On mechanical fuel-injection systems, an "idle-bypass circuit" with an adjustable screw controls idle speed. On most late-model applications with electronic fuel injection, idle speed is controlled by an "idle-air control (IAC) valve" (Fig. 9-10). A computer controlled electric stepper motor opens and closes a valve to regulate idle speed. Failure of the IAC motor may cause stalling. Air leaks downstream of the throttle body can also play havoc with the system's ability to regulate idle speed accurately.

Fuel Pump

All vehicles with electronic fuel injection (EFI) have electric fuel pumps. This includes all new vehicles

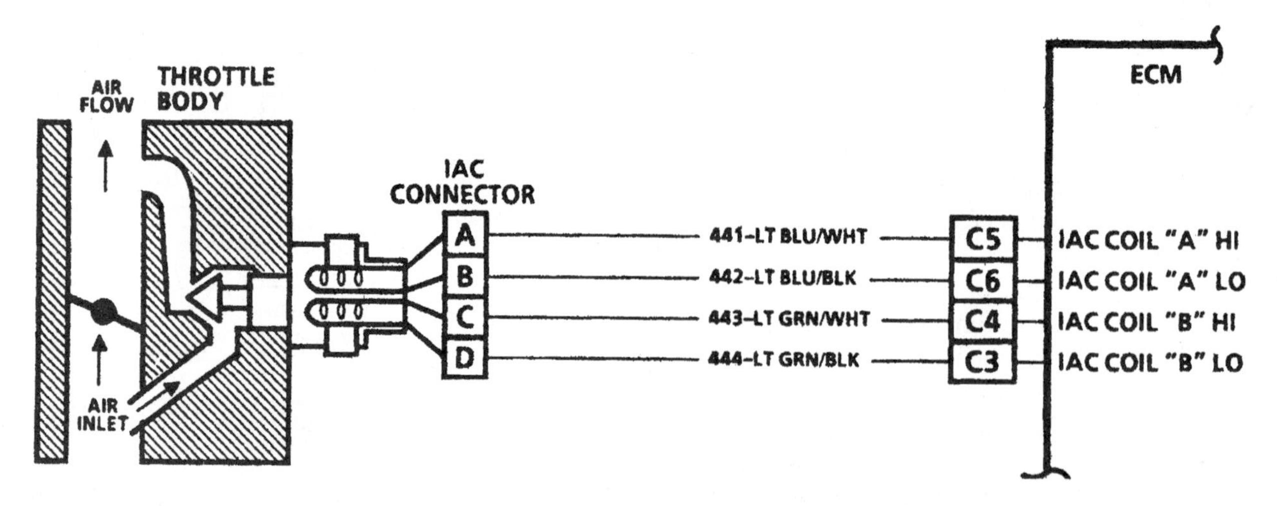

Fig. 9-10 Idle air control motor (courtesy General Motors).

today as well as most of the cars and light trucks that have been built since the mid-1980s.

Electric fuel pumps are required with fuel injection because the fuel must be supplied to the injectors at much higher pressure than required for carbureted vehicles. High pressure is also needed to create a cone-shaped spray pattern that promotes proper fuel atomization. What's more, the pressure must be maintained within a preset range to provide the correct air/fuel ratio under various operating conditions. This means anything that affects fuel delivery to the engine or system pressure can have an adverse affect on engine performance and emissions.

Carburetors, on the other hand, rely on intake vacuum and a venturi effect to draw fuel into the engine from a nonpressurized fuel bowl. So a carburetor only needs enough pressure in the fuel line to keep the bowl full. That's why carbureted engines have either mechanical fuel pumps or low-pressure electric pumps.

Any one of a variety of different pump designs may be used with electronic fuel injection: a single- or double-vane, roller-vane (Fig. 9-11), turbine-, or gerotor-style pump. The pump that's used in a given application depends on the vehicle manufacturer's requirements and preferences.

The pump is usually mounted inside the fuel tank in a hanger or module assembly. Often attached to this are the float arm and sending unit for the fuel gauge. Mounting the pump inside the tank makes access much more difficult, but also provides additional protection and pump cooling. Externally mounted pumps are usually located under the vehicle near the tank. Accessibility is easier, but failure rates can be higher because of increased vulnerability to road hazards, moisture, and corrosion (a problem which can afflict the wiring on in-tank pumps, too).

On some applications, two pumps are used: a "transfer" or "feeder" pump inside the tank and a main pump or "pusher" pump located outside the tank.

Regardless of the type of pump that's used or it's location, the pump's function is to push fuel from the fuel tank to the engine and create pressure for the injectors. In most applications, the pump runs continuously as long as the engine is cranking or running. Most pumps also contain a one-way check valve to maintain residual pressure when the engine is shut off. This makes starting easier and helps keep the fuel in the injectors and fuel rail from boiling when a hot engine is shut off.

The pump receives voltage through a relay

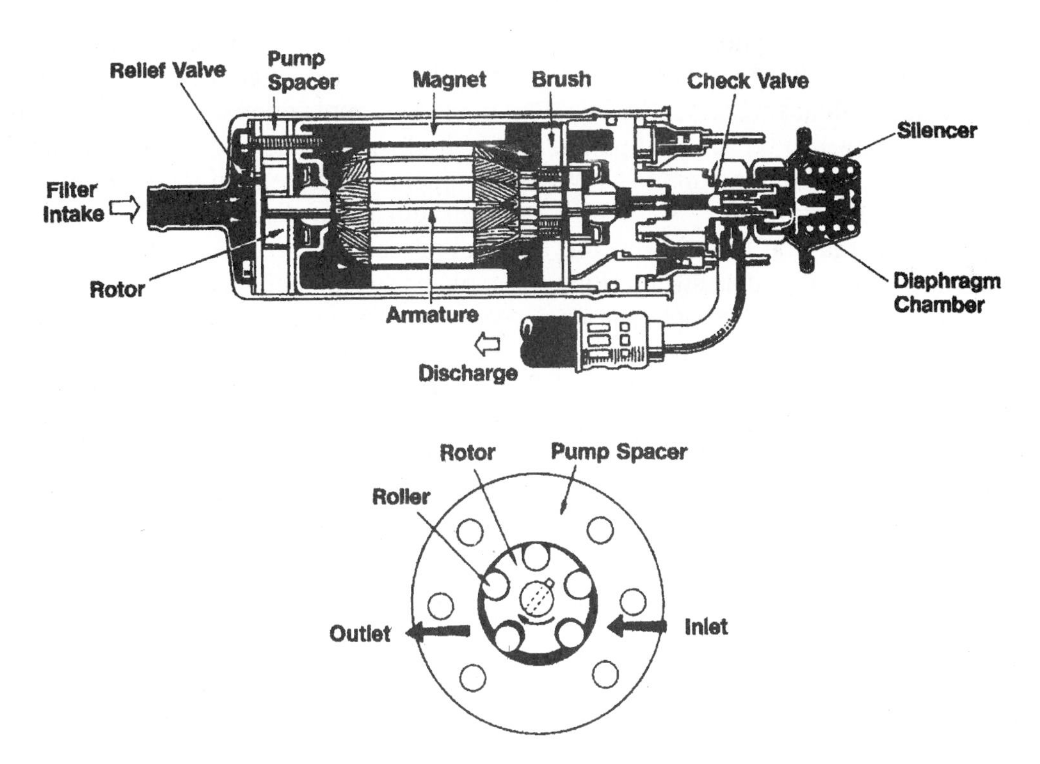

Fig. 9-11 Electric fuel pump (courtesy Toyota).

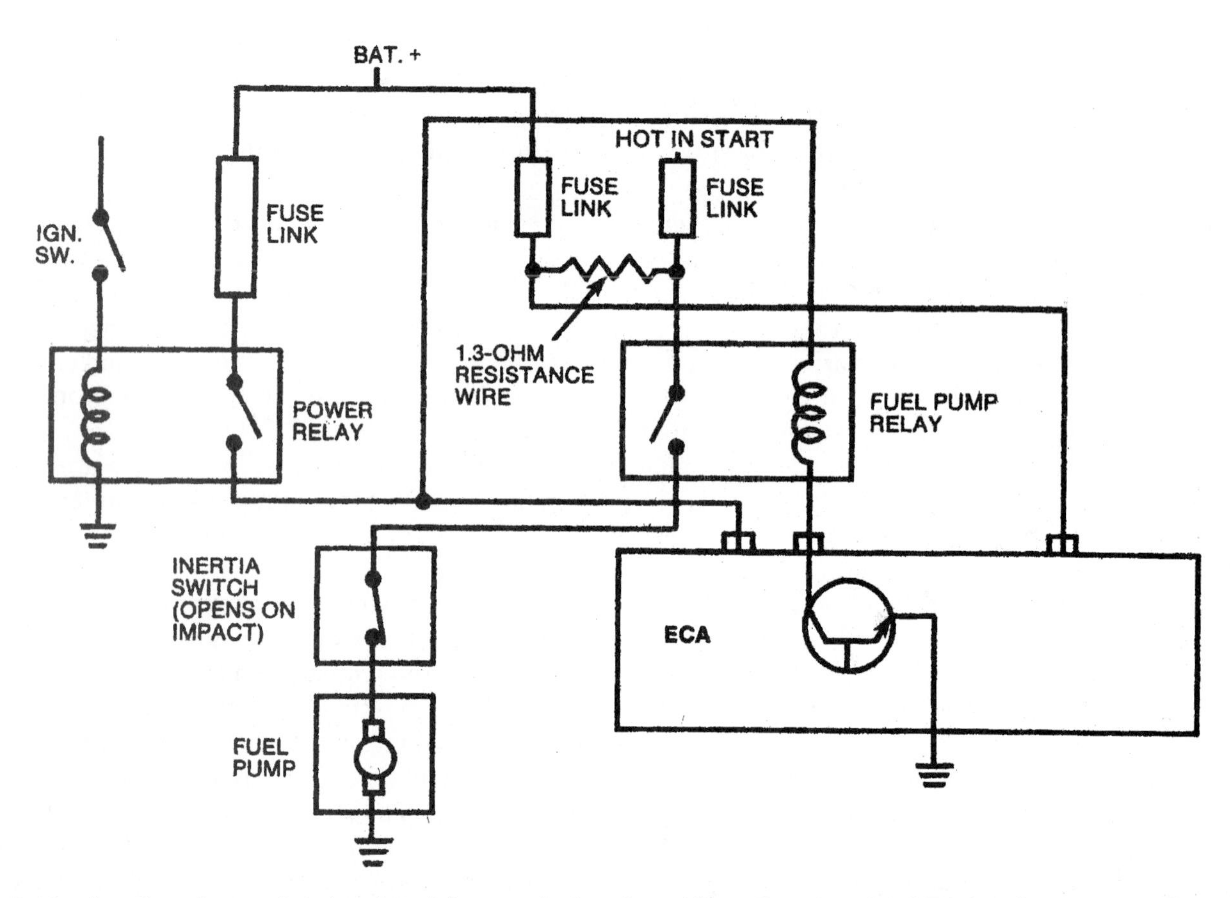

Fig. 9-12 Inertia safety switch (reprinted, by permission, from King, *Computerized Engine Control, 4E.* Copyright 1993 by Delmar Publishers Inc.).

when the ignition is turned on, or when the engine-control module receives a cranking signal from the distributor pickup or crankshaft sensor. On GM vehicles, the relay circuit may also be wired in parallel to the oil-pressure switch or sending unit to allow starting, should the relay fail.

Another element that may be included in the pump wiring circuit is an "inertia" safety switch (Fig. 9-12). Located in the trunk, under the back seat or behind a rear kick panel, the switch turns the voltage to the pump relay off in a severe impact. Sometimes jolts and bumps encountered in normal driving may trip the safety switch, causing the fuel pump to quit and the engine to stall. The pump won't run until the inertia switch is reset by pushing its reset button.

Fuel Pressure And The Regulator

Though the fuel pump creates pressure for the injectors, the "fuel-pressure regulator" (Fig. 9-13) keeps system pressure within a preset range. The regulator is usually mounted on the fuel rail in multiport EFI applications, or the throttle body assembly in throttle body injection (TBI) applications.

Inside the regulator is a spring-loaded diaphragm and relief valve. The regulator controls fuel pressure by releasing excess fuel back to the tank via a return line when a certain pressure is exceeded. Depending on the application, this may vary from 9 psi up to 80 psi or higher.

On many applications, the regulator also varies system pressure in response to changes in engine load and manifold vacuum. At idle and cruise when manifold vacuum is high, less fuel pressure is used to force fuel through the injectors. (This provides less fuel flow during the time the injector is open.) At wide-open throttle or when the engine is under load and vacuum is low, more fuel pressure is needed. More fuel pressure may also be required to overcome increases in manifold pressure in turbocharged or supercharged engines during periods of high boost. This is accomplished by connecting

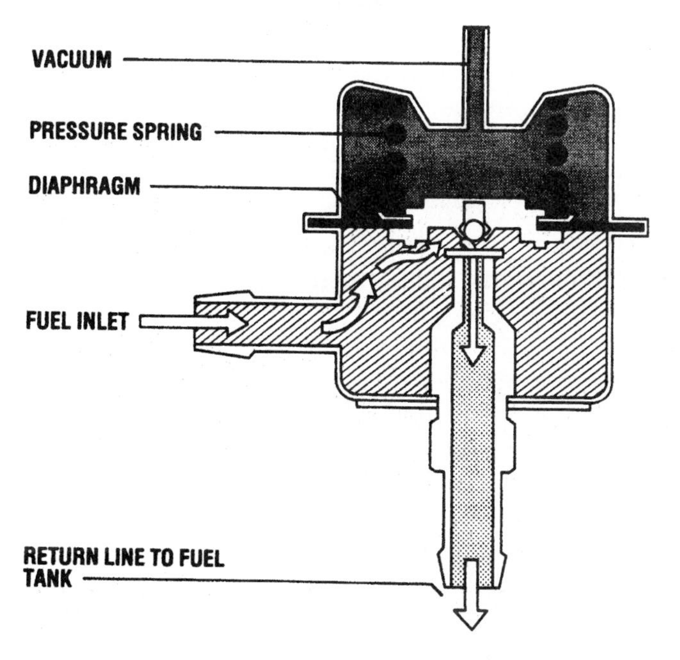

Fig. 9-13 Fuel pressure regulator (courtesy General Motors).

a hose from a manifold vacuum source to the regulator's diaphragm housing. Vacuum pulls the diaphragm and modifies the point at which the regulator opens and closes. This allows the regulator to "fine-tune" fuel pressure to changing loads and the engine's requirements.

When the regulator diverts fuel to keep pressure within range, the excess fuel is routed back to the fuel tank through a return line. (Restrictions or kinks in the return line can cause excessive pressure and a rich-fuel condition.)

On some newer Chrysler vehicles, a "returnless" constant-pressure system is used. There is no return line from the engine to the fuel tank, because the regulator is mounted on top of the fuel pump module assembly. Even so, the basic operating principle is still the same. The pump creates pressure, and the regulator controls it by returning excess fuel to the tank.

The important thing to remember about fuel pressure is that it affects the air/fuel mixture in a fuel-injected engine. Too much pressure causes the engine to run rich (excessive emissions, poor fuel economy, surging, etc.). Too little pressure causes it to run lean (hard cold-starting, stumbling, hesitation, lack of high-speed power, lack of acceleration, lean misfire, etc.). It's also important to remember that other problems such as air and vacuum leaks, inaccurate, misadjusted or defective engine sensors, a leaky EGR valve, or dirty injectors can also upset the engine's air/fuel ratio and cause driveability symptoms. These may be misdiagnosed as a weak fuel pump or faulty fuel-pressure regulator. That's why everything that affects the air/fuel ratio should be inspected when a fuel-delivery problem is suspected.

WHY PUMPS FAIL

Electric fuel pumps can fail for any number of reasons, the most common of which are:

- *Loss of current or low voltage*—The pump can't run without electricity, so anything that prevents current or voltage from reaching the pump will make it stop. Corroded, loose, or broken wiring is one major cause of electric fuel pump "failure." The pump may still be good, but it won't run unless the wiring problem is identified and repaired. A bad relay, inertia switch, oil-pressure sending unit, or bad ground or engine-control module are all problems that can keep a good pump from working.
- *Dirt*—Dirt, sediment, or other debris in the tank can clog the pickup strainer, accelerate pump wear, damage the pump, and/or cause the pump's check valve to stick open (An open check valve can cause a hard-starting condition due to loss of pressure when the engine is shut off.) The use of alcohol-enhanced gasoline or alcohol fuel additives may be a contributing factor because these are strong solvents that can loosen deposits in the tank.

  When dirt has caused a pump failure, or if there appears to be a lot of dirt or sediment in the tank, the tank should be thoroughly cleaned to prevent a repeat failure.
- *Rust*—Corrosion inside the tank produces rust which can flake off and plug up the pickup strainer and have the same damaging effects on the pump as dirt. Rust is caused by condensation during cool, humid weather when the fuel tank is low. Keeping the tank full will minimize the formation of condensation. If the tank is badly rusted or leaking, it should be replaced.

- *Wear*—Most pumps are capable of going 100,000 miles or more, but depend on lubrication and cooling provided by the fuel itself. Frequent driving with a low fuel level, however, may occasionally starve the pump for lubrication and cooling. This can lead to accelerated wear or even pump damage. If a vehicle experiences a momentary hesitation when cornering, for example, it may be because the fuel is sloshing away from the pump and allowing it to suck air.

  Wear can also be caused by running at excessive pressure. On some vehicles, a faulty regulator, check valve, or crimped line can cause blockages that force the pump to run at a higher pressure. So too can a clogged fuel filter. This accelerates pump wear and can cause a replacement pump to fail unless the faulty return valve is diagnosed and replaced.

  Noise may be an indication of excessive pump wear—but not always. Some pumps are inherently noisier than others, often because of the way in which they're mounted inside the tank. Noise can also be caused by a loose or missing rubber noise insulators around the pump or physical contact with the bottom of the tank or tank baffles. So always determine what's causing noise before condemning the pump.

PUMP REPLACEMENT

Because fuel pressure has such an influence on performance, driveability, and emissions in an EFI-equipped engine, a replacement fuel pump (and regulator, too!) must be the right one for the application. A fuel system that requires a working pressure of 30 to 39 psi, for example, won't run right with a pump or regulator that's designed for a 9 to 13 psi application. Always follow the replacement pump listings in the pump supplier's catalog, and refer to a shop manual for specific pressure ratings.

To protect in-tank replacement pumps against possible contamination, make sure the fuel tank is clean and free from rust, dirt, scale, and debris. Then either filter the old fuel before it is returned to the tank, or refill it with fresh gasoline.

Most in-tank pumps require dropping the fuel tank to replace the pump. (*Before removing the fuel tank, disconnect the battery!*) On some vehicles, though, a small access panel is provided in the floor under the back seat or in the trunk for this purpose.

While the tank and pump are out of the vehicle, inspect all fuel lines for age cracks, chaffing, weakening, leaks, or other damage. Replace any hoses or lines found to be defective. Use replacement hose that's approved for high-pressure fuel-injection applications only! Using low-pressure fuel hose for a carbureted engine or hose that is not approved to carry fuel is very dangerous.

Follow the replacement instructions carefully that come with the replacement pump. On some applications that have a pulsator as part of the pump assembly, the pulsator may not be needed with the replacement pump. Note the position of the strainer when removing the original pump so that it may be reinstalled correctly. Always replace the strainer.

Pay attention to the positioning and placement of pump grommets, noise isolators, and brackets. The pump may be noisy if these components are not reinstalled correctly.

Place a new tank seal on the tank opening and carefully place the pump assembly back into the tank. Make sure it is positioned correctly before locking it in place.

Use care when reattaching the fuel filler pipe so dirt doesn't get into the tank. Then once the lines have been reconnected and the tank reinstalled, install a new fuel filter and refill the tank with clean gas. Reconnect the battery, then start the engine and check for leaks and proper system operation.

MECHANICAL INJECTION SYSTEMS

To better understand the basics of fuel injection, we'll take a closer look at a Bosch CIS mechanical injection system that's used on many import cars. We'll take a closer look at some electronic fuel-injection systems (Bosch L-Jetronic, Chevrolet TPI, etc.) in Chapter 17.

Though often referred to as a "mechanical" fuel-injection system, Bosch prefers to call CIS a "hydraulic" fuel-injection system. Early Porsche 911s had a true mechanical injection system that used throttle linkage and governor speed to regulate delivery. But the Bosch CIS system is different.

"CIS" stands for continuous injection system (Fig. 9-14). It's also called "K-Jetronic." The "K"

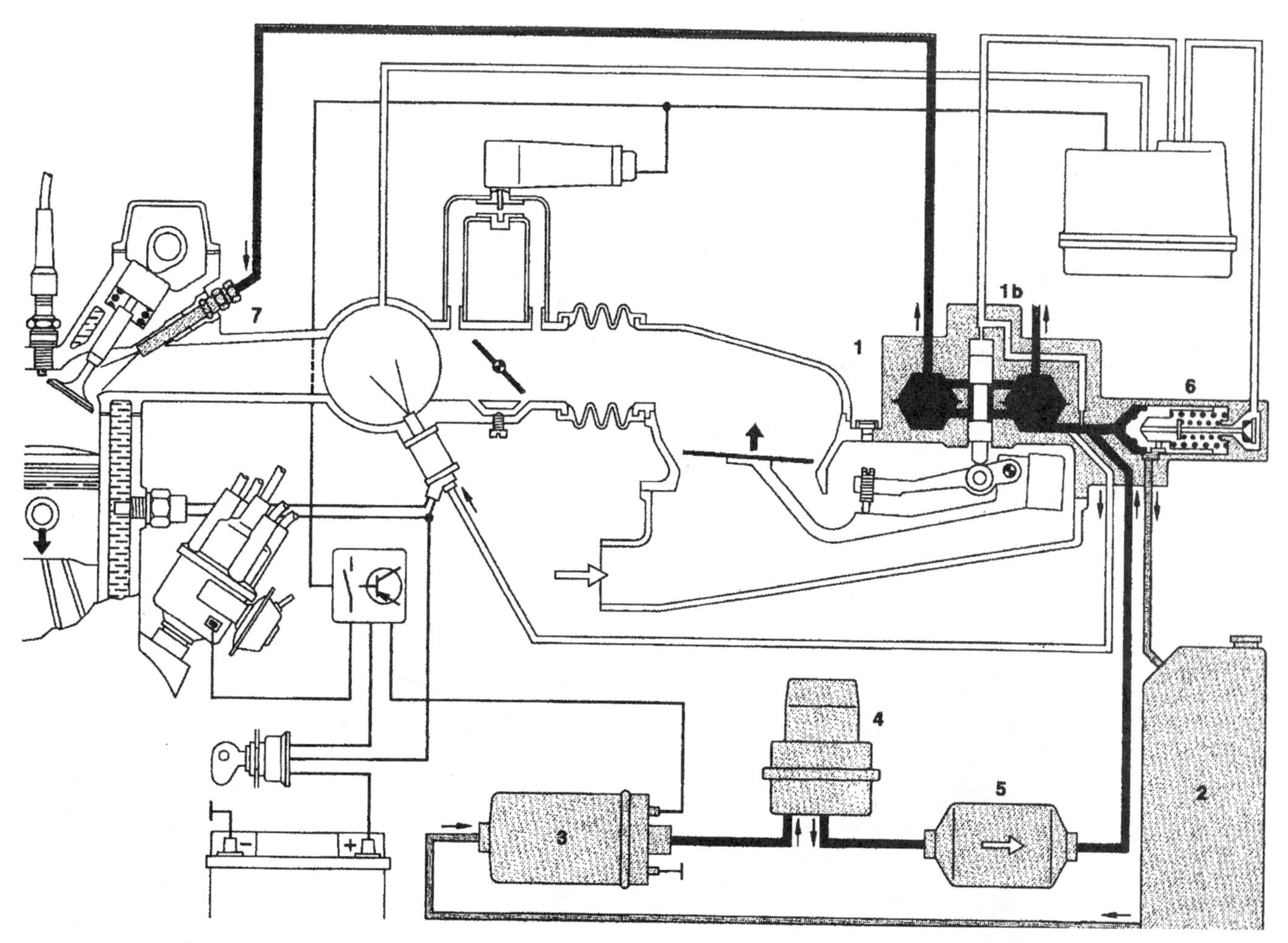

Fig. 9-14 Continuous injection system (courtesy Robert Bosch).

stands for *kontinuierliche,* the German word for "continuous." But regardless of the name it goes by, CIS is essentially a hydraulic injection system because fuel pressure controls fuel delivery. Here's how it works.

Unlike a diesel injection system, there is no mechanically driven pump or governor to control fuel delivery. Nor is it like electronic injection systems such as D- and KE- and L-Jetronic, where electronics are employed to regulate fuel delivery.

The CIS system is a relatively simple system that uses an electric fuel pump to deliver a continuous supply of fuel to the injectors. The pump is mounted under the car near the fuel tank. It runs as long as the key is on and the engine is cranking or running. It's a high-pressure pump, capable of producing upwards of 80 psi and pumping up to 35 gallons of fuel per hour.

When the engine is first started, the pump is energized by a relay. It also provides voltage to the warm-up regulator, auxiliary air device, thermo-time switch, and cold-start valve. The relay is also wired to terminal 1 on the ignition coil so it can pick up a tach signal. The ignition signal tells the relay when the engine is running. If the engine stops while the key is still on, the relay shuts off the pump as a safety precaution. This minimizes the fire hazard that would be created by a burst or torn fuel hose in an accident.

Fuel flows from the pump to an "accumulator" (Fig. 9-15), which muffles pump noise and maintains pressure in the fuel line when the engine is shut off. A check valve in the fuel pump also helps keep pressure from escaping back into the tank. The accumulator, nothing more than a spring-loaded diaphragm with a reed valve, makes restarting easier, especially when the engine is hot.

From here, the fuel flows through a fuel filter.

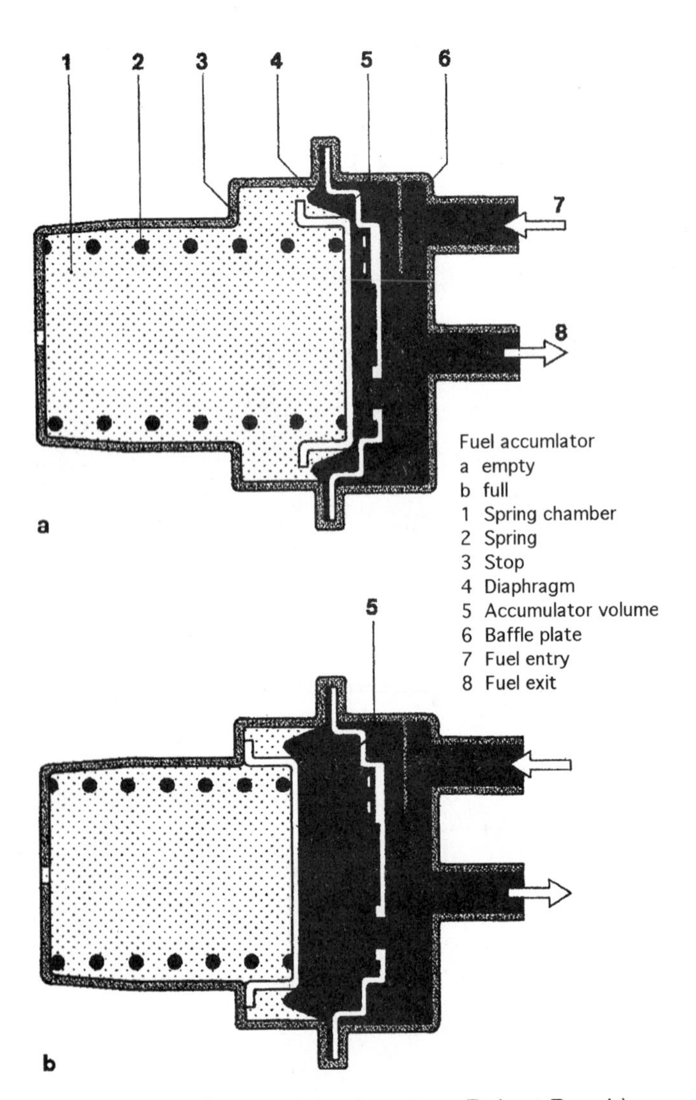

Fig. 9-15 Accumulator (courtesy Robert Bosch).

The original-equipment filter is designed to trap anything larger than 4 microns in size.

After leaving the filter, fuel flows on to the primary pressure regulator in the fuel distributor. The regulator's job is to maintain a constant supply pressure (usually about 65 to 75 psi or 4.5 to 5.2 bar in most applications). The incoming fuel pushes open a spring-loaded plunger valve. As pressure builds, the plunger moves further and further, until it uncovers a port to bleed off the excess fuel. The excess fuel then returns to the fuel tank through a nonpressurized return line. Primary fuel pressure, therefore, depends on the amount of spring resistance behind the plunger. Changing shims under the plunger spring changes the primary fuel pressure of the system.

The primary pressure regulator also acts like a check valve. When the engine is shut off, the plunger closes the fuel return port to hold pressure in the supply line.

Warm-up Regulator

The primary pressure regulator is attached to a "warm-up regulator" (Fig. 9-16), also called a control pressure regulator), which functions something like an electric choke. It provides a richer mixture when the engine is first started, then gradually leans

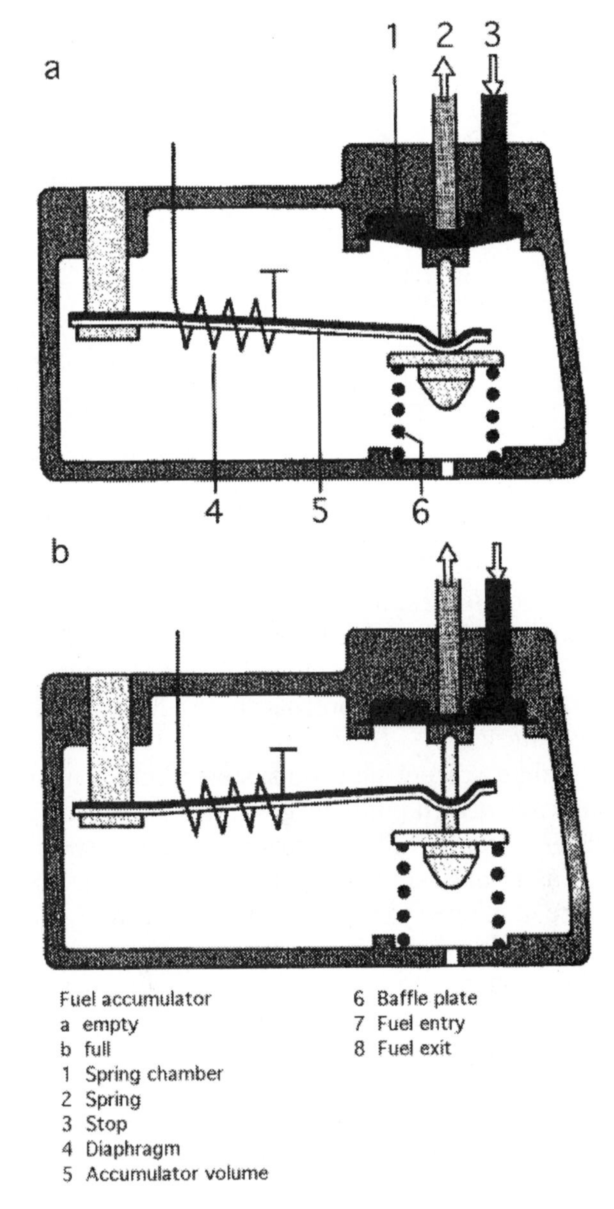

Fig. 9-16 Warmup regulator (courtesy Robert Bosch).

out the mixture as the engine warms up. But instead of choking off the air supply to enrichen the fuel mixture, it causes the pressure regulator to increase the control pressure of the fuel system as the engine warms up.

The warm-up regulator has an electrically heated bimetallic spring that pushes against a diaphragm to uncover a small port. This port routes fuel to the backside of the plunger in the primary pressure regulator. The amount of pressure pushing against the backside of the plunger creates more or less resistance against the fuel entering the fuel distributor. This modifies the operating pressure in the fuel distributor, which changes the amount of fuel delivered to the injectors.

Cold-start Valve

Another temperature compensation device in the CIS system is the "cold-start valve," an extra fuel injector that sprays additional fuel into the intake manifold when a cold engine is first started. The cold-start valve is actuated electrically through the "thermo-time switch (Fig. 9-17)." This is nothing more than a temperature-sensitive, electrically heated switch designed to open above 95 degrees F. The thermo-time switch is usually screwed into the head or engine block. Its operating time is affected

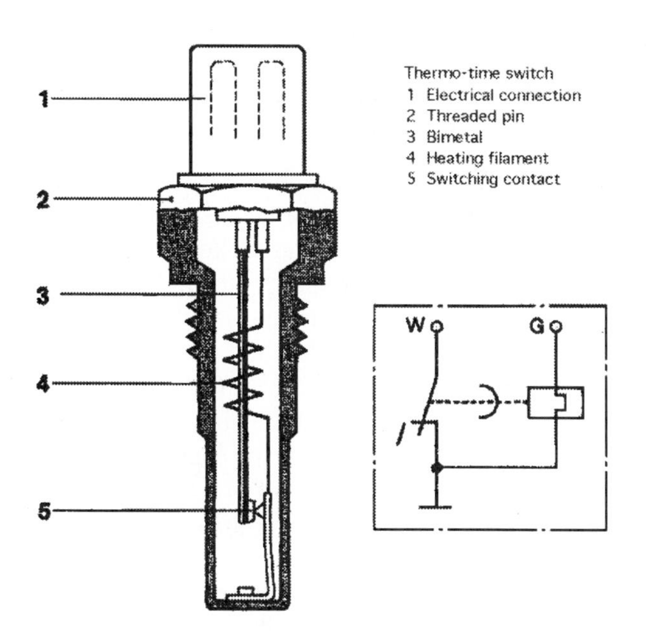

Fig. 9-17 Thermo-time switch (courtesy Robert Bosch).

both by ambient air temperature and the temperature of the coolant in the engine.

FUEL DISTRIBUTOR

The real heart of the CIS system is the "fuel distributor." Its job is to regulate the amount of fuel delivered to the injectors, which in turn determines the engine's air/fuel ratio. Now things get even more complicated—but bear with us.

The amount of fuel that's passed to the injectors is determined by the position of the "control plunger" (Fig. 9-18) inside the fuel distributor. The control plunger is merely a metering barrel with precision-machined slits in its side. As the plunger moves up, it exposes more slit area to the fuel-delivery ports. This allows more fuel to flow through the plunger and to the injectors.

Also within the fuel distributor are "differential-pressure valves" (one for each injector). These help maintain a constant pressure drop across the metering slits (0.1 bar) regardless of the volume of fuel flow.

The control plunger in the fuel distributor is moved up and down by a lever attached to the "airflow sensor" (Fig. 9-19). The tip of the plunger rests atop the lever, and is held against it by gravity and fuel pressure. As air enters the engine and lifts the airflow sensor plate, the lever on the sensor arm pushes the plunger up. This increases fuel delivery in proportion to the volume of air entering the engine and allows the system to maintain a properly balanced air/fuel ratio regardless of speed or load.

The shape of the funnel in which the airflow sensor plate sits also helps determine the air/fuel ratio. The taper of the funnel changes the amount of sensor plate movement for a given increase in airflow. The plate moves further for a given increase in airflow at both idle and wide-open throttle.

A couple of things need to be noted here. One is that the control plunger inside the fuel distributor is a precision-fit assembly. Tolerances are within millionths of an inch, so it doesn't take much rust or dirt to cause problems. A blocked metering slit, for example, can cause a cylinder to run lean or receive no fuel at all. And if the plunger itself binds, the engine can run dangerously lean as load and rpm increase—or run too rich as the engine slows back down to an idle. There's no easy way to clean or re-

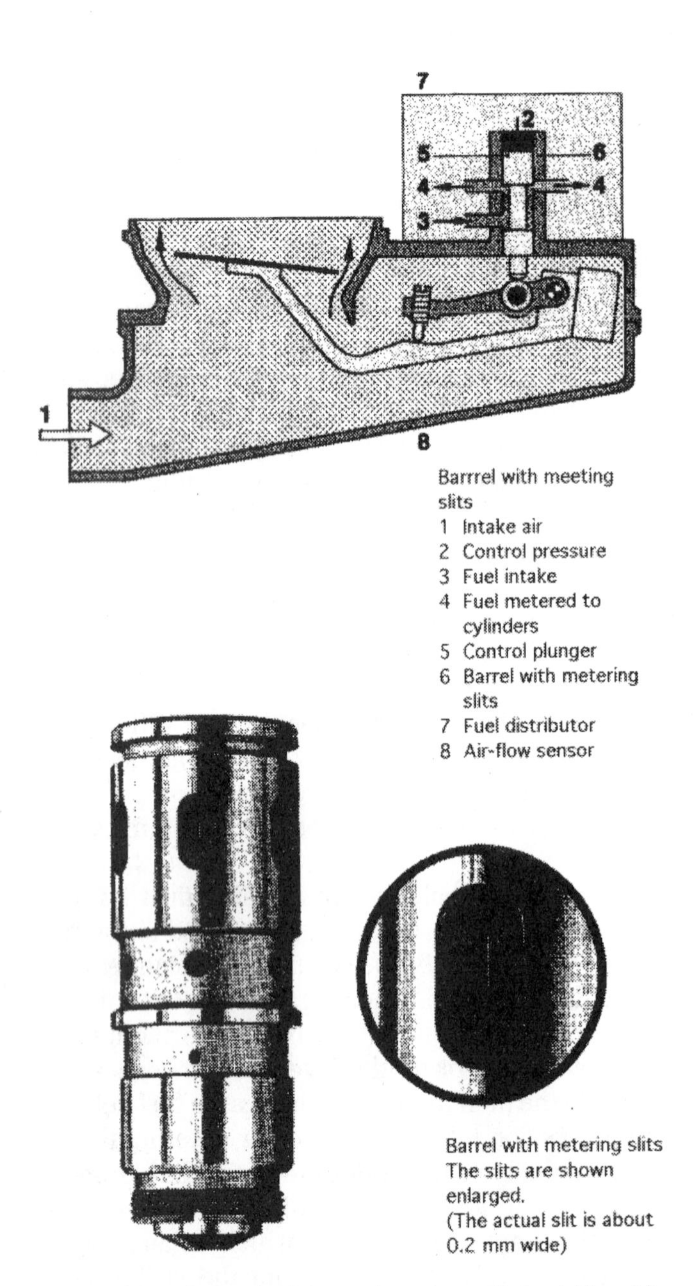

Fig. 9-18 Control plunger (courtesy Robert Bosch).

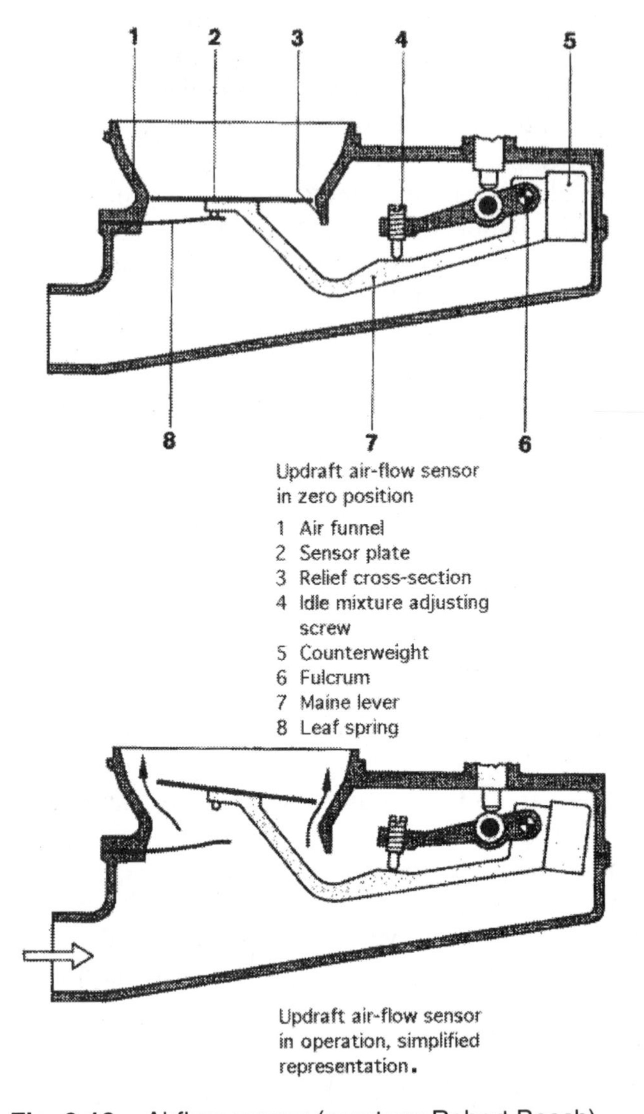

Fig. 9-19 Airflow sensor (courtesy Robert Bosch).

pair the fuel distributor. So if there's rust or dirt inside a new or remanufactured fuel distributor, replacement is the only repair option.

Another point to note is that the airflow sensor can't tolerate vacuum leaks downstream of the throttle, or air leaks in the plumbing between itself and the throttle. Any air leakage between the sensor and the intake ports in the cylinder head allows unmetered air to upset the carefully balanced air/fuel ratio. This leans out the fuel mixture and can cause hot-starting problems and detonation.

There's also an adjustment screw on the lever arm (under an anti-tamper plug) that allows for precise adjustment of the air/fuel ratio. The screw is extremely sensitive and only a small adjustment (one-eighth turn or less) can make a big difference in carbon monoxide readings at idle. It's also important to note that this adjustment screw not only changes the idle mixture, but the air/fuel ratio throughout the engine's operating range.

The injectors in the CIS system are little more than spray nozzles with spring-loaded pintles designed to open above a preset pressure (Fig. 9-20). Like any injector, they can become clogged with dirt and varnish.

Idle speed on the CIS system is adjusted by an idle-adjustment screw that allows-air to bypass the throttle. The throttle also has an "auxiliary-air" device that serves the same purpose as a fast-idle cam on a carburetor. It allows extra air to bypass the

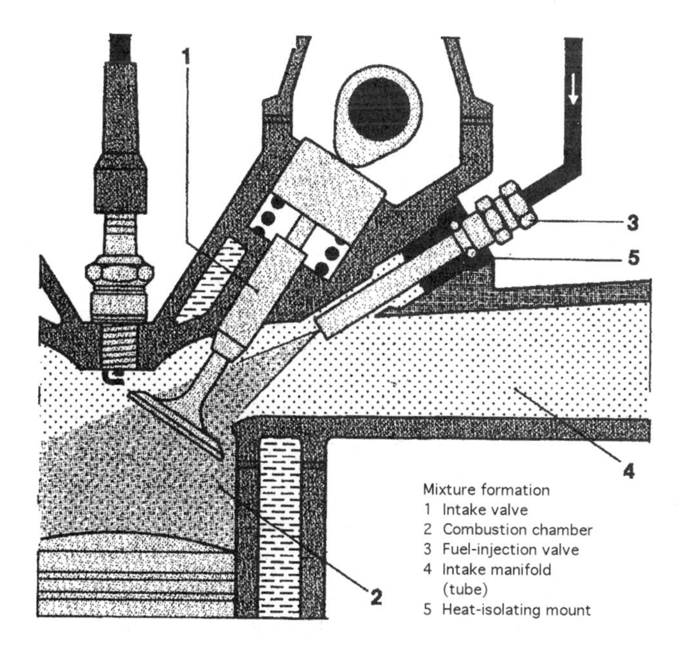

Fig. 9-20 K-Jetronic injector (courtesy Robert Bosch.)

throttle when a cold engine is first started. This helps to prevent stalling until the engine warms up. The auxiliary-air device has a spring-loaded gate valve. The valve is gradually pulled shut by an electrically heated bimetallic strip once the engine has been started.

Pressure Checks

When a CIS fuel-delivery problem is suspected, step one should be to visually inspect the airflow sensor. The plate should be centered in the housing, and even with the bottom of the venturi or within .020 to .030 in. below it. A wire loop under the plate holds it in position. Bending the wire can raise or lower the plate. The plate can be recentered if necessary by loosening the center bolt. The plate should also move up and down freely, with only the resistance of the control plunger in the fuel distributor against it.

If the air sensor appears to be okay, the next step is to check the operating pressure of the system. For this, you'll need a pressure gauge capable of reading up to 100 psi. You'll also need a three-way valve and hoses to connect the gauge to the line that runs between the center of the fuel distributor and the control-pressure regulator. Make sure the engine is off before attaching or disconnecting the fuel-pressure gauge.

There are actually four pressure checks involved:

(1) *System pressure check.* This test checks the operation of the fuel pump and the primary pressure regulator. The test is done by closing the three-way valve so fuel can't flow to the fuel distributor, then energizing the fuel pump with a jumper wire, or by bridging terminals L14 (pump) to L13 (battery) on the pump relay. The electrical connections on the warm-up (control-pressure) regulator and auxiliary-air device should be disconnected for this test so these units will not heat up.

System pressure will depend on the application, but will generally be 65 to 75 psi (4.5 to 5.2 bar). If the fuel pressure is only a little out of range, it can be adjusted by replacing the shims under the pressure-regulator plunger spring. Thicker shims increase pressure, while thinner shims reduce it.

If the pressure is unusually low, the fuel pump may be weak. Pump output can be checked by measuring its fuel delivery. In 30 seconds, a good pump should deliver at least 3/4 qt. (750 cc) of fuel.

(2) *Cold control pressure.* If the system pressure is okay, the cold-control pressure can be checked next. For this test, the engine must be at room temperature (cold) with the electrical connector on the warm-up (control-pressure) regulator disconnected. Open the three-way valve, start the engine, but don't let it idle for more than one minute.

The cold-control pressure should match the line on the accompanying graph. At 68 degrees F., for example, the cold-control pressure should be around 21 psi (1.5 bar). If the readings are out of range, the warm-up (control-pressure) regulator is defective and needs to be replaced.

(3) *Warm control pressure.* This checks whether or not the control-pressure regulator gradually increases fuel pressure as the engine warms up. Reconnect the electrical connector on the warm-up (control-pressure) regulator, start the engine, and allow it to idle.

The gauge reading should increase and stabilize after several minutes at 49 to 57 psi (3.4 to 3.8 bar). If the pressure fails to go up, use a test light or ohmmeter to check for voltage at the warm-up (control-pressure) regulator. If there's voltage but no pressure increase, replace the regulator.

(4) *Rest (leakdown) pressure.* This checks the system's ability to maintain pressure after the engine has been shut off. With the engine warm and the gauge reading at its normal warm pressure (49 to 57 psi), close the three-way valve and turn the engine off.

After five seconds, the system should be no more than 37 psi (2.6 bar) and no less than 26 psi (1.8 bar). It should hold the minimum pressure for 10 minutes and leakdown to no more than 23 psi (1.6 bar) within 20 minutes.

If the pressure drops quickly, restart the engine and pinch off the rubber hose between the fuel tank and fuel pump. Then shut the engine off again. If the pressure now holds within specs, the one-way check valve in the fuel pump is leaking. If the pressure still leaks down, check the plunger in the primary pressure regulator for a damaged O-ring, rust, or dirt. The cold-start valve may also be leaking as might one or more of the fuel injectors. A bad accumulator can also allow the loss of residual pressure.

Injector Checks

The cold-start valve can be checked by removing it from the intake manifold and placing it in a container. Energize the fuel pump. There should be no fuel leakage from the cold-start valve while its electrical lead is disconnected. Reconnect the electrical lead and turn the ignition on, or use a jumper wire to feed it battery voltage. You should see a steady, cone-shaped spray pattern.

If one or more cylinders appear to be running lean or are short on power, try switching the injector lines at the fuel distributor. If there's no change, the injector(s) are probably dirty and need to be cleaned or replaced. But if the problem changes cylinders, the fuel distributor needs to be replaced.

Bosch as well as others make test equipment for bench-testing mechanical injectors. Three tests should be done here: leaks (no drips at 0.2 bar less than the injectors' opening pressure); spray pattern (should be cone-shaped mist); and delivery (the volume of fuel delivered into a graduated cylinder by each injector should vary no more than 10-15% from the others). If flushing with solvent can't salvage a weak injector, it will have to be replaced.

CHAPTER 10

BASIC EMISSION CONTROLS

When most people think of automotive-related pollution, they think only about what comes out the tailpipe. But cars and trucks can actually emit pollutants three different ways:

- Gasoline vapors from the fuel tank and carburetor, or "evaporative emissions."
- Combustion byproducts and vapors from the engine's crankcase, or "blowby" emissions.
- Exhaust gases produced by combustion or "tailpipe" emissions. These include unburned hydrocarbons (HC), carbon monoxide (CO), carbon dioxide (CO2), oxides of nitrogen (NOX), water vapor (H2O), particulates (soot or chunks of carbon), various sulfur compounds and other substances.

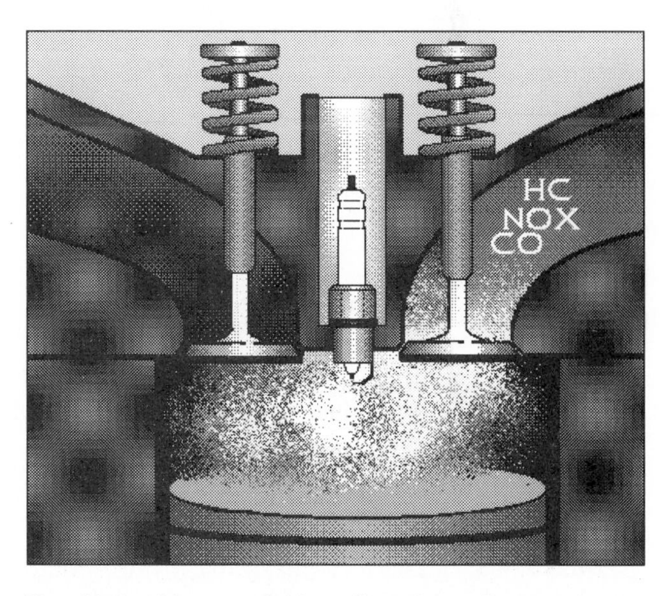

Fig. 10-1 Many variables affect the pollutants that are produced by internal combustion.

EXHAUST EMISSIONS

Exhaust emissions are the most difficult to control because there are so many variables that affect their formation (Fig. 10-1). The most important factors are the air/fuel ratio (see Chapter 8), ignition timing and advance (Chapter 5), and mechanical considerations. Some of these factors are the design of the combustion chamber, camshaft timing, valve duration and overlap, intake manifold design and temperature, engine compression, piston-to-cylinder wall clearances, the type of valve seals used (and their condition), etc. (Chapter 7).

Two things typically contribute to excessive exhaust emissions. One is incomplete combustion. The other is the presence of unwanted substances in the combustion process.

When gasoline is burned inside an engine, combustion is never totally complete. There is always a tiny amount of fuel that fails to burn, or is only partly burned. When gasoline is drawn through the carburetor venturis or sprayed out of a fuel injector, it is broken up into tiny droplets. To burn properly, the tiny droplets of fuel must be mixed with oxygen in

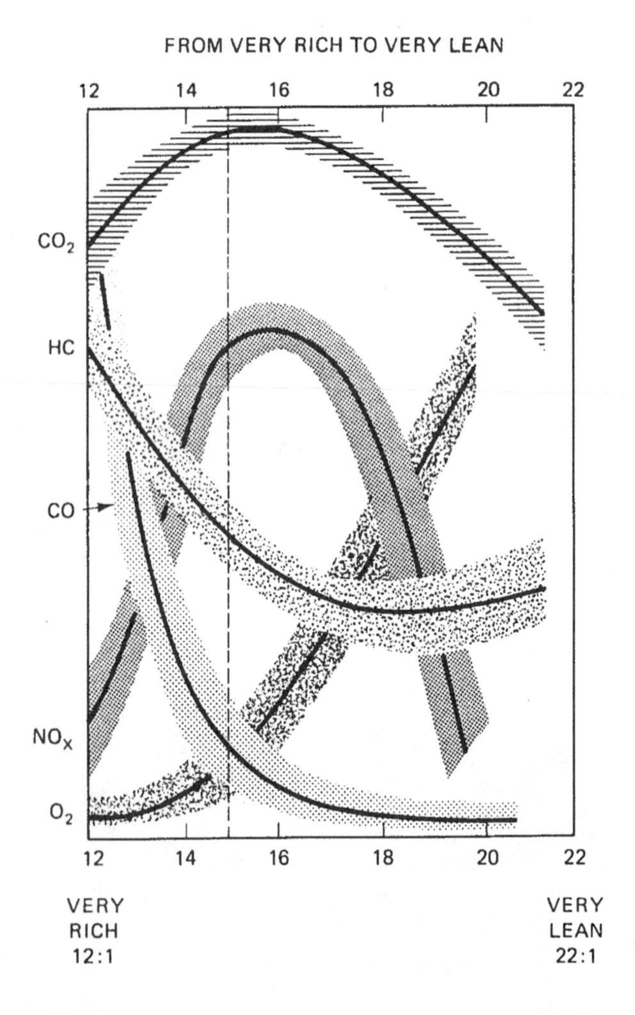

Fig. 10-2 The composition of the various exhaust gases varies with the air/fuel ratio. The dotted vertical line is at the ideal 14.7:1 ratio (courtesy Chrysler).

the right proportions. But air contains only about 21% oxygen, so it takes a lot of air to provide enough oxygen for combustion. The ideal, or stoichiometric, air/fuel ratio for gasoline is 14.7: 1 (Fig. 10-2).

The air/fuel mixture is drawn into the combustion chamber, compressed, and ignited by a spark. For the combustion process to be totally complete. Then the air and fuel must be mixed in the correct proportions. Then all the oxygen would combine with all the gasoline to produce heat energy, water vapor, and carbon dioxide.

Water vapor and carbon dioxide are harmless byproducts of combustion. Therefore, we would have no air pollution problems if combustion inside a real engine were this simple. But it isn't. For one thing, the relatively cool surfaces of the piston, cylinder walls, and cylinder head have a quenching effect on the tiny droplets of fuel. Some of the fuel will cling to these metal surfaces and not burn. Some fuel will also not burn when it gets trapped down between the piston and cylinder wall just above the top compression ring. This results in unburned fuel (hydrocarbons) in the exhaust.

Intake manifold design can also interfere with combustion. It interferes by allowing the tiny droplets of fuel to separate from the airflow in engines that have carburetors or throttle body injection. This causes an uneven mixing of the air and fuel which hinders complete combustion.

Too Rich

An engine needs a richer than usual air/fuel ratio when it is first started because the fuel does not vaporize very easily in a cold engine. A richer mixture is initially provided by the choke on a carbureted engine, or a "cold start valve," or increased injector duration on engines with electronic fuel injection. As the engine warms up, the mixture is gradually leaned until it is balanced. On late model engines with oxygen sensors, and either feedback carburetion or electronic fuel injection, the engine computer keeps the fuel mixture balanced (computerized fuel management is covered in Chapter 17).

But sometimes things get out of whack. A choke may be stuck or misadjusted. The fuel level inside the carburetor bowl may be too high due to a misadjusted, leaky, or fuel saturated float. There may be excessive fuel pressure or a leaky needle valve that's flooding the carburetor with fuel. Someone may have replaced the main metering jets in the carburetor with ones that are too large, or screwed up the idle mixture adjustment screws. A plugged up PCV valve or hose can also contribute to a rich mixture. On late model engines with electronic feedback carburetors, the carburetor mixture control solenoid may be defective or misadjusted. Or the oxygen sensor may be bad. On a fuel injected engine, a leaky cold start valve can leak extra fuel into the intake manifold. A bad fuel pressure regulator may be raising the fuel pressure too much. A computer or sensor problem may be throwing the mixture off. Any of these conditions can make the fuel mixture run rich.

When the air/fuel mixture is too rich (too much fuel, not enough oxygen or an air/fuel ratio lower than 14.7:1), there will not be enough oxygen to burn the fuel completely. This results in incomplete combustion and the formation of carbon monoxide (CO) in the exhaust:

insufficient 02 + HC = H20 + CO2

If the air/fuel mixture is excessively rich, it may even cause the formation of soot particles in the exhaust:

insufficient 02 + HC = H20 + C

Or, it may not ignite at all, allowing the entire mixture to pass through the combustion chamber unburned and into the exhaust. This can be especially damaging to a catalytic converter because raw fuel in the exhaust makes the converter overheat. If the converter gets too hot, the ceramic or metallic substrate that holds the catalyst can melt, creating an obstruction or blockage in the exhaust.

Too Lean

Sometimes the air/fuel mixture may be too lean. Any number of things might cause this: a vacuum leak in engine's vacuum plumbing, carburetor, or intake manifold is a common cause. Others include a float level in the carburetor that's too low, jets that are dirty or too small, misadjusted idle mixture screws, and low fuel pressure in a fuel injected engine. Also, obstructions in the fuel line and clogged fuel injectors, or injectors with insufficient flow capacity for the application (wrong injectors).

A lean mixture (too much oxygen and not enough fuel) may create an emissions problem if the air/fuel ratio is leaner than about 18:1, or too lean for what the engine needs to handle its present load. Under these circumstances, the mixture may fail to ignite. This condition is known as "lean misfire" and typically causes a rough idle. It can also cause misfiring at high speed. If lean misfiring occurs momentarily while accelerating, it can cause a hesitation, stumble or bog, when the throttle opens. This kind of driveability problem is often caused by a weak or plugged accelerator pump in a carburetor. Dirty fuel injectors, a faulty throttle position sensor, or an accumulation of heavy carbon deposits on the intake valves are problems as well.

Lean mixtures may also cause other problems because they elevate combustion temperatures. If things get too hot, the engine may go into preignition and/or detonation and suffer expensive damage (like a burned piston).

If an engine is suffering from a lean misfire condition, it will have high HC emissions. The high HC emissions occur because unburned fuel is passing through the combustion chamber and entering the exhaust. And as we said earlier, that may overheat and damage the catalytic converter.

Too Hot

Excessively high combustion temperature inside an engine is not a desirable situation. Air is almost 80% nitrogen. Normally, nitrogen is inert and does not do much of anything. But at temperatures above 2500 degrees F, nitrogen and oxygen combine to form oxides of nitrogen. The abbreviation "NOX" is used to describe the various nitrogen compounds that they form.

The higher the combustion temperature, the greater the tendency to form NOX. In an engine without emission controls, combustion temperatures can easily exceed 2500 F. Therefore, some means of lowering the temperatures must be used to minimize the formation of NOX. This system is the "Exhaust Gas Recirculation (EGR) system, which is covered later in this chapter.

CARBON MONOXIDE (CO)

Of the three main pollutants, carbon monoxide is the deadliest because you can't see it or smell it. An exhaust analyzer is used to measure the percent of carbon monoxide in the exhaust in percent (%). A concentration of only half a percent (0.5%) CO in the air can render a person unconscious—and kill within 10 to 15 minutes! Even concentrations as small as four hundredths of a percent (0.04%) can cause headaches and be life threatening after several hours exposure.

Carbon monoxide is deadly because it displaces the oxygen in your bloodstream. A molecule of carbon monoxide has 210 times the affinity with joining red blood cells as does oxygen. Breathing CO reduces the amount of oxygen that reaches your brain. This starves the brain for oxygen and rapidly leads to unconsciousness and death. When the CO concentration in a person's blood reaches 2 to 5 percent, vision begins to blur and reaction time starts

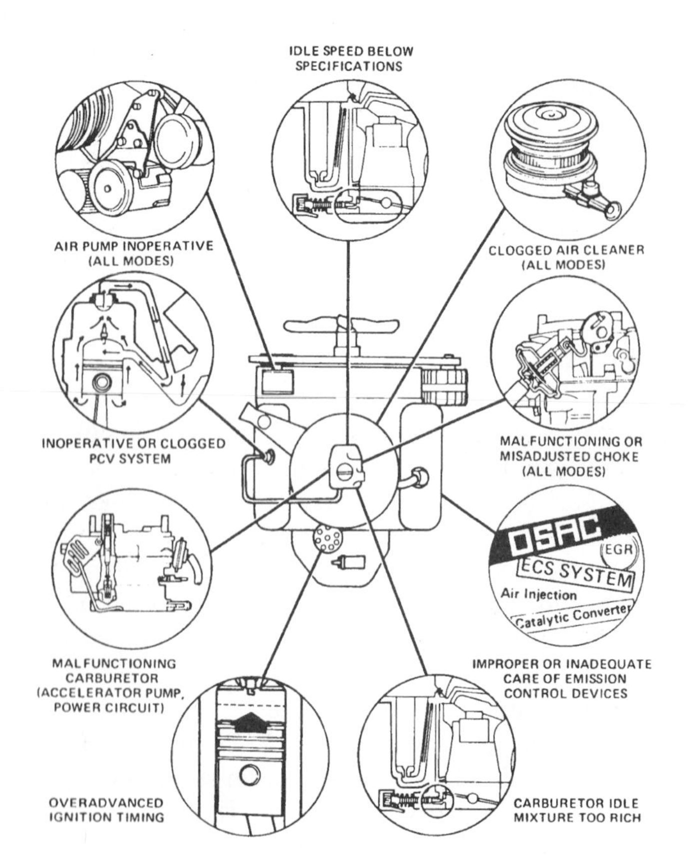

Fig. 10-3 Causes of high carbon monoxide (CO) emissions (courtesy Chrysler).

to drop. It can also cause headaches, dizziness, chest pains, and breathing difficulty. Carbon monoxide concentrations of 20 ppm are not uncommon in heavy traffic.

Carbon monoxide is formed when the fuel mixture is rich and there is insufficient oxygen to completely burn all the fuel (Fig. 10-3). The richer the fuel mixture, the greater the quantity of CO produced. That makes CO a good diagnostic indicator of incomplete combustion, carburetor maladjustment and similar problems. Some of these problems are clogged air filter, sticking choke, defective heated air intake system, and leaky cold start injector or fuel injectors, etc..

Carbon monoxide emissions are highest when an engine is first started. The fuel mixture is richer than normal during this time, and the catalytic converter has not yet reached operating temperature. On 1994 and newer cars, the converters are designed to reach operating temperature more quickly to reduce carbon monoxide emissions while the engine is warming up.

Carbon monoxide formation is minimized in the engine by leaning out the mixture as quickly as possible as the engine warms up. Then a balanced fuel mixture (which depends heavily on a good oxygen sensor) must be maintained. The carbon monoxide that is formed during combustion is changed into carbon dioxide (CO2) in the catalytic converter. Carbon dioxide is not considered a pollutant, but it is a "greenhouse gas" that may contribute to global warming.

A well-tuned engine with a good converter will produce CO levels in the exhaust that are practically zero—compared to as much as 2% in an engine without a converter.

HYDROCARBONS (HC)

Hydrocarbon emissions are unburned gasoline and oil vapors. Though not directly harmful, hydrocarbons are a major contributor to the formation of atmospheric smog and ozone. Hydrocarbons react with sunlight, and break down to form these other chemical compounds that irritate the eyes, nasal passages, throat and lungs.

Ozone is probably one of the most toxic and dangerous air pollutants known. Ozone is formed when an extra atom of oxygen attaches itself to the normal oxygen molecule: 02 + 0 = 03 (ozone). The extra oxygen atom causes the molecule to be very toxic to other materials. It irritates the eyes and lungs. It causes a variety of symptoms including coughing, headaches, choking, and a feeling of weariness, in concentrations as low as 0.5 ppm. Ozone attacks rubber products and is toxic to many types of plants and microbes. Health experts say people should not be exposed to ozone concentrations of more than 0.1 ppm during an 8-hour period.

Elevated HC emissions (Fig. 10-4), are measured in "parts per million" (ppm) with an exhaust analyzer. The elevated HC emissions can be caused by ignition problems (a fouled spark plug or bad plug wire), lean misfiring, loss of compression (burned exhaust valve or leaky head gasket), or engine wear that causes the engine to burn oil (worn valve guides, seals and/or rings).

Hydrocarbon formation is minimized by maintaining a balanced air/fuel ratio, making sure the engine has reliable ignition (proper plug gap, clean plugs, good wires and distributor cap, etc.) You

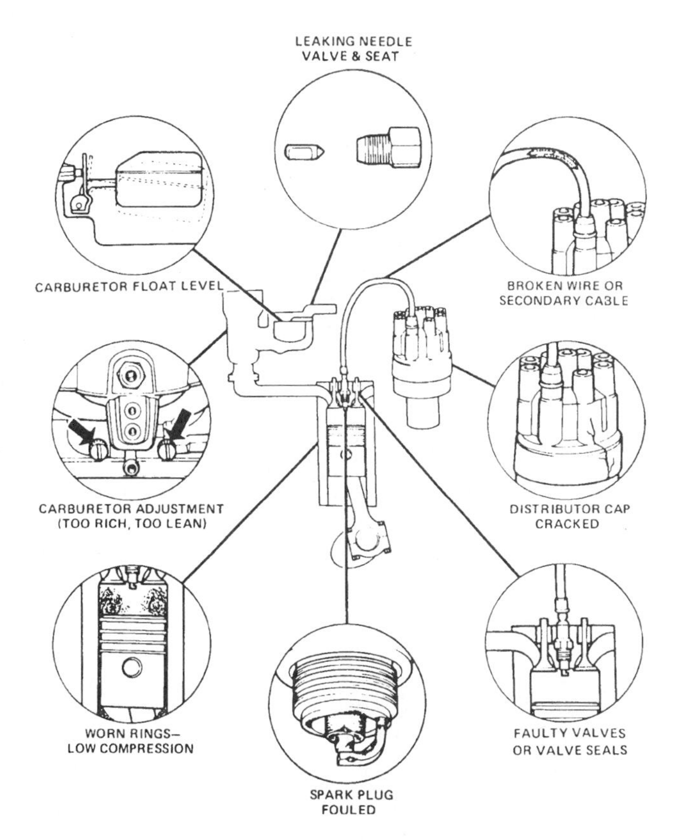

Fig. 10-4 Causes of high hydrocarbon (HC) emissions (courtesy Chrysler).

must also have good compression and close tolerances in the engine (piston rings that are fully seated and seal properly, valve guides and seals that don't leak oil, etc.).

The hydrocarbon's that do pass into the exhaust are "reburned" in the catalytic converter and transformed into water vapor and carbon dioxide. A late model engine in good running condition, properly tuned and with a good converter should produce HC exhaust readings of less than 50 ppm. Compare this to several hundred ppm HC for an engine without a converter.

OXIDES OF NITROGEN (NOX)

Nitrogen makes up almost 80% of the atmosphere. Though normally inert and not directly involved in the combustion process itself, flame temperatures above 2500 degrees F cause nitrogen and oxygen to combine. This combination forms various compounds called "oxides of nitrogen" or NOX. This typically occurs when the engine is under load and combustion temperatures soar.

Most of the NOX that comes out the tailpipe is in the form of nitric oxide (NO), a colorless poisonous gas. It then combines with oxygen in the atmosphere to form nitrogen dioxide (NO2), which creates a brownish haze in badly polluted areas.

NOX is a nasty pollutant, both directly and indirectly. In concentrations as small as a few parts per million, it can cause eye, nose and lung irritations, headaches and irritability. It has an odor that becomes noticeable in concentrations as small as 1 to 3 ppm. When levels reach 5 to 10 ppm, NOX causes eye and nose irritation in some people. Higher concentrations can cause bronchitis and aggravate other lung disorders. Prolonged exposure to 10 to 40 ppm can have serious health consequences. Once in the atmosphere, it reacts with oxygen to form ozone (which is also toxic to breathe) and smog (Fig. 10-5).

To minimize the formation of NOX in the engine, exhaust gas recirculation (EGR) is used. Recirculating a small amount of exhaust gas back into the intake manifold to dilute the air/fuel mixture has a "cooling" effect on combustion. This process keeps temperatures below the NOX formation threshold.

On many 1981, and later, engines with computerized engine controls, a special "three-way" catalytic converter is also used to further reduce NOX in the exhaust. The first chamber of the converter contains a special "reduction" catalyst that breaks NOX down into oxygen and nitrogen. The second chamber in the converter contains the "oxidation" catalyst that reburns CO and HC.

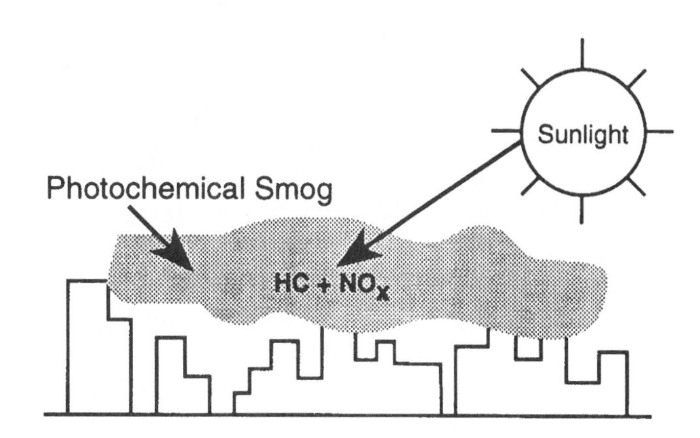

Fig. 10-5 Photochemical smog forms when NOX and HC react with sunlight. NOX also reacts with oxygen to form ozone.

Until the arrival of the new "enhanced" I/M 240 emission inspection program, the only way to measure exhaust NOX was in a laboratory. Ordinary emissions test equipment could not measure it. You could tell if an engine was producing excessive NOX emissions, by visually inspecting the EGR system to see that it was working (not an easy task on many engines). Or another way was to listen for a change in idle quality when vacuum was applied to the valve. But the new I/M 240 test equipment is capable of measuring NOX emissions. Now it can be read directly to determine if the EGR system is functioning properly.

OTHER POLLUTANTS

In addition to the terrible trio (HC, CO & NOX), other pollutants may also be found in the exhaust. Particulates are tiny particles of carbon soot that result from incomplete combustion and excessively rich air/fuel ratios. Diesel engines, which produce little carbon monoxide, are notorious for their sooty particulate emissions—especially when accelerating or pulling a load. Particulate emissions are typically measured with an "opacity" test that measures the darkness of the smoke that's coming out the tailpipe. Accurate injection timing can minimize diesel particulate emissions. Newer diesel engines with electronic direct injection produce much less soot.

Scientists estimate that as much as 100 tons of particulates fall to the ground, or are breathed in by the public in most large American cities every day. Most of these come from trucks and buses with diesel engines. Many particulates are smaller than 1 micron in diameter (39 millionths of an inch). Particles this small are called "aerosols" because they tend to remain suspended in air, rather than settle to the ground. Such particles pose a significant health hazard because they are too small to be filtered out by the mucous membranes in our nose and windpipe. They penetrate deep into our lungs and remain there, accumulating with each passing year. This can increase the odds of developing emphysema and cancer, especially if a person smokes.

Sulfuric acid is another pollutant that's formed by exhaust pollution in catalytic converter equipped vehicles. All crude oil contains a certain amount of sulfur. Crude oil from the western United States contains much more sulfur than do eastern crude oils. When the crude is refined to make gasoline, some of the sulfur remains in the fuel. The concentrations are small, but enough to cause pollution problems.

When gasoline containing sulfur is burned, the sulfur combines with oxygen to form sulfur oxides. When sulfur oxide concentrations reach 8 to 12 ppm, most people's eyes will water. Such levels also cause coughing, breathing difficulty, and can increase the likelihood of heart disease over time. Sulfur dioxide (SO2) produces the familiar rotten-egg odor.

When exhaust sulfur emissions combine with water, it forms sulfuric acid. The sulfuric acid contributes to "acid rain," a corrosive brew that eats away at buildings and statues, ruins paint jobs, and kills trees and fish in lakes.

NO MORE LEAD

Lead pollution is not a problem with cars that burn unleaded gasoline. But it was in vehicles that used leaded regular gasoline. Tetraethyl lead was long used in gasoline for two reasons. One was to raise the octane rating of the fuel. This allowed refiners to take a lower grade of gasoline and boost its performance to acceptable levels by adding lead as an octane enhancer. Federal regulations limited the amount of lead that could be added by refiners. But many secondary refiners, or blenders, added up to five times the maximum permissible amount of lead to marginal gasoline to make the product salable. The other reason why lead was used was because it had a lubricating effect on exhaust valves which helped prolong valve life.

If leaded gasoline is used in a car with a catalytic converter, the lead fouls the catalyst and renders it useless. This, in turn, causes a significant increase in HC and CO emissions. It can also foul up the oxygen sensor in computer controlled engines. Because of these problems, many emission test programs include a visual check of the restricter in the fuel tank filler neck. The visual check shows if the restrictor has been punched out to accept a pump nozzle for regular leaded gasoline.

Lead is considered a serious threat to the environment because it literally lasts forever. Lead is ingested into the body by breathing in lead-polluted air, or by drinking water from lead-polluted wells, or by eating plants contaminated by lead dust or lead-polluted soil. The metal accumulates in the body and eventually produces lead poisoning. There is ample evidence to suggest that lead from automobile exhaust in metropolitan areas adversely affects the learning abilities of inner city school children. Lead is absorbed into nerve cells in the brain, where it inhibits normal functions. Tests have shown that school children who live near areas of traffic congestion score lower on intelligence tests. These children are more prone to behavior problems than those who live away from lead-polluted environments. As a result of this, school children in many urban areas must now have their blood tested for excessive lead levels when they start kindergarten.

Fortunately, leaded fuels are now a thing of the past. Lead has been gradually phased out over the years and is no longer available in pump gasoline.

POSITIVE CRANKCASE VENTILATION SYSTEMS

The first emission control device to be required by law was one that prevented crankcase blowby vapors from escaping into the atmosphere. It was called "Positive Crankcase Ventilation" (PCV), and today you'll find it under almost every hood you open.

Prior to the introduction of PCV systems, crankcase blowby gases were dumped into the atmosphere through a "road draft tube." Fresh air entered the crankcase through an open breather cap on the oil filler tube. The fresh air then circulated through the engine, and exited through the road draft tube along with moisture and blowby gases from the crankcase. It was not a very efficient method of venting the crankcase, nor did it help the growing air pollution problem. In fact, prior to PCV, nearly 20% of the pollutants vehicles dumped into the atmosphere came from open crankcases.

Positive crankcase ventilation (Fig. 10-6) prevents crankcase blowby vapors from escaping into the atmosphere by siphoning the vapors back into the intake manifold. Then the crankcase blowby vapors can be reburned in the engine. The open road draft tube is replaced with a hose that reroutes the vapors back to a vacuum port on the intake manifold or under the carburetor. Fresh air is pulled through the engine by intake vacuum (Fig. 10-7), and along with it the blowby vapors and moisture from the crankcase. Airflow through the PCV system is metered by the PCV valve. This PCV valve is a spring-loaded variable orifice valve that changes the flow rate according to engine load and throttle position.

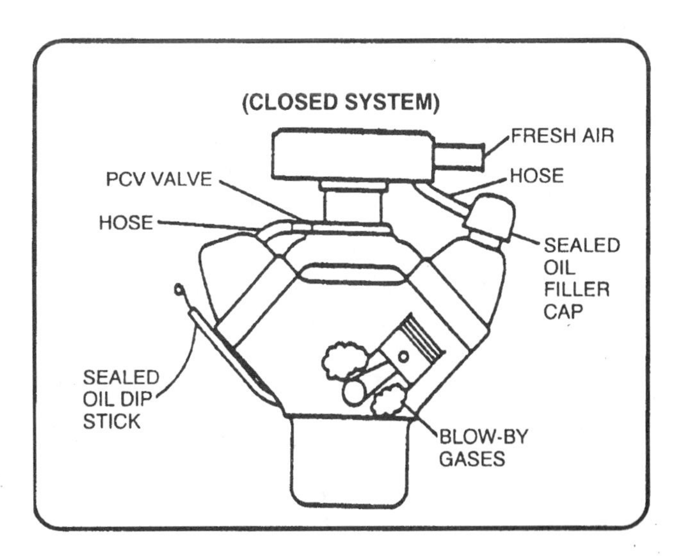

Fig. 10-6 Positive crankcase ventilation (PCV) system (reprinted, by permission, from New York State Department of Motor Vehicles, *Systems Training in Emissions and Performance.* Copyright 1993 Delmar Publishers Inc.).

The earliest PCV systems were an "open" style system. Air entered the crankcase through a breather cap on the valve cover or oil filler tube. But some blowby gases could still escape through the vent. So in 1968, "closed" PCV systems became standard. The air inlet vent in a closed system is positioned inside the air cleaner housing. If blowby gases build up faster

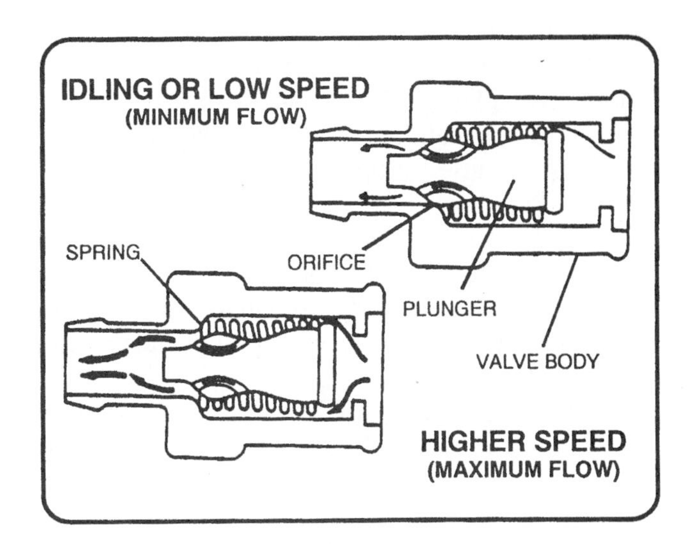

Fig. 10-7 Airflow through the PCV valve (reprinted, by permission, from New York State Department of Motor Vehicles, *Systems Training in Emissions and Performance.* Copyright 1993 Delmar Publishers Inc.).

than the PCV valve can extract them, the gases back up into the air cleaner. They are then sucked into the carburetor and reburned.

On some vehicles, a small PCV air filter is mounted inside the air cleaner. This filter should always be inspected, and cleaned or replaced as needed, on a regular basis. On other vehicles, the crankcase vent is located on the inside of the air filter, so a separate filter is not used.

Besides eliminating crankcase blowby vapors as a source of air pollution, PCV also reduces the buildup of moisture in the crankcase. The PCV prolongs the life of the oil as well. If the PCV valve or a hose becomes clogged, moisture can rapidly accumulate in the crankcase, leading to sludge formation. And if the oil isn't changed often enough, sludge can ruin the engine! That's why regular inspection of the PCV valve and plumbing system is so important.

A clogged PCV system will also allow pressure to build within the crankcase, possibly forcing oil to leak past gaskets and seals. Oil leaking past valve covers or crankshaft seals, or oil vapor backing up into the air cleaner, might by symptoms of a clogged PCV system. Regular inspection of the PCV valve, and periodic replacement will keep the system functioning properly.

The other problem to watch out for with PCV systems is vacuum leakage. Because the PCV valve acts like a calibrated vacuum leak, any additional air leaking past the valve can lean out the air/fuel ratio. This results in lean misfire, hesitation, hard starting and stalling. Watch for cracked or loose hoses, or poorly fitting valves. The PCV valve must also be the correct one for the application. The wrong valve may flow too much or too little air, upsetting the fuel calibration.

The PCV valve can be tested several ways. You should feel strong vacuum when you hold your thumb over the end of the valve at idle. This should also cause a momentary rise in exhaust carbon monoxide readings. The valve should also rattle when shaken.

Some engines have no PCV valve. A couple of examples include the Ford Escort and General Motors Quad 4. These engines use a calibrated orifice in a breather box to siphon blowby vapors back into the intake manifold. Like a PCV valve, an orifice ventilation system can plug up, allowing pressure to build inside the crankcase. The orifice can be tested by feeling for vacuum on the inlet side when the engine is running. You can test the orifice by trying to blow air through it with the unit disconnected as well. If a plugged orifice cannot be cleaned with solvent, the breather box has to be replaced.

EVAPORATIVE EMISSIONS CONTROL

Gasoline fuel vapors, or evaporative emissions, as they are called, contain a variety of hydrocarbons (HC). The lighter elements in gasoline evaporate easily, especially in warm weather. These include aldehydes, aromatics, olefins, and higher paraffins. These substances can react with air and sunlight (called a photochemical reaction) to form smog. Aldehydes are often called instant smog because they can form smog without undergoing photochemical changes.

Evaporative emissions can also account for about 20 percent of a vehicle's total emissions. A parked car can pollute the air with hydrocarbon emissions even though the engine is not running, so you should appreciate the importance of controlling evaporative emissions. The problem becomes especially bad on hot summer days when a vehicle is parked in the sun on an open parking lot. The gasoline vapors can literally spew out of the vehicle's fuel system, unless they are prevented from doing so by an evaporative emission control system.

Evaporative emissions are eliminated by sealing off the fuel system from the atmosphere (Fig. 10-8).

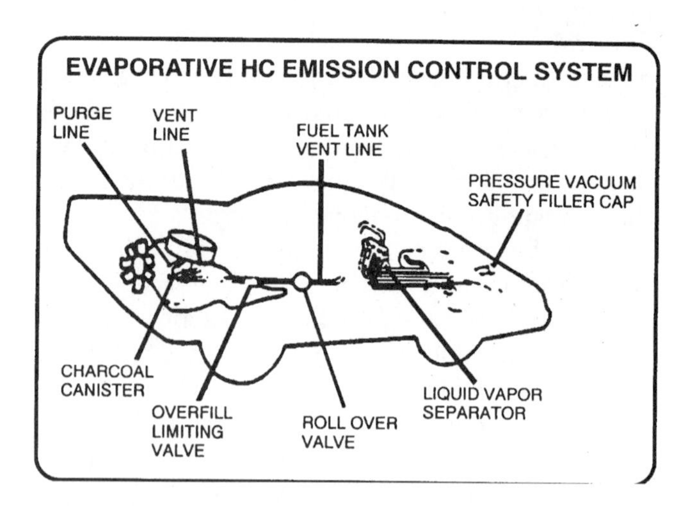

Fig. 10-8 Evaporative emission system components (reprinted, by permission, from New York State Department of Motor Vehicles, *Systems Training in Emissions and Performance.* Copyright 1993 Delmar

This prevents the gasoline vapors from escaping from the fuel tank or carburetor bowl. Vent lines from the fuel tank and carburetor bowl route vapors to a charcoal canister. They are trapped and stored there until the engine is started. The vapors are then drawn into the intake manifold and burned.

With fuel injection, there are no evaporative emissions from the engine compartment because the injectors are part of a sealed system. Unlike a carburetor, there is no vented fuel bowl to leak vapors. The fuel is all contained within the pressurized fuel rail and injectors. But the fuel tank must still be sealed to prevent vapors from escaping out the filler pipe.

Evaporative emission controls were first required on cars sold in California in 1970, and on all other cars since 1971.

SEALED SYSTEM

Sealing the fuel tank and carburetor is a simple way to prevent the escape of fuel vapors into the atmosphere. But doing so is not as simple as it sounds. For one thing, a fuel tank must be vented so that air can enter to replace fuel as the fuel is used up. If a tank were sealed tight, the fuel pump would soon create enough negative suction pressure inside the tank to cause the tank to collapse. Or it would also restrict the flow of fuel to the engine. It is something like trying to pour oil out of a can with a single small hole in it. The vacuum created inside the can slows the flow of oil to a trickle. Punch a vent hole in the can and the oil gushes out. A fuel tank needs the same kind of ventilation so that the fuel pump can suck the fuel out. The job of venting the gas tank is usually performed by the gas cap.

A fuel tank must also allow for a certain amount of fuel expansion. Gasoline, like other liquids, expands as it gets warmer. If you fill a tank on a cool morning, rising temperatures later in the day can cause the fuel to expand. If there is not sufficient reserve capacity built into the tank to handle the added fuel volume, the tank will overflow. Like the fuel tank, the carburetor bowl must also be vented to function properly. If the bowl were sealed tight, negative pressure inside the bowl would decrease the flow of fuel through the metering circuits and venturis. This negative pressure causes a leaning of the air/fuel mixture, or possibly, fuel starvation. Therefore, some means of venting the bowl must be provided.

Evaporation of fuel from a carburetor increases with temperature. The hotter the bowl, the faster the rate of evaporation. While the engine is running, fresh fuel entering the bowl has a cooling effect. This helps to minimize evaporation somewhat. But as soon as the engine is turned off, the carburetor begins to soak up heat like a sponge. Evaporation increases dramatically, and on especially hot days the fuel can literally boil in the bowl.

To keep the gasoline in the fuel system and out of the atmosphere, the evaporative emission control system must allow for fuel expansion, tank venting, and carburetor bowl venting. The evaporative emission control system must be able to store gasoline fumes for extended periods of time.

EFE COMPONENTS

The major components of the evaporative emission control system include:

- FUEL TANK— All fuel tanks in today's cars are designed to allow for fuel expansion (Fig. 10-9). The expansion space is usually 10 to 12% of the total tank volume. For example, a tank designed to hold 12 gallons of fuel, when filled, would need at least an additional 1 gallon capacity for expansion. There are several ways expansion space can be designed into a fuel tank. The easiest way is to locate the filler neck. An air space is then created at the

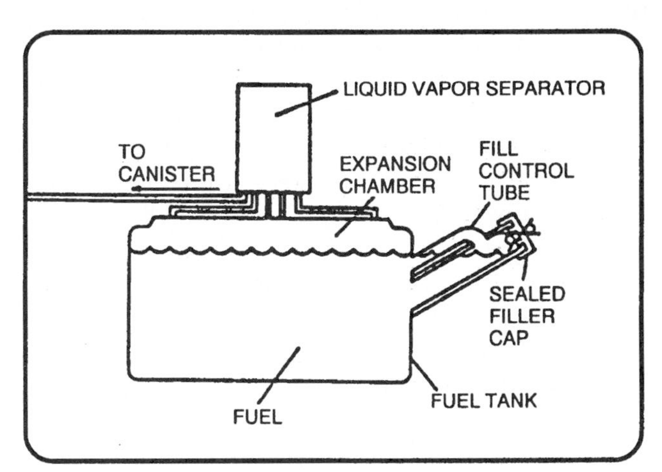

Fig. 10-9 Sealed fuel tank with liquid-vapor separator (reprinted, by permission, from New York State Department of Motor Vehicles, *Systems Training in Emissions and Performance.* Copyright 1993 Delmar Publishers Inc.).

top of the tank when it is filled. Designing a bulge or dome on the top of the tank serves the same purpose. The air pocket absorbs the increase in volume, as the fuel expands.

Another way to create an air space in the top of the tank is to connect a fill control tube to the filler neck. When the tank reaches a certain level as it is being filled, gasoline begins to flow back through the fill control tube into the filler neck. This causes the gas nozzle to kick off and prevents overfilling the fuel tank. The remaining air space at the top of the tank then serves as the expansion reserve.

Some vehicle manufacturers solve the expansion problem by using a small external expansion tank on top of the main fuel tank. The expansion tank has a capacity of about 1 to 2 gallons, and is connected to the main tank with vent lines and a fill control tube. With this approach, the main tank can be filled to capacity. The expansion tank then handles any resulting fuel expansion. It is something akin to the expansion tank on a radiator.

Yet another approach to controlling fuel expansion is to use the tank-within-a-tank method. A little fuel tank, with several small orifices punched in the sides, is located inside the main fuel tank. The orifices limit the speed with which the smaller tank can fill with fuel. When the main tank is being filled, it will reach capacity long before the smaller tank. So when the gas nozzle kicks off, indicating a "full" tank, gasoline will continue to seep into the smaller tank from the big one. This creates an air space in the top of the main tank for fuel expansion.

The only problems any of these expansion control techniques can create are complaints about slow filling. Many motorists quickly discover that such fuel tanks fill slowly, or that they never seem to be quite full. That is because the tanks are designed that way. Overfilling, by continually squeezing in a few more cents' worth of gasoline after the nozzle has kicked off, defeats the design purpose of expansion control.

- GAS CAP— Most people don't realize the gas cap is an emission control device, but it is. In precontrol days, the gas cap's main job was to keep gasoline from sloshing out of the tank, and dirt from getting in the tank. It was equipped with a small vent hole so that the tank could breathe. Air entered through the cap to make up for fuel as it was used, and fuel vapors exited through the cap as internal pressure rose on warm days.

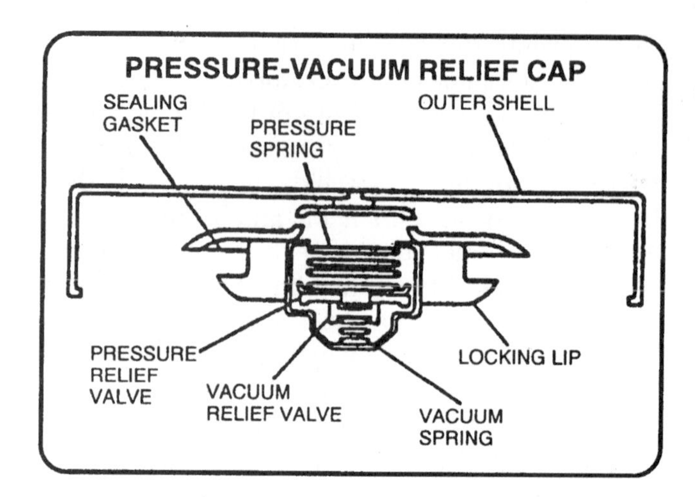

Fig. 10-10 Pressure-vacuum relief cap (reprinted, by permission, from New York State Department of Motor Vehicles, *Systems Training in Emissions and Performance.* Copyright 1993 Delmar Publishers Inc.).

Today's emission control gas caps are considerably different. They are either of solid construction (venting is provided by other means) or they contain a pressure/vacuum valve (Fig. 10-10). The valve-type cap will vent tank pressure if it exceeds 1 psi. It will also allow air to enter the tank if a vacuum exists within the tank. In other words, the valve-type cap can vent pressure or relieve vacuum as the situation warrants without allowing gasoline vapors to pollute the environment.

The valve itself is a simple double-spring arrangement similar to a radiator cap. One spring reacts to internal pressure while the other reacts to external pressure. A plate or diaphragm between the two springs opens and closes to allow air to pass through the valve in the direction needed.

Internal fuel tank pressure can also be vented by means of a three-way valve in the vapor line to the charcoal canister. Some Ford vehicles do not have a pressure-vacuum relief gas cap. Instead they use a three-way valve in the fuel tank vent line to control internal tank pressure. The valve vents tank pressure to the charcoal canister. When there's a vacuum in the tank, the upper diaphragm will allow air to be drawn into the vent line to the tank. The lower diaphragm serves as a safety vent for excessive tank pressure in case the main vent line becomes clogged.

If a gas cap has to be replaced on a vehicle, the replacement must be the same type as the original (sealed or vented).

• LIQUID-VAPOR SEPARATOR—On top of the fuel tank or as part of the expansion tank is a device known as a liquid vapor separator (Fig. 10-9). The purpose of this unit is to prevent liquid gasoline from entering the vent line to the charcoal canister (located in the engine compartment). You do not want liquid gasoline going directly to the charcoal canister because it would quickly overload the canister's ability to store fuel vapors.

The liquid-vapor separator works on the principle that vapors rise and liquids sink. The vapor vent lines from the fuel tank, that go to the separator are positioned vertically inside the unit with the open ends near the top. This allows the vapors to rise to the top of the separator. Any liquid that enters the separator through the vent lines dribbles down the sides of the vent tubes and collects in the bottom of the separator. (Examples of this are fuel sloshing around inside the fuel tank as a result of hard driving, parking on a steep hill, excessive fuel expansion, etc.) A return line allows the liquid gasoline to dribble back into the fuel tank. The vapors then exit through an opening in the top of the separator. This oening which usually has an orifice restriction to help prevent any liquid from getting into the canister vent line.

The liquid-vapor separator is a simple device that is relatively trouble-free. One problem that can develop is if the liquid return becomes plugged with debris, such as rust or scale from inside the fuel tank. Another problem is if the main vent line becomes blocked or crimped. A third problem is if a vent line develops an external leak due to rust, corrosion, or metal fatigue from vibration.

Some liquid-vapor separators use a slightly different approach to keeping liquid fuel out of the canister vent line. A float and needle assembly is mounted inside the separator. If liquid enters the unit, the float rises and seats the needle valve to close the tank vent.

Another approach sometimes used is a foam-filled dome in the top of the fuel tank. Vapor will pass through the foam, but liquid will cling to the foam and drip.

If a blockage occurs in the liquid-vapor separator, or in the vent line between it and the charcoal canister, the fuel tank will not be able to breathe properly. Symptoms include fuel starvation, or a collapsed fuel tank on vehicles with solid-type gas caps. If you notice a whoosh of pressure in or out of the tank when the gas gap is removed, suspect poor venting.

You can check tank venting by removing the gas cap and then disconnecting the gas tank vent line from the charcoal canister. If the system is free and clear, you should be able to blow through the vent line into the fuel tank. Blowing with compressed air can sometimes free a blockage. If not, you will have to inspect the vent line and possibly remove the fuel tank to diagnose the problem.

Some Chrysler and Ford cars have an overfill limiting valve in the vent line between the fuel tank and charcoal canister. The valve's purpose is to prevent any liquid that might have passed through the separator from reaching the canister. It consists of a simple float valve that closes if liquid fills the small chamber around it.

• CHARCOAL CANISTER—The charcoal canister is a small round or rectangular plastic or steel container mounted somewhere in the engine compartment (Fig. 10-11). On some late model vehicles, where underhood space is a premium, the canister may be hidden behind a fender splash panel.

The canister's job is to store gasoline vapors from the fuel tank, so that the fumes do not pollute the atmosphere. The canister contains activated charcoal (about 1-1/2 lbs.). Charcoal acts like a sponge to soak up the gasoline vapors, holding up to twice its weight in fuel.

The vapors are stored in the canister until the engine is started. The vapors are then drawn into the air cleaner, through a vacuum port in the carburetor or intake manifold, or are siphoned into the engine through the PCV plumbing.

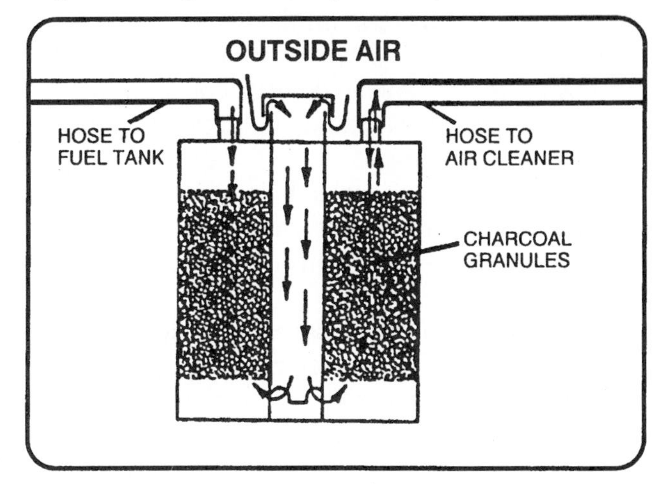

Fig. 10-11 Charcoal canister (reprinted, by permission, from New York State Department of Motor Vehicles, *Systems Training in Emissions and Performance.* Copyright 1993 Delmar Publishers Inc.

Some early Chrysler evaporative control systems did not use a charcoal canister. Instead, the fuel vapors were routed into the engine's crankcase for storage. When the engine was started, the PCV system would draw the fumes out of the crankcase and into the intake manifold. This approach had its drawbacks, though. For one thing, the gasoline vapors tended to dilute the crankcase oil. The vapors also formed an explosive mixture that could literally blow the valve covers right off the engine. Because of such problems, the approach was dropped in favor of the charcoal canister method of storing fuel vapors.

The charcoal canister is connected to the fuel tank via the tank vent line, and to the carburetor bowl with another vent hose. The pressure, created by evaporating fuel, drives the vapors through the lines and into the canister. Here they stay until the canister is purged by starting the engine.

The auto manufacturers have been quite clever in coming up with various ways to purge the charcoal canister of its contents. Some canisters in older vehicles have an open bottom with a small, flat air filter across the opening. The filter is there to keep dirt out of the canister, while it is being purged. On applications that use such a filter, the filter should be inspected periodically and replaced according to the manufacturer's recommendations. A good rule of thumb is to replace the filter every two years.

On canisters with the open bottom, fresh air is drawn in through the filter by connecting a purge line from the top of the canister to the air cleaner, a carburetor vacuum port, or the PCV plumbing. Airflow through the canister ("purging") is regulated by a purge control valve on the canister. The valve opens in response to a ported vacuum signal. Others use an electric solenoid purge control valve. On these systems, the solenoid is regulated by the engine control computer.

On canisters with sealed bottoms fresh air is circulated through the canister via a center tube. Air is sucked down the tube, up through the charcoal, and out the purge line.

Depending on the design of the canister, purging may be controlled in different ways. Those that use ported vacuum use a technique called constant and demand purge. The purge valve on the canister allows constant purging at a restricted rate, through an orifice, until a certain level of vacuum exists at the canister outlet. When ported vacuum is applied to the purge control valve, it allows a higher purging rate. The reason for having constant and demand purging is because the engine cannot handle a large flow of air through the canister at idle or slow speeds. The additional air and fuel vapor would upset the air/fuel ratio, causing a rough idle and increased tailpipe emissions. At higher rpm rates, however, the engine can digest the canister's contents without problem. Purging is calibrated to match the engine's ability to handle it. A vapor feed rate of around 12 cubic feet per hour might be average for a small V8 cruising down the highway.

Under normal circumstances, the charcoal canister causes few problems. Since the charcoal does not wear out, it should last the life of the vehicle. Problems can result, though, when the canister filter is neglected, or when a purge control valve malfunctions. A problem also occurs or if someone gets the vent, purge, and control vacuum lines mixed up (the connections are usually labeled to avoid such mistakes).

If vapor is not being purged from the canister, the purge valve may be defective or the canister filter may be plugged. The purge valve can be tested with a hand vacuum pump. it should hold vacuum for at least 15 to 20 seconds without leaking down. Vacuum connections should be inspected to make sure that they are tight and properly routed.

On units equipped with computer-controlled solenoid purge valves, refer to the manufacturer's shop manual for the specs on when, and under what conditions, the solenoid is supposed to open. Generally speaking, such systems will not purge the canister until the engine reaches operating temperature. The coolant sensor monitors temperature. When the computer reads the appropriate temperature, it sends a command to open the canister purge solenoid.

On General Motors products with the Computer Command Control system, for example, the computer energizes the canister solenoid when the engine is operating in the "open loop" mode. Open loop means that there is no feedback computer control over the air/fuel mixture. It is set at a fixed value until the engine warms up to improve cold idle. This prevents vacuum from reaching the purge valve on the canister. When the engine enters the "closed loop" mode of operation (when the oxygen sensor is hot enough to produce a signal and the engine is at operating temperature), the solenoid is deenergized. Then, vacuum is allowed to open the canister purge valve and purge the fuel vapors.

• ANTI-PERCOLATOR VALVE—To prevent fuel evaporation from the fuel bowl during engine operation, some carburetors have an anti-percolator valve, where the canister vent line connects to the fuel bowl. The valve seals off the vent line while the engine is running. The valve is connected to the throttle linkage. It will be closed when the throttle is open, and open when the throttle is closed. The valve opens when the engine is off (the throttle closed), so that the hot fuel vapors can boil out through the vent line and into the canister.

HEATED AIR INTAKE SYSTEMS

The heated air intake, or early fuel evaporation system's purpose, is to preheat the air before it enters a carburetor when the engine is first started. It does this by routing air through a "stove" around the exhaust manifold before it enters the air cleaner. A temperature sensor inside the air cleaner controls a vacuum diaphragm in the air cleaner inlet (Fig. 10-12). When cold, the temperature sensor passes vacuum to the diaphragm. This closes a flap to outside air so heated air will be drawn into the air cleaner. Warm air improves fuel vaporization while the engine is warming up. This helps the engine idle more smoothly. It also improves cold driveability by reducing the tendency to hesitate or stumble when the throttle is opened.

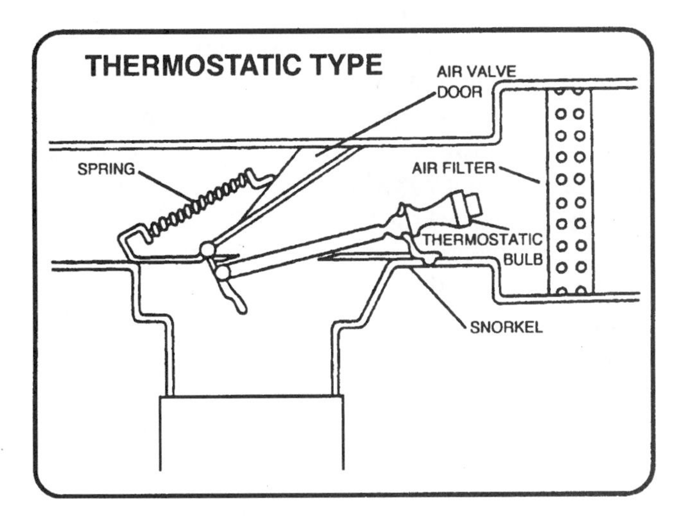

Fig. 10-12 Heated air intake systems: thermostatic type (top) and vacuum motor type (bottom)(reprinted, by permission, from New York State Department of Motor Vehicles, *Systems Training in Emissions and Performance.* Copyright 1993 Delmar Publishers Inc.).

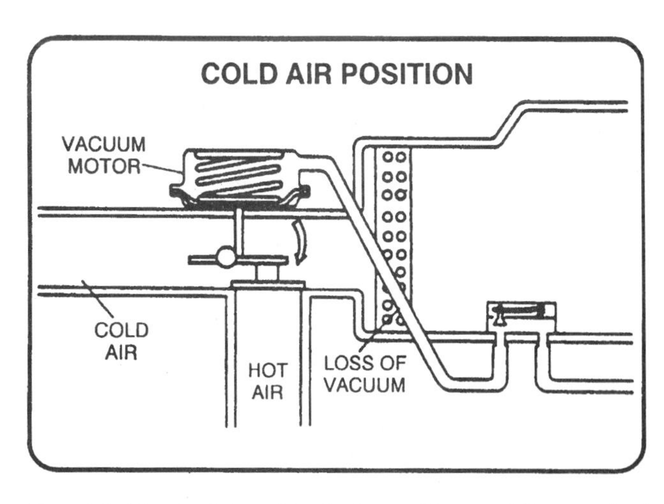

Fig. 10-13 Cold and hot air positions of the control door (reprinted, by permission, from New York State Department of Motor Vehicles, *Systems Training in Emissions and Performance.* Copyright 1993 Delmar Publishers Inc.).

As the engine warms up, the air doesn't have to be heated as much. The temperature sensor reacts to the rising temperatures and begins to bleed air. Spring pressure overcomes the vacuum diaphragm, and the air flow control door opens to admit more unheated air into the air cleaner (Fig. 10-13). By reacting to changes in the incoming air temperature, the temperature sensor is thus able to keep the carburetor supplied with warm air. This makes it easier for the carburetor to maintain a consistent air/fuel mixture. That's because air density changes with temperature. Cold air is more dense than warm air. When the air outside is cold (more dense), it has to be heated (made less dense) to prevent the fuel mixture from becoming too lean.

Driveability problems can arise when the heated air intake system fails to do its job properly. A missing, loose, or damaged heat riser tube connects the air cleaner to the stove on the exhaust manifold. This will prevent warm air from entering the air cleaner. A defective temperature sensor, vacuum leak, or faulty vacuum diaphragm in the air flap control motor, will likewise keep the system from functioning as it should. And the result can be rough idle, hesitation, and poor cold driveability, while the engine is warming up (especially during cold weather).

If the air control door is stuck in the closed position, a continuous supply of heated air is fed to the carburetor. The overheated air can then cause detonation (spark knock or pinging) during warm weather. The fuel mixture will also run rich because the air is too hot. This will have a negative effect on fuel economy and emissions, especially on older engines that don't have electronic feedback carburetion.

A check of the heated air intake system should begin with a visual inspection of the components. Is the tubing between the air cleaner and stove on the exhaust manifold intact and tight? Is the air duct that routes outside air into the air cleaner properly mounted and free from obstructions?

Next, remove the air inlet duct from the air cleaner and look inside. The air door should be in the open position when the engine is off. It should also be open when the engine is idling at normal operating temperature during warm weather. The air door should close to outside air when a cold engine is first started. The air door should then remain closed until the incoming air reaches a temperature of about 95 to 100 degrees F.

If the air door doesn't close when a cold engine is first started, check for a leak in the vacuum plumbing, a defective vacuum motor, or a faulty temperature sensor. If the door closes, but fails to open once the engine reaches normal operating temperature, the temperature sensor may be bad.

You can check the operation of the air control door by applying vacuum, with a hand pump, directly to the vacuum motor on the air cleaner inlet. If the door fails to move, or if the motor can't hold vacuum, the diaphragm inside the vacuum motor is leaking and the motor needs to be replaced.

The temperature sensor can also be checked with a hand vacuum pump. Leave the hose from the temperature sensor to the vacuum motor connected. Disconnect the vacuum supply hose to the sensor and apply vacuum directly to the sensor. The temperature sensor should pass vacuum on to the air door motor, as long as it is below about 95 to 100 degrees F. If a cold sensor leaks vacuum, replace it.

AIR INJECTION

No matter how well an engine is designed to minimize emissions, a certain amount of unburned hydrocarbons and carbon monoxide will always be left in the exhaust as it exits the engine's cylinders. So some means of cleaning up the exhaust is needed to reduce or eliminate these pollutants once they've exited the engine.

Air injection is just such a system. It first appeared 1968. Called AIR (Air Injection Reaction) by GM and Thermactor by Ford, the name that stuck was "smog pump."

The basic idea behind air injection is to pump extra oxygen into the exhaust. It can then combine with the pollutants before they exit the tailpipe and enter the atmosphere. In the early days of pollution control (before catalytic converters), one approach that was taken was the thermal reactor. The thermal reactor was a very large and well-insulated exhaust manifold used in conjunction with a high-capacity air pump. This combination provided a furnace-like environment in which the HC and CO was given a good opportunity to combine with oxygen to complete the combustion process. In the early 1970's, it looked as if this might be the answer to the emissions problem. It was an inexpensive technology, but it wasn't very efficient. It raised underhood temperatures drastically. So along came catalatic converters.

The wholesale adoption of the two-way catalytic converter by U.S. auto makers in 1975 changed the purpose of air injection. It became the means for supplying extra oxygen to the exhaust, so the pollutants could be reburned inside the converter.

The mixture of oxygen from the air pump and HC and CO flares up in the presence of the catalyst. Temperatures as high as 1600 degrees F are generated, eliminating most of the harmful pollutants by oxidizing them into C02 and water.

Some catalyst-equipped cars have been built that don't need air injection. These have engines calibrated in such a way that there's enough 02 in the exhaust to support the reaction in the converter. But these are exceptions. Most vehicles today have either an air injection system or an aspirator-valve setup to provide the extra fresh air the catalyst needs to do its job. (This will be explained later in this chapter).

You'll find the following components in a typical air injection system (Fig. 10-14):

- A belt-driven vane pump.
- A vacuum-operated diverter valve. This valve vents pump output to the atmosphere during decel, so the combination of a rich mixture and extra oxygen doesn't cause backfiring.

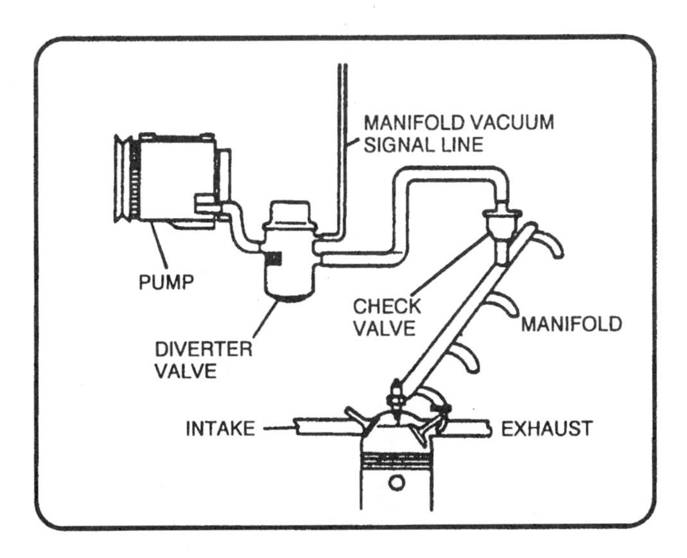

Fig. 10-14 Air injection system components (reprinted, by permission, from New York State Department of Motor Vehicles, *Systems Training in Emissions and Performance.* Copyright 1993 Delmar Publishers Inc.).

- A pressure relief valve. This valve allows excess pump output to escape.
- A one-way check valve. This valve allows air flow into the exhaust manifold, but keeps exhaust out of the pump if the belt breaks.
- The plumbing and nozzles necessary to distribute and inject the air.

Air Pump

When the engine is started, a belt causes the pump pulley to rotate. Inside the pump, vanes riding against the walls of a cylindrical chamber start moving air. The rotor turns on an axis that's different from that of the pump bore. The vanes slide in and out of slots in the rotor (Fig. 10-15). This causes a pumping action that moves air through the pump. Usually, intake is through a centrifugal filter mounted behind the drive pulley, but separate intake filters have been used on some applications. Most systems use a relief valve that allows excess pump pressure to escape. It may be in the pump itself, or incorporated into the diverter valve. The pressurized air exits into a large-diameter hose that routes it to this valve.

During all modes of engine operation except deceleration, air from the pump flows into the hose to the check valve. This is a simple one-way device, which lets air enter the air injection manifold. This one-way device also keeps exhaust from backing up

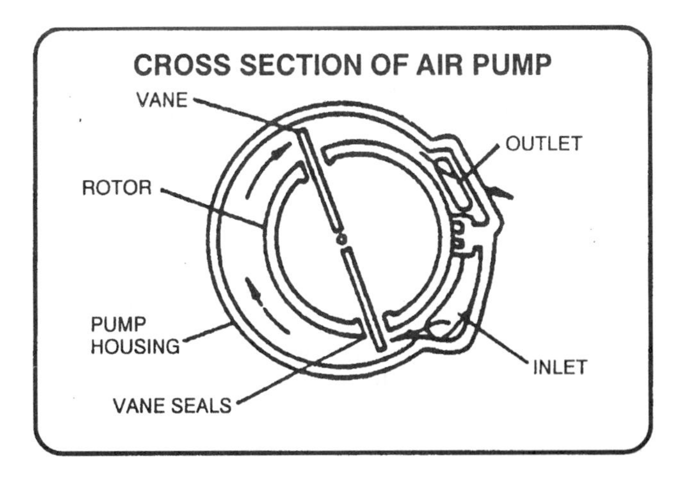

Fig. 10-15 Air pump (reprinted, by permission, from New York State Department of Motor Vehicles, *Systems Training in Emissions and Performance.* Copyright 1993 Delmar Publishers Inc.).

into the pump if a belt should break, or if the pump should otherwise stop working (normally the pump's pressure is high enough to overcome exhaust pressure). From the check valve, the air flows into the air injection manifold, which directs it into each exhaust port.

Diverter Valve

The diverter valve is the most complicated part of the air injection system (Fig. 10-16). The most common type receives a vacuum signal through a hose that runs to the intake manifold, or the base of the carburetor or throttle body. During closed-throttle deceleration when engine vacuum is strongest, it di-

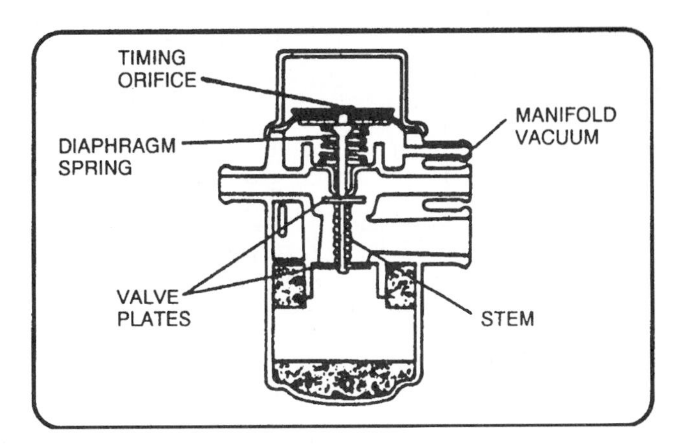

Fig. 10-16 Diverter valve (reprinted, by permission, from New York State Department of Motor Vehicles, *Systems Training in Emissions and Performance.* Copyright 1993 Delmar Publishers Inc.).

rects air flow through a small muffler, which is usually mounted on the pump.

The valve has a vacuum chamber, diaphragm, and spring arrangement. This valve moves the stopper from one of its seats to the other, and so controls the switching operation. During deceleration, the diaphragm in the valve's chamber overcomes the force of the spring, and dumps air flow to the muffler. This prevents backfiring in the exhaust system. If the the pump provides extra air, it would allow the rich mixture, that is present in the exhaust stream during deceleration, to ignite explosively. To put it simply, the diverter valve directs the air pump's output away from the exhaust system during deceleration.

Gulp and VDV

Another anti-backfire device is known as the gulp valve (Fig. 10-17). It has a diaphragm chamber, a spring, and a normally-closed valve inside. During deceleration when intake manifold vacuum reaches 20 to 22 in. Hg, the vacuum signal pulls the diaphragm against spring pressure. This action opens the valve. It also allows some of the air pump's output to flow into the intake manifold to dilute the rich mixture present in this mode. Thereby the possibility of extra gasoline vapor exploding in the exhaust system is eliminated. Usually, the addition of pump air to the intake stream lasts from I to 3 seconds.

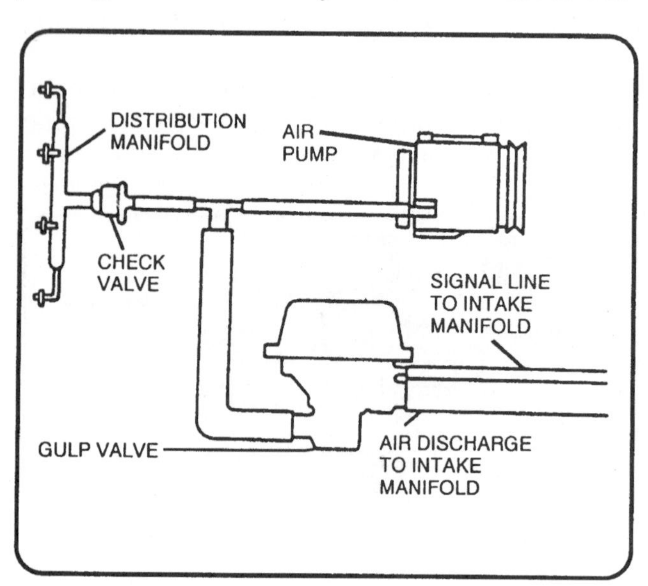

Fig. 10-17 Gulp valve in the air injection system (reprinted, by permission, from New York State Department of Motor Vehicles, *Systems Training in Emissions and Performance.* Copyright 1993 Delmar Publishers Inc.).

Some systems use a gulp valve with no air pump. They simply open a passage between the air cleaner and the intake manifold on deceleration, and let the engine draw as much as it can.

There is yet another means of preventing backfiring: the Vacuum Differential Valve (VDV). This valve works together with an air bypass valve. The VDV shuts off the vacuum signal to the bypass valve momentarily, whenever intake manifold vacuum rises or drops sharply. In normal operation, while the bypass valve is receiving vacuum, it allows air pump output to flow freely to the air injection manifold. When the VDV stops the vacuum signal, a spring inside the bypass valve opens a vent port. Pump flow is then directed to the atmosphere.

If there is excessive pump pressure, or a restriction in the check valve or air injection manifold, a relief valve inside the bypass valve opens. By opening this valve, some of the pump's output is dumped, and the rest of it to take the normal route.

AIR INJECTION WITH THREE-WAY CONVERTERS

In a typical late-model system, air flow goes to the exhaust manifold only while the engine is cold (Fig. 10-18). Very little NOx is produced then anyway, so the efficiency of the reduction process isn't

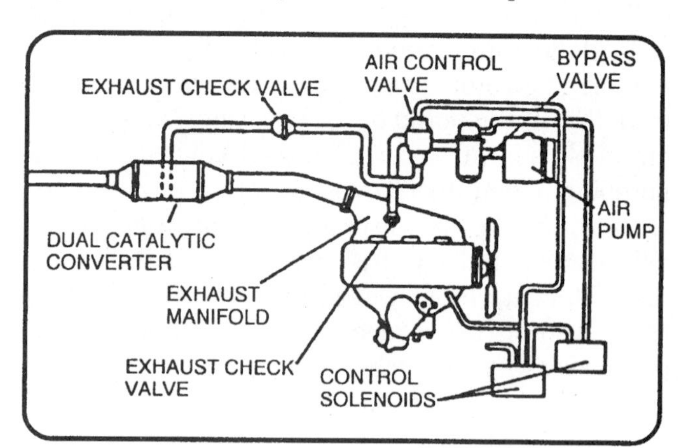

Fig. 10-18 Routing of air from the pump is directed either to the exhaust manifold or converter by the computer-controlled solenoids (reprinted, by permission, from New York State Department of Motor Vehicles, *Systems Training in Emissions and Performance.* Copyright 1993 Delmar Publishers Inc.).

important. Once normal operating temperature is reached, however, air pump output is switched downstream of the three-way section of the converter. This boosts the performance of the oxidation catalyst.

Some cars employ a computerized system that switches air injection flow from the exhaust ports to a line that runs directly into the catalytic converter. An early Ford Electronic Engine Controls (EEC) Ill system is an example of this. The ECA (Electronic Control Assembly) is Ford's name for what is more commonly known as the "Electronic Control Unit," or "ECU." The ECU makes decisions on where to send Thermactor (air pump) output on the basis of input from various engine sensors, notably that for coolant temperature, and according to certain time calibrations. It works in conjunction with a pair of solenoid valves: the bypass solenoid, which directs air pump flow to the atmosphere when energized, and the diverter solenoid, which switches air flow to either the exhaust ports or the catalytic converter.

During normal engine operation, air pump output is routed to the catalytic converter (there's a mixing chamber inside the shell between the reduction catalyst and the conventional two-way oxidation catalyst, and this is where the air enters). The computer energizes the bypass solenoid when time at closed throttle exceeds a specified time in its memory, when the interval between a lean and a rich signal from the oxygen sensor is longer than a set time value, and during wide-open throttle. This protects the converter from overheating because of an overly rich mixture, and guards against backfiring.

The computer energizes the diverter solenoid when the coolant sensor tells it that the engine is cold. This directs air pump flow upstream to the exhaust manifold during engine warmup, giving HC and CO more time to oxidize. This essential air switching is accomplished on some Chrysler cars without the help of electronics. An ordinary diverter valve is used to dump air pump output into the atmosphere during deceleration. But the directing of flow from the exhaust ports downstream to the exhaust pipe, ahead of the catalytic converter, is done by an air switching valve and a coolant control engine vacuum switch (CCEVS).

The purpose of this setup is different from that of the Ford system described above. When the engine is cold, fresh air directed at the exhaust ports is very helpful in reducing HC and CO emissions. But after normal operating temperature is reached, the high heat levels generated here start producing NOx. If the air is injected downstream in the exhaust system where it's cooler, HC and CO will still be oxidized. NOx emission won't be adversely affected, however.

The switching action is as follows: As long as there's a vacuum signal from the CCEVS to the air switching valve, the valve stays open. It allows air pump output to flow to the exhaust ports. When the engine warms up, the CCEVS shuts off the vacuum signal. The switching valve then directs most of the air flow to the downstream injection point. A bleed hole in the valve allows a small amount of air to be routed to the exhaust ports at all times. This assists in the reduction of HC and CO, but isn't enough to promote the formation of NOx.

AIR INJECTION WITHOUT A PUMP

Vacuum is present in the exhaust manifold momentarily after each cylinder's exhaust stroke (refer to Figs. 10-19 through 10-21). This is the result of the exhaust valve's closing, and the inertia of the column of spent gases as it speeds through the exhaust system. After the valve has closed, the column continues to travel away from the exhaust port, leaving a negative pressure area behind it. This is easier to understand if you picture the exhaust, not as a steady stream, but as a series of pulses.

This little-known phenomenon is the basis of

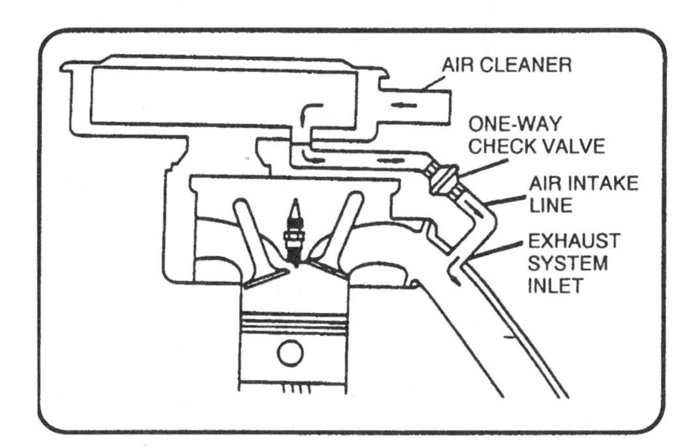

Fig. 10-19 An aspirator system is far simpler than an air pump system (reprinted, by permission, from New York State Department of Motor Vehicles, *Systems Training in Emissions and Performance.* Copyright 1993 Delmar Publishers Inc.).

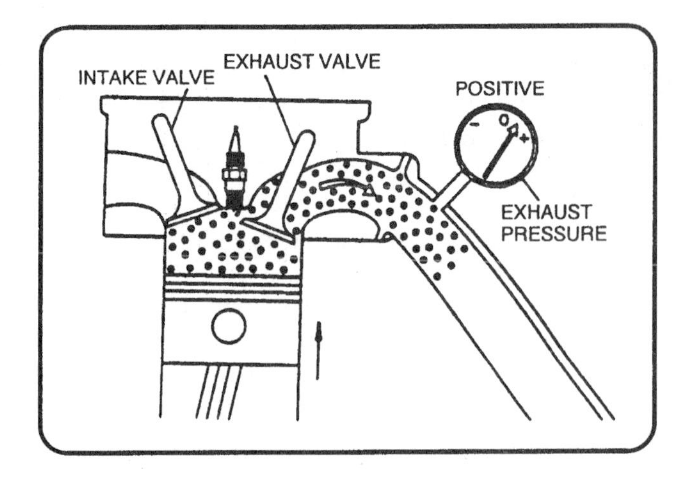

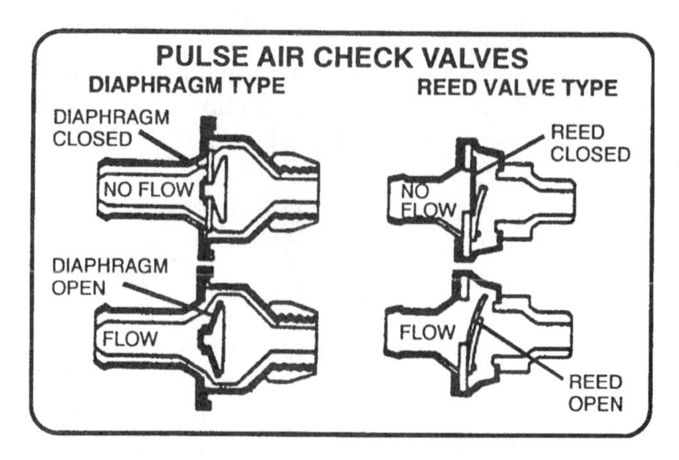

Fig. 10-21 Pulse air check valves (reprinted, by permission, from New York State Department of Motor Vehicles, *Systems Training in Emissions and Performance.* Copyright 1993 Delmar Publishers Inc.).

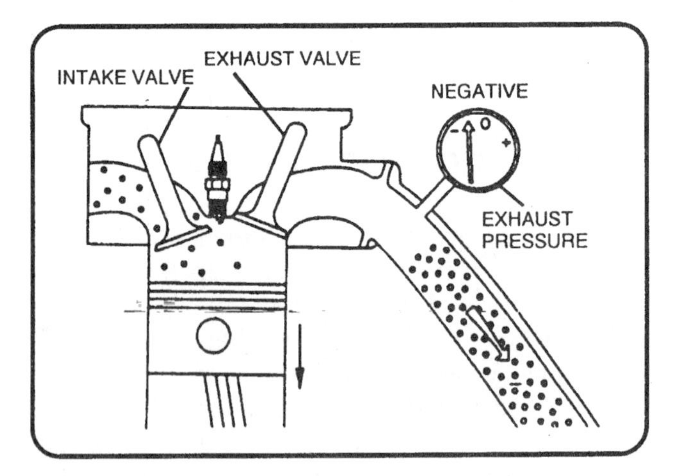

Fig. 10-20 Exhaust pulses create a momentary vacuum in the exhaust as they exit the exhaust port (reprinted, by permission, from New York State Department of Motor Vehicles, *Systems Training in Emissions and Performance.* Copyright 1993 Delmar Publishers Inc.).

pumpless air injection systems, generally known as aspirators. The main component is a one-way valve (similar to the check valve in pump-type air injection). This valve that allows fresh air picked up from the air cleaner to enter the exhaust stream whenever vacuum opens it. The valve stops the hot gases from escaping by closing when it encounters pressure (Fig. 10-21). There may be one aspirator valve for the whole engine, or one for each cylinder.

Aspirators have been used to eliminate the air injection pumps because they reduce complication, and save space in that crowded engine compartment. They're quite effective at idle and low-speed operation, but much less so at high rpm.

GM's Pulsair is a common aspirator system (Fig. 10-22). It comprises a housing that contains one valve for each cylinder, a tube that goes into each exhaust manifold runner, and a hose that picks up fresh air from the air cleaner.

CATALYTIC CONVERTERS

Catalytic converters have been in use since 1975 when most cars were required to burn unleaded gasoline. The converter was added to the exhaust system as a sort of "afterburner" device to reduce the levels of certain pollutants in the exhaust. There are essentially three basic types:

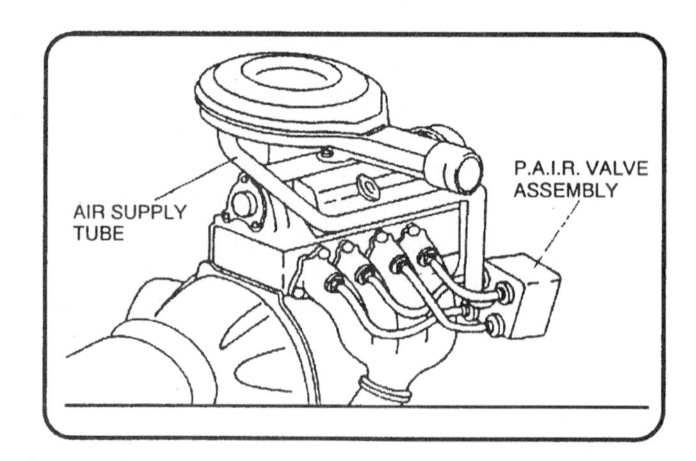

Fig. 10-22 General Motors Pulsair system (reprinted, by permission, from New York State Department of Motor Vehicles, *Systems Training in Emissions and Performance.* Copyright 1993 Delmar Publishers Inc.).

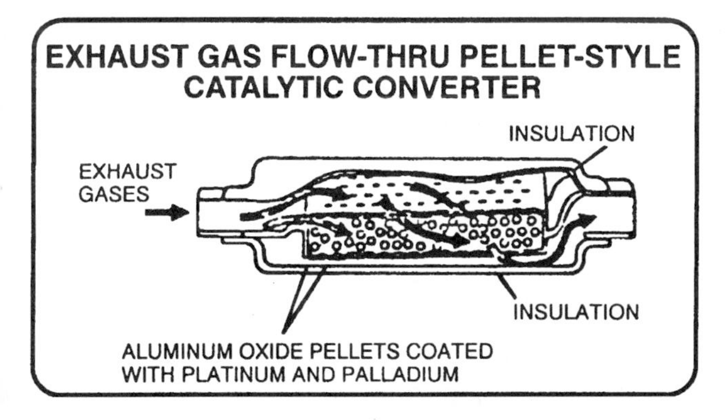

Fig. 10-23 A GM two-way pellet-style converter (reprinted, by permission, from New York State Department of Motor Vehicles, *Systems Training in Emissions and Performance.* Copyright 1993 Delmar Publishers Inc.).

- Two-way converters control unburned hydrocarbons (HC) and carbon monoxide (CO). They are used primarily on pre-1980 vintage vehicles (Fig. 10-23).
- Three-way converters (TWC) that control HC, CO and oxides of nitrogen (NOX) used on 1980-85 vehicles without computerized engine controls or air injection.
- Three-way plus oxidation converters which are used on 1980 and later vehicles with computerized engine controls and air injection (Fig. 10-24).

The catalytic converter reduces the amount of HC and CO in the exhaust by converting them into carbon dioxide (CO2) and water vapor. Additional oxygen (provided by the air pump or an aspirator valve on vehicles without an air pump) is required.

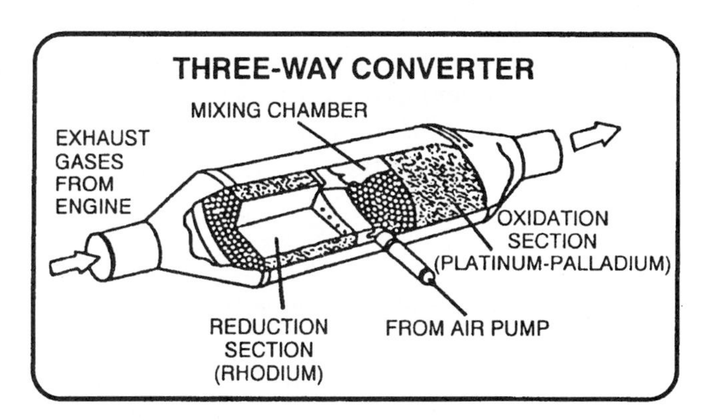

Fig. 10-24 A monolithic three-way converter (reprinted, by permission, from New York State Department of Motor Vehicles, *Systems Training in Emissions and Performance.* Copyright 1993 Delmar Publishers Inc.).

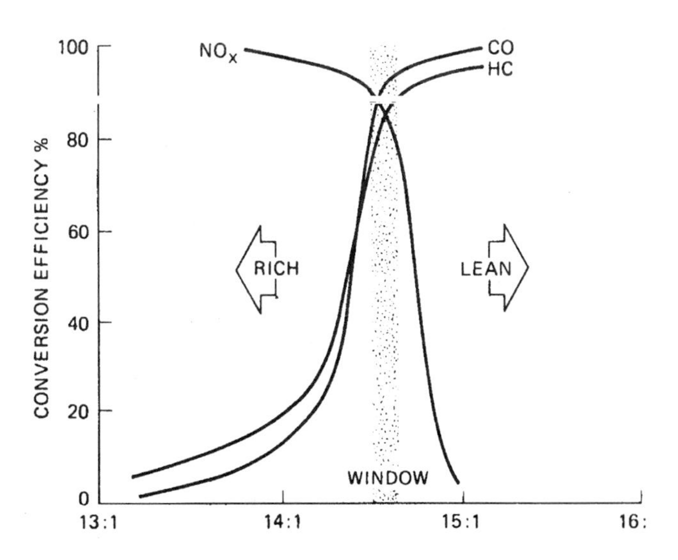

Fig. 10-25 The efficiency of a three-way converter is dependent upon the air/fuel ratio (courtesy Volkswagen).

Two other requirements are heat (provided by the exhaust itself), and a balanced air/fuel mixture. Converter efficiency drops significantly if the air/fuel mixture is too rich or too lean (Fig. 10-25).

The reburning of pollutants occurs when the hot exhaust and extra air pass over the catalyst inside the converter. The catalyst consists of a thin coating of platinum and palladium metal on a ceramic honeycomb, or pellets. The catalyst ignites the mixture and keeps the reaction going as long as there are pollutants to burn.

In three-way converters, an additional catalyst (rhodium) is used in a forward compartment to break down NOX. Extra oxygen isn't needed for this reaction to occur. In fact, oxygen hinders the conversion of NOX. The extra air the converter needs to burn HC and CO is routed through a pipe that feeds it into the converter, just aft of the first compartment. When the engine is cold, NOX emissions are low. This enables the extra air to be pumped directly into the exhaust manifold to get the converter going. But once things start to warm up, the diverter valve (also called an air control valve) reroutes the extra air directly to the converter.

The most common reasons for converter failure are overheating and fuel poisoning (from using leaded gasoline). A misfiring spark plug, leaky exhaust valve or head gasket, can allow unburned fuel to pass through the engine and into the exhaust. The overload of raw fuel can send the converter's temperature soaring. The high temperature results in a partial or complete meltdown of the converter's innards, and a severe or complete exhaust restriction.

If an engine doesn't run right or seems sluggish and there are no obvious ignition, fuel or compression problems, then be sure to check either intake vacuum or exhaust backpressure. Then see if the exhaust is obstructed. A low vacuum reading or high backpressure reading would tell you something is wrong, and that further diagnosis is needed. You should disconnect or remove the converter, and look inside it with a trouble light. If you can't see through it, and it isn't a GM pellet type converter, then you've found the problem.

In cases where the converter has suffered a complete meltdown, the engine may start. Then after a few seconds it may die, and refuse to start again. Again, a peek inside the converter will usually confirm the problem.

A "thunk" test on the outside of the converter with a soft rubber mallet can reveal a loose catalyst inside. But unless the noise is objectionable or the damaged catalyst is causing a blockage, replacement may not be necessary.

A poisoned converter will not burn the pollutants in the exhaust. Higher than normal HC and CO readings on your exhaust analyzer would be a good indication of a dead converter. Check the restricter plate under the gas cap. If it's been enlarged or knocked out, the converter has probably been poisoned with lead. But before you condemn the converter, check its air supply. This can be done by monitoring the oxygen level in the exhaust, while switching the diverter valve solenoid on and off. No change in the oxygen reading would indicate a defective diverter valve or air pump.

Another item to inspect is the air pump check valve. The valve is located between the air pump and catalytic converter. It's one way diaphragm allows air to travel in only one direction, preventing hot exhaust from backing up into the air pump. A check valve can be damaged by severe back pressure caused by a clogged converter or muffler. A buildup of carbon deposits in the valve can also cause it to stick open or closed. Corrosion, however, is the most common reason for failure.

Converters can also die of old age. After so many tens of thousands of miles, the catalyst inside gets "tired" and gradually loses its ability to do its job. The process can be accelerated by the accumulation of phosphorus, lead or silicone deposits. These deposits can come from an engine that burns oil or from using the wrong fuel. Such deposits coat the catalyst and render it ineffective.

When a converter gets too old to do its job efficiently, it no longer reduces the level of pollutants in the exhaust as it once did. Consequently, HC and CO emissions go up—and if the vehicle is subject to an emissions test, it may flunk because of excessive emissions.

The converter is covered by the vehicle manufacturer's emissions warranty. The EPA has strict rules governing its replacement. Only new car dealers can replace a converter that's still under warranty. A converter that's out of warranty can be replaced by anyone. A legitimate need for replacing it must be established and documented (such as a blockage, failing an emissions test, or the converter is missing). When a converter is replaced, the shop performing the work must obtain the customer's written authorization. The shop must keep the required paperwork for six months and the old converter for 15 days. The replacement converter must be the same type as the original, and installed in the same location in the exhaust system. Violators can be fined.

EXHAUST GAS RECIRCULATION (EGR)

The exhaust gas recirculation (EGR) system's purpose is to reduce NOX emissions. As long as the EGR system is functioning properly, it should have no noticeable effect on engine performance. But when something goes amiss with the EGR system, it can cause a variety of symptoms. These symptoms include detonation (knocking or pinging when accelerating or under load), a rough idle, stalling, hard starting, elevated NOX emissions, and even elevated HC emissions in the exhaust.

Why EGR?

Exhaust gas recirculation reduces the formation of NOX by allowing a small amount of exhaust gas to "leak" into the intake manifold. The amount of gas leaked into the intake manifold is only about 6 to 10% of the total. It's enough to dilute the air/fuel mixture just enough to have a "quenching effect" on combustion temperatures. This keeps temperatures below 2500 degree F, which is the threshold point at which nitrogen reacts with oxygen to form NOX.

The need for some type of engine control technology to lower or reduce NOX emissions became apparent when scientific studies proved the link between NOX emissions and smog. The Environmental Protection Agency decided to add NOX emission standards to its list of things to regulate. In the early 1970s, the auto makers were designing engines with later ignition timing, leaner carburetion, and higher combustion temperatures to lower HC and CO emissions in compliance with EPA regulations. Unfortunately, the changes that reduced HC and CO emissions tended to increase NOX emissions. Figuring out how to lower HC and CO without affecting NOX (or even lowering it) seemed like an impossible task —until engineers discovered that slowing down the combustion process slightly, and lowering combustion temperatures, did the trick. They could achieve lower HC and CO emissions, and lower NOX emissions too, by recirculating a small amount of "inert" gas into the air/fuel mixture to dilute it slightly. It worked great in a laboratory, but the problem was finding an inexhaustible supply of inert gas. The answer was found in the exhaust itself. Exhaust is mostly carbon dioxide and water vapor, with nitrogen, small amounts of unburned hydrocarbons, carbon monoxide and other trace gases. There's very little free oxygen in the exhaust because most of it is burned in the engine. So for all practical purposes, exhaust is essentially inert.

The first production EGR system appeared on 1972 Buicks. It was then added to most passenger car engines in 1973 to meet federal NOX emission standards.

How EGR Works

To recirculate exhaust back into the intake manifold, a small calibrated "leak" or passageway is created between the intake and exhaust manifolds. Intake vacuum in the intake manifold sucks exhaust back into the engine. But the amount of recirculation has to be closely controlled. Otherwise it can have the same effect on idle quality, engine performance, and driveability as a huge vacuum leak. As mentioned earlier, most EGR systems restrict maximum exhaust gas recirculation to no more than 6 to 10% of the total air/fuel mixture. Higher rates of flow can dilute the air/fuel mixture excessively, causing the fuel mixture to misfire. This, in turn, can create a very rough idle, a stumble or hesitation upon acceleration, and contribute to hard starting. Engine misfire can also allow unburned fuel to pass through into the exhaust, creating elevated HC readings in the exhaust.

Some of the early V8 EGR systems had small jets in the base of the intake manifold to regulate exhaust flow into the manifold, from the exhaust crossover passage. Though simple in design, the "floor jet" EGR system had several drawbacks. One was that there was no way to "fine tune" exhaust flow. It varied in direct proportion to intake vacuum. Under some driving conditions (such as wide open throttle under load) there was not enough exhaust gas recirculation to keep NOX under control. At idle, there sometimes tended to be a bit too much exhaust gas recirculation which made for a rough idle. Another problem with the EGR jets in the bottom of the intake manifold was that they often plugged up with carbon and varnish.

A better approach proved to be that of using a vacuum regulated valve to control exhaust gas recirculation. The "EGR valve" thus became the heart of the EGR system. It made possible a significant improvement in NOX reduction and overall system performance. The EGR valve meant that EGR could now be regulated more in sync with the demands of the engine: no EGR at idle, and maximum EGR when the engine was working hard under load.

In addition to EGR, other changes in engine design and operation also contributed to minimizing NOX. These included increasing camshaft valve overlap, redesigning combustion chambers and modifying ignition advance curves. With the addition of three-way catalytic converters that could also reduce NOX in the exhaust, it even became possible on some engines to eliminate the EGR system altogether.

Nox or Knocks?

Diluting the air/fuel mixture with exhaust is the right prescription for making horsepower. The quenching effect it creates in the combustion chamber helps the engine resist detonation (spark knock). That means the engine can tolerate more spark advance and compression on lower octane regular fuel.

If an EGR system is rendered inoperative the quenching effect that was formerly provided by the

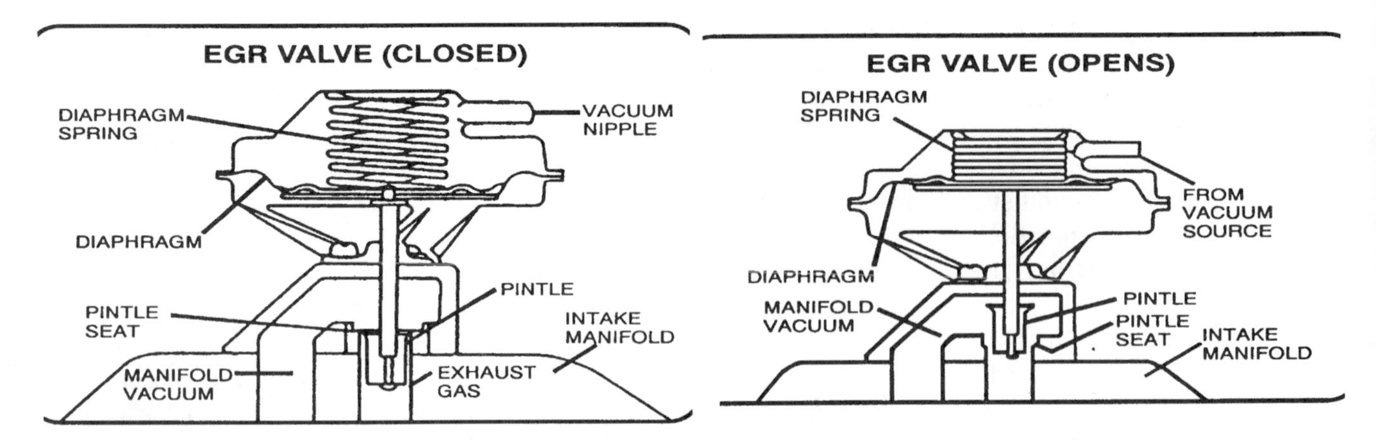

Fig. 10-26 EGR valve in the closed and open positions (reprinted, by permission, from New York State Department of Motor Vehicles, *Systems Training in Emissions and Performance.* Copyright 1993 Delmar Publishers Inc.).

EGR system will be lost. One way this can happen is when someone illegally unhooks or plugs the vacuum hosé to the EGR valve. It can also happen if someone removes the EGR valve, or installs an aftermarket intake manifold that lacks the required EGR connections. Without EGR, the engine will likely knock and ping during hard acceleration, or when the engine is heavily loaded. If ignition timing is retarded, relative to stock specifications in an attempt to "offset" the loss of EGR, performance and fuel economy usually suffer.

EGR Valves

The typical EGR valve consists of a vacuum diaphragm connected to a poppet or tapered stem flow control valve (Fig. 10-26). The valve opens a small passage between the exhaust and intake manifolds. The EGR valve itself is usually mounted either on a spacer under the carburetor, or on the intake manifold. A small pipe from the exhaust manifold, or an internal crossover passage in the cylinder head and intake manifold, carries exhaust to the valve.

When the EGR valve opens, intake vacuum pulls exhaust into the engine, where it mixes with and dilutes the incoming air/fuel mixture. This normally occurs when the engine has reached normal operating temperature (little NOX is formed when the engine is cold so EGR isn't needed until the engine is warm). It occurs also only when EGR is needed to reduce combustion temperatures (as when accelerating under load or driving at part-throttle to full-throttle).

The amount of exhaust that enters the intake manifold is determined by (1) the size of the EGR valve orifice, (2) how far the valve opens, and (3) how long it is held open.

EGR, remember, is not a full-time thing. Because it has the same effect on driveability as a vacuum leak, it is used only during part-throttle operation, when the engine can tolerate a diluted fuel mixture. EGR is not used at idle (Fig. 10-27) because it would cause the engine to run rough and possibly stall (a classic symptom of an EGR valve that is stuck in the open position). Nor is EGR al-

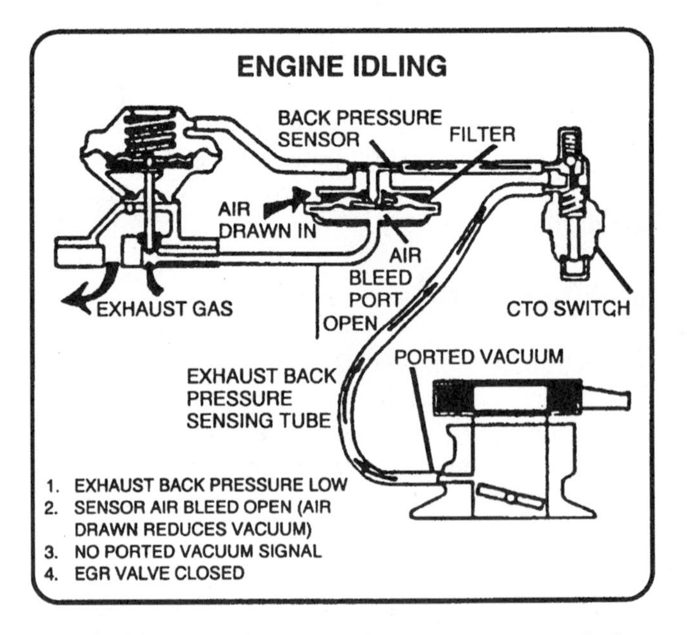

Fig. 10-27 No EGR at idle (reprinted, by permission, from New York State Department of Motor Vehicles, *Systems Training in Emissions and Performance.* Copyright 1993 Delmar Publishers Inc.).

lowed when the engine is being cranked, because it would act like a vacuum leak and make the engine hard to start.

How does the EGR valve know when to open and close? In most applications, it operates in response to engine vacuum either from a ported vacuum source above the throttle plates or from venturi vacuum, at the carburetor throat.

With ported vacuum systems, engine vacuum pulls the EGR valve open. This occurs when the throttle plates are cracked open far enough to expose the vacuum port to intake vacuum. The use of ported vacuum prevents EGR at idle. A spring inside the EGR valve pushes the valve shut when there's no vacuum (as when the throttle plates are closed), or when engine vacuum drops below a certain level.

EGR systems, using venturi vacuum, require a vacuum amplifier to boost the relatively weak vacuum signal at the carburetor venturi. Reading vacuum from this point allows precise measurements between EGR action and airflow through the carburetor. The amplifier is connected to a vacuum reservoir in many such applications. The amplifier contains a check valve to maintain an adequate vacuum supply regardless of variations in engine vacuum. A relief valve may also be used to dump or cancel the output EGR signal when ever the venturi vacuum signal is equal to, or greater than, intake vacuum. This allows the EGR valve to close at wide open throttle on some applications, when maximum power is required.

Positive and Negative Backpressure EGR Valves

Another feature that has been incorporated into EGR systems is the ability to change EGR flow in response to changes in exhaust system backpressure (Fig. 10-28). NOX formation increases when the engine is under load and during acceleration. EGR flow should also increase during these times to compensate for higher combustion temperatures. Exhaust backpressure is a good measure of engine load. For this reason, a pressure-sensitive diaphragm that reacts to changes in backpressure is an effective means of regulating EGR operation. The backpressure diaphragm may be located inside the EGR valve itself, together with the main control diaphragm, or in a separate housing. The backpressure diaphragm opens and closes a small vac-

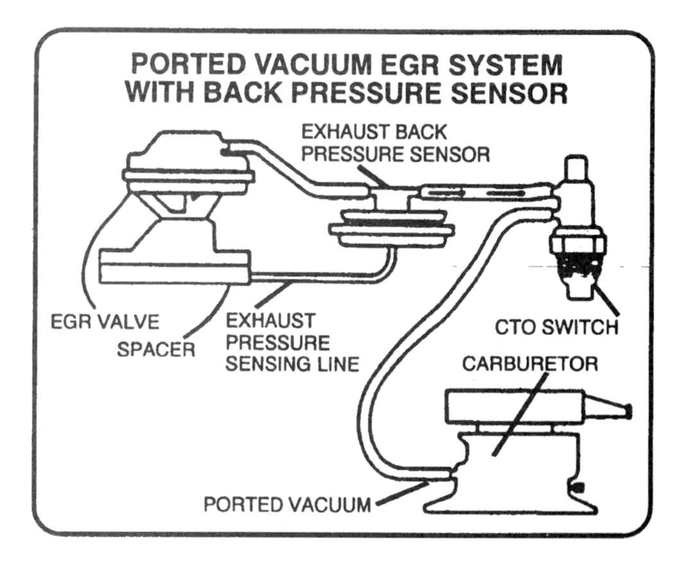

Fig. 10-28 Ported-vacuum EGR system with an external backpressure sensor (reprinted, by permission, from New York State Department of Motor Vehicles, *Systems Training in Emissions and Performance.* Copyright 1993 Delmar Publishers Inc.).

uum bleed hole in the main EGR vacuum circuit or diaphragm chamber. Opening the bleed hole reduces the vacuum to the EGR valve, and prevents it from opening fully. Closing the bleed hole allows full vacuum and maximum EGR flow. This allows the backpressure diaphragm to increase or decrease EGR vacuum in direct proportion to changes in exhaust backpressure.

There are two types of backpressure EGR valves: positive and negative. The positive type (Fig. 10-29) uses positive exhaust backpressure to regulate EGR flow. As pressure increases in the exhaust, the valve begins to open, allowing increased EGR into the intake manifold. This reduces backpressure somewhat, allowing the backpressure diaphragm to bleed off some control vacuum. The EGR valve begins to close, and exhaust pressure rises again. The EGR valve oscillates open and closed, with changing exhaust pressure, to maintain a sort of balanced flow. The negative backpressure type of EGR valve (Fig. 10-30) reacts in the same way. However, it reacts to negative or decreasing pressure changes in the exhaust system to regulate EGR action. Decreasing backpressure signals decreased engine load, and the backpressure diaphragm opens a bleed hole to reduce EGR flow. It's the same principle as with the positive type, except that the control function reacts to decreasing pressure rather than increasing pressure.

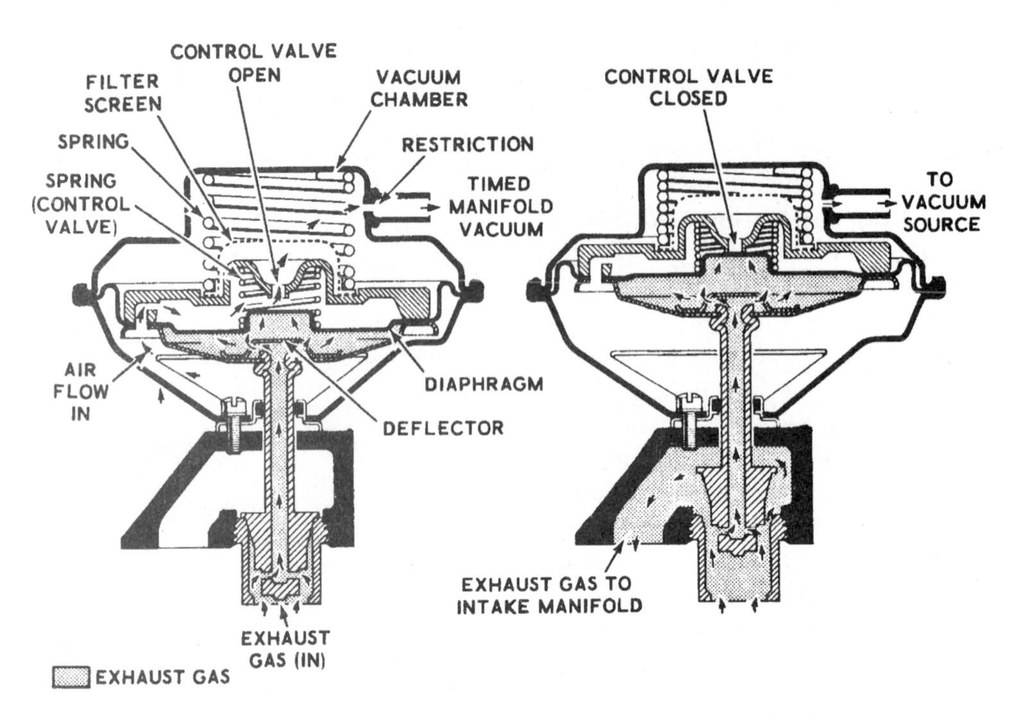

Fig. 10-29 Positive backpressure EGR valve (courtesy AC Delco).

Backpressure EGR valves, both positive and negative, provide better control over EGR flow. This is because they have the ability to react directly to changing engine loads. There are a couple of drawbacks, however. For one, the small bleed hole can become clogged. On backpressure EGR valves with the backpressure diaphragm inside the valve itself, a hollow valve stem is used to carry exhaust pressure to the diaphragm. This stem can also become plugged quite easily.

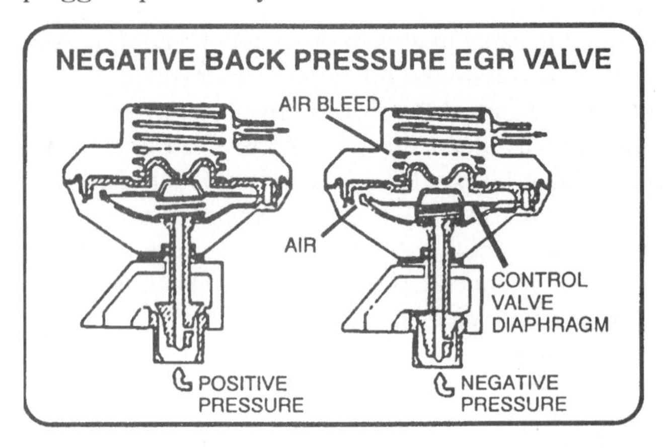

Fig. 10-30 Negative backpressure EGR valve (reprinted, by permission, from New York State Department of Motor Vehicles, *Systems Training in Emissions and Performance.* Copyright 1993 Delmar Publishers Inc.).

In respect to engines with backpressure EGR valves, the original equipment EGR valve is calibrated to the backpressure in the stock exhaust system. If the stock muffler has been replaced with a low restriction aftermarket muffler, the change may reduce exhaust backpressure enough to have an adverse affect the operation of the EGR valve. Pinging (detonation) when accelerating would be a good clue that reduced backpressure is preventing your EGR valve from opening. Reduced EGR because of a change in exhaust backpressure may cause the vehicle to fail an I/M 240 emissions test because of elevated NOX emissions.

TVS Switch

Most precomputer EGR systems have a temperature vacuum switch (TVS) or ported vacuum switch between the EGR valve and vacuum source. This TVS prevents EGR operation until the engine has had a chance to warm up (Fig. 10-31). The engine must be relatively warm before it can handle EGR. If an engine runs rough or stumbles when cold, it may indicate a defective TVS that is allowing EGR too soon after starting. A TVS stuck in the closed position would block vacuum to the EGR, and prevent any EGR operation. The symptom here would

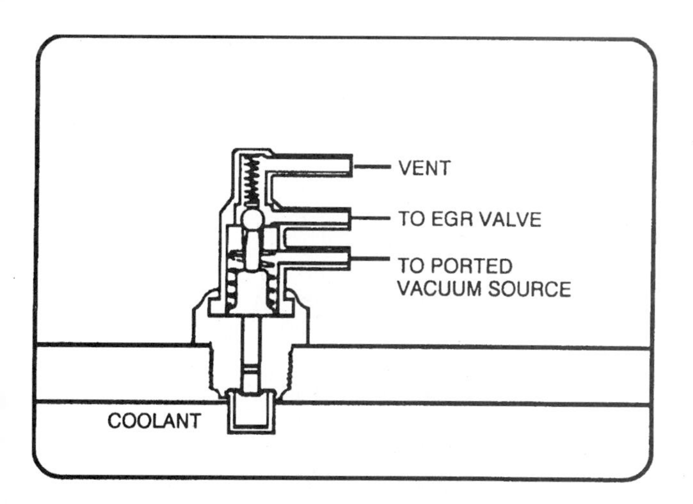

Fig. 10-31 The TVS switch blocks vacuum to the EGR valve when the engine is cold to improve cold driveability (reprinted, by permission, from New York State Department of Motor Vehicles, *Systems Training in Emissions and Performance.* Copyright 1993 Delmar Publishers Inc.).

be excessive NOX emissions and possible pinging or detonation.

Wide Open Throttle Switch or Valve

Many EGR systems use a wide open throttle (WOT) switch or valve to eliminate EGR action during those times when maximum power and acceleration are needed. On some systems, a diaphragm vents EGR vacuum to the atmosphere when intake manifold vacuum drops to zero. On later model engines with computerized engine controls, a throttle switch or throttle position sensor (TPS) signals the computer when the throttle is wide open. It can then temporarily turn off the EGR system.

Other Means of Modulating EGR Action

Some EGR systems use an air bleed orifice, or solenoid, to modulate EGR action. According to engine operating conditions, the air bleed may be opened to reduce EGR vacuum. Opening the air bleed reduces how far the EGR valve opens. On some General Motors applications, for example, the amount of EGR is reduced when the automatic transmission torque converter clutch (TCC) is engaged. The same thing may be used on vehicles with manual transmissions when running in high gear. The reason for doing this is to provide smoother engine operation. Too much EGR flow can cause roughness and hesitation.

With diesel engines, there is no intake vacuum for EGR operation or control, so vacuum is usually created by an auxiliary vacuum pump. The pump provides a steady amount of vacuum for opening the EGR valve. Operation is regulated by computer-controlled vacuum solenoids, and input from an electrical backpressure sensor in the exhaust system.

EGR & Computer Controls

On most engines with computerized engine control, a temperature vacuum switch is not used, because EGR vacuum is controlled by the computer. The computer monitors engine temperature through the coolant sensor. When the programmed operating temperature is reached, the computer opens the EGR vacuum solenoid, allowing intake manifold vacuum to pass through to the valve. On some systems, the control solenoid is normally closed. Energizing it opens the solenoid and allows vacuum to reach the EGR valve. On other systems, the EGR solenoid may be open in the normal position. It is energized (closed) only when EGR is not wanted, as when the engine is cold, during cranking, or at wide-open throttle. In either case, once the engine warms up, EGR flow is controlled as usual by ported vacuum and exhaust
backpressure.

Some computer systems use air bleeds or vents in conjunction with the EGR solenoid to regulate EGR flow. Both Ford and General Motors do this on certain systems. Early Ford Electronic Engine Control systems (EEC-II & EEC-III) have two EGR control solenoids, and an EGR valve position sensor mounted on top of the EGR valve (Fig. 10-32). The EEC computer monitors various engine functions, as well as EGR position to determine how much EGR is needed. Opening the normally closed EGR vacuum control solenoid (EGRC) allows manifold vacuum to pass to the EGR valve. The normally open EGR vent solenoid (EGRV) vents some air into the vacuum line, to reduce vacuum and limit how far the EGR valve opens. If more EGR is needed, the computer energizes (closes) the EGRV to stop the air leak, so that full vacuum can reach the EGR valve. The computer can modulate EGR action by "dithering" (opening and closing) the two solenoids to achieve the amount of EGR needed. Energizing both

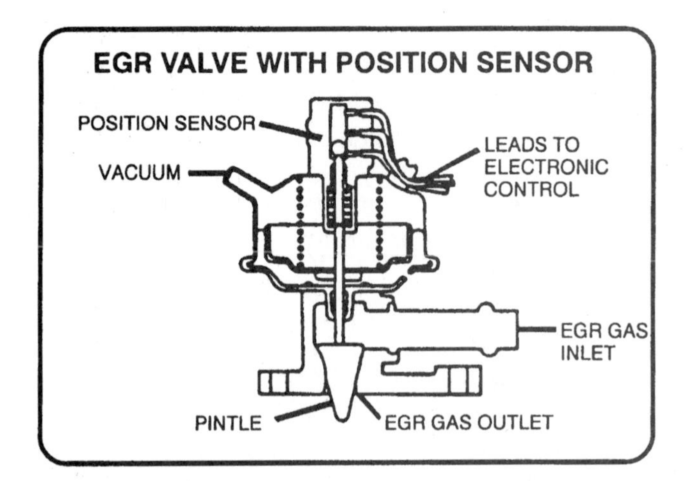

Fig. 10-32 EGR valve with position sensor (reprinted, by permission, from New York State Department of Motor Vehicles, *Systems Training in Emissions and Performance.* Copyright 1993 Delmar Publishers Inc.).

provides maximum EGR (full vacuum). Energizing only the EGRC, but not the EGRV, provides a sort of "midrange" EGR (part vacuum). Not energizing either solenoid prevents any EGR (no vacuum).

Another approach to EGR control (which was first introduced in 1984 by General Motors) is to use a pulse width-modulated EGR control solenoid. With this technique, the engine control module cycles the EGR vacuum control solenoid rapidly on and off. This creates a variable vacuum signal that can regulate EGR operation very closely. The amount of "on" time versus "off" time for the EGR solenoid ranges from 0 to 100 percent. The average amount of "on" time, versus "off" time, at any given instant determines how much EGR flow occurs.

On some applications, a "digital" EGR valve is used. This type of valve also uses vacuum to open the valve, but regulates EGR flow according to computer control. The digital EGR valve has three metering orifices that are opened and closed by solenoids (Fig. 10-33). By opening various combinations of these three solenoids, different flow rates can be achieved to match EGR to the engine's requirements. The solenoids are normally closed, and open only when the computer completes the ground to each.

The latest innovation in EGR systems is a "linear" EGR valve. This valve uses a small computer-controlled stepper motor to open and close the EGR valve instead of vacuum. The advantage of this approach is that the EGR valve operates totally independent of engine vacuum. It is electrically operated

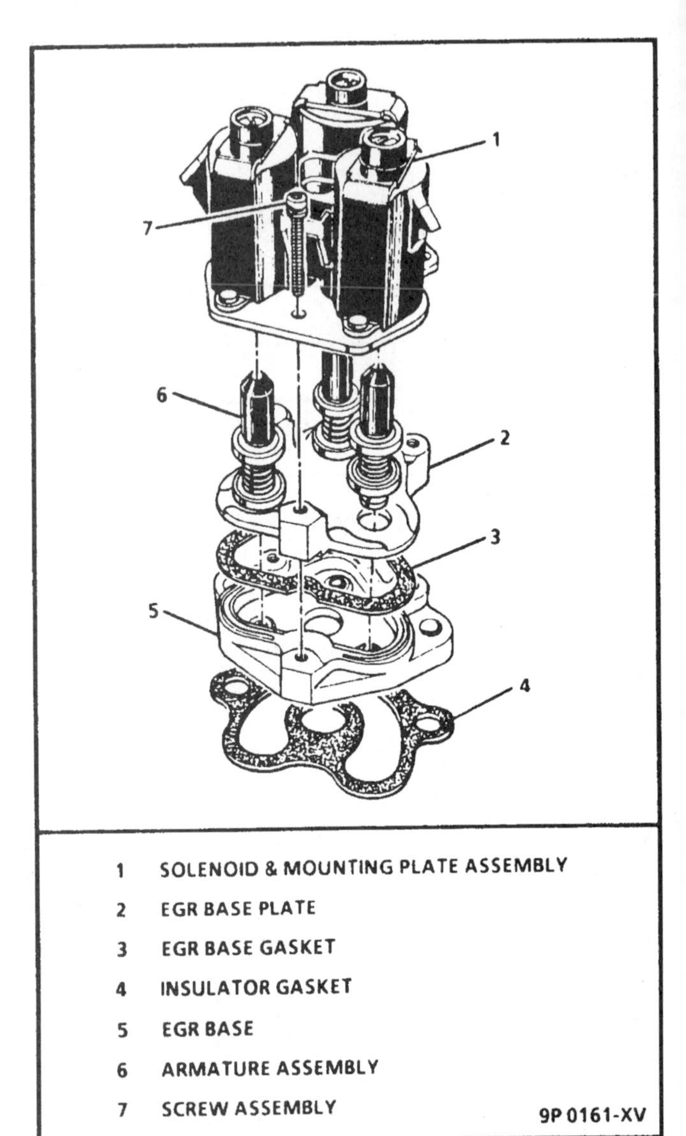

Fig. 10-33 Digital EGR valve (courtesy General Motors).

and can be opened in various increments. The increments depend on what the engine control module determines the engine needs at any given moment in time. GM started using this type of valve on many of its passenger car engines in 1992.

EGR Problems

If you think an EGR system isn't working properly it will be necessary to diagnose the EGR system to find out what's wrong. Perhaps the engine is experiencing engine detonation under load, a rough idle or hesitation problem, Another problem might be that it failed the NOX part of an I/M 240 emissions test.

There are many different EGR systems in use today. The first step in troubleshooting a suspected EGR problem is to find out what type of EGR system is used on the engine (refer to a service manual).

Does the EGR system use ported vacuum, venturi vacuum, or is it computer-controlled? Does it have a temperature vacuum switch, a computer-controlled EGR vacuum, or vent solenoid, or a wide open throttle (WOT) switch or valve? Is the EGR valve the backpressure type, and if so which type, positive or negative? Is there an EGR valve position sensor? Are there other systems, such as canister purge, plumbed into the EGR vacuum circuit? These are some of the things you should know before troubleshooting the system.

There are several ways to troubleshoot an EGR system. You can follow the EGR troubleshooting procedure that's listed in a service manual for the engine. On late model computer controlled engines, there may be trouble codes that relate to the EGR system. On such an application, the first step would be to read out the code, or codes, using a special diagnostic procedure or scan tool. You would then refer to the specific diagnostic charts in a service manual that tell you what to do next.

On late model GM applications, for example, a code 32 indicates an EGR problem. The logic by which the onboard diagnostics detects trouble follows one of two routes. On some applications, a code 32 is set when the computer detects a richer fuel mixture off idle (indicating no EGR). On others, a code is set if the computer energizes the EGR vacuum solenoid, but does not detect a corresponding drop in intake vacuum.

On late model Fords, a code 31 indicates a problem with the EGR valve position sensor (EVP). It works like a throttle position sensor, going from high resistance (5500 ohms) when the EGR valve is closed, to low resistance (100 ohms) when it is open. You'll find these EVP sensors mostly on Ford EEC-IV V6 and V8 engines. Other codes include a code 32, which indicates the EGR circuit is not controlling. A code 33 is triggered when the EVP sensor is not closing, and a code 34 indicates no EGR flow. Any of these codes could indicate a faulty EGR valve as well as a problem in the EGRC or EGRV vacuum solenoids. Other codes include a code 83 (EGRC circuit fault) and code 84 (EGRV circuit fault). Both indicate an electrical problem in one of the solenoid circuits. The solenoids should have between 30 and 70 ohms resistance.

EGR Diagnosis

The following "generic" procedure may help you troubleshoot EGR problems.

1. Does the engine have a detonation (spark knock) problem when accelerating under load? Refer to the timing specs for the engine and check ignition timing. The timing may be overadvanced. If the timing is within specs, check the engine's operating temperature. A cooling problem may be causing the engine to detonate. If the temperature is within its normal range and there are no apparent cooling problems, you should investigate other possibilities. These possibilities include spark plugs that are too hot for the engine application, a lean air/fuel mixture, and low octane fuel. Too much compression, due to a buildup of carbon in the combustion chambers, or because of pistons or heads that have too much compression for the fuel you're using, is another possibility. Be sure you've ruled out all the other possibilities before focusing on the EGR system.

2. Use a vacuum gauge to check the EGR valve vacuum supply hose for vacuum at 2000-2500 rpm. There should be vacuum if the engine is at normal operating temperature. No vacuum would indicate a problem, such as a loose or misrouted hose, a blocked or inoperative ported vacuum switch or solenoid, or a faulty vacuum amplifier (or vacuum pump in the case of a diesel engine).

Sometimes loss of EGR can be caused by a failed vacuum solenoid in the EGR's vacuum supply line. Refer to a vacuum hose routing diagram in a service manual, or the hose routing information on the vehicle's emission decal for the location of the solenoid. If the solenoid fails to open when energized, jams shut or open, or fails to function because of a corroded electrical connection, loose wire, bad ground, or other electrical problem, it will obviously affect the operation of the EGR valve. Depending on the nature of the problem, the engine may have no EGR, EGR all the time, or insufficient EGR. If bypassing the suspicious solenoid, with a section of vacuum tubing, causes the EGR valve to operate, find out why the solenoid isn't responding before you replace it. The problem may be nothing more than a loose or corroded wiring connector.

3. Inspect the EGR valve itself. Because of the valve's location, it may be difficult to see whether or not the valve stem moves when the engine is revved to 1500 to 2000 rpm, by slowing opening and closing the throttle. The EGR valve stem should move if the valve is functioning correctly. A hand mirror may make it easier to watch the valve stem. Be careful not to touch the valve because it will be hot! If the valve stem doesn't move when the engine is revved (and the valve is receiving vacuum), there's probably something wrong with the EGR valve.

Another way to "test" the EGR valve, on some engines, is to apply vacuum directly to the EGR valve. Depending on the type of EGR valve that's used, vacuum should pull the valve open, creating the equivalent of a large vacuum leak. This should cause a sudden change in the engine's idle quality, causing a noticeable increase in roughness and/or a drop at least 100 rpm. But this test doesn't work on many engines with backpressure EGR valves.

Backpressure type EGR valves are more difficult to check. This is because there must be sufficient backpressure in the exhaust before the valve will open when vacuum is applied. One trick that's sometimes used is to create an artificial restriction, by inserting a large socket into the tailpipe, then applying vacuum to the valve to see if it opens. Don't forget to remove the restriction afterwards.

4. Remove and inspect the EGR valve if you suspect a problem. Most failures are caused by a rupture or leak in the valve diaphragm. If the valve is not a backpressure type, it should hold vacuum when vacuum is applied with a hand-help pump. If it can't hold vacuum, it needs to be replaced.

This test doesn't work on a backpressure type EGR valve, however, because you also have to have some here means of simulating exhaust backpressure at the same time vacuum is applied. A special tester that hooks up to a shop's air supply can be used for this type of test.

Backpressure EGR valves sometimes fail if the hollow valve stem becomes clogged with carbon or debris. This you can see for yourself. It's almost impossible to remove such a clog, so replace the EGR valve.

Carbon accumulations around the base of the EGR valve can sometimes interfere with the opening or closing of the valve. These can be removed by careful brushing, or by soaking the tip of the valve in solvent. Do not soak the entire valve in solvent or allow solvent to get anywhere near the diaphragm. The solvent will attack and ruin the diaphragm.

5. Inspect the EGR passageway in the manifold for clogging. Use a pipe cleaner or small piece of wire to explore the opening for a blockage. Sometimes you can dislodge material that's clogging the opening by carefully poking at it. Other times, it may be necessary to remove the manifold and have it professionally cleaned.

Vacuum Amplifier Diagnosis

On EGR systems with vacuum amplifiers, there are two types of amplifier design. Early-model units typically use a single connector, while late-model amplifiers have two connectors (Fig. 10-34). To test the system:

1. With the engine at operating temperature, perform the throttle test. Slowly open and close the throttle while keeping the engine under 2500 rpm. If the EGR valve opens and closes, everything is okay. If the valve fails to open, proceed to the next step.

2. Pull the vacuum hose off the EGR valve and connect the hose to a vacuum gauge. Repeat the throttle test. If a strong vacuum signal (10 or more inches) is reaching the EGR valve, the amplifier is okay and the problem is a bad EGR valve diaphragm. If no vacuum or a weak vacuum is detected, move on to step 3.

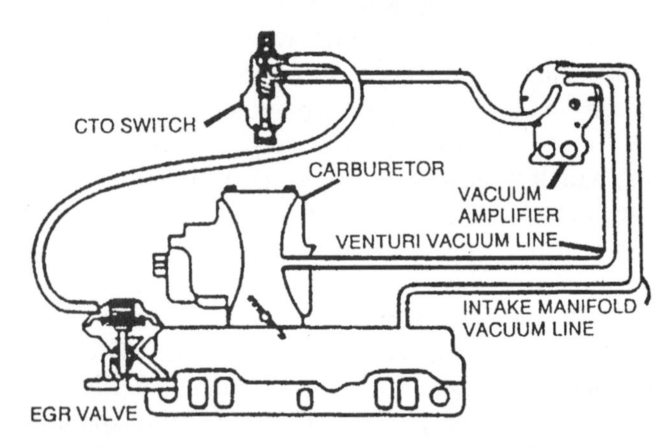

Fig. 10-34 Testing a vacuum-amplified EGR system.

3. If the system has a temperature vacuum switch or solenoid between the amplifier and EGR valve, pull the hose off the TVS that leads to the amplifier. Then connect your vacuum gauge. Repeat the throttle test. If vacuum is reaching the TVS, the amplifier is okay and the problem is in the switch. If still no vacuum or only a weak signal is detected, go to step 4.
4. Reconnect the TVS hose. Pull the vacuum hose from the carburetor venturi signal port. Temporarily plug the port and connect a hand vacuum pump to the signal hose. Start the engine. Send a 1- to 3-inch vacuum signal through the line to the amplifier, using your hand pump. If the engine starts to idle rough, indicating that the EGR valve is opening, the venturi vacuum port may be clogged or obstructed inside the carburetor. A thorough cleaning will be necessary to clear the path. If nothing happens, and the vacuum supply line to the amplifier is okay, the amplifier is defective and should be replaced.

It should be noted that on some cars, the vacuum amplifier is calibrated to generate a small vacuum (2 inches or so) at all times. The calibration is done even when there is no carburetor venturi vacuum signal. This is to maintain vacuum in the system for quicker response. This small amount of vacuum is not enough to cause the EGR valve to open (most require at least 10 inches of vacuum). Therefore, so do not think that the amplifier is defective. If you discover that full manifold vacuum is reaching the EGR valve at idle, the amplifier is leaking vacuum internally and should be replaced.

The thing to remember, when troubleshooting EGR systems, is that the basic principle of operation is the same, regardless of how many components and gizmos are incorporated into the system. There should be no EGR at idle or when the engine is cold. And there should always be EGR in a warm engine under part-throttle operation. If the EGR valve is not working, start at the valve and trace backward to see why vacuum is not getting through. The same applies to those situations where vacuum is reaching the valve when it is not supposed to. The most likely cause here is a temperature vacuum switch, or solenoid that is open all the time, or a defective vacuum amplifier.

One other thing that should be checked, when inspecting any EGR system is to make sure that all the vacuum plumbing is correctly routed. With all the vacuum hoses under the hood, it is easy to get things mixed up.

EGR Valve Replacement

There are many variations from one vehicle application to the next in emission control systems and calibration. Therefore, it is extremely important that you get the correct replacement EGR valve for the application. Two EGR valves may look identical, but be calibrated differently in terms of flow and the amount of vacuum and/or backpressure it takes to open the valve. Therefore, you may have to refer to the vehicle's VIN number as well as year, make, model and engine size when ordering a replacement EGR valve. It may also be necessary to refer to the OEM part number on the old EGR valve (if possible) when ordering a replacement. If so, don't throw the old EGR valve away until you have, installed the new one and made sure it's working correctly.

Many aftermarket EGR valves are "consolidated," so fewer part numbers are necessary to cover a wider range of vehicle applications. Some of these valves use interchangeable restricters to alter their flow characteristics (Fig. 10-35). Follow the suppliers instructions as to which restricter to use for the correct calibration.

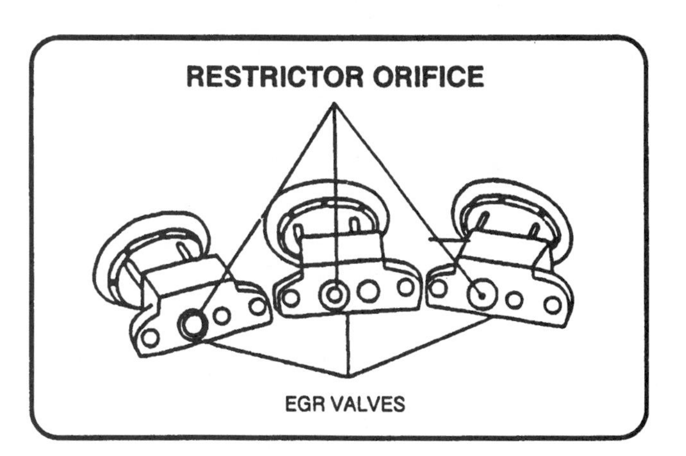

Fig. 10-35 Some replacement EGR valves have replaceable restrictor orifices so one valve can fit a broader variety of applications (reprinted, by permission, from New York State Department of Motor Vehicles, *Systems Training in Emissions and Performance.* Copyright 1993 Delmar Publishers Inc.).

CHAPTER 11

Exhaust Analysis

There are many good reasons for using a 4-gas infrared exhaust analyzer to diagnose carburetion, ignition, and emissions problems. Exhaust readings are necessary for verifying emissions levels. In areas that require periodic emissions testing, you must be able to confirm that a vehicle meets the required emission standards. Exhaust analysis enables you to do that. Without exhaust analysis you have no way of knowing whether or not a vehicle's emissions are within specs or not.

Another reason for using the infrared is to assure consistent quality when a tune-up or preventative maintenance is performed. Checking tailpipe emissions at idle and 2500 rpm is a good way to make sure readings are within specifications. This assures a consistent level of quality in the work performed, and serves as a double check that all adjustments have been performed correctly.

Exhaust analysis is also a very useful diagnostic tool for isolating fuel delivery, ignition, and even mechanical problems. It simplifies troubleshooting and helps you find the source of a driveability problem more quickly.

An emissions analyzer will tell you the concentration of unburned hydrocarbons (HC), carbon monoxide (CO), carbon dioxide (CO_2) and oxygen (O_2) in the exhaust. Ignition, fuel delivery, and mechanical problems can produce similar driveability symptoms and similar individual gas readings. Comparing the various gas readings against one another and/or against oscilloscope ignition patterns and injector patterns can help you narrow down the list of possibilities and find the problem more quickly.

WHAT THE READINGS MEAN

In Chapter 8, we covered the basics of carburetion. But to better understand what exhaust analysis is all about and the chemistry behind it, let's briefly review a couple of things that affect combustion. A balanced or "stoichiometric" air/fuel ratio of 14.7 lbs. of air to 1 lb. of fuel results in near-perfect combustion (Fig. 11-1). There is just enough oxygen to convert all of the hydrocarbon fuel to carbon dioxide and water vapor. Under these conditions, CO_2 levels will be high while HC, CO, and O_2 will be low:

Stoichiometric mixture: $O_2 + HC = CO_2 + H_20$ (low HC and CO)

If there is insufficient oxygen (a rich fuel mixture), there won't be enough oxygen to make carbon dioxide which takes two atoms of oxygen per molecule. The shortage of oxygen will yield carbon monoxide (CO). There will be lower levels of carbon dioxide in the exhaust and very little oxygen left over. Hydrocarbon levels will also be elevated.

Rich mixture: Insufficient $O_2 + HC = CO + H_2O$ (low O_2 and CO_2)

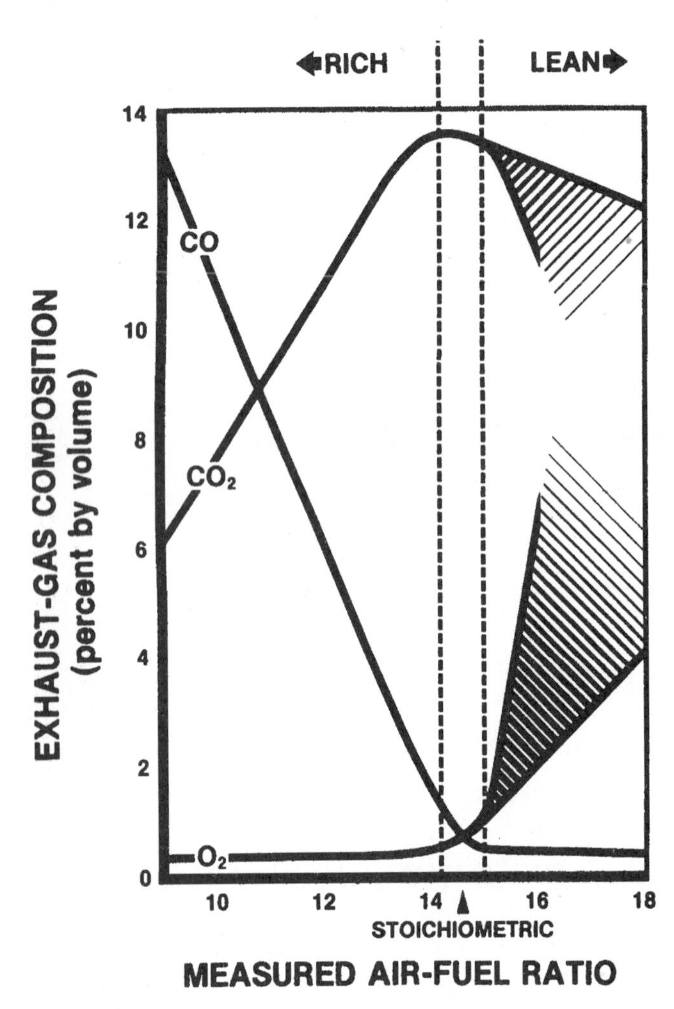

Fig. 11-1 Stoichiometric air/fuel ratio (courtesy Sun Electric Corp.).

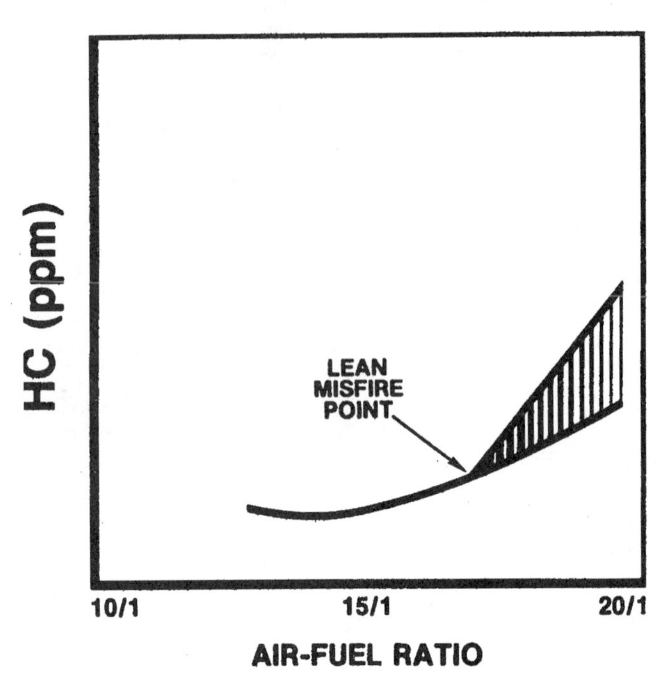

Fig. 11-2 HC and air/fuel ratio (courtesy Sun Electric Corp.).

If there is too much oxygen and not enough fuel (lean mixture), the fuel mixture may not ignite or burn completely, a condition known as "lean misfire." In this instance, the amount of unburned hydrocarbons in the exhaust will go up along with oxygen. Carbon dioxide and carbon monoxide will be low (Fig. 11-2).

Lean mixture: Excessive O_2 + HC = O_2 + HC (low CO and CO_2)

A problem known as "density misfire" that is very similar to lean misfire can sometimes occur with a defective EGR valve. If the EGR valve is stuck open and admits too much exhaust into the intake manifold, the exhaust gases can dilute the air/fuel mixture to the point where the mixture simply isn't dense enough to ignite. The exhaust readings in this instance would be similar to those of ignition misfire: High HC and O_2, and low CO_2 and CO.

THE FOUR INDICATORS

The four gases that a standard 4-gas exhaust analyzer can measure are as follows:

Hydrocarbons (HC)

HC is an indicator of misfiring, usually ignition related. One misfiring spark plug, for example, can raise HC readings to the 1600 to 2000 ppm range! You should look at the HC readings first and then the other three gases to see what the relationship is between them. If HC is low, it's usually means the fuel mixture, ignition system, and compression are all okay. But when HC is high, it indicates a major problem—probably a fouled plug or bad plug wire, or a leaky exhaust valve that's allowing unburned fuel to enter the exhaust.

One very important thing to keep in mind is that the catalytic converter reduces HC and CO levels to almost nothing when the air pump is operating, assuming the converter is in good working condition. Sniffing the tailpipe without disconnecting the air pump, therefore, should show very low HC and CO readings, unless there is a gross problem (such as a fouled spark plug), or unless there is a problem with the converter or air pump. A "minor" problem such as a slightly rich fuel mixture raises HC emissions, but the converter will mask such problems by effectively burning the HC out of the exhaust. That's why the air pump or aspirator needs to be disabled if a major problem is suspected.

HC levels are displayed in parts per million (ppm). "Acceptable" HC levels will vary depending on the model year vehicle, but generally speaking HC should be within the following ranges:

1967 and earlier vehicles	300–500 ppm
1968–1969	200–300 ppm
1970–1972	150–250 ppm
1973–1974	100–200 ppm
1975–1978*	100–150 ppm
1979–1980*	100 ppm
1981 and up*	50 ppm

* Converter-equipped vehicle readings are taken with the air-injection system disabled.

On newer vehicles with the air pump or aspirator system operating, HC levels should only be a few parts per million or almost zero.

Remember, HC is unburned fuel leaving the combustion chamber. If the cause is ignition related, anything that prevents the spark plug from firing properly could be the source of the trouble. This includes the absence of a spark, or the spark occurring at the wrong time or with insufficient duration. Things to look for would include a fouled or worn spark plug, a bad plug wire, arcing around the plug wire or spark plug boot, arcing in the distributor cap, excessive rotor air gap, or low firing voltage (weak coil, low primary voltage, etc.).

If you suspect an ignition problem, an oscilloscope should be used to check ignition patterns and to perform a cylinder balance test (where possible). By shorting out cylinders one at a time, the one that does not increase HC is the problem cylinder because it is already putting HC into the exhaust. If HC is okay at idle but increases with rpm, it means the engine is misfiring because ignition is breaking down at higher speeds.

NOTE: Do not short out any given cylinder for more than a few seconds, as doing so allows a lot of unburned fuel to pass through the engine and into the exhaust. This may cause the catalytic converter to overheat.

If the cause of high HC levels is fuel related, look for anything that might cause an extremely lean fuel mixture or fuel starvation: air or vacuum leaks, an EGR leak, a lean idle adjustment or restricted idle circuit, low float level, clogged fuel filter, weak fuel pump, or dirty fuel injectors. A leaky fuel injector can also cause elevated HC emissions in a given cylinder.

If the cause of high HC levels is compression related, the source of the trouble is usually a leaky exhaust valve. To isolate the offending cylinder, a cylinder balance test can be performed. The cylinder which causes the least change in rpm when the spark plug is grounded is the weak cylinder. A compression check will likewise show the weak cylinder as reading substantially lower than the rest, and probably close to zero. Compression loss can also be caused by a leaky head gasket, "sticky" lifters, or insufficient valve lash.

Oil consumption can also elevate HC readings. Oil leaking past worn valve guides, rings, and/or cylinders may be the cause.

Carbon Monoxide (CO)

CO is a rich indicator. On most converter-equipped vehicles, CO levels should be very low unless the fuel mixture is running very rich. If the fuel system is running lean, the level of CO produced remains relatively flat above 15:1 air/fuel mixtures (Fig. 11-3), but below about 14.4:1, the level of CO rises dramatically as the fuel mixture becomes richer.

If high CO levels are noted, the engine is starved for oxygen. On a carbureted engine, this could be due to a rich idle adjustment, a high float level (or heavy carburetor float that has sunk in the bowl), or a leaky needle valve or power valve in the carburetor. It could also be an indication of restrictions in the air supply such as a dirty air cleaner, clogged air bleeds in the carburetor, or a sticky or misadjusted choke. On a fuel-injected engine, high carbon monoxide readings may indicate a leaky cold-start injector, leaky injectors, or excessive fuel pressure due to a defective fuel-pressure regulator.

A rich fuel mixture can also be created by overheated air entering the carburetor. The lower the density of the air, the richer the relative fuel mixture. Causes here might include a heat riser airflow

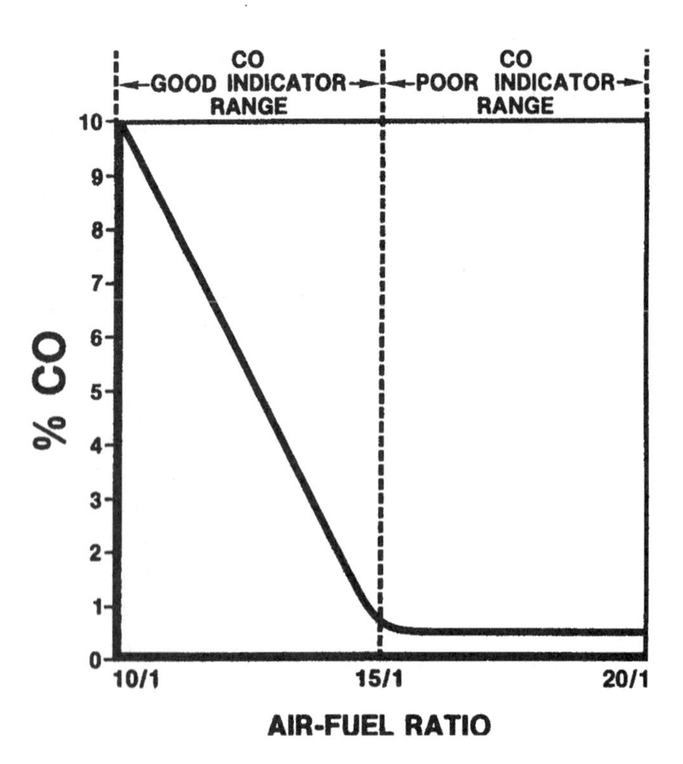

Fig. 11-3 CO and air/fuel ratio (courtesy Sun Electric Corp.).

control door stuck in the closed position, an overheated engine, or a heat riser valve in the exhaust manifold stuck shut.

On late-model, computer-controlled cars with feedback carburetion, a defective oxygen sensor and/or coolant sensor (see Chapters 14 and 15) can prevent the system from going into closed loop, causing the engine to run rich in an open-loop condition.

CO readings are displayed in percent. "Acceptable" CO levels will vary depending on the model-year vehicle, but generally speaking CO should be within the following ranges:

1967 and earlier	2.5–3.0%
1968–1969	2.0–2.5%
1970–1972	1.5–2.0%
1973–1974	1.0–1.5%
1975–1978*	0.5–1.0%
1979–1980*	0.5%
1981 and up*	0.0–0.5%

* Converter-equipped vehicle readings are taken with the air-injection system disabled.

As with HC readings, carbon monoxide readings on a late-model car with the air pump or aspirator and converter intact should be almost zero.

Carbon Dioxide (CO_2)

CO_2 is an indicator of combustion efficiency. CO_2 readings are highest when the air/fuel mixture is at the ideal stoichiometric ratio (Fig. 11-4). On either side of 14.7: 1, the level of CO_2 in the exhaust drops sharply. The higher the CO2 reading, therefore, the closer the air/fuel mixture is to the ideal ratio, and the more completely the fuel is being burned.

Carbon dioxide is not considered a pollutant (though it is a "greenhouse gas" that may contribute to global warming). It is a natural byproduct of combustion along with water vapor (H_20). As such, the percentage of CO_2 in the exhaust is not reduced by the catalytic converter the way carbon monoxide is. In fact, the converter actually increases the amount of CO_2 in the exhaust by converting carbon monoxide (CO) to carbon dioxide (CO_2), and by burning unburned hydrocarbons to make water vapor and more carbon dioxide. Thus, CO_2 readings can give you a direct window past the converter through which you can determine combustion efficiency and air/fuel mixture adjustments.

CO_2, like CO, is expressed as a percentage of exhaust volume. The exact percentage of CO_2 in the exhaust of a properly tuned engine can vary from 8-15%, but generally speaking it should be in the 12-14% range for 1978 and earlier vehicles, and 13-15% for 1979 and later vehicles.

You can use the CO_2 reading to determine if a problem is fuel delivery related by creating a mo-

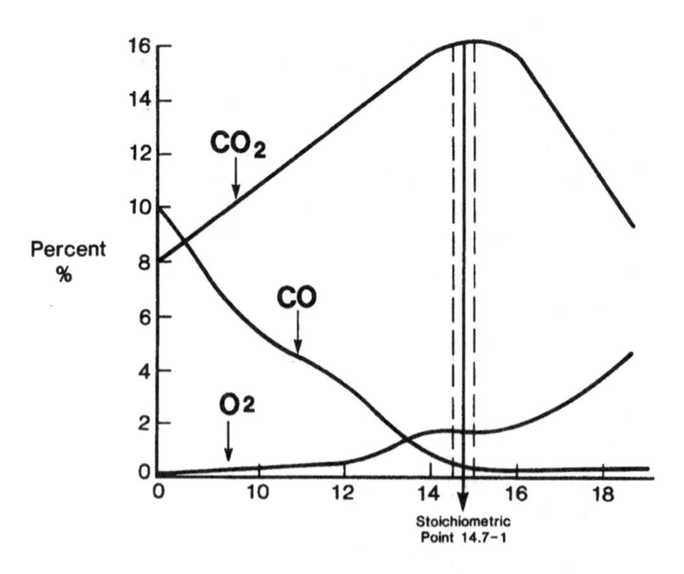

Fig. 11-4 Relationship between CO_2 and O_2 (courtesy Sun Electric Corp.).

mentary rich or lean condition and noting any change that takes place in the level of CO_2 emissions. If CO_2 is low because the fuel mixture is rich, the level should drop even more when you make the mixture richer. If the fuel mixture is rich and you create a lean condition, CO_2 will rise until the engine starts to lean-misfire. If artificially enrichening or leaning the mixture fails to cause a change in the CO_2 level, the problem is not fuel related. The engine has an ignition misfire, a mechanical, or compression problem.

Oxygen (O_2)

O_2 is a lean indicator because oxygen levels change very dramatically when the air/fuel mixture becomes lean. Rich fuel mixtures of less than 14.7: 1 leave very little oxygen in the exhaust after combustion. As such, the level of O_2 changes very little regardless of how rich the mixture is, unless the mixture is so rich that it won't ignite. In that case, the level of O_2 in the exhaust would rise because unburned air and fuel would be passing through the cylinder and into the exhaust. Rich mixtures, therefore, produce very low oxygen readings—but high CO readings. (Fig. 11-5) A low oxygen reading combined with a high CO reading would be a definite indication of a rich-fuel condition.

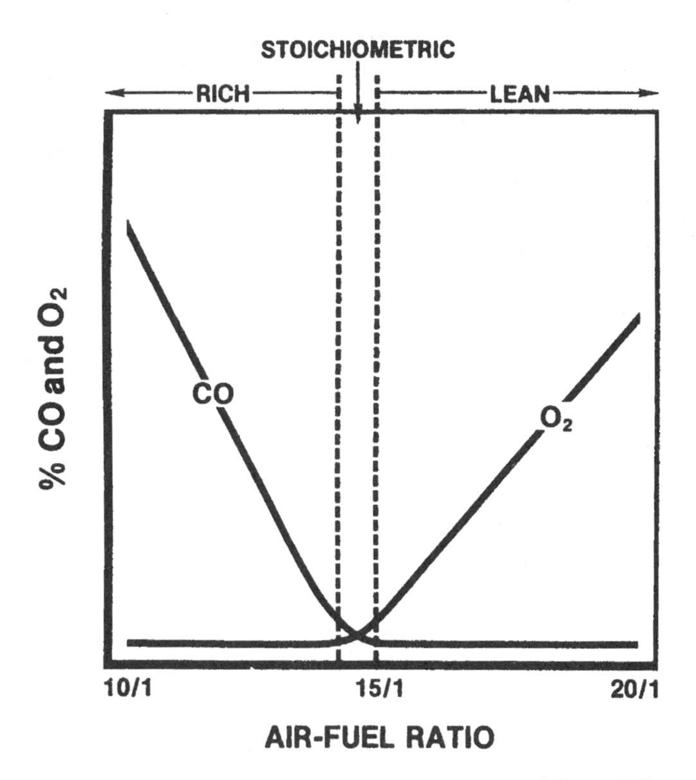

Fig. 11-5 Rich and lean indications with CO and O_2 (courtesy Sun Electric Corp.).

Where oxygen really shows its stuff is when the fuel mixture goes lean. At air/fuel ratios above 14.7: 1, the amount of oxygen left in the exhaust rises sharply. At the same time, a rising oxygen level would reduce the CO readings. An engine that showed high O_2 readings and low CO readings, therefore, could be assumed to be running lean.

Several things unrelated to the fuel mixture itself can influence the amount of oxygen in the exhaust. Because the catalytic converter needs extra oxygen to reburn the exhaust pollutants, the air pump adds oxygen to the exhaust. To get an accurate indication of how much oxygen is leftover after combustion, therefore, the air pump or aspirator may have to be temporarily disabled. If the air pump is not disabled, O_2 readings will be higher than normal, showing about 2-5% O_2 content in the exhaust. The same situation can be caused by air leaks in the exhaust system.

Oxygen can also enter the engine through a vacuum leak. A nonfeedback carburetor or mechanical fuel-injection system may be adjusted correctly. But if a vacuum leak is allowing "unmetered" air to enter the manifold, the mixture will be made artificially lean. If the mixture goes so lean as to cause lean misfire, you'll see high O_2 readings and high HC readings simultaneously.

Like carbon monoxide and carbon dioxide, oxygen content is displayed as a percentage of volume. "Acceptable" O_2 levels will vary depending on the model-year vehicle, but generally speaking O2 readings should be from 1-2% An O_2 reading of over about 3-4% would indicate excessively lean carburetion.

OTHER EMISSION TESTS

A 4-gas exhaust analyzer can be used to perform a variety of emissions-related tests:

PCV VALVE

The purpose of the PCV system is to purge the crankcase of blowby gases. A plugged or restricted PCV system can cause high CO readings. High CO readings can also result if there is excessive blowby and/or the engine oil is heavily diluted with gasoline, because the vapors will be drawn through the

PCV system back into the intake manifold. CO levels from this kind of problem will increase as the engine temperature increases. This condition occurs most commonly in vehicles that are used for short trips and stop-and-go driving. The oil doesn't get hot enough to boil off the contaminants. Infrequent oil changes may also be a factor (more frequent oil-change intervals would be a good idea if this is found to be the problem).

A PCV system that leaks vacuum has too large a PCV valve (wrong part number for the application), or has a PCV valve with a weak or broken spring, causing elevated HC readings because of lean misfire.

To test the PCV valve, use the following procedure:

1. With the engine idling, remove the PCV valve from the engine leaving it connected to the hose from the carburetor, throttle body, or intake manifold. CO should read slightly lower when the valve is removed. If the CO reading drops by more than 0.5%, the oil is diluted with fuel and should be changed. The extra fuel that's being pulled in through the PCV is making the engine run slightly rich.
2. Cover the end of the PCV valve with your thumb. You should feel strong vacuum when you do this and note a change in idle speed if the hose and valve are not restricted. At the same time, you should see a higher CO reading than before.

EGR VALVE

The EGR valve is designed to leak a metered amount of exhaust gas into the intake manifold when the engine is warm and running at speeds above idle. If the valve fails to close at idle, the result will be a rough idle caused by "density misfire," a condition similar to a lean misfire. It can also cause an off-idle stumble. Since the exhaust gas contains little unburned oxygen, it doesn't actually lean out the mixture but dilutes it. The air and fuel is less closely spaced causing a density misfire.

To test the EGR with an exhaust analyzer:

1. With the engine idling, note the HC readings. If the EGR valve is leaking, the HC readings will be slightly higher at idle and off-idle.
2. Artificially enrich the fuel mixture by partially obstructing the air intake. If idle quality improves and HC decreases, look for an air leak. If idle quality and HC remain the same, the EGR valve is not closing, the passage is not being sealed by the valve because it has burned through, or the EGR vacuum supply is connected to an intake rather than ported vacuum source.

If you suspect the EGR valve, disconnect the EGR valve vacuum hose and see if there's any change in idle quality or HC readings. If there is, reroute the vacuum hose correctly. If no change, try to force the EGR valve shut by pushing down on the stem using needlenose pliers. If idle quality and HC improve, you've found the problem. Remove the EGR valve and clean or replace as needed.

VACUUM LEAKS

Vacuum leaks will almost always cause fluctuating HC readings. Two types of problems can be diagnosed here. The first kind is a general vacuum leak (PCV hose, brake booster, etc.) that leans out the mixture and causes a very low CO reading and only a slightly higher fluctuating HC reading. The O_2 reading will be high. The second kind of vacuum leak is a "point" leak that affects only one or two cylinders (a leaky intake manifold gasket, fuel injector O-ring, etc.). This will be indicated by a normal or low CO reading combined with high fluctuating HC readings. O_2 will again be high.

A good way to track down such leaks is with propane. Carburetor cleaner also works, but most exhaust analyzer manufacturers caution not to use any type of solvent when checking emissions because the solvent may be drawn into the analyzer and damage it (Fig. 11-6).

With a piece of hose connected to a propane bottle, feed propane at suspected leak points until you note an improvement in idle quality and/or a change in the HC/CO/O_2 readings. When you've found the leak, the idle should smooth out, HC and O_2 should drop and CO rise.

It's important to note that an overly lean idle mixture will also cause a fluctuating HC reading the same as a vacuum leak. To tell one from the other, there is another trick you can use. Momentarily enrich the idle mixture to 1.5-2.0% CO by placing a clean shop rag over the top of the carburetor or throttle body. If the engine smooths out and HC drops and remains stable, the problem is a lean idle

Fig. 11-6 Caution: May cause damage (courtesy Hennessy Industries).

mixture adjustment. If HC still fluctuates, however, the engine is still too lean in one or more cylinders, indicating a vacuum leak.

CARBURETOR BALANCE

On V6 and V8 engines with two- or four-barrel carburetors, it is important to adjust the idle mixture evenly on both "sides" of the engine. To do this with an exhaust analyzer:

1. Run the engine at idle.
2. Observe the CO_2 readings.
3. Adjust the idle mixture screws in or out until the highest possible CO_2 reading is obtained. Then turn the idle mixture screws individually until the desired mixture is obtained. This is done by leaning out the mixture until O_2 indicates lean misfire. (When the O_2 meter readings begin to fluctuate, it indicates lean misfire.)
4. Back the adjustment screws out slightly to richen the mixture.

IDLE MIXTURE ADJUSTMENT

To set the idle mixture adjustment on a carbureted engine with catalytic converters, use the following procedure:

1. Disable the air pump or aspirator.
2. Remove the PCV valve from the valve cover.
3. Disconnect and plug the charcoal canister purge hoses.
4. Lightly bottom all idle mixture screws and back them out two full turns.
5. Start the engine, bring it up to normal operating temperature, and observe CO, HC, and CO_2 readings. All should be stable before proceeding.
6. Adjust the curb idle speed to specifications.
7. Turn the idle mixture screws equally (same direction and amount) 1/8 turn at a time. Wait until the infrared has responded before making any further adjustments.
8. Continue to enrichen or lean the mixture until CO_2 reaches its highest value (12% or higher).
9. Reset idle speed as required.
10. Adjust the mixture screws equally to the leanest setting possible while still maintaining maximum CO_2 readings and good idle quality (both HC and CO should be stable and within specs for the vehicle being adjusted).
11. Reconnect the PCV valve, canister purge hoses, and air-injection system. Idle speed should barely change.

CRUISE MIXTURE

To check consistency of the air/fuel mixture under cruise conditions:

1. Run the engine at 3000 rpm no load.
2. Observe the O_2 readings.

On vehicles with catalytic converters and conventional carburetors, O_2 readings should be from 1-4%. Over 4% indicates an excessively lean condition. Less than 1% O_2 means the carburetor is allowing too much fuel to enter the engine.

Excessively rich mixtures will show CO readings higher than 1%. Electronic feedback carburetors should show no more than 1% O2 or 1% CO if operating properly. HC and CO readings should be the same or lower than idle. CO_2 should be the same or higher than idle.

If HC readings show a rise at 3000 rpm, it indicates a weakness in the ignition system is causing a misfire. If O_2 readings shoot up, it signals a lean-fuel condition possibly as a result of fuel starvation.

CARBURETOR ACCELERATOR PUMP CIRCUIT

This test measures the response of the accelerator pump. A leaky pump plunger or pump diaphragm, defective check ball, or plugged discharge nozzle can cause stumbling on takeoff or rapid acceleration.

1. With the engine idling, snap the throttle open momentarily then return to idle.
2. Note the O_2 readings on the infrared. On converter-equipped vehicles, O_2 is the best indicator of accelerator pump response. You should see a 0.1-5% decrease before it rises. If there is no decrease in O_2 and the readings rise, the mixture is leaning out indicating a weak accelerator pump circuit.
3. CO should decrease no more than 0.5% and increase by at least 1%. CO and HC should both increase at about the same time with CO leading slightly if the accelerator pump is delivering the right amount of fuel. If too little fuel is delivered, HC will rise before CO which may barely increase at all. CO_2 will generally decrease by about 1%.

AUTOMATIC CHOKE ADJUSTMENT

An exhaust analyzer can also be used for choke adjustments. The thing to watch for here is any indication of the mixture going lean too quickly (as indicated by a sharp rise in O_2 readings in the exhaust) as the choke opens up. If the choke opens too rapidly, the fuel mixture will lean out, leading to lean misfire and hesitation problems while the engine is cold.

If the engine leans out as soon as it starts, check the adjustment of the choke pull-off to see that it is not opening the choke plate too far (use an angle gauge or appropriate-size drill bit or rod gauge). Adjust the choke pull-off if necessary by bending the linkage arm, or by turning the adjustment screw on the diaphragm.

If the initial mixture appears to be correct, but the choke opens up too quickly as the engine warms up, reset the choke to the factory index marks (if someone has adjusted it too lean). Or increase the richness of the choke setting one or two notches (or as needed to maintain proper fuel balance).

An excessively rich fuel mixture on start-up may be due to an inoperative or misadjusted choke pull-off, too much tension on the choke plate (choke set too rich), or sticking. Check the choke pull-off with a hand-held vacuum pump to see that it holds vacuum and opens the choke plate the specified amount. On GM carburetors that have a temperature vacuum switch (TVS) in the air cleaner, check to see that the TVS does not pass vacuum when cold.

If the start-up mixture is okay but starts to go rich as the engine warms up, the choke is not opening quickly enough. Check the choke setting and the operation of the choke heating element. On older vehicles that use a heat riser pipe to siphon hot air from near the exhaust manifold up to the choke housing, the pipe is often plugged or corroded. On V8 and V6 engines where the thermostat coil is mounted in a well in the intake manifold rather than on the carburetor, an inoperative heat riser valve on the exhaust manifold or a plugged heat crossover passage in the intake manifold can slow the rate at which the choke opens. On carburetors that rely on hot coolant circulating by the thermostat housing to warm the choke, a defective or missing cooling thermostat, low coolant, or coolant circulation problems can affect the opening of the choke.

CHAPTER 12

Diagnosing Emission Problems

Emissions testing is an essential element of making sure vehicles meet applicable emission standards. After all, if vehicle emissions were not actually measured, how could compliance with clean air standards be enforced? The Environmental Protection Agency, as well as most state regulatory and enforcement agencies that are responsible for overseeing clean air programs, are committed to mandatory emissions testing as a means of assuring general compliance.

Mandatory emissions testing places a responsibility on vehicle owners to maintain their vehicles. An emissions test confirms whether or not a vehicle is in compliance. As long as a vehicle is within the state's defined emissions cut points, it is considered to be in compliance with the standards and requires no special attention. But if a vehicle exceeds the emissions cut points, then it is not in compliance and must be repaired before the vehicle's owner can receive the required smog certificate or inspection sticker.

A vehicle that fails an emissions test is a vehicle that requires attention. Unless the emissions problem is corrected, the vehicle's owner will not receive the required certificate or sticker to renew his vehicle registration. Consequently, he won't be able to legally drive his vehicle after the current certificate, sticker, or registration expires. Violators may be subject to fines and/or suspension of driving privileges.

Most inspection programs include a "waiver" provision. This allows vehicles that do not meet emission standards to pass anyway, provided a fixed dollar amount has been spent in a good-faith attempt to correct an emissions problem. All repairs must be documented (receipts, repair orders, etc.), however, to receive credit toward the applicable waiver amount. Waiver limits vary from state to state, but generally range from $50 to $300 or more. The waiver limit may be fixed (one amount for all vehicles), or it may vary depending on the model year of your vehicle. In some programs, the waiver limit for older vehicles is set lower than that for newer ones. In some states, a motorist may be able to qualify for a waiver, if the cost of "anticipated" repairs will obviously exceed the waiver limit. In areas that have adopted the new federal I/M 240 emissions testing program, waiver limits of up to $450 may be imposed on newer vehicles!

WHAT TO LOOK FOR

Each state determines its own cut points for emissions testing, as well as the specific tests that must be performed on the vehicles being tested. Most emissions test programs currently check for two pollutants: unburned hydrocarbons (HC) and carbon monoxide (CO). Most also measure carbon dioxide (CO_2), but for diagnostic purposes only since CO_2 is not a pollutant (though it is a "greenhouse gas" that may contribute to global warming).

Some inspection programs also require visual checks of various emissions-related equipment for evidence of tampering. These include:

- Checking the restrictor in the fuel tank filler neck to make sure it hasn't been knocked out or enlarged to accept regular leaded gasoline.
- Inspecting the gas cap to make sure it is the correct type for the application and seals tight.
- Checking under the hood to make sure the engine has all the required emissions control components.
- Looking under the car to see that the catalytic converter is in place.
- Checking the instrument panel to see if a "check engine" or "sensor" warning light is illuminated, possibly indicating a performance problem that could affect emissions.
- Checking any nonstock aftermarket parts on the engine to make sure they are emissions certified.
- If an engine has been swapped or replaced, making sure it has all the required emissions equipment for the original model year and application.
- Safety checks. Some states also include various "safety checks" in addition to the emissions test. Safety checks may include inspecting the brakes, tires, glass, exhaust system, lights, horn, wiper blades, etc. Such items have nothing to do with emissions, but can affect the safe operation of a vehicle.
- The new I/M 240 emissions test program adds oxides of nitrogen (NOX) to the tailpipe checks, as well as "loaded mode" dyno testing that simulates actual driving conditions. Also required is a performance check of the vehicle's evaporative emissions control system (charcoal canister and purge valve) to make sure evaporative emissions are being controlled.

TEST STANDARDS

The cut points for acceptable HC and CO levels are generally based on the emissions standards that a vehicle was required to meet when it was new. Older vehicles do not have to meet the same standards as newer ones, which means the test requirements will vary depending on the model year of the vehicle. The actual cut points for a given model year of vehicle will usually be somewhat "looser" than the original emissions standards for that year to compensate for wear. The cut points, therefore, may actually be 20-30% higher than the same-year new car standards. Even so, a good emissions testing program will catch the majority of vehicles that have an emissions problem.

How clean do the tailpipe emissions have to be to pass an emissions test? As we said, it varies from state to state and depends on the model year of the vehicle. But generally speaking, cut points fall within the following ranges:

Model year	**Typical Cut Points**		**Well-tuned engine**	
	CO%	HC ppm	CO%	HC ppm
pre-1968	7.5–12.5	750–2000	2.0–3.0	250–500
1969–70	7.0–11.0	650–1250	1.5–2.5	200–300
1971–74	5.0–9.0	425–1200	1.0–1.5	100–200
1975–79	3.0–6.5	300–650	0.5–1.0	51–100
1980	1.5–3.5	275–600	0.3–1.0	50–100
1981–93	1.0–2.5	200–300	0.0–0.5	10–50
1994 & up	1.0–1.5	50–100	0.0–0.2	02–20

Notice that the actual emissions produced by the average well-tuned engine are substantially less than the typical cut points required for an official emissions test. That's because the goal of emissions testing is to identify only those vehicles with serious emissions problems so the vehicles can be brought back into compliance.

HOW EMISSIONS ARE TESTED

The primary tool for measuring emissions testing is an infrared exhaust analyzer, which we covered in the last chapter. Exhaust analyzers operate on the principle that different gases absorb different wavelengths of infrared light. The analyzer reads HC and CO content by splitting a beam of infrared light with a mirror. It passes half the beam through a sample of the exhaust and the other half through a reference gas. The beam is chopped on and off by a spinning wheel. The beam then passes through a series of optical filters and strikes a light detector. The detector senses the difference between the two beams and generates an electrical signal that is processed electronically to produce the meter readings. Unburned hydrocarbons are displayed on one meter in parts per million (ppm), and carbon monoxide is displayed on a second meter in percent.

Most analyzers also have the ability to measure carbon dioxide (CO_2) and/or oxygen (0_2) in the exhaust for diagnostic purposes. Neither gas is a pollutant, but are good indicators of what's going on inside the engine. Using HC and CO alone to measure combustion efficiency is difficult because the catalytic converter "masks" many problems by significantly lowering HC and CO emissions. In other words, the converter cleans up the exhaust pollutants so well that some alternative means of measuring combustion efficiency must be used. That's where the ability to read oxygen and/or carbon dioxide helps. The relative proportions of these two gases in the exhaust can reveal whether the air/fuel ratio is correct or not, as well as other problems that affect engine performance and emissions.

As combustion efficiency decreases, the oxygen content in the exhaust rises and carbon dioxide falls. An engine that is running at a nearly ideal air/fuel ratio of 14.5:1 will show about 14.5% carbon dioxide and 2.5% oxygen in the exhaust. Carbon dioxide readings of less than about 13% and oxygen readings greater than about 4-5% indicate poor combustion efficiency. This translates to an over-rich or over-lean air/fuel ratio, poor compression, or an ignition problem.

High HC readings result when a lot of unburned gasoline and/or oil passes through the engine. This could be due to a misfiring plug (which can increase the HC reading 10 times over normal), worn piston rings or valve guide seals (which allow oil to be sucked into the combustion chamber), or a super-rich air/fuel mixture.

High CO readings indicate incomplete combustion caused by insufficient air to burn the fuel completely. This is usually due to an over-rich air/fuel mixture.

To get accurate readings with an exhaust analyzer, the analyzer must be properly calibrated (which is automatically done and self-checked on most equipment today). The vehicle must also be at normal operating temperature, and the exhaust system must be leak-free.

WHY SOME VEHICLES THAT SHOULD PASS AN EMISSIONS TEST DON'T

Most vehicles that are in good running condition and properly maintained should pass an emissions test. In some cases, though, "minor" problems can cause the vehicle to fail. These include:

- Engine and/or converter not at operating temperature. If a vehicle is only driven a short distance to the test facility, it may not be warm enough for the engine to be at normal operating temperature and/or the converter at light-off temperature. This will affect the emissions of the engine and may cause it to fail. Excessive idling while waiting in a test lane may also cause the catalytic converter and/or oxygen sensor to cool down enough to not control emissions properly, causing higher-than-normal readings.
- Idle speed too high. A few hundred rpm can sometimes make the difference between passing and failing an emissions test, if an engine's emissions are marginal to begin with.
- Dirty air filter. A restricted air filter will choke off the engine's air supply, causing higher than normal CO readings.
- Worn or dirty spark plugs. Excessive plug gap can create ignition misfire, resulting in excessive HC emissions.
- Dirty oil. The oil in the crankcase can become badly contaminated with gasoline, if a vehicle has been subject to a lot of short trip driving, especially during cold weather. These vapors can siphon back through the PCV system and cause elevated CO readings.

WHEN A VEHICLE FAILS AN EMISSIONS TEST

Every state has its own emissions test report format, so we'll stick to generalities in describing what's on the form itself. As a rule, most reports are computer printed and contain the following:

- Vehicle identification (year, make, and model, license number and/or VIN number).
- Test information (date, time, test location, lane number, operator identification, etc.).
- Emissions test results (the vehicle's tailpipe readings).
- A checklist of visually inspected items (if applicable).

- The applicable emissions standard cut points that the vehicle must meet to pass the test.
- A statement or notation indicating that the vehicle either passed or failed the test.
- If it failed, a possible explanation as to why it failed (diagnostic information).
- Instructions on what the vehicle's owner must do to correct an emissions problem, including requirements for waivers and reinspection.

An emissions test failure usually means a vehicle either has a emissions problem (excessive HC, CO, or NOX emissions), or an emissions "violation" (missing or disconnected emissions control devices, the incorrect emissions equipment for the engine, or noncertified aftermarket components on the engine). Either way, the problem must be corrected to bring the vehicle into emissions compliance. There is usually a time limit (30 days typically) on how long the vehicle's owner has, to get the problem corrected.

If a vehicle failed the test because of excessive exhaust emissions and not a technical violation, the diagnostic information that's provided with the test form can help guide you in repairing the problem. If the problem seems minor, a few simple adjustments or replacing the spark plugs, air filter, PCV valve, etc. may be all that's necessary to bring the emissions down to acceptable levels.

On vehicles with computerized engine controls, feedback carburetion, or electronic fuel injection, however, troubleshooting an emissions problem can be much more complex. It's not always easy to determine which component in the system is causing the problem. It may be something as simple as a fouled spark plug or bad plug wire. Or it might be a bad oxygen sensor or other engine sensor. Then again, it might be something as complicated as an intermittent computer malfunction. Tracking down the underlying cause, therefore, may take more time and expertise, and in many instances special test equipment.

Once the cause of the emissions failure has been accurately identified and the necessary adjustments or repairs made, the vehicle is ready to be retested.

EMISSION VIOLATIONS

Emission violations are more serious for three reasons:

1. An emissions violation for tampering may also result in a fine, if there's obvious evidence that someone intentionally removed, disconnected, or otherwise rendered inoperative any emission control device.

2. Missing emissions control equipment must be replaced regardless of cost. Either the parts are replaced, or the vehicle is not considered legal for street use.
3. The cost of replacing missing emissions control parts cannot be applied toward a waiver limit for emissions repairs. This means the vehicle's owner may end up spending more money on any additional repairs or adjustments that might be needed to bring emissions down to acceptable levels.

If a vehicle failed an emissions inspection because some minor piece of emissions control equipment was missing (such as the hose that connects the heat riser stove on the exhaust manifold to the air cleaner snorkel, or the wrong gas cap, etc.), correcting the violation should be no big deal. On the other hand, if the violation occurred because a major emissions component was missing (such as the catalytic converter, air pump, no EGR system, etc.), or because the stock carburetor and intake manifold have been replaced with a noncertified aftermarket carburetor and manifold, replacing the missing or incorrect parts can get mighty expensive. A original equipment replacement converter can cost several hundred dollars. Even an aftermarket "universal" type of replacement converter can cost well over a hundred dollars by the time it's installed.

With an older vehicle, it may be very difficult, if not impossible to replace missing emissions control components if the components are no longer available through a car dealer or automotive parts stores. The only source for such parts may be a salvage yard, which means the chances of finding all the correct components may be poor, unless the vehicle is fairly common or a "collector" car.

In some instances, the cost to repair or replace missing emissions control components may be more than the vehicle is worth. It makes no sense to spend a lot of money replacing missing or defective emissions control parts on a vehicle that itself may only be worth a few hundred dollars.

Warranty Coverage

Newer vehicles are covered by an emissions performance warranty and emissions defect warranty. These cover 100% of the cost of parts and labor for repairing any emissions-related part or system during the covered period. The coverage is for 5 years or 50,000 miles, whichever comes first, for vehicles outside California. In California, 1990 and newer cars and light trucks (under 8500 GVW) are covered by a 3-year, 30,000-mile warranty. Emissions-related parts costing more than $300 are covered by a 7-year, 70,000-mile warranty. To get free repairs under this warranty, a vehicle must be taken back to an authorized new car dealer. It doesn't have to be the same dealer that sold the vehicle, but it does have to be a dealer that sells the same make (Ford, Chevy, Toyota, etc.). Most dealers will not honor warranty claims for work done by anyone else, except under unusual circumstances.

Every vehicle that's been built to federal emissions standards since 1981 has had this 5-year/50,000-mile emissions warranty. The list of items covered is specific to each vehicle, and varies somewhat from one model to another and from one make to another. The following list describes the items that are generally covered. Note that not all of these items are found on every vehicle. The vehicle owners manual or warranty papers should include a comprehensive list that defines exactly what is, and is not covered:

Items Generally Covered By The Federal 5/50 Emissions Defect Warranty

- Air/fuel feedback control system and sensors.
- Air pump and related plumbing.
- Altitude compensation system.
- Catalytic converter.
- Cold-start enrichment system.
- Cold-start fuel injector.
- Distributor or distributorless ignition system.
- Electronic ignition module.
- Electronic engine control computer, wiring harness, and sensors.
- Electronic fuel-injection system, including fuel injectors, fuel-pressure regulator, and fuel-rail assembly (but not the pump).
- Exhaust head pipe from the manifold to converter.
- Exhaust manifolds.
- Fuel filler cap and neck restrictor.
- Fuel vapor storage canister (charcoal canister), purge valve, and related plumbing.
- Fuel tank.
- Fuel tank pressure-control valve.
- Ignition coil and/or control module.
- Intake airflow meter or sensor.
- Intake manifold and gaskets.
- Malfunction Indicator Light (check engine/sensor warning light).
- Oxygen sensor.
- PCV valve & related plumbing.
- All spark control components.
- Spark plugs and wires (plugs must be replaced at specified mileage interval).
- Supercharger and related hardware and plumbing.
- Throttle body assembly.
- Turbocharger, intercooler, wastegate, and related plumbing.

All hoses, clamps, brackets, tubes, gaskets, seals, wires, bulbs, and connectors related to any of the above items or systems are also covered.

To qualify for free repairs under the emissions defect warranty, a motorist must maintain the vehicle properly (according to the vehicle manufacturer's recommended maintenance schedule).

TROUBLESHOOTING PROBLEMS RELATED TO EMISSIONS CONTROL SYSTEMS

The following covers specific tests you can perform to troubleshoot emissions problems:

PCV

A common test of a typical system is to remove the PCV valve from its grommet in the valve cover and shake it to make sure it rattles. This will let you know whether or not the valve is clogged or jammed, but it can't tell you if the spring has broken or lost its tension.

Another check is to feel for strong vacuum at the end of the valve while the engine is idling (Fig. 12-

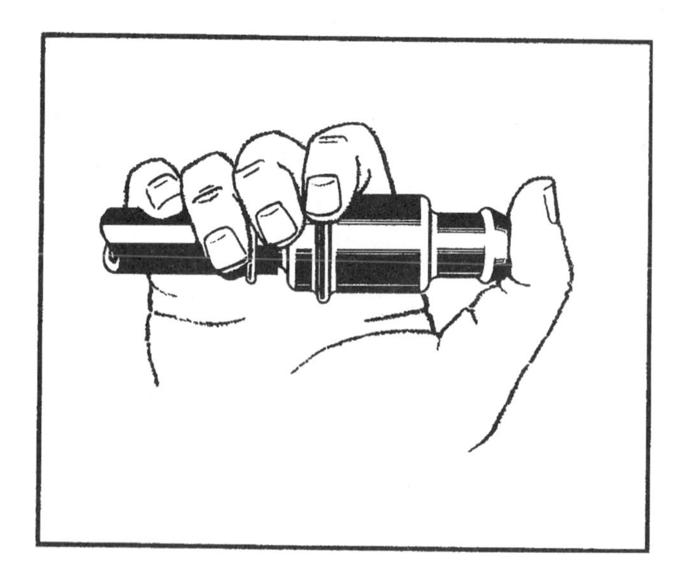

Fig. 12-1 PCV functional check (courtesy AC Delco).

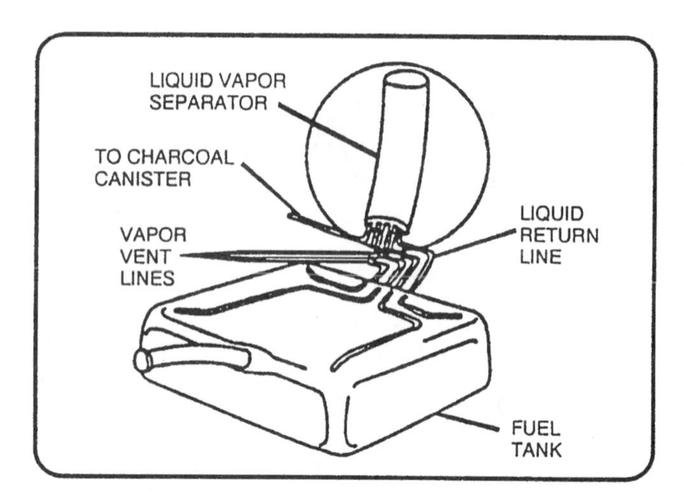

Fig. 12-2 Liquid vapor separator (reprinted, by permission, from New York State Department of Motor Vehicles, *Systems Training in Emissions and Performance.* Copyright 1993 by Delmar Publishers Inc.).

1). If there isn't any, the valve, hose, or passages are blocked. Even if vacuum is present, you still won't know if the valve's spring is weak or broken. Testers are available that attach to the oil filler hole and measure the vacuum or airflow the system produces.

EVAPORATIVE EMISSIONS

One of the most common problems with evaporative emissions control systems is the smell of gasoline inside the vehicle. This is usually due to a disconnected or improperly routed vapor line. Other possibilities are fuel starvation, and even a collapsed tank caused by a restriction in the system or the installation of the wrong gas cap.

The liquid/vapor separator (Fig. 12-2) is mostly trouble-free, although it's possible for the liquid return to become plugged with debris such as rust or scale from inside the tank.

Any blockage between the tank and the charcoal canister can

cause trouble. Check tank venting by removing the gas cap, disconnecting the vent line from the canister, then blowing into the line—it should be clear.

Charcoal canisters are usually good for the life of the car. Problems with purge control valves and clogged filters do occur, however. You can check a typical purge valve with a vacuum pump (Fig. 12-3). It should hold vacuum for 15-20 seconds.

Also, there is always the chance that somebody may have mixed up the vent, purge, and control vacuum lines.

On units equipped with computer-controlled solenoid purge valves, you'll have to refer to the manufacturer's service literature for information on when the solenoid is supposed to open.

Heated Air Intake

For some unexplainable reason, the heated-air intake system is frequently ignored as the cause of driveability complaints. Many supposedly incurable hesitation problems are cured simply by replacing a missing hot air duct, reconnecting a vacuum hose, or installing a new bleed valve.

The symptoms associated with a malfunction in the heated-air intake system are exactly what you'd expect: (a) hesitation, stalling, and rough idling, if too little hot air is ducted to the carburetor or throttle body; and (b) the opposite, reduced power output and fuel efficiency, if the flapper valve keeps the unheated air passage closed.

A leak or other interruption in the vacuum signal to the vacuum motor, or a perforated diaphragm in the vacuum motor itself, will normally result in the flapper valve closing off heated air (in other words, always open to unheated air). So, the first set of problems just mentioned is the most common. Only if there's some sort of binding condition, or if the bleed valve jams closed, will the flapper valve stay up and close off unheated air.

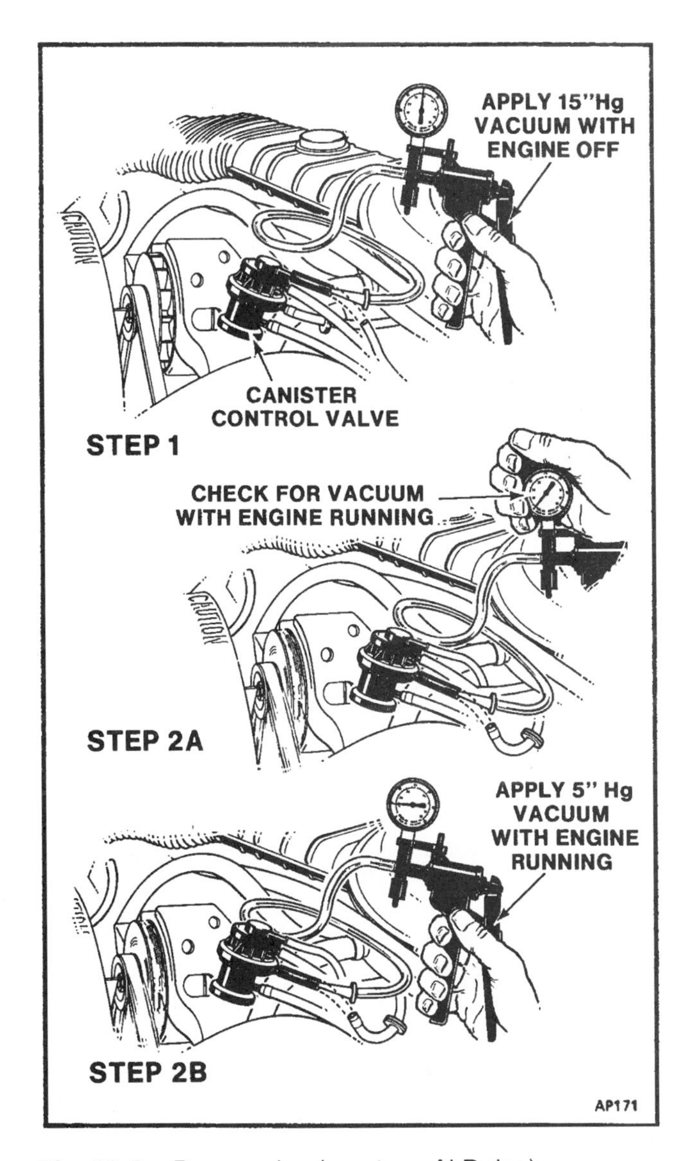

Fig. 12-3 Purge valve (courtesy Al Delco)

Diagnosis is essentially a process of making sure the valve closes below the specified temperature, and opens above it, and that the vacuum motor raises the door when you hook it to a manual vacuum pump.

First, give it a visual examination. Make sure the vacuum lines are not cracked, crushed, or broken, and that they're properly connected. Next, see that the heat tube that runs from the exhaust manifold stove to the air cleaner snorkel is intact. Look inside the snorkel (you may have to remove the air cleaner or use a mirror) to see if the flapper valve is in the down position. Or, you can insert a pencil or screwdriver to find out the position of the valve.

Start the engine. Providing the temperature is below that at which the bleed valve opens, the flap-

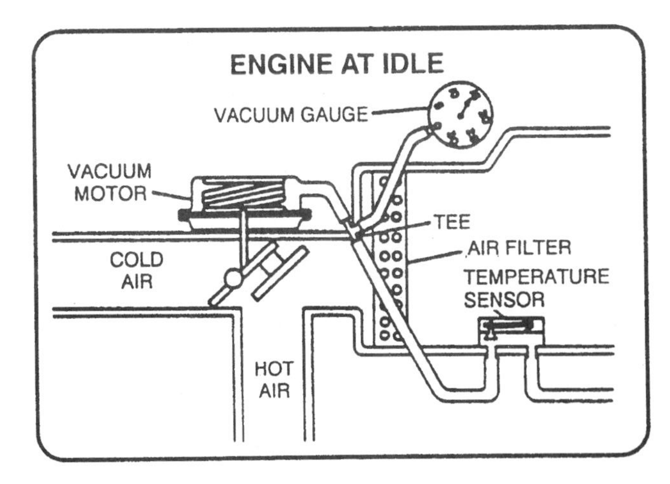

Fig. 12-4 Vacuum testing of bleed valve (reprinted, by permission, from New York State Department of Motor Vehicles, *Systems Training in Emissions and Performance.* Copyright 1993 by Delmar Publishers Inc.).

per valve should rise to the heat-on position, then drop gradually as the engine warms up (Fig. 12-4).

If you don't get these results, there are several ways to proceed to uncover the fault. One way is as follows:

1. Attach a spare line to a known source of manifold vacuum or to a manual vacuum pump.
2. Remove the standard vacuum line from the nipple of the vacuum motor.
3. Connect your source of vacuum to the vacuum motor.
4. If the flapper valve rises now, the trouble is probably in the bleed valve or the line that feeds it vacuum.
5. the flapper valve doesn't rise when given direct vacuum, there's most likely a leak in the vacuum motor diaphragm or possibly the flapper valve or linkage is jammed.
6. If you're using a manual vacuum pump, and the flapper valve does rise when vacuum is applied to the vacuum motor, see that the motor holds vacuum for a reasonable length of time. (A common limit is less than a 10-in. Hg (mercury) drop in 5 minutes.) If not, the diaphragm is leaking.
7. To check the bleed valve, first look up the temperatures at which it should close and open. A common Chrysler system, for example, will raise the flapper valve with a cold engine and the ambient temperature below 50 degrees. F (Fig. 12-5).

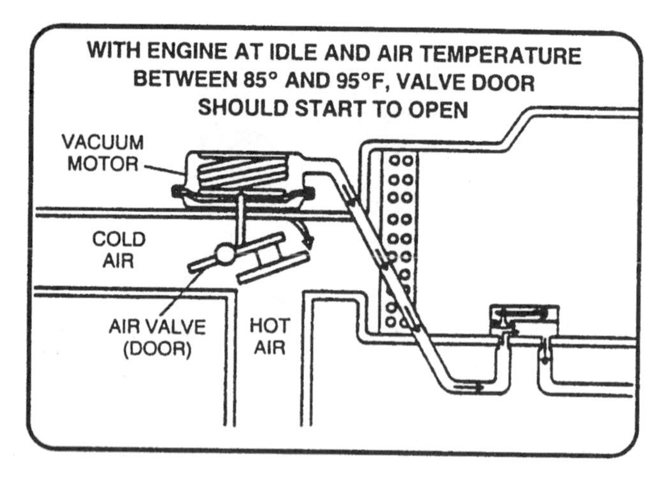

Fig. 12-5 TAC operation (reprinted, by permission, from New York State Department of Motor Vehicles, *Systems Training in Emissions and Performance.* Copyright 1993 by Delmar Publishers Inc.).

8. Remove the air cleaner and cool the bleed valve below the closing temperature. This can be done by covering it with ice cubes (Fig. 12-6).
9. With the hand pump, apply vacuum to the inlet side of the bleed valve. (Make sure the outlet side is properly connected to the vacuum motor.)
10. If the flapper valve doesn't rise, the bleed valve is probably defective. It's relatively inexpensive and very easy to replace.

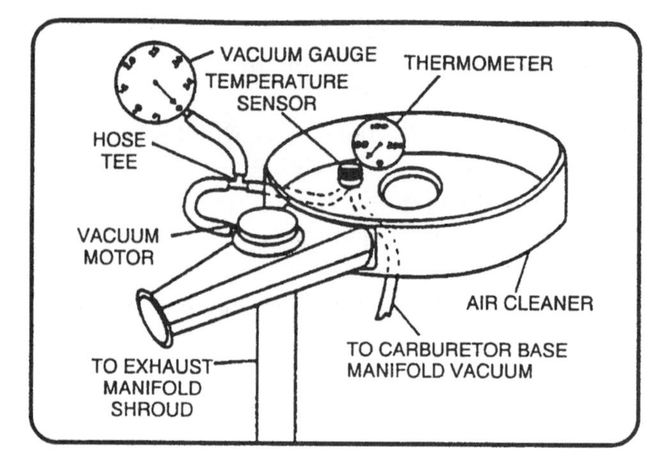

Fig. 12-6 Temperature testing of bleed valve (reprinted, by permission, from New York State Department of Motor Vehicles, *Systems Training in Emissions and Performance.* Copyright 1993 by Delmar Publishers Inc.).

Early Fuel Evaporation

Testing mechanical systems that control early fuel evaporation is just a matter of applying vacuum to the motor to find out if it closes the heat riser (on a garden-variety GM system, 10-in. Hg level shouldn't bleed off in less than 20 seconds), and feeling for vacuum at the line below the specified temperature. Of course, make sure the heat riser shaft isn't frozen in the manifold.

On engines where there's an electrically heated grid under a carburetor, current must be present at the heating element's lead when the engine is cold. If not, check the fuse, the thermostatic switch, and the wiring and connections. On the computer-controlled type, check for a trouble code. The element itself should have the specified resistance between its hot lead and ground, typically no more than three ohms.

Air Injection

The following is a list of common problems and diagnostic tips associated with pump-type air-injection systems:

- Pump noise—Remove the belt and spin the pulley by hand. Don't expect silence, but reject loud squealing and heavy turning resistance. Sometimes the cause of pump failure is a leaking check valve that allows exhaust to enter the pump.
- Backfiring—With an ordinary diverter valve, see if vacuum from a hand-operated pump applied to the valve's nipple causes air to flow from the muffler. If not, either the valve isn't functioning, or the pump isn't producing flow.
- Failure of an emissions test—Disconnect the pump outlet hose at the check valve to make sure there's sufficient pressure and volume at idle (Fig. 12-7).
- Exhaust leakage noise—Especially on GM vehicles, the air-injection manifold tends to rust through, causing a noisy exhaust leak under the hood.

With aspirator valves, about all you can do is listen for exhaust noise and see if there's pressure at the intake instead of vacuum.

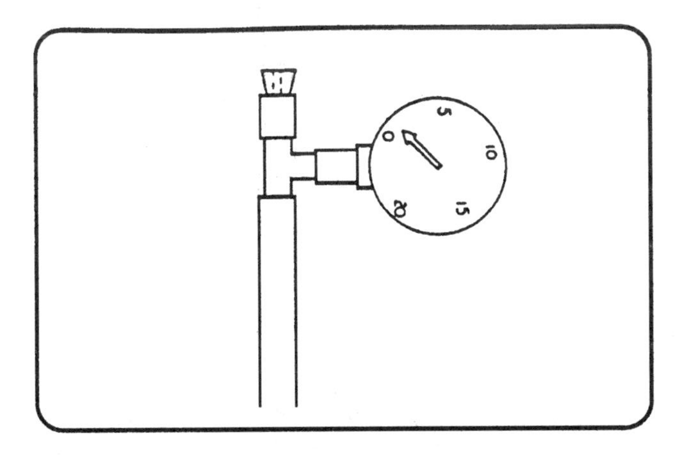

Fig. 12-7 AIR (air injection reaction) pump pressure test (reprinted, by permission, from New York State Department of Motor Vehicles, *Systems Training in Emissions and Performance.* Copyright 1993 by Delmar Publishers Inc.).

Catalytic Converters

Fouling, clogging, melt-down, breakage of the ceramic substrate, etc. can cause a converter to stop doing its job, and/or plug it and raise backpressure to the point of serious performance, driveability, fuel-mileage degradation, or even of stalling and no-starts.

Both monolithic- and pellet-type converters are subject to plugging from a melt-down (overheating can occur if there's too much gasoline in the exhaust from a rich mix or a misfiring cylinder), or disintegration of the ceramics (GM dual-bed units, for example, are prone to this).

The most common cause of failure is an engine that pumps out too much unburned fuel, which can overheat or carbon-clog the catalyst. The excess gasoline in the exhaust may be due to a bad spark plug or valve, but an overly rich air/fuel mixture is certainly a good possibility. With older cars, you might attribute it to something like a heavy carburetor float. But on later models with computerized engine controls, most authorities point to continuous running in "open loop." In other words, the ECU (electronic control unit) doesn't have the opportunity to tailor the air/fuel mixture to conditions, so it falls back on the limited operating strategies in its PROM (Programmed Read-Only Memory). These emergency calibrations are usually on the rich side.

A failed oxygen sensor or coolant sensor are others reasons why a computer-controlled engine may not go into closed loop. (See Chapter 17 for more information on computerized fuel management.) It makes sense to check out the oxygen and coolant sensors any time you encounter a bad converter—and not to replace the converter until you have corrected whatever condition caused it to fail.

Another cause of clogged and contaminated catalysts is oil burning. A set of bad valve seals, for instance, can cause a great deal of carbon formation, and metals present in the oil will coat the catalytic agents.

The easiest test for converter plugging is done with a vacuum gauge. Note the reading at idle, then hold rpm at 2500. The needle will drop when you first open the throttle, then stabilize. If the reading then starts to fall (for example, Chrysler says from 15 down to 10 in. Hg), excessive backpressure is the probable cause. So, suspect a blockage somewhere in the exhaust system.

The next step is to check backpressure directly (Fig. 12-8). If the car has air injection, disconnect the check valve from the distribution manifold, and plug in a pressure gauge with a low scale. Or, remove the oxygen sensor and take your reading at its hole in the manifold or headpipe. There's some dispute over how many psi is normal. Check the specifications for the vehicle at hand, but if you see over 1.25 psi at idle, or more than 3 psi at 2000 rpm, there's a restriction in the exhaust system.

Another direct kind of test is to open a connection to relieve pressure, then see how the engine runs. Take another vacuum reading under these conditions.

In cases where you still aren't sure whether or not the converter is restricting exhaust flow, you can resort to one of the inexpensive kits now available that allow you to check pressure both fore and aft of the unit. Typically, these comprise a hole-punch for your air chisel, a self-tapping hollow nipple, and some screw-in plugs for the holes you'll be making. If there's any noticeable difference between the readings, you've found a restriction.

The "thunk" test can be very useful. Raise the car securely, then put on a heavy glove or use a rubber mallet to give the converter a few solid blows. With the pellet type, you should hear some rattling. If not, the pellet bed is probably clogged or heavily contaminated. On the monolithic type, a rattle means the ceramic honeycomb substrate is at least

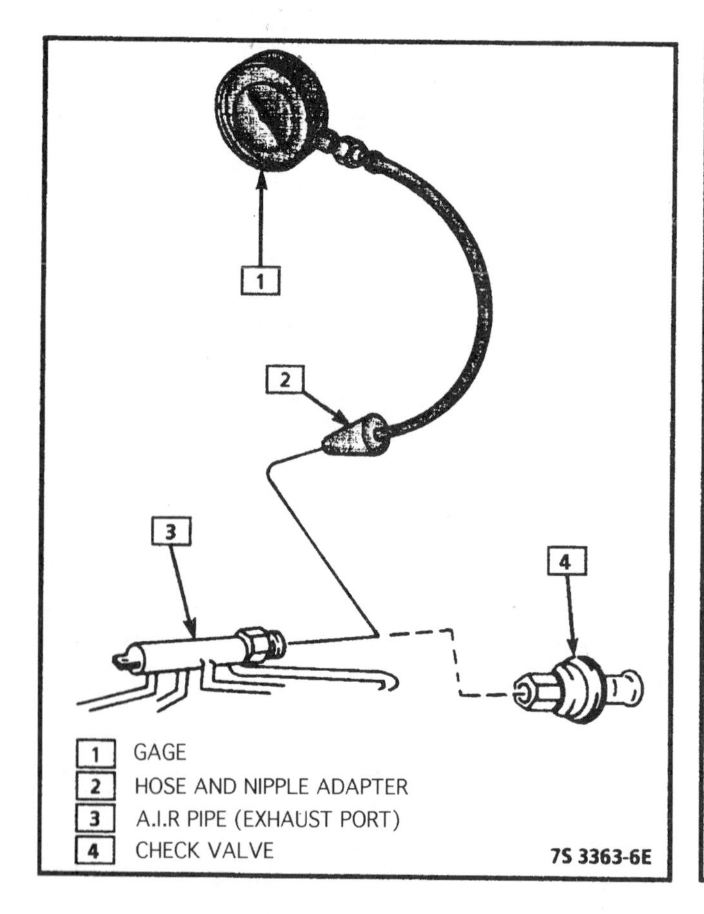

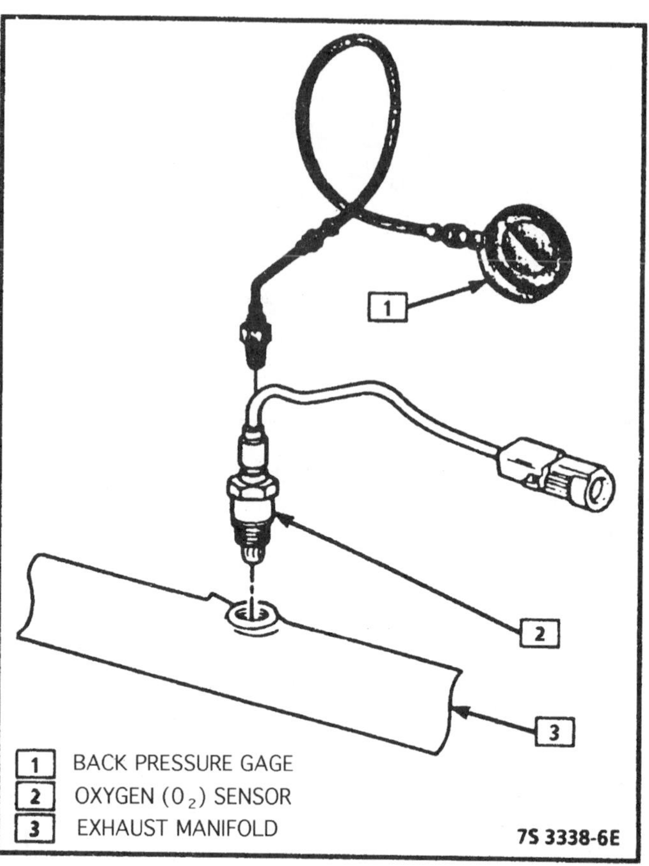

Fig. 12-8 Exhaust backpressure test (courtesy AC Delco)

partially broken, although the situation may not be bad enough to require replacement.

Temperature comparison can be used to find out if the catalyst is working. Using a pyrometer, check the temperature of the pipe just ahead of the converter with the engine fully warmed up and running. Then take a reading on the outlet pipe. If the outlet temperature is at least 100 degrees. F. higher than the inlet temperature, the catalyst is operating. If both readings are nearly the same, or the outlet is cooler than the inlet, the rapid oxidation reaction is not occurring. This, however, may be due to lack of air injection rather than to catalyst contamination.

If emissions are high, or there's little temperature differential between the inlet and outlet pipes, you still can't be sure if the catalyst itself is bad, or if it's just not getting enough air to support oxidation. Check the air-injection system or aspirator valve. See that the air-pump drive belt is okay and properly adjusted, and examine the check valve to see if it's corroded through. You may hear an exhaust leak, or the pump may start making excessive noise. If you suspect that the check valve is allowing exhaust to flow backwards, remove it and blow through both ends. It should let air pass in one direction, but not in the other.

Examine the air-injection manifold, another component that tends to rust out. Find out if the diverter valve is dumping pump output when the throttle is opened and then allowed to snap closed. You should feel and hear air escaping from the little muffler on the valve. If not, backfiring may occur, which can shatter the catalyst. Before condemning the diverter valve itself, make sure its vacuum line is intact and properly connected.

With aspirator valves, you should be able to hear and/or feel the fluttering of the internal flapper.

EXHAUST GAS RECIRCULATION

Troubles associated with a malfunctioning EGR system can be reduced to recirculation occurring when it shouldn't, or not occurring when it should. Typical symptoms include:

- Detonation, if the valve fails, the vacuum signal is interrupted, or the exhaust passages become plugged, since EGR's cooling effect keeps the charge from exploding.
- Stalling or roughness at idle, if the valve jams open or the signal is constant due to a hose mixup.
- Poor cold driveability, if the signal reaches the valve too soon after startup.

It's always a good idea to take EGR into account when troubleshooting any of these symptoms. With all the controls that have been used over the years (see Chapter 10), EGR diagnosis can be complicated. So make sure you have the specific information for the vehicle. However, the basic tests included here are applicable to most types.

As a preliminary, see that all the vacuum lines involved are intact and properly routed. (Use the diagram under the hood, or look it up in a suitable manual.) You may have seen cars with hoses so scrambled that nothing was working properly—not to mention those on which they've been plugged intentionally to disable the EGR valve or other emissions control devices.

A related point: On vehicles which have the automatic transmission modulator hose teed into the EGR vacuum line, anything that disrupts vacuum to the EGR will cause shifting problems and eventually a ruined transmission.

The following procedures will uncover many common EGR problems:

1. If the EGR valve stem is accessible, push it against spring pressure. It should move freely and return fully. If not, remove the valve for cleaning or replacement.

2. With the engine at normal operating temperature, open the throttle enough to reach at least 2500 rpm while watching the EGR valve stem (use a mirror if necessary). It should move, then return. If it doesn't, remove the hose and feel for vacuum as you speed up the engine again. If you find vacuum, the valve is at fault. If there's no vacuum present, check out the controls (see step #3). Note that if you're dealing with a Ford pressure-operated unit, you should feel pressure instead.
3. If you find no vacuum at the valve, and there's a thermostatic vacuum switch (Fig. 12-9) in the hose, pull off the source line and feel for vacuum above idle. If you find it, but none gets to the valve with the engine warm, the switch is faulty.
4. On all but the positive backpressure and pressure-operated types, you can use a manual vacuum pump to test valve action. The engine should roughen and perhaps die when vacuum is applied at idle. This is especially useful on valves with an enclosed stem.

Remember, don't condemn a backpressure-type valve until you're sure the exhaust system is stock, has no leaks, and isn't clogged. Also, don't try to get the positive backpressure type to hold vacuum with the engine off or idling.

GM says never to use a solvent to dissolve deposits in an EGR valve, but Chrysler tells you it's okay providing you're careful not to get any on the diaphragm. With most specimens, you'll be cleaning the pintle and valve seat with a dull scraper or wire brush, and knocking out loose carbon by tapping the pintle. But these parts are expensive, so a more satisfactory means of cleaning them is desirable. Some GM versions, for instance, can be disassembled for this purpose. Just make sure you don't damage the seating surface in the process, and that

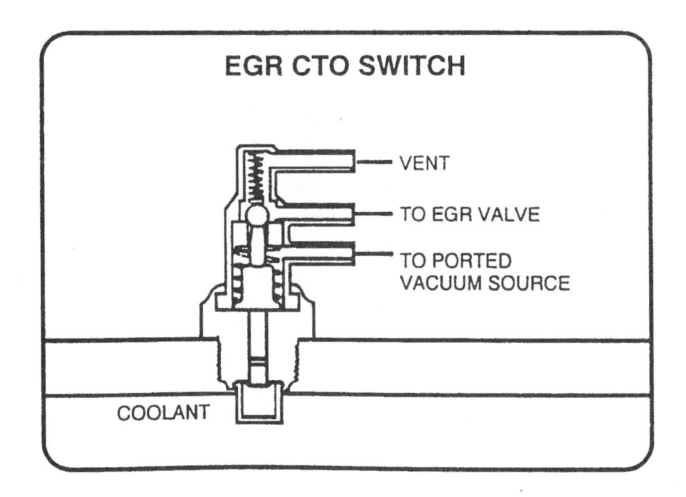

Fig. 12-9 EGR temperature switch (reprinted, by permission, from New York State Department of Motor Vehicles, *Systems Training in Emissions and Performance*. Copyright 1993 by Delmar Publishers Inc.).

you scribe marks so you can get it back together in the proper alignment.

EGR manifold passages often need cleaning, too. Frequently, the cause of a detonation condition can be traced to carbon buildup in these passages. If the EGR valve is indeed opening, remove the valve and look into the passages (use a mirror if necessary). Deposits can usually be broken loose with a stiff wire or an awl.

Mechanical Fuel Injection

A list of the most common CIS symptoms and their probable causes is a logical beginning for troubleshooting:

- Difficult cold starting may be due to a malfunctioning cold-start valve or thermo-time switch.
- Rough running or stalling during warm up should cause you to suspect the auxiliary air regulator, system pressure, and the idle speed and mixture adjustments.
- Surging or missing when the engine is warm points to a problem in system pressure, warm control pressure, the fuel distributor and injectors, and the idle speed and mixture settings.
- Hard starting when the engine is hot may be caused by a faulty hot-start relay (if present), improper air sensor plate height, incorrect idle speed and mixture settings, or faulty leak-down/cut-off pressure.

A fuel pressure gauge of sufficient capacity, connecting fittings, and specific service information for the vehicle at hand will be necessary for checking fuel pressure at various points in the system. If you find a deviation from specifications, first be certain the pump is delivering the proper pressure. Then you can make adjustments by changing shims in the pressure relief valve.

Ideally, this system supplies all the cylinders with the same amount of fuel. If not, poor idle quality and part-throttle performance will be the symptoms. To find out if all the injectors are providing equal amounts of gasoline, you can use a set of graduated measuring tubes and a special fixture that depresses the air sensor plate—or move the air sensor by hand.

The two emissions-related adjustments you can perform on K-Jetronic are those of the idle speed and the air/fuel mixture. To set speed, turn the screw next to the throttle plate linkage, which varies the size of a bypass drilling. Counterclockwise equals faster. Mixture is adjusted by turning a screw which bears on the air sensor lever. You'll find a rubber plug between the fuel distributor and the air sensor cone, which must be removed for access to the screw. Use a long, 3-mm Allen wrench to adjust. Clockwise richens the mixture.

Electronic Fuel Injection

Whenever you have a problem with an EFI-equipped car, heed this list of symptoms and their most probable causes before jumping to any unfortunate conclusions:

- No start—Lack of spark, fuel pump circuit, fuel pump, fuel filter, or defective control unit.
- Hard starting—Cold-start valve, thermo-time switch, intake air sensor, fuel pressure regulator, or temperature sensor.
- Stalls during warm-up—Auxiliary air regulator, or temperature sensor.
- Runs poorly when cold—Auxiliary air regulator, or temperature sensor.
- Too lean—Fuel pressure regulator (pressure too low), restricted injectors, vacuum leak, or clogged fuel filter.
- Too rich—Fuel pressure regulator (pressure too high), or leaking cold-start valve.
- Flooding—Thermo-time switch, or cold-start valve.
- Low power—Fuel pressure regulator (pressure too low), or restricted injectors.
- Poor fuel economy—Intake air sensor, or fuel pressure regulator (pressure too high).
- Erratic performance—Vacuum leak or intake air sensor.
- Rough idle when warm—Vacuum leak, fuel pressure regulator, leaking cold-start valve, or bad injector.

There are some good quick checks that'll tell you quite a bit about the condition of the system. First, listen to each individual injector using a mechanic's stethoscope or a long rod held to your ear. They should all make an identical clicking sound while the engine is idling. If one sounds different from the others, or makes no noise at all, suspect it as the cause of missing, rough running, or other performance problems.

In cases where one or more injectors are silent, or if the engine won't start, disconnect the injector harness plug and check for 12 volts across the plug terminals using a test light. If you get no flashing while the engine is being cranked, the trouble is in the electrical circuit, not the injector itself.

AC Rochester Central Point Injection (CPI)

The following diagnostic notes pertain to CPI (Fig. 12-10):

- The top of the manifold will have to come off for any service, but that's just a matter of removing 10 Torx bolts and stud nuts. At reinstallation, torque them to 124 in.-lbs. Starting with the one at the front of the driver's side, continue clockwise around the perimeter. This joint is sealed with a reusable gasket, not a chemical compound as you might have expected.
- It's easy to remove the injector nozzles. Squeeze the two plastic prongs together as you might on an electrical connector.
- Resistance across the central injector should be 1.5 ohms.
- The TPS is non-adjustable.
- The O_2 sensor heating element gets current through the IGN/GAUGES fuse.
- Key-on/engine-off fuel pressure should be 54-62 psi.
- Just as with many other GM systems, long cranking time may mean the fuel pump relay isn't working, as the pump has to get its complete circuit by means of the oil pressure switch.

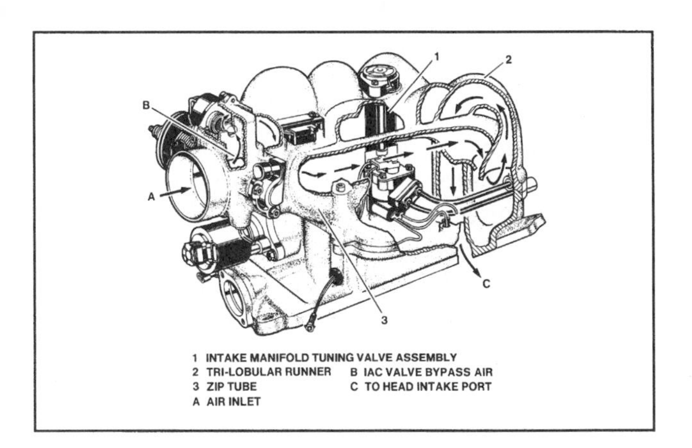

Fig. 12-10 4.3L central point injection (courtesy AC Rochester)

Engines With Computerized Controls

This is definitely a case where the specific diagnostic information for the model at hand will be necessary if you're going to nail down a subtle problem. You'll need the proper test procedures and voltage or resistance specifications in order to be sure your efforts result in an accurate diagnosis. Also, if a self-diagnostic mode is included, you'll have to determine how to generate, read, and interpret fault codes.

First and foremost, make sure all the traditional basics are okay. Just because a computer system is present, don't jump to the conclusion that it's the cause of the trouble. Check compression, ignition, and coolant temperature, and look for vacuum leaks. In many cases, the fault will lie in one of these, rather than in the electronics.

Next, when symptoms or trouble codes lead you to suspect a particular sensor, make sure you check its circuit before replacing it. Look for frayed or broken wires and loose or corroded connectors. Then, check the sensor itself with a digital volt/ohmmeter to see if it's doing what it's supposed to.

On early feedback/closed-loop systems, the recommended replacement interval for the oxygen sensor was generally 30,000 miles. Later vehicles typically have no recommended replacement interval—meaning you supposedly don't have to replace it unless it fails. Several things can wreck an oxygen sensor's ability to provide accurate exhaust gas oxygen readings. Mechanical damage in the form of a broken element or wire happens, but the most common killer is contamination. Lead, carbon, and silicon (from RTV or anti-freeze—expect O_2 sensor problems whenever you replace a blown head gasket) can all coat that precious platinum and make the unit sluggish or altogether inoperative.

Contamination isn't always the kiss of death. When you get poor readings, try running the engine at 3000 rpm for a few minutes, then retest. You may have burned off whatever was interfering, plus you'll be sure the sensor's hot enough to make voltage.

The most basic check of both the oxygen sensor and feedback system operation begins with running the engine at fast idle until everything's hot, then tapping your tester into the lead. (With electrically-heated units, make sure you use the output wire.) You should see rapidly changing readings as the computer keeps adjusting the blend. Deciding whether or not response is slow enough to justify replacement requires some judgment. A common rule of thumb for minimum oxygen sensor activity is 8 trips across the rich/lean line in 10 seconds. Sometimes you can find specs for this under the heading "cross-counts."

Next, try pulling the vacuum hose off the brake booster. This should send the oxygen sensor's reading to zero, then it'll start working its way back up. Don't expect it to go very high, because the feedback carb or injectors probably won't be able to supply enough fuel to make a rich mix with that much extra air. Thumbing the hose should bump the voltage to about .9, then it'll gradually come down to the midpoint. Some technicians like to blow a generous amount of acetylene or propane into the air intake to see what happens at the rich end—a rise in voltage is normal. Shutting off the gas suddenly should produce a momentary zero reading, then the ECM will start to compensate.

O_2 Sensor Checks

To test the O_2 sensor all by itself, disconnect its pigtail (Fig. 12-11) and attach the meter's positive lead to it (the other lead to ground, of course), so that the computer is cut off from this source of data. Cause an artificial lean condition (pull a vacuum hose or the PCV valve), and voltage should drop. Then, cause a rich condition (with a carb, close the choke partially), and you should see the reading rise. If the O_2 sensor doesn't respond, it's either too cold, or it's dead.

Knock Sensor Checks

The most obvious trouble you can have with a knock/detonation sensor system is pinging. There shouldn't be anything beyond a trace of this noise. A more subtle malady is less-than-sparkling performance and fuel mileage, which may be the result of the computer believing that detonation is present when it really isn't.

The most straightforward means of testing the knock sensor circuit is to simulate the sound of detonation by tapping on the area nearest the sensor with a wrench or extension while you hold rpm at 2000 or so (the system's disabled at idle) and seeing what happens. If you're using a scan tool, you'll read degrees of retard, get a yes/no display, and/or see a continuous 0-255 counter loop (no change in the number means no knock signal).

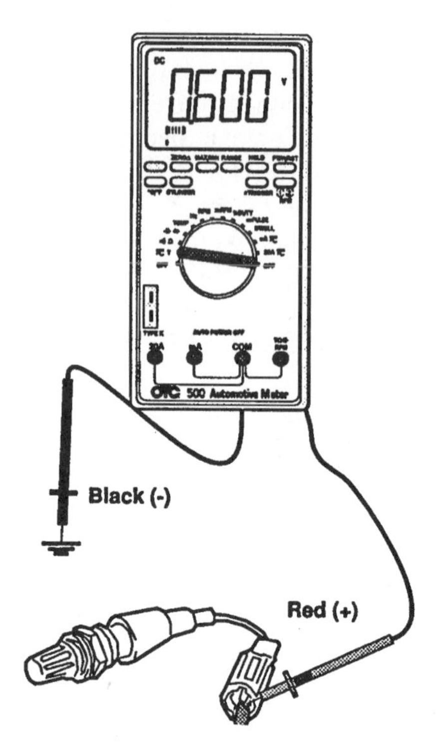

Fig. 12-11 CTS resistance test (courtesy OTC)

But you can observe the action without a scanner. Just hook up your timing light, shine it at the marks, hold fast idle, and start tapping. If you don't see retard occurring, first change to a different tool and modify the force you're using. If that doesn't produce a change in ignition timing, check out the wiring and connections before condemning either the detonation sensor itself or the computer.

If you're getting retard all the time, suspect an overly sensitive sensor, or an internal engine noise that mimics spark knock. Although the crystal assembly is calibrated to respond only to the exact frequency that detonation causes, it can be fooled. After all, that's what you do to test it.

Throttle Position Sensor Checks

When the TPS fails, as it often does, driveability problems can be expected, especially tip-in hesitation. If it's hanging up, it can cause high idle speed because there is more voltage than the specified minimum in the return wire. Combined with high vacuum, this will make the computer think you're in decel, so it'll back out the AIS (automatic idle stabilizer) motor.

Another possible glitch if the sensor doesn't return completely is hard starting. For example, if the voltage signal is significantly above the closed-throttle value while cranking, the ECM will read it as full throttle and engage the clear-flood mode—the injectors won't get the necessary pulses and fuel won't flow.

The direct DVOM test of the TPS involves pulling the sensor's plug, then connecting all three terminals to the harness again using jumpers with contact taps. Hook your meter between the output and ground wires, switch on the ignition, then look for the minimum voltage spec with the throttle at curb-idle position. Open the throttle slowly to do the sweep test (Fig. 12-12)—you want to see an even rise to about 5 V. Voltage should go up smoothly. If it drops back or jumps ahead at any point in its travel, the TPS needs to be replaced. Of course, you can also use your scan tool to do this check. If you want to back-up your findings, repeat the sweep test using an ohmmeter across the sensor's supply and output terminals.

Many throttle position sensors are adjustable, but must be set with a high degree of accuracy. A setting that's as little as 0.2 volts off can cause trouble.

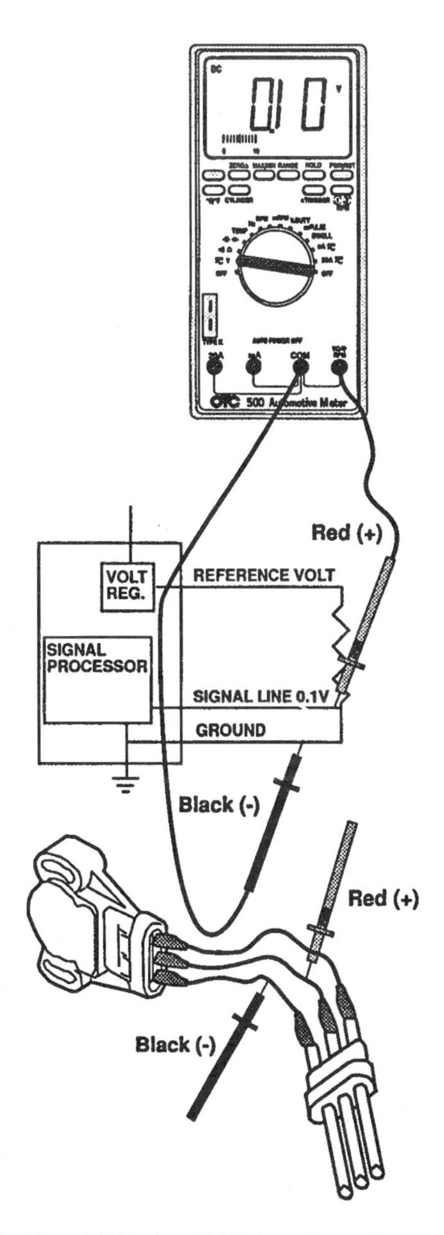

Fig. 12-12 AC Delco MAF (courtesy General Motors)

Coolant Sensor Checks

There are several ways of checking a coolant sensor. With a scan tool, for instance, you can read coolant temp directly. A cold engine should obviously show about ambient temperature, and a hot engine should produce a display of between 190 and 220 degrees F.

If the basic readings are off, you can usually find specs in factory diagnostic info for the resistance the sensor is supposed to have at various temperatures. Check the actual coolant temperature with a thermometer or digital pyrometer, switch your DVOM to "Ohms," then take a reading across the sensor's terminals to see if you get the correct resistance (Fig. 12-13).

Next, start the engine and watch the reading. If you don't see at least a 200-ohm change within a minute, unscrew the sensor to see if it's coated with crud. Clean it and try again. If you don't get that rapid resistance change, or your readings don't match specs, the sensor needs to be replaced.

There are several ways of finding out how the system is responding to the coolant sensor's output. Using a scan tool, unplug the sensor's lead. This open circuit will simulate the high resistance of super-low temperatures, so you should see the minimum temp reading, typically –40 degrees F. Go to the other extreme by jumping the two connector terminals, which should display the max temperature (250 degrees F. and up).

Another procedure involves setting up to read

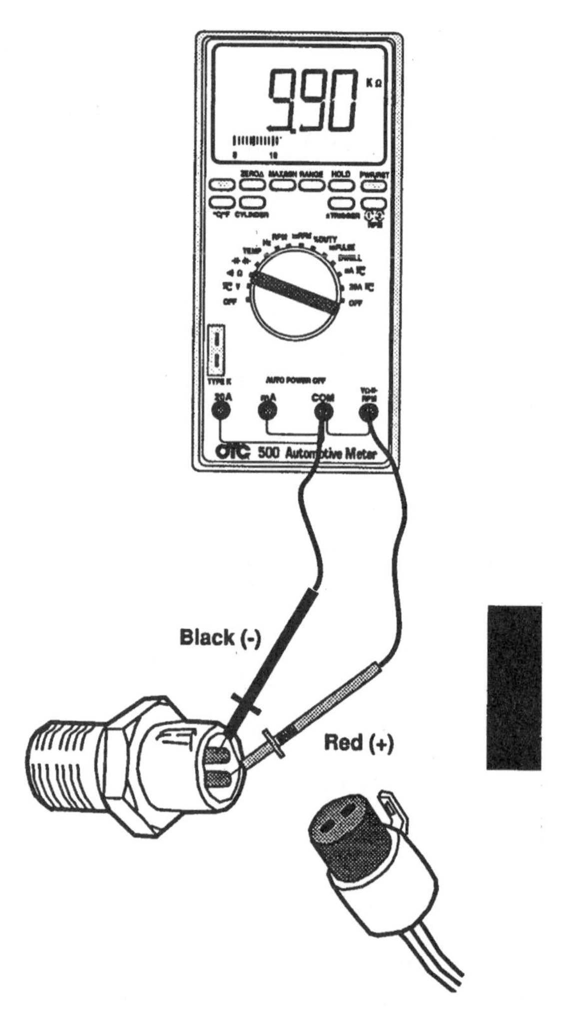

Fig. 12-13 Meter functions-frequency (HZ) (courtesy Owatonna Tool Co.)

pulse width (a scanner or scope maybe, or you could even use the 6-cylinder scale of a digital dwell meter tapped into an injector's output wire). Then put a 50 K-ohm pot and a 100-ohm resistor (to provide some resistance when the pot's turned to full hot) in series in the sensor wire. If the dwell/pulse width varies as you change resistance (you'll have to keep adjusting idle speed to keep it constant), the computer is indeed using input on coolant temp in its calculations.

MAP SENSOR CHECKS

The biggest clue to a bad MAP sensor is a lean condition. Using your scan tool, look at BARO. It should be about what the weatherman tells you (prevailing barometric pressure). Get both BARO and MAP readings, which should be the same with the engine off. With the engine running, subtract MAP from BARO to get actual in. Hg, which you can compare to a mechanical gauge reading.

As with the TPS, most MAP sensors can be checked using jumpers with contact taps so the component can stay in the system. On a garden-variety GM car with a feedback carb, terminal A is ground, B is sensor-output voltage, and C is reference voltage. With the key on, first see that you've got pretty close to 5 volts between A and C. Then, look for 4.6 V between A and B. Using a hand pump, apply vacuum to the sensor's nipple, and make sure 2.3 V is available at 10 in. Hg, and 1.0 V at 20 in. Hg.

For a dynamic test, use your favorite method of reading pulse width and apply vacuum to the sensor using a hand pump. You should see pulse width decrease as vacuum is increased. Commonly with EFI, 23 in. Hg or so at idle will kill the engine because it simulates decel, a mode in which the computer shuts down the injectors which can cause the injectors to cut out intermittently at idle. This happens when somebody has advanced timing beyond specs, which results in a smaller-than-normal throttle opening for proper idle speed. That, in turn, produces enough vacuum at idle to trigger the injector cut function.

A possible source of trouble you'd probably never think of is frozen condensation in the MAP's vacuum line (reroute it to eliminate low spots).

MASS AIRFLOW SENSOR CHECKS

When an airflow/mass sensor or its wiring or ducting goes awry, what kind of symptoms can you expect? The answer depends on whom you ask. GM guys will tell you the engine will crank up and die. Bosch talks about starting problems both hot and cold, hesitation, stalling (especially under load), rough idle, and low power output. Nissan gives stalling, poor idle, black smoke, and switching to the fail-safe mode as evidence of airflow meter problems. (In some models, this mode will be manifested by the inability to exceed 2000 rpm.) Generally, contamination of a hot wire or film-sensing element, which slows response, will result in stumble.

That's all fine, but the most prominent logical effect of a bad signal or lack of any signal is trouble at transient throttle: stalling, sagging, missing. If the signal is far enough out of range to cause the electronics to shift to LOS (limited operating strategy), overall performance and driveability will be lousy.

Unfortunately, other things can cause many of those same symptoms, some of them very basic. Before you jump to conclusions, you've got to think about ignition, compression, fuel supply, etc. And a simple problem that's commonly overlooked is a hole or rip in the duct between the sensor and the throttle body, which admits unmeasured or "false" air and leans out the mixture. An open PCV can do the same thing, and plugged air filter can mean trouble, too.

The common GM mass airflow sensor (Fig. 12-14) has a pretty poor reliability record. But the higher-frequency 10-kHz Hitachi unit used on late-model GM cars has a much lower failure rate.

One way to test a GM MAF sensor is to tap the sensor while the engine is idling. A bad sensor will not only produce a dramatic change in frequency, it may also cause the engine to stumble or stall. This is certainly a convenient check, and it's almost 100% accurate. You might even want to try it before pulling codes.

But there's another quick check that's almost as fast. With the key off, unplug the MAF's harness connector, then start the engine. If it runs appreciably better now, it's time for a new sensor.

If you have a digital multi-meter that can measure frequency (Fig. 12-15), you can use that mode to check AC, Hitachi, and any other unit you run into that produces a frequency signal. Set the meter to read Hz or kHz, and connect its leads to the sensor's signal and ground wires. An ordinary AC MAF as found on a 2.8L Chevy V6 should show you about 45 Hz at 1000 rpm, and 72 Hz at 3500 rpm, whereas the high-frequency type of a late-model 3800 will read 2.9 kHz and 5.0 kHz at those same speeds. Record the readings at various rpm and compare them to specs. You should see a linear frequency rise with no dips or jumps as speed increases.

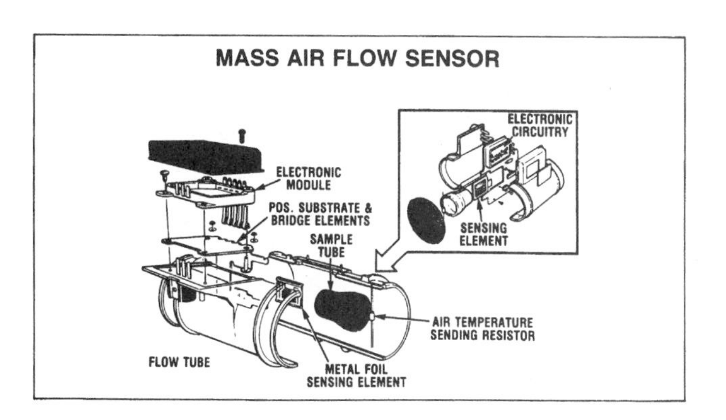

Fig. 12-14 O_2 sensor test (courtesy OTC)

⇨ Select the **Frequency (Hz)** setting with the rotary switch.

Insert:

- Black lead in **COM** terminal.
- Red lead in VΩ* RPM terminal.

Connect the Black test probe to ground.

Connect the Red test probe to the "signal out" wire of the sensor to be tested.

Note:
For frequencies below 1 Hz, the display will show 00.00 Hz.

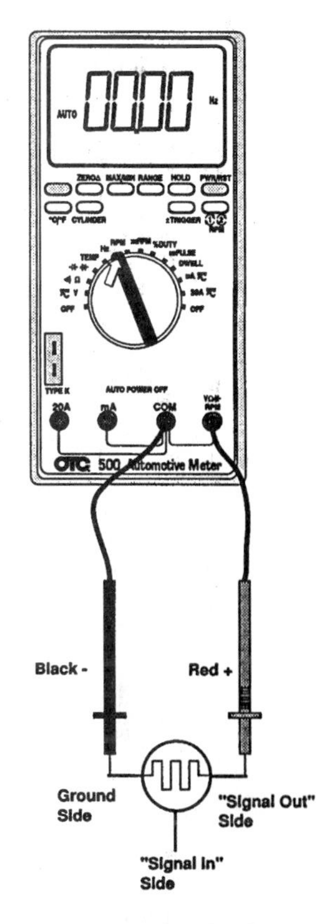

Fig. 12-15 TPS voltage test (courtesy OTC)

To check out a Bosch hot-wire air-mass sensor, first look for battery voltage at the appropriate terminal, then measure output. A typical unit should read about 2 V at idle, rising to almost 3 V at 3500 rpm. To give you some more ballpark references, common output specs for Ford hot wire units are 0-0.5 V key-on/engine-off, 0.5-1.0 V at hot idle, 1.5-2.5 V at hot cruise, and 3.0-4.7 V at WOT. Also, you should see the voltage change when you blow air through the sensor.

Vane Airflow Meter Checks

On the rotating vane type air meter, the first thing to do is reach inside the air box and move the flap through its range by hand. You should feel no binding or roughness. If it's a version that incorporates a fuel pump switch, make sure you hear the pump start when you push the vane (key on).

Using your favorite connection method (back-probing, jumpers, wire piercing, or break-out box) and a digital volt-ohmmeter, look for reference voltage input (that's 5 V). Switch to the output contact, and you should see the reading change smoothly as you push the flap to the fully open position. (In most cases, this will be a falling reading, but on Ford EEC-IV units it will rise—look for .25 V closed, and 4.50 V open.) Some carmakers give you direct resistance specs, too. For instance, on a Mazda 323, the reading between sensor terminals E2 and VS should be 20 ohms with the vane closed, and 1,000 ohms with it open. Just as with a TPS (throttle position sensor), you're checking the condition of the resistive strip or track. Any jumps in either voltage or ohms readings mean it's time for a new sensor.

CHAPTER 13

Servicing the Fuel System

Carburetor service involves making any or all of the following adjustments:

1. Curb idle.
2. Idle mixture.
3. Fast idle.
4. Choke spring tension.
5. Choke pull-off.
6. Choke unloader.

In addition, other adjustments may be necessary, especially when a carburetor has been overhauled. These include setting the float level and float drop, checking the accelerator pump stem height or linkage adjustment on certain carburetors, making bowl vent adjustments, setting the throttle stop screw, adjusting a throttle position sensor, and so on.

WHY ADJUSTMENTS ARE SO IMPORTANT

Curb idle is one of the first and most basic carburetor adjustments because it determines airflow past the throttle plate as well as engine speed (Fig. 13-1). The volume of air flowing through the carburetor in turn affects the idle mixture and intake vacuum. Curb idle is usually set at the same time as the idle mixture (adjusting either one affects the other). Idle speed also affects ignition timing, so timing changes likewise usually require curb idle speed adjustments. Ignition timing should be set prior to making the final carburetor idle mixture and curb speed adjustments.

Once the idle mixture and curb idle speed have been set, the next item to be adjusted is usually fast idle. Fast idle speed affects the way the engine runs when it is first started and while it is warming up. A cold engine needs a wider throttle opening to idle smoothly. A stepped cam on the throttle linkage provides this function. The fast idle cam is operated by the choke linkage, and the two work together to give the engine both fuel enrichment and more

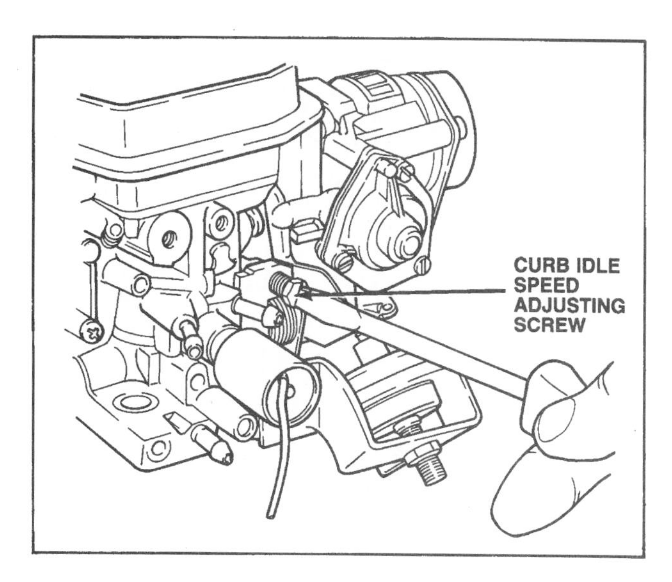

Fig. 13-1 Curb idle screw (courtesy Motorcraft - Ford).

throttle during engine warm-up. A proper fast idle speed adjustment, therefore, also requires an accurate choke adjustment.

The basic operation of the choke is determined by the tension of a bimetal coil spring inside the choke housing (Fig. 13-2). Increasing spring tension by rotating the choke housing in the opposite direction of the spring enrichens the starting mixture, and causes the choke to open more slowly. The engine receives a richer fuel mixture and more throttle during warm-up. Decreasing spring tension by rotating the housing in the same direction of the spring inside does just the opposite. It leans the starting mixture and causes the choke to open more quickly. Now the engine receives a leaner fuel mixture and less throttle during warm-up.

Most hard-starting and cold-driveability problems, such as hesitation or stumbling when accelerating, are usually caused by automatic choke problems (misadjustment, sticking, or inoperative parts). The rate at which the choke opens is determined by the rate at which the bimetal spring is heated. When there is a problem with the heating mechanism (heat riser pipe corroded or blocked, electric heating element not working or not receiving voltage, etc.), the choke may be slow to open or not open at all. The engine may not idle down as it warms-up, or it may chug and puff black smoke out the tailpipe. If dirt, corrosion, or varnish have built up on the choke linkage or choke plate shaft, sticking may prevent the choke from closing or opening properly. If the bimetal spring inside the choke housing is broken or binding, the choke may not work at all. The engine may not start or only start with repeated pumping of the gas pedal, and then idle roughly or stall until it warms up.

When the engine first starts, it needs air to keep running. A vacuum-actuated "choke pull-off" (also called a "vacuum break" or "choke pull-down") attached to the choke plate linkage cracks the choke open slightly as soon as there is manifold vacuum (Fig. 13-3). The adjustment of the choke pull-off is important. If it fails to open the choke wide enough, the engine can flood and stall. If it pulls the choke open too wide, the fuel mixture will lean out before the engine can handle it, causing lean misfire, cold hesitation, and a rough idle.

To make sure the engine is not choked when the throttle is wide open, a "choke unloader" adjustment is provided in the choke linkage. The choke unloader performs the same basic function as the choke pull-down, except that it is mechanical rather than vacuum. The choke unloader is needed because engine vacuum drops to near zero when the throttle is floored. Since the choke pull-down needs vacuum to pull the choke open, it can't do its job at wide-open throttle because there isn't enough vacuum. The mechanical choke unloader then takes over to pull the choke open, so the engine will receive sufficient air and not flood-out.

The importance of making accurate carburetor adjustments can't be overemphasized because of

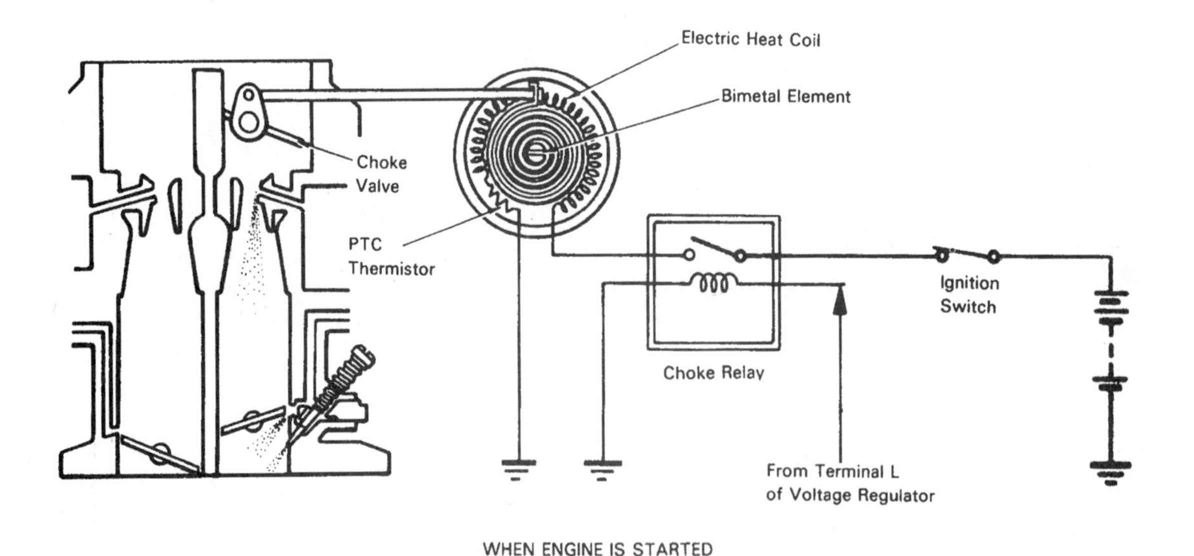

Fig. 13-2 Choke circuit (courtesy Toyota).

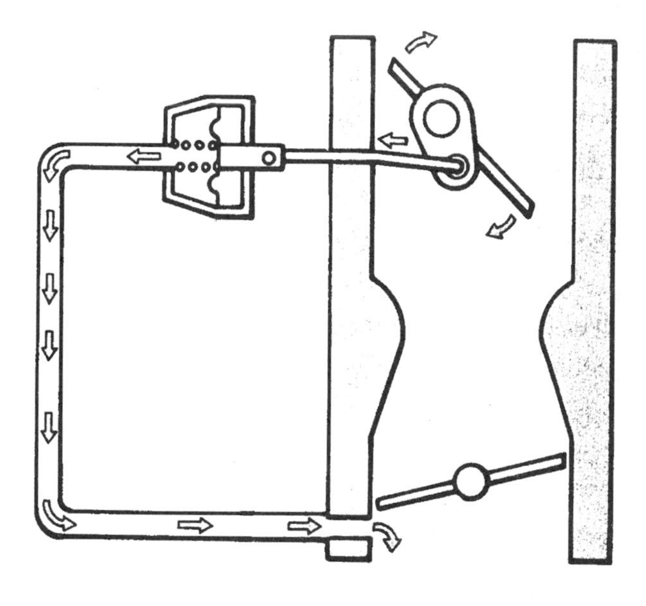

Fig. 13-3 Vacuum break/choke pull-down (courtesy Toyota).

the way in which the adjustments affect each other, as well as engine performance and emissions.

CURB IDLE SPEED

Don't underestimate the importance of an accurate idle speed adjustment. Close enough is not good enough, and you should never set idle by ear alone. Idle speed affects idle quality, the idle mixture, idle emissions, ignition timing, engine smoothness, and driveability.

When the idle speed is too slow, the engine may run roughly, or stall when brought to a stop or when accessory loads are placed on the engine (air conditioning, electric rear-window defroster, power steering assist, etc.). The volume of air flowing through the carburetor may not be great enough to maintain a properly balanced air/fuel mixture. This can lead to lean misfire and increased emissions.

When the idle speed is too high, it may be impossible to set the idle mixture correctly. Idle vacuum is also affected (the higher the idle speed, the lower the intake vacuum). This affects not only fuel metering but also the performance of vacuum-actuated emissions control devices and ignition timing.

Increasing idle speed can advance ignition timing, so even before you can set the timing accurately, the engine has to be idling at close to its specified rpm. Too fast an idle speed can also make an engine engage harshly when the automatic transmission is shifted into gear. This puts added strain on the transmission, driveshaft joints, and differential gears. Too fast an idle speed can also cause an engine pull or strain against an automatic transmission when the vehicle is stopped in gear. The vehicle will try to creep unless the driver maintains firm pressure on the brake pedal. This can create a potentially dangerous situation, as well as shortening the life of both the transmission bands and brake linings. Even with a manual transmission, too fast an idle speed can diminish the effectiveness of engine braking when decelerating.

The curb idle speed specified on the emissions decal takes into account the normal loads the engine experiences at idle. In most instances, the curb idle speed is to be set with the engine at operating temperature. If an electric cooling fan is cycling on and off, or if the upper radiator hose feels hot to the touch, you can be reasonably sure the engine is at normal temperature.

Curb idle specs may specify that the transmission be in neutral or park, though some require an automatic transmission to be placed in drive (Fig. 13-4). This is a rather dangerous procedure to attempt without a helper. It is strongly recommended that you have an assistant sit behind the wheel with one foot on the brake, if the curb idle needs to be checked with the transmission in gear. Under no circumstances should you rely on the parking brake alone to hold the car if the transmission is in gear with no one behind the wheel.

Most curb idle specs also require that all power accessories be off (air conditioner, lights, heater, rear defroster, radio, etc.). Any accessory that places a load on the engine will lower the idle speed. If someone has left the rear defroster on, for example, the amp load on the alternator can pull down the idle speed noticeably. The same is true if you're trying to check curb idle on a vehicle with a low battery. The extra charging load on the alternator can lower the idle speed. With air conditioning, the same holds true—with one exception. Most vehicles that are equipped with A/C also have an idle kicker solenoid or vacuum diaphragm to increase idle speed when the A/C is on. This compensates for the added load placed on the engine by the air-

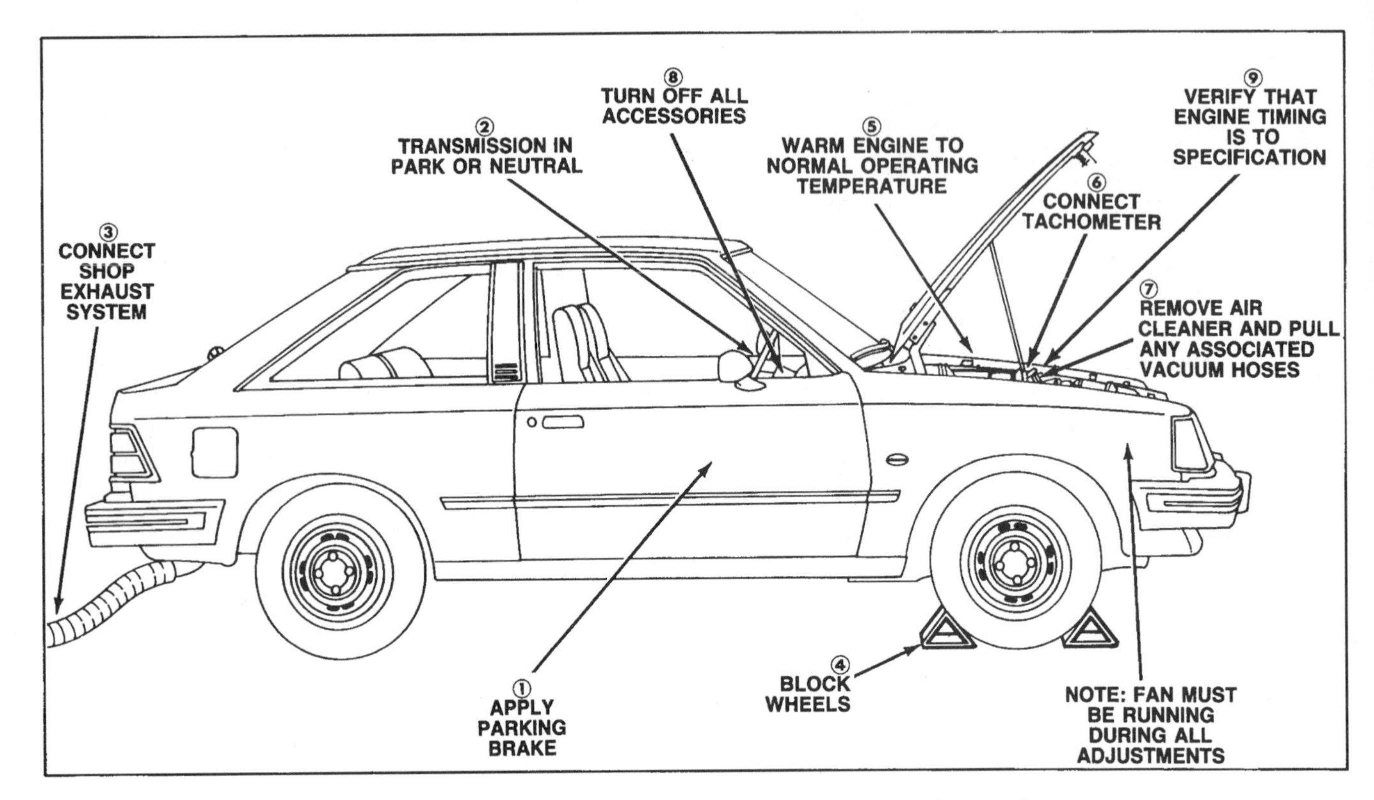

Fig. 13-4 Vehicle preparation (courtesy Motorcraft - Ford).

conditioning compressor. On A/C systems where the compressor cycles on and off (as opposed to those that run continuously), the idle kicker solenoid may cycle on and off with the compressor. Or it may stay on as long as the heater control is set to A/C. This can sometimes cause a noticeable fluctuation in the idle speed, which is normal for this type of arrangement. Where problems arise is when the idle kicker fails to work, and the engine lugs down or stalls when the A/C is switched on. To check the performance of the idle kicker solenoid, therefore, the A/C should be switched on momentarily to check the effect on idle speed. With ""A/C ON," a separate idle speed adjustment is required to set the idle kicker "A/C ON" idle speed.

CURB IDLE SPEED ADJUSTMENT PROCEDURE

Before adjusting curb idle speed, engine timing must have already been set to specs. Changing the ignition timing will affect idle speed. Advancing the timing usually increases idle speed, while retarding it decreases idle speed.

The basic procedure for setting the idle speed on a carbureted engine that does NOT have idle speed control (ISC) goes as follows:

1. With the engine at normal operating temperature, the transmission in neutral or park, all accessories off and the air cleaner removed (be sure to plug any vacuum lines), check to see that the choke is wide open and that the fast idle cam is not engaged with the throttle linkage. Also check the throttle linkage to see that it is not binding or being held open by a short or sticking throttle cable, misadjusted transmission kickdown linkage, interference with the manifold or a gasket, etc.
2. Check the emissions decal or your shop manual to see if any special procedures are required (disconnecting and plugging any vacuum lines, turning the A/C or headlights on, putting the transmission in drive, etc.). Enlist a helper if the idle speed must be checked in drive.
3. Observe engine rpm on your tachometer. If the idle speed is not within plus or minus 25 rpm of the curb idle speed specified on the emissions decal, it will have to be adjusted.

4. Locate the correct adjustment screw on the throttle linkage. Don't confuse it with the fast idle speed screw, the throttle stop screw, or a separate dashpot or idle kicker adjustment screw. On some carburetors, the idle speed is adjusted by turning a screw on the idle kicker, while on others it is not. The only way to tell which is which is to carefully examine the linkage to see which screw is carrying the tension of the throttle. If the throttle linkage is resting against an idle kicker, see the special procedure that follows for setting Ford VOTM and other idle kicker systems.
5. Turn the curb idle adjustment screw until your tachometer reads within 25 rpm of the speed specified. Turning the screw counterclockwise (backing it out) closes the throttle opening and lowers idle speed. Turning the screw clockwise (in) opens the throttle and increases idle speed.

FORD VOTM AND VACUUM IDLE SPEED CONTROL

Some Fords (typically 1980 and later models) use a VOTM (vacuum-operated throttle modulator) to regulate idle speed. Others use a slightly different system called "mechanical vacuum idle speed control." Both have a vacuum diaphragm with a screw adjustment and plunger that rests against the throttle linkage.

The basic idle speed adjustment procedure for carburetors that have a VOTM that retracts when vacuum is applied is to disconnect and plug the VOTM vacuum hose. Then apply full vacuum to the unit with a slave hose from an intake vacuum source or with a hand-held pump. Vacuum should pull the throttle plunger away from the linkage so the basic idle speed adjustment can be made by turning the screw on the top of the VOTM.

NOTE: if the VOTM doesn't hold vacuum, it means the diaphragm is leaking and the unit needs to be replaced.

On other Ford applications, a somewhat different style of VOTM is used. On engines without air conditioning, there is no VOTM, but there is an idle tracking switch on the throttle linkage to inform the engine computer when the engine is at idle. On all engines with air conditioning, a "reverse" type of VOTM is used wherein the plunger extends and pushes against the throttle linkage to increase idle speed when vacuum is applied. The VOTM is supposed to receive vacuum when the air conditioner is running. This increases the throttle opening to maintain idle speed and to compensate for the extra load of the A/C compressor to prevent stalling. When adjusting the curb idle speed on carbs with the reverse style VOTM, make sure the A/C is off.

An easy way to tell the difference between the two types of VOTMs is to apply vacuum and see if the throttle linkage plunger retracts or extends. If it retracts, then it is the type that needs to be fully retracted with full vacuum before adjusting curb idle. If the VOTM plunger extends when vacuum is applied, then it must not receive vacuum while the idle speed adjustment is made.

On Fords that have a vacuum idle speed control system, the basic idle adjustment is made by removing and plugging the ISC vacuum hose and applying vacuum to the ISC diaphragm. The ISC adjusting screw is then turned until the ISC plunger is clear of the throttle lever. The curb idle speed can then adjusted in the usual manner, except that the throttle stop screw is used instead of the ISC screw. The ISC is then reconnected to its vacuum hose, the transmission is put in drive, and the ISC adjustment screw is then adjusted to obtain the specified idle speed in drive.

A/C-"ON" IDLE SPEED WITH VOTM

Idle speed should also be checked with the A/C running to make sure the VOTM is functioning correctly and that the A/C-"ON" idle speed is within specs (fast enough to prevent stalling, but not so fast that it causes the engine to race or strain against the transmission when it is in drive).

Ford recommends that this adjustment be made BEFORE setting curb idle speed.

To make the adjustment, make sure the vacuum hose to the VOTM is connected in the normal manner, then turn on the air conditioner to maximum cool and high blower speed. The increase in idle speed is usually anywhere from 50 to 200 rpm when the VOTM is activated with the A/C running. Adjust idle speed as needed to bring it within specs.

If the idle speed drops because the VOTM failed to move the throttle linkage further open, first check to see that the VOTM is properly adjusted so that it makes contact with the throttle linkage.

If a VOTM plunger does not move when the A/C is on, check the VOTM vacuum hose for the presence of vacuum. If there is vacuum, but the VOTM fails to move, the VOTM is defective and needs to be replaced. You should confirm your diagnosis by applying vacuum to the VOTM directly with a handheld pump. No vacuum at the VOTM means there's a problem in the vacuum supply line. Refer to the vehicle vacuum hose routing diagram to trace down the problem (such as a leaky or misrouted hose, a faulty vacuum solenoid, etc.).

IDLE KICKER ADJUSTMENTS

On carburetors that use an idle kicker solenoid (also called an throttle kicker solenoid or TKS) or a throttle solenoid positioner (TSP) to boost idle speed when the air conditioning compressor is engaged, a separate idle spec may be listed on the emissions decal or in the manual (usually 50 to 200 rpm higher than the curb idle speed).

To check it, simply turn on the air conditioning and note the engine rpm. Readjust it if it is not within 25 rpm of specs. The adjustment is made by turning a screw on the unit, or by loosening an adjustment nut on the solenoid bracket and repositioning the unit in or out with respect to the throttle linkage.

If the idle kicker solenoid or TSP plunger fails to move when the A/C is switched on, sometimes the pressure of the return spring is enough to prevent the solenoid from kicking out. Try opening the throttle slightly to relieve spring pressure. If the idle kicker solenoid or TSP plunger still fails to move, check the electrical connector for the presence of voltage with a volt-meter or test lamp. You should see battery voltage if the control circuit if functioning properly. If voltage is reaching the idle kicker or TSP, but nothing is happening (no clicking or extension of the plunger), you can assume the component is bad and needs to be replaced. If you don't get voltage at the connector when the A/C is on, then the problem is in the wiring harness or electrical circuit.

IDLE STOP SOLENOID

Do not confuse an idle kicker solenoid or TSP with an idle stop solenoid (Fig. 13-5). Although they both

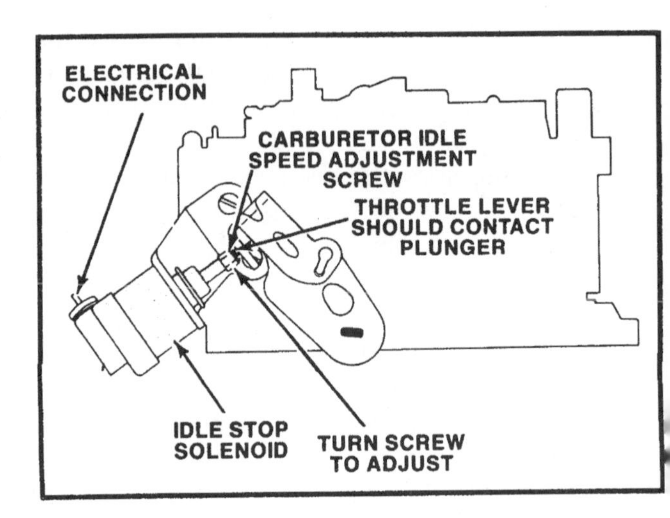

Fig. 13-5 Idle stop solenoid (courtesy General Motors).

look alike, they have completely different purposes. An idle stop solenoid is designed to hold the throttle open and maintain a constant idle speed as long as the engine is running. But when the ignition is switched off, the solenoid retracts allowing the throttle to close completely. This shuts off the engine's supply of air and prevents the engine from running on or "dieseling" when hot.

The idle stop solenoid should be adjusted so that it releases the throttle completely when the ignition is turned off. If an engine with an idle stop solenoid has a problem with run-on, it means the solenoid is not adjusted correctly (not releasing the throttle fully), is not working, or the throttle stop screw is holding the throttle too far open. (See the throttle stop adjustment procedure that follows.)

There are a couple of ways to tell an idle stop solenoid from an idle kicker or TSP. (**NOTE**: some engines may have both!) One is to check for voltage at the solenoid when the engine is idling and the A/C is off. If you see voltage, then you have an idle stop solenoid. If there's no voltage, then it's an idle kicker solenoid (or TSP). The other way is to turn the A/C on and note what happens. An idle kicker solenoid or TSP will extend while an idle stop solenoid will remain extended. You can also check for voltage while switching the A/C on and off. If the voltage at the solenoid cycles on and off with the A/C, then it is an idle kicker or TSP, not an idle stop solenoid.

The curb idle speed adjustment screw usually rests against the tip of the idle stop solenoid plunger or is part of the plunger itself. In either case, the curb idle speed is set with the idle stop solenoid energized. The solenoid should extend fully when the ignition switch is turned on (listen for a click and watch for plunger movement), and retract when the ignition is switched off. You can check for the presence of voltage at the solenoid with a voltmeter or test lamp.

THROTTLE STOP ADJUSTMENT

A separate throttle stop screw is usually located on the base of the carburetor or on the throttle linkage. This screw is used in conjunction with an idle stop solenoid to adjust the throttle so it will close completely when the solenoid is off (retracted). The closing of the throttle is adjusted by backing the stop screw out until the throttle is completely shut, then turning the screw in a fraction of a turn so the throttle doesn't stick shut.

The throttle stop should not require adjustment unless the engine is experiencing a run-on problem when turned off. A run-on problem would indicate that either the throttle is remaining too far open, or that the idle stop solenoid is not working or misadjusted. Check the operation and adjustment of the idle stop solenoid first before making any adjustments to the throttle stop screw.

On engines with idle speed control (ISC), the throttle stop screw is used to set minimum idle speed (see section on ISC that follows). It is turned in like a curb idle screw to set the minimum idle speed. Adjustment is not normally needed if the ISC motor is functioning correctly and is properly adjusted. But if the ISC motor is replaced or needs adjustment, then the throttle stop screw must be reset.

IDLE SPEED CONTROL MOTORS

On carburetors that have ISC (idle speed control) motors, a small electric stepper motor is connected to the throttle linkage to regulate idle speed. The ISC motor receives its commands from the engine control computer, which monitors engine rpm through distributor pickup reference pulses. By counting these pulses, it knows how fast the engine is running. A separate idle switch on the throttle linkage closes at idle and tells the computer when the engine is idling, so it can start monitoring and regulating the idle speed. The computer also monitors the relative position of the ISC motor, and changes the ISC plunger position too by running the motor in or out to increase or decrease idle speed as required. The direction the motor turns is determined by the polarity of the voltage to its terminals.

Adjustment of the ISC motor is critical because if the plunger is too far extended (a common condition), the linkage may never close the idle contact switch to activate the ISC system.

Do not use the ISC plunger to adjust curb idle because you won't accomplish anything. The computer will simply reposition the plunger when the ISC is reconnected (which must be done with the key off to prevent a voltage spike that could damage the ISC motor).

For curb idle adjustments on carburetors with ISC, the basic procedure is to disable the automatic idle speed control motor by unplugging its connector (the ignition key must be off) and to then set a minimum idle speed using the throttle stop screw. You'll also need a pair of fused jumper wires to run the ISC motor plunger in and out.

ISC IDLE SPEED ADJUSTMENT (GM)

The idle speed check procedure goes as follows:

1. On GM applications, check the ISC plunger for an identification letter (you'll find it stamped on the side of the plunger shaft near the head). If it has a letter, proceed to step 3. If not, unscrew the plunger and measure the distance from the back of the plunger head to the end of the plunger. Record this as dimension "A," because you'll need it to find the correct adjustment.
2. Reinstall the plunger so the distance between the back of the plunger head and the ISC motor is less than dimension "B" again.
3. With the engine at normal operating temperature, turn the ignition off and disconnect the ISC motor connector. As mentioned earlier, the ignition key must be off when this is done. Otherwise you can damage the ISC motor.
4. Use a fused jumper wire to apply battery voltage to terminal "C" on the ISC motor connector to retract the plunger. Terminal "D" must also

be grounded. Do not leave battery voltage connected to the ISC motor any longer than is necessary to retract the plunger, and do not accidentally apply battery voltage to terminals "A" or "B," as doing so can damage the ISC motor.

5. Restart the engine. GM recommends running the engine until the M/C (mixture control) solenoid on the feedback carburetor is producing a fluctuating dwell signal (which you can read with a dwell meter set to the 6-cylinder scale connected to the green M/C dwell lead, or by looking for a "closed loop" indication on a handheld scan tool. This makes sure the engine is at normal operating temperature and idle conditions.
6. On vehicles with automatic transmissions, the idle adjustment procedure calls for the transmission to be placed in drive. (Be sure to have a helper behind the wheel while you do this.)
7. With the ISC plunger fully retracted, adjust the idle speed to the minimum rpm specified for the application. In most cases this will be 400 to 500 rpm.
8. Now put the transmission back into neutral and reverse your jumper connections, applying voltage to terminal "D" and ground to "C". This will cause the ISC motor plunger to fully extend. Again, be careful not to leave the jumpers connected any longer than necessary.
9. Set the maximum idle speed to the rpm specified for the application. For manual transmissions and automatics in neutral, this will be around 1200 to 1500 rpm. If the maximum rpm spec specifies the transmission be in drive, then the maximum idle speed is usually between 900 and 1300 rpm.
10. Measure the distance between the back of the plunger head and the ISC motor (dimension "B"). It should not exceed the maximum specification shown in the shop manual.
11. Reverse your jumper connections once more to fully retract the ISC plunger. Then turn the engine off and reconnect the ISC wiring harness.
12. Disconnecting the ISC motor on a GM application will set a fault code in the computer memory and cause the "Check Engine" light to come on. When the ISC motor is reconnected, the light will go out, but the fault code will remain in memory. The final step of the adjustment procedure, therefore, is to clear the computer memory by pulling the ECM (electronic control module) fuse for 10 seconds. Don't just disconnect the battery cable to erase the memory, because you'll also erase all the radio and climate control settings.

ISC IDLE SPEED ADJUSTMENT (FORD)

The Ford ISC motor works on the same principle as GM, but the adjustment procedure is somewhat different. If the idle speed doesn't match the specs after the engine has idled for 60 seconds, and is higher than the decal specs, check to see if the ISC plunger is making contact with the throttle linkage. If it is not, and the throttle is being held open by the throttle stop screw, the ISC plunger must be retracted using the following procedure.

1. Shut the engine off and locate the diagnostic self-test connector and pigtail located in the engine compartment.
2. Connect a jumper wire between the signal return terminal in the diagnostic connector and the pigtail connector.
3. Turn the ignition on, but do not start the engine. This should cause the ISC plunger to retract. Wait until the plunger is fully retracted (about 10 seconds). If the plunger fails to retract, the ISC motor could be defective, or there could be a problem in the wiring harness or control system.
4. Turn the ignition key off and remove the jumper wire.
5. The throttle stop screw should now be adjusted to achieve the desired throttle opening. Ford says to remove the old screw and install a new one. Turn the screw in until there is .005-in. clearance between the tip and throttle linkage; then turn it 1½ turns more.
6. To perform the fast idle adjustment, remove the rubber dust cover from the tip of the ISC motor and push the plunger back toward the motor to remove any lash. **NOTE**: The tip of the plunger contains an idle contact switch. This switch must not be held in when checking the clearance, so don't push on the tip of the plunger. Use gentle pressure from needlenose pliers on the plunger shaft instead to remove any lash. There should

be 9/32 inch clearance (use a .281 drill bit) between the tip of the ISC motor and the throttle linkage, if the position of the ISC motor is correct. If not, loosen the ISC bracket adjusting screw, and reposition the ISC motor as required. Then retighten the adjusting screw, and replace the rubber dust cover on the plunger.

IDLE SPEED WITH ELECTRONIC FUEL INJECTION

Idle speed on a fuel-injected engine is controlled by the amount of air allowed to bypasses the throttle plates. A computer-controlled idle air control (IAC) motor moves a valve that varies airflow through the bypass circuit. If idle speed is below specs, the computer will signal the motor to increase airflow. If idle speed is above specs, the computer will signal the motor to decrease airflow.

The operation of the IAC motor is nonadjustable. If idle speed is incorrect, the vehicle may have a vacuum leak, defective IAC motor, loose or corroded IAC motor wiring connections, a computer problem, or the wrong PCV valve. You should also check the computer for any applicable fault codes (a Code DTC 35 on newer GM products, for example). The IAC motor itself can be tested on many applications with a scan tool. By using the scan tool to drive the IAC motor to change airflow through the bypass circuit, you can check for a corresponding change in idle rpm.

CARBURETOR IDLE MIXTURE ADJUSTMENT

The relative proportions of air and fuel at idle (called the "idle mixture") are adjusted by a tapered "idle mixture adjustment screw." On older carburetors, the idle mixture screw extends through the cavity behind the idle port (Fig. 13-6). Turning the screw in decreases the size of the opening to reduce the flow of fuel, which leans out the idle mixture. Backing the screw out allows more fuel to flow through the idle port, which enrichens the mixture.

On late-model carburetors with "sealed" idle mixture adjustment screws (to discourage tamper-

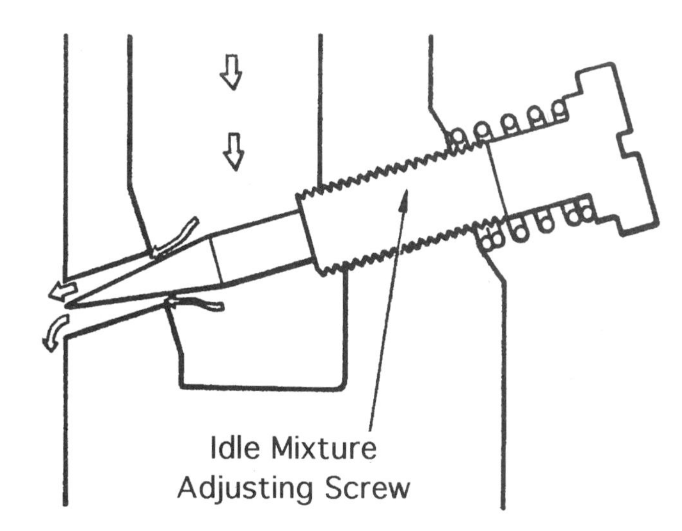

Fig. 13-6 Idle mixture screw (courtesy Toyota).

ing), a second adjustment screw is provided to control the amount of air bleeding into the idle circuit. Thus the idle mixture screw is really an air bleed screw. The effect of turning the air bleed screw in and out are reversed compared to an ordinary idle mixture screw: turning the screw in (clockwise) enrichens the mixture, while turning it out (counterclockwise) leans the mixture.

"Idle limiter caps" are used on many carburetors to limit the range of adjustment in the idle mixture. The reason for limiting the adjustment is to prevent overly rich or lean idle mixtures that could adversely affect exhaust emissions. Tabs that project from the plastic caps restrict the amount of adjustment to about half a turn. The caps can be pried or pulled off, but even with the caps removed, the range of adjustment is limited by the internal restriction in the idle tube.

As just mentioned, the idle mixture screw is often hidden beneath a metal plug to discourage tampering (Fig. 13-7). If a satisfactory idle cannot be obtained by adjusting the idle air bleed screw, it may be necessary to remove the plug and readjust the factory-set mixture screw. (Which may require removing the carburetor from the engine.) The plug can be removed by drilling, or by punching a small hole in it and then using a screw-in slide puller to pull it out. Care must be taken not to damage the adjustment screw. Once the adjustment has been made, the screw should be resealed by inserting a new plug or covering the hole with RTV silicone sealer.

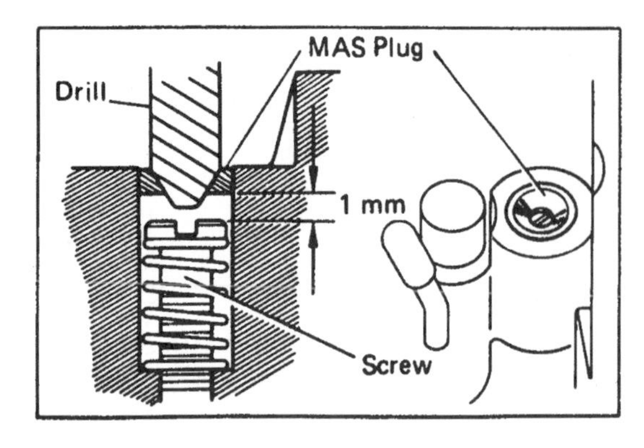

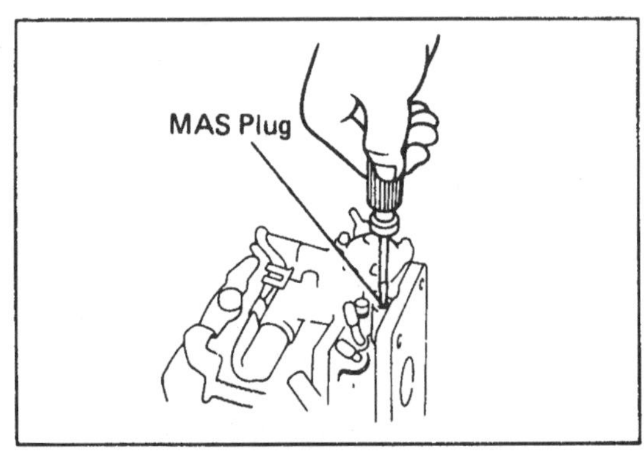

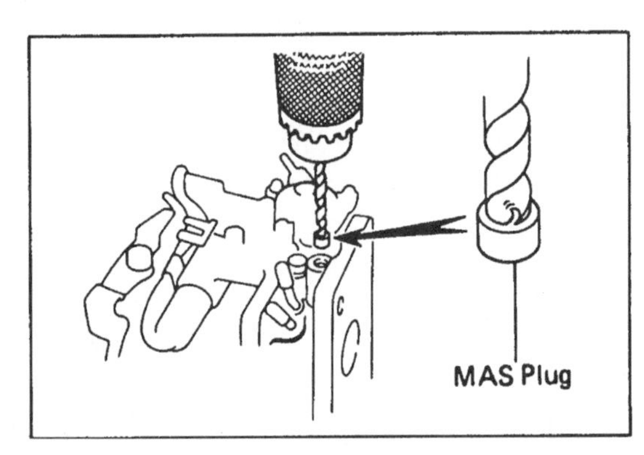

Fig. 13-7 Anti-tamper plug removal (courtesy Toyota).

LEAN DROP IDLE ADJUSTMENT

Ideally, the idle mixture should be adjusted to achieve the smoothest idle, the highest idle vacuum reading, and the lowest exhaust emissions. This can usually be accomplished by using the "lean drop" technique. After the idle speed has been adjusted to minimum rpm, the idle mixture screw is turned in until the rpm begins to drop, then it is readjusted out until the minimum specified idle speed is achieved.

When using an exhaust analyzer for idle mixture adjustments, carbon monoxide (CO) readings will tell you when the mixture is too rich, and oxygen (O^2) will tell you when the mixture is too lean.

PROPANE ENRICHMENT

On some carburetors, a procedure called "propane enrichment" can be used to set the idle mixture. Because the carburetor is calibrated so lean to reduce carbon monoxide emissions, the traditional lean drop technique may not give the best results.

The propane enrichment procedure involves feeding propane vapor into the engine through a vacuum fitting on the carburetor or intake manifold to enrich the idle mixture and increase the idle speed. The flow of propane should be increased until maximum idle speed is achieved. (When too much propane is added, rpm will decrease.)

To make the check, hold the propane bottle in the upright position. Then slowly open the main valve on the propane bottle until maximum engine rpm is reached. When the idle speed starts to drop, back-off the amount of propane slightly until the highest idle is achieved.

NOTE: If the idle mixture is extremely rich, idle speed will drop when any propane is used. In this case, turn the idle mixture in about half a turn and try adding propane again.

The maximum idle speed the engine reaches while being fed propane is the "propane enriched" idle speed. You obtain it by fine-tuning the metering valve on the propane bottle until the highest rpm is achieved, and then readjusting the idle speed to the "enriched" idle specs. The propane enriched rpm is usually specified on the underhood emissions decal.

If the peak propane enriched idle speed is higher than the one specified on the decal, it means the idle mixture is too lean and needs to be made richer.

If the peak propane enriched idle speed is less than the one specified, it means the mixture is too rich and it needs to be made leaner.

Once the idle speed has been adjusted to the propane enriched specification, turn off the main propane valve and allow the engine speed to stabilize. Now slowly adjust the mixture screw until the specified idle rpm is reached. Then turn the propane back on again and fine-tune the metering valve to get the highest engine rpm.

If the maximum enriched speed differs by more

than 25 rpm from the specified enriched speed, you'll have to repeat the adjustment procedure. If the enriched idle speed is within 25 rpm of the decal specs, the idle mixture is properly adjusted and no further adjustments are necessary.

FAST IDLE SPEED ADJUSTMENT

The fast idle speed is typically adjusted with the engine warm and the fast idle screw (Fig. 13-8) resting on the specified step of the fast idle cam (which may be the highest or lowest step on the cam). A separate screw is usually provided for fast idle adjustment, which should not be confused with the curb idle screw or a separate A/C kicker solenoid screw (if provided). Fast idle is always checked after first setting curb idle.

AUTOMATIC CHOKE ADJUSTMENT

The basic adjustment is made by loosening the three screws on the metal retaining ring that holds the plastic choke housing to the carburetor airhorn. Then rotate the choke housing to change the tension on the choke plate. Increasing tension reduces the choke plate opening for a richer mixture, while decreasing tension opens the choke plate further for a leaner mixture.

The most accurate way to achieve proper choke adjustment is to use an exhaust analyzer. Watching the fuel enrichment with the exhaust analyzer will tell you if the carburetor is leaning out too quickly during engine warm-up.

CAUTION: A small adjustment goes a long way on most automatic chokes. Changing the setting by more than a couple of index marks on the choke housing can significantly alter the way the choke behaves.

On many late-model carburetors, the choke housing is designed to be "tamper-proof." Rivets are used in place of screws on the retaining collar. To make an adjustment, the rivets must be drilled out using a ⅛-inch (No. 30) drill bit (Fig. 13-9). Do not try to chisel the rivet heads off, as doing so can easily damage the soft metal housing. The rivets can be replaced with self-tapping screws, or pop-rivets once the choke adjustment has been made.

To adjust the choke, proceed as follows:

1. With the choke coil cold (choke closed), loosen the choke coil cover screws. **NOTE**: The choke will take awhile to cool down once the engine is shut off.
2. Work the throttle linkage so the choke can snap shut. Check the tension of the choke spring by pressing against the choke plate with your finger. See that it moves with even spring resistance and does not stick or bind.
3. On many chokes, an index mark on the plastic choke cover is aligned with a scale or series of marks on the air horn. Refer to a manual for the

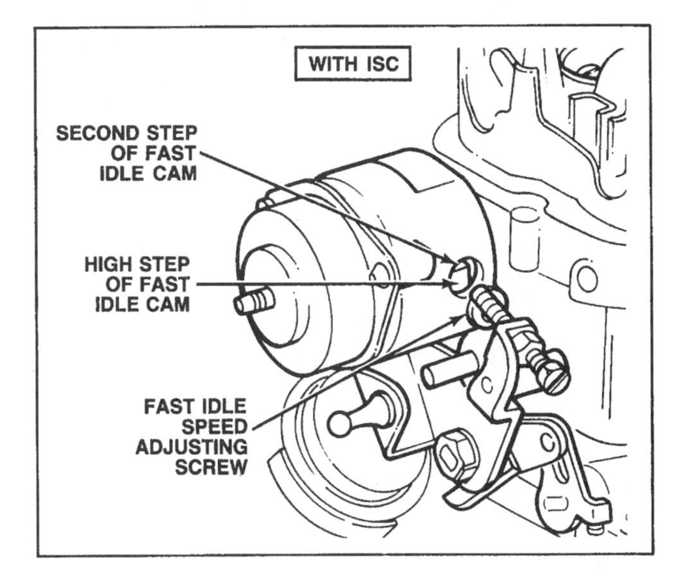

Fig. 13-8 Fast idle speed screw (courtesy Motorcraft-Ford).

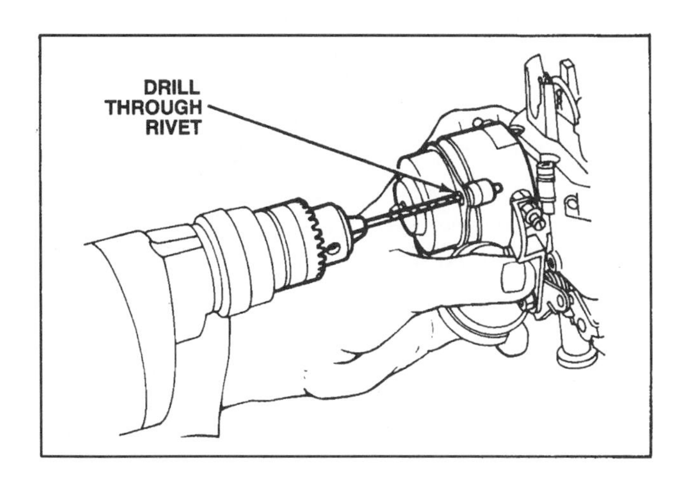

Fig. 13-9 Drilling rivets (courtesy Motorcraft - Ford).

manufacturer's recommended choke setting. Then turn the choke so the index mark aligns with the appropriate mark on the casting. If the specs call for "2 lean," for example, the mark would be aligned with the second mark from center towards the lean position.

NOTE: The recommended choke setting is a basic starting point only. The choke may require additional adjusting one way or the other for best performance.

CHOKE PULL-OFF ADJUSTMENT

Making the choke pull-off adjustment requires the use of (a) an angle gauge, (b) a drill bit or rod gauge, and/or (c) an exhaust analyzer. An angle gauge is used to measure the angle at which the choke plate is positioned in the air horn. The measured angle of the choke plate is compared to the specs listed in your manual so the necessary adjustment can be made. Some manufacturers specify the choke plate opening according to what sized drill bit or rod gauge can be inserted between the edge of the plate and the air-horn. The exhaust analyzer can be used to tell you if the fuel mixture is too rich or too lean when the engine is first started. This enables you to fine-tune the choke pull-off adjustment for best performance.

To adjust the choke pulloff, proceed as follows:

1. With the choke cold, and after making the initial automatic choke adjustment, work the throttle linkage so the choke plate snaps shut.
2. Disconnect the vacuum hose from the pull-off, and apply vacuum with a hand-held pump to pull the choke open. The choke pull-off should pull down on the choke plate and open it slightly. If the pull-off diaphragm does not hold vacuum, it is defective and needs to be replaced.

   NOTE: On some two- and four-barrel carburetors (Rochester), two choke pulloffs are used. The primary choke pull-off handles the initial vacuum break, while the second holds the choke open during warm weather to prevent a rich mixture when hot-starting the engine. The primary choke pull-off works the same as a single pull-off, but the secondary pull-off usually has a tiny vacuum bleed hole it (so it won't hold vacuum unless the hole is first plugged).
3. Using either an angle gauge or the specified-size drill bit or rod gauge, check the choke opening for the proper dimensions.
4. If adjustment is needed, bend the linkage rod between the choke pull-off and choke plate until the choke opening matches the specifications. On other carburetors (Holley 5200, 6510, etc.), the adjustment is made by turning an adjustment screw (some are allen-head screws) on the pull-off chamber to change the movement of the diaphragm. A hardened steel anti-tampering plug in the pull-off housing may have to be removed to get to the adjustment screw on some carburetors.

   NOTE: On some four-barrel carburetors, a second choke pull-off is provided for the secondaries. Repeat the same adjustment procedure for it, too.
5. If the carburetor does not seem to work properly with the factory-specified choke pull-off setting, additional adjustment may be needed using the 4-gas analyzer to monitor the initial start-up fuel mixture. If the mixture is running too rich, open up the choke a little more (a couple of degrees) by bending the rod slightly. If the mixture is too lean, close the choke opening slightly by again bending the linkage rod.
6. The choke pull-off on some GM cars is controlled by a thermal vacuum switch (TVS) inside the air cleaner. It can be distinguished from an air cleaner temperature sensor by a hose extending between the switch and the pull-off. If the engine is exhibiting cold-start problems that indicate the choke pull-off isn't working, make sure the hoses between the TVS, pull-off and vacuum source are tightly connected and not leaking or cracked.

   To test the pull-off TVS, put a cold rag or ice cube on it for several minutes to cool it down. Disconnect the hoses and connect a hand-held vacuum pump to the TVS vacuum inlet port. Place your finger over the outlet port and apply vacuum. If you feel any pull on your finger, the TVS is bad and needs to be replaced.

FLOAT ADJUSTMENTS

There are a number of ways in which the float level can be set. On some carburetors (certain Holley models, for example), it can be set without having to remove the top of the carburetor. An adjustment screw is provided that changes the pivot height of the float arm. The fuel level in the bowl is observed by removing an inspection plug on the side of the bowl. The level is then adjusted until the fuel level inside the bowl is even with the bottom of the inspection hole.

On other carburetors, the float level can be set by inserting a depth gauge through a bowl vent and adjusting the float arm (by turning a screw) until the proper height is obtained.

On many carburetors, though, the top of the carburetor must be removed to adjust the float. On some, the level is adjusted by bending the float arm so a point on the top of the float is at a specified depth with respect to the top of the bowl. With others, the level is checked by holding the top of the carburetor upside down and measuring the height of the float with respect to the housing. Float level settings are usually specified within 1/32 of an inch, so "close enough" isn't good enough for accurate fuel calibration (Fig. 13-10).

"Float drop," or the amount by which the float itself can sink in the fuel bowl is another adjustment that may be required, though it is not as critical as the float level adjustment. Float drop is adjusted by bending a tang on the float arm to limit how far the float can move.

BOWL VENT

On some carburetors (Holley Model 1946 one-barrel carb used on Ford 6-cylinder engines, for example), a bowl vent adjustment screw is provided to regulate the volume of air being siphoned through the vent passageway.

To make the adjustment, proceed as follows:

1. Disconnect the canister vent hose from the bowl vent tube.
2. Connect a hand-held vacuum pump to the bowl vent tube.
3. Remove the small metal bowl vent cover, gasket, and spring to gain access of the adjustment screw.
4. Rotate the vent adjustment screw clockwise until it protrudes 1/8 inch or less above the vent arm.
5. Apply vacuum to the vent tube and slowly turn the adjusting screw counterclockwise in 1/8-inch increments until the vacuum reading indicates that the vent is closed. Stop applying vacuum and turn the screw clockwise 1/2 turn. The vent should now be properly adjusted.

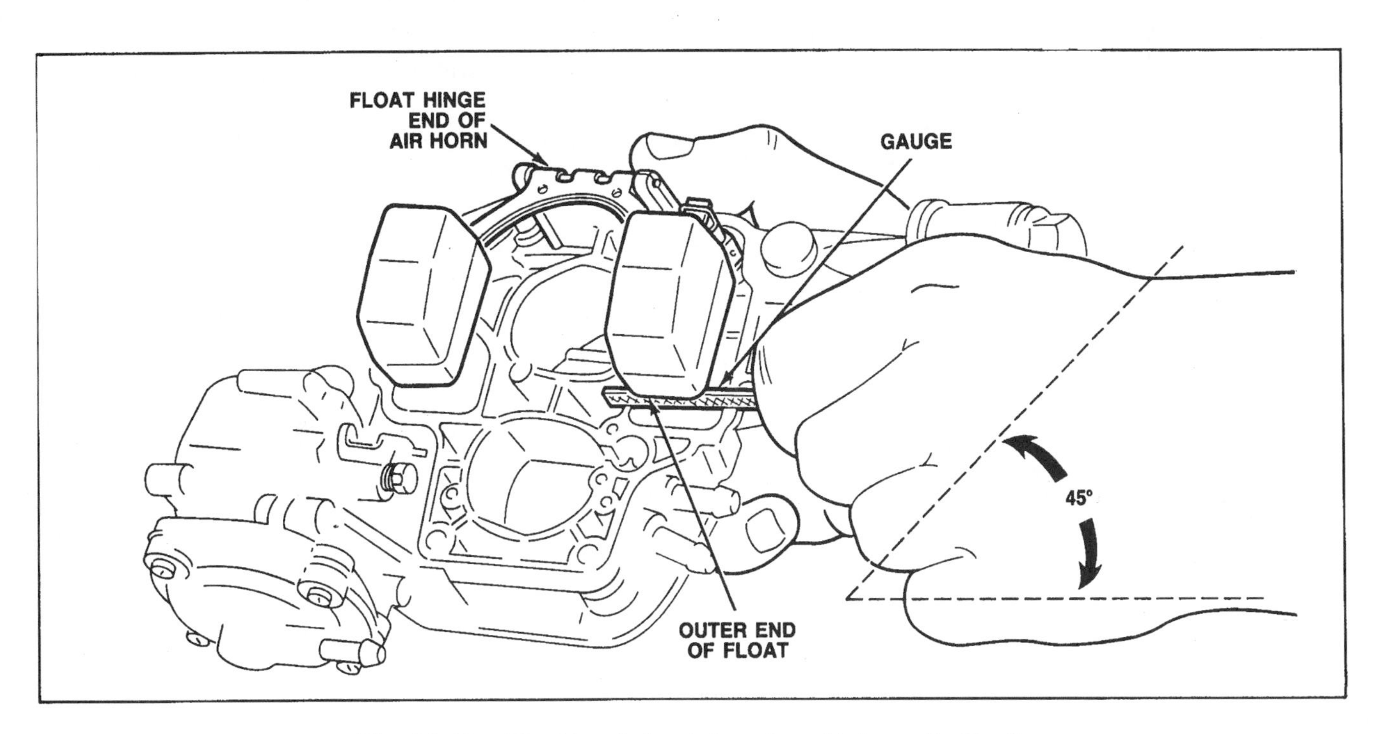

Fig. 13-10 Float level check (courtesy Motorcraft - Ford).

ACCELERATOR PUMP ADJUSTMENT

On some carburetors, the stem height of the accelerator pump must be set to specs when the carburetor is overhauled. This is done by assembling the pump linkage, measuring the height and then bending the linkage arm if the height does not match the specs.

On some Holley four-barrel carburetors, the accelerator pump action can be modified by replacing the plastic pump cam on the throttle linkage. Changing the profile of the pump cam changes the rate at which fuel is delivered when the throttle is opened. This type of adjustment is limited primarily to performance applications, where other engine modifications require more fuel delivery at lower rpm to prevent hesitation.

TPS AND THROTTLE SWITCH ADJUSTMENTS

On late-model engines with feedback carburetion or electronic fuel injection and computerized engine controls, a "throttle position sensor" (TPS) is used to inform the computer about the rate of throttle opening and throttle position. A separate idle switch and/or wide-open throttle switch may also be used.

On some carburetors, such as the Rochester E2SE, E2ME, E2MC, E4ME and E4MC, the throttle position sensor is located inside the carburetor body (Fig. 13-11). An adjustment screw on top of the carburetor is provided, but it is covered by an anti-tampering plug. External throttle position sensors are used primary on fuel-injection throttle bodies.

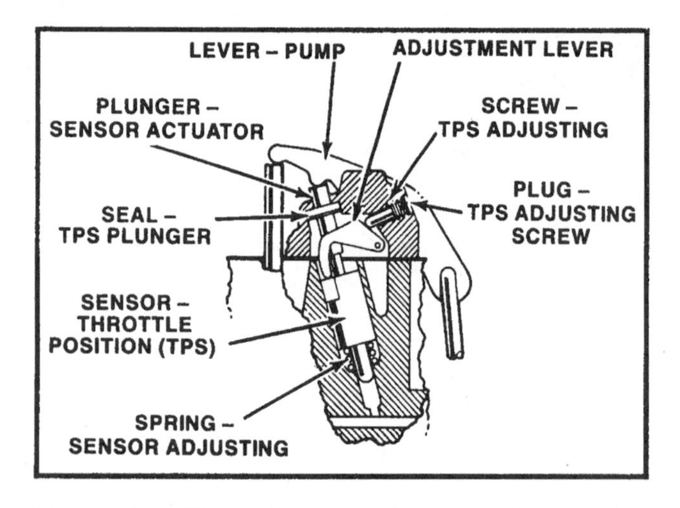

Fig. 13-11 TPS adjustment (courtesy General Motors).

Under normal circumstances, the TPS should not require adjustment. But if a diagnosis reveals a problem with the TPS, if the TPS is replaced, or if a carburetor or throttle body is replaced, then adjustment is needed.

The TPS is essentially a variable resistor that changes resistance as the throttle opens. The initial setting of the TPS is very important. The voltage signal the computer receives back from the TPS tells the computer the position of the throttle. The initial setting, therefore, must be set as closely as possible to the factory specs. This is accomplished by reading the TPS voltage at a specific throttle position using a 10-megohm impedance digital voltmeter or a hand-held scan tool that plugs into the vehicle diagnostic connector.

The adjustment procedures goes as follows:

1. Remove any anti-tampering plug by drilling a small hole ($\frac{5}{64}$ inch) it in, threading in a sheet-metal screw, and popping or prying the plug out.
2. Connect a digital voltmeter to the TPS output and ground terminals—or plug in a scan tool to the vehicle's diagnostic connector.
3. Turn the ignition on. With the engine off, adjust the TPS screw with the throttle in the specified position (idle, high step of fast idle cam, or resting against the throttle stop screw with the ISC plunger fully retracted) until the proper voltage reading is obtained. Always refer to a shop manual for the exact specs, because they vary greatly from one application to another.
4. Turn the ignition off, disconnect the voltmeter or scan tool, and reseal the TPS adjustment screw with a new plug or silicone sealer.

SERVICING FUEL-INJECTED ENGINES

Service is limited to adjusting idle speed (if adjustable), replacing the fuel filter periodically, and cleaning or replacing components such as the injectors or pressure regulator (when necessary). Even so, a variety of "diagnostic checks" may be necessary if a fuel-delivery problem is suspected.

Electronic problems will usually trigger some kind of trouble code in vehicles where the EFI control module or engine computer has self-diagnostic capability. This, in turn, can help you decide where's the best place to start your diagnosis. But when there are no fault codes, you'll have to start with the injection system itself.

Pressure Checks

Checking fuel pressure is an important element of troubleshooting fuel injection performance because fuel pressure affects fuel delivery. Too much pressure will make an engine run rich. Not enough pressure, and it will run lean. Little or no pressure, and the engine won't run at all.

To check the operating pressure of the system, you'll need to connect a pressure gauge to the test fitting or schrader valve on the fuel rail (Fig. 13-12). If there is no test fitting, you'll have to plumb in a T-fitting to accommodate your gauge.

Start the engine and note the fuel pressure reading at idle. A pressure reading that's below specs tells you either the fuel pump is weak or the pressure regulator is defective. Too much pressure means a defective pressure regulator.

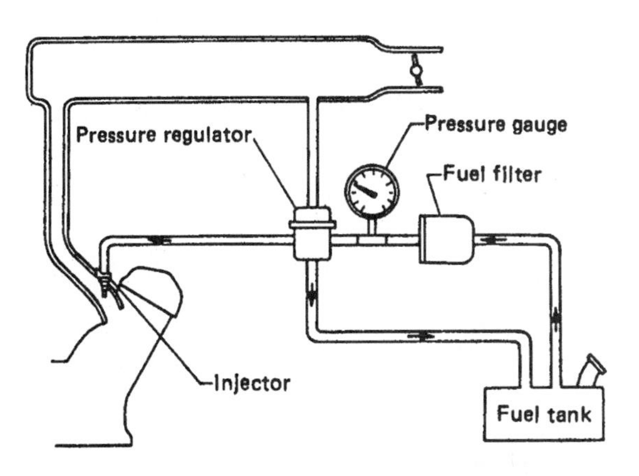

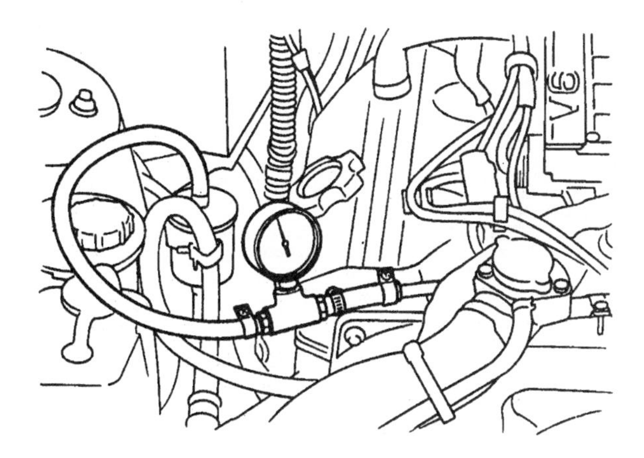

Fig. 13-12 Fuel pressure testing (courtesy Nissan).

Some manuals call for checking pump output volume at this point. If the pump is putting out a sufficient quantity of fuel, but pressure is low, the regulator may be bleeding off too much fuel and routing it back to the tank. A new regulator would be needed here.

A quick check for testing a pressure regulator is to momentarily disconnect the vacuum hose to the regulator while the engine is running. Fuel pressure should rise if the unit is working correctly.

Injectors

With throttle body fuel injection, it's easy to see whether or not an injector is delivering fuel properly. A good injector should produce a cone-shaped spray pattern with no "streamers" or large droplets. But with multipoint fuel injection, you can't observe the injector spray patterns without removing the injectors. So you have to use other means to weed out the bad ones from the good ones.

If an injector isn't delivering it's normal dose of fuel because of clogging or nozzle deterioration, the cylinder it feeds will run lean and will probably misfire. A steady miss, rough idle, misfiring on acceleration or when the engine is under load, poor fuel economy, loss of power, or high HC emissions in the exhaust can all be symptoms caused by one or more bad injectors.

A power balance test will usually identify the weak or misfiring cylinders. The challenge then is to determine if the problem is ignition-, compression-, or fuel-related. If the spark plugs aren't misfiring and compression is good, the problem is in the injector.

Another way to check injector fuel delivery is to energize one injector at a time, while noting the drop in fuel pressure with the engine off (Fig. 13-13). If all the injectors are good, the pressure drop will be approximately equal for each injector. But if one injector is clogged or not working, you'll see less of a pressure drop (or no drop).

The first check should be to see if the injector is receiving a voltage signal. Attach a 12-volt test light or analog voltmeter to the injector terminal. If voltage is getting through, the light should flicker on and off (or the needle jump) while the engine is running (or cranking). No voltage? Then turn the igni-

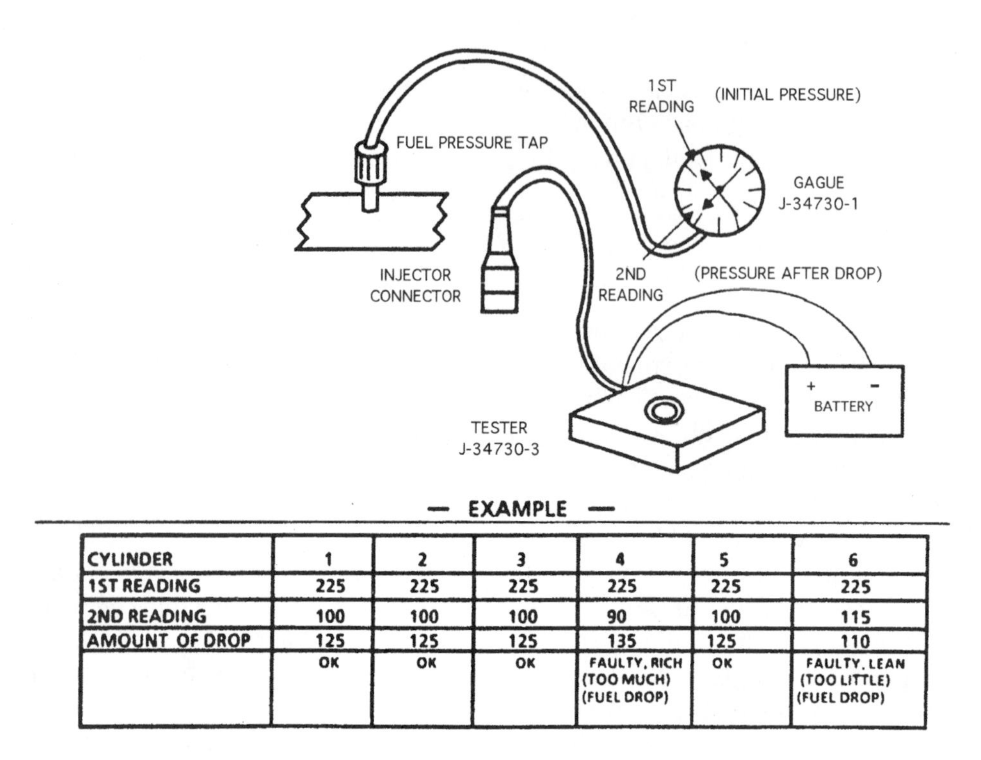

CYLINDER	1	2	3	4	5	6
1ST READING	225	225	225	225	225	225
2ND READING	100	100	100	90	100	115
AMOUNT OF DROP	125	125	125	135	125	110
	OK	OK	OK	FAULTY, RICH (TOO MUCH) (FUEL DROP)	OK	FAULTY, LEAN (TOO LITTLE) (FUEL DROP)

Fig. 13-13 Injector balance TES (courtesy General Motors).

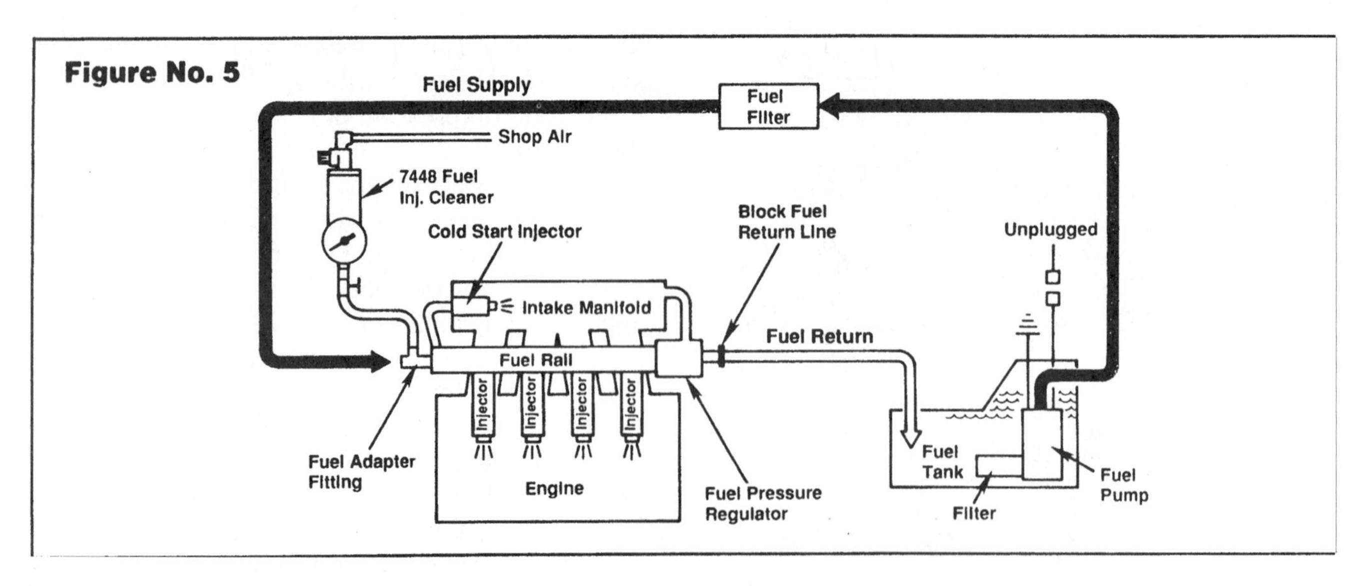

Fig. 13-14 Fuel injector cleaning (courtesy OTC).

tion off and use an ohmmeter to check continuity and resistance at the injector's wiring connector, in the wiring harness, and at the computer connector to find out where the open or short is.

If the injector on a weak or dead cylinder is receiving a pulsating voltage signal, but isn't buzzing (listen to or feel the injector), the injector solenoid is probably open or shorted. Either way, the injector needs to be replaced. If the injector is buzzing, on the other hand, then clogging is the problem, which calls for either cleaning or replacement.

Clogged Injectors

"Dirty" injectors are actually quite rare. As long as there's a good fuel filter in the fuel line, dirt shouldn't get through. Sometimes tiny pieces of rubber can flake off the inside of ancient fuel hoses and plug up an injector inlet. But injector clogging due to varnish buildup is a common problem. Varnish builds up in the injector nozzle over time and eventually restricts the orifice. This reduces the flow of fuel and disrupts the cone-shaped spray pattern, causing the fuel mixture to lean out and the cylinder to misfire.

Detergent fuel additives help keep injectors clean, but there's still a lot of "cheap" gas that contains little if any detergent. Adding a can of injector cleaner to the fuel tank every 5,000 miles or so may help lessen the dangers of varnish buildup. But some engines will experience injector clogging anyway, because of heat-soak and frequent short trip stop-and-go driving.

Cleaning Dirty Injectors

One way to clean dirty injectors is to flush them with solvent by feeding pressurized solvent directly into the fuel rail. If on-car cleaning doesn't do the trick, however, then it will be necessary to either remove the injectors for off-car cleaning using special cleaning equipment, or replace the injectors (Fig. 13-14).

Some technicians prefer off-car injector cleaning in spite of the extra labor that's involved to remove and reinstall the injectors because of the superior results they can achieve. Removing the injectors and flushing them can often open up clogged injectors that don't respond to normal on-car cleaning procedures.

Another reason for pulling the injectors is that you can flow test the injectors after cleaning to check both fuel delivery and spray pattern. (This requires special cleaning and test equipment.)

Some off-car cleaning equipment has the capability to "reverse-flush" injectors. By forcing solvent to flow backwards though the injector, many deposits that might not be removed otherwise are dislodged and flushed out. What's more, deposits on the inlet filter screen are carried away from the injector rather than flushed into it.

For really stubborn deposits, the tip of the injector may have to be immersed in an ultrasonic bath. The combination of solvent and sound waves can usually dislodge even the most difficult-to-remove deposits given enough time.

MISMATCHED INJECTORS

Another source of performance problems can be "mismatched" injectors. The volume of fuel that two supposedly identical injectors deliver to the engine can vary significantly. A difference of more than 7-10% in flow rates between the injectors on an engine equipped with multiport injection can cause noticeable driveability problems. Some cylinders will run too rich, while others will run too lean. Ideally, the injectors should flow within 3-5% of one another. For performance applications, they should be matched within 1-2%.

There are only three ways to tell if injectors are mismatched. One is to remove the injectors and flow-test each one. By measuring the volume of fuel each injector sprays into a graduated cylinder during the same time interval, you can see how closely matched the injectors are.

Another method is to use on-car test equipment that measures the pressure drop in the fuel rail when each injector is energized for the same amount of time. Though not as accurate as off-car flow testing, the technique can be used to spot weak injectors.

The third (and least accurate) method is to check the factory color code on the injector itself. Some manufacturers flow-test and color-code injectors by placing a small paint dot at the base of the injector's electrical connector. It doesn't make any difference what the color of the dot is just as long as all the injectors in an engine have the same color code. If somebody has replaced an injector with one that carries a different color code, it may flow differently enough from the others to create a drive-ability problem.

LEAKY INJECTORS

Another problem you're going to encounter is leaky injectors. Wear in the injector orifice and/or accumulated deposits can sometimes prevent the injector valve from seating completely, allowing fuel to dribble out of the injector nozzle. The extra fuel causes a rich fuel condition which can foul spark plugs, contribute to hard starting, increase emissions, and cause a rough idle. A carbon-fouled spark plug in one cylinder of a multiport injected engine usually indicates a leaky injector. If cleaning fails to eliminate the leak (which it can if dirt or varnish are responsible), replacement will be necessary.

Most injectors are pressed in place and held with O-ring seals (Fig. 13-15). Replacing one is simply a matter of pulling out the old one and dropping in a new one. Always replace the O-ring seals and make sure the unit is sealing properly.

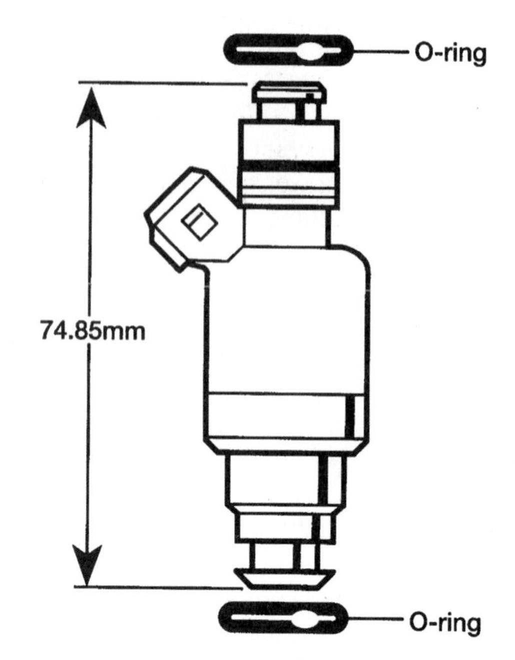

Fig. 13-15 Injector O-rings (courtesy AC Rochester).

CHAPTER 14

Introduction to Computerized Engine Controls

Computerized engine controls have come about as a means of meeting tighter exhaust emissions and fuel economy standards. Ignition, carburetion, and emissions control functions have been interwoven within the engine control computer (the "electronic control module" or ECM). This serves as a sort of glorified electronic calculator. The vehicle's onboard computer receives inputs from a variety of engine sensors to determine ignition timing, fuel enrichment, when certain emissions control functions should occur, and the lockup of the torque converter in the automatic transmission.

The introduction of computerized engine controls has taken automotive technology a giant step forward and added hundreds of dollar's worth of electronics to the average vehicle. But it has also proved to be quite a challenge for technicians who find themselves confronted with complex driveability and emissions problems.

WHAT COMPUTERS DO

If you understand the basics of how one computerized engine control system works, you have a framework for understanding how they all work, whether you're talking General Motors, Ford, Chrysler, or any of the imports. So let's take a look at the basics they all share.

The engine control computer (ECM, control module, microprocessor, or whatever you choose to call it) is the brains of the system (Fig. 14-1). Its primary functions are to regulate fuel enrichment, ignition timing, idle speed, and various emissions control functions. On many vehicles with automatic transmissions, it controls the lockup of the torque converter. And on newer vehicles with "electronic" transmissions, it even controls the shifting of the gears by opening and closing solenoids in the transmission valve body. What's more, many vehicles have additional computers for chassis functions such as electronic ride control (both shock-dampening and ride-height adjustments), antilock brakes, variable rate power steering, climate control system, and the supplemental restraint (air bag) system. Some operate more or less independently of the others, while some are tied together via a common bus network and communicate back and forth.

You don't have to know a lot about electronics or the inner workings of a computer to understand what it does. An engine control computer is essentially a black box that takes in information, processes the information according to an internal program, and then outputs a control signal. The information that goes into the box comes from various sensors and switches. The control signals that come out of the box regulate various devices that in turn control fuel delivery, ignition timing, emissions, and drivetrain functions.

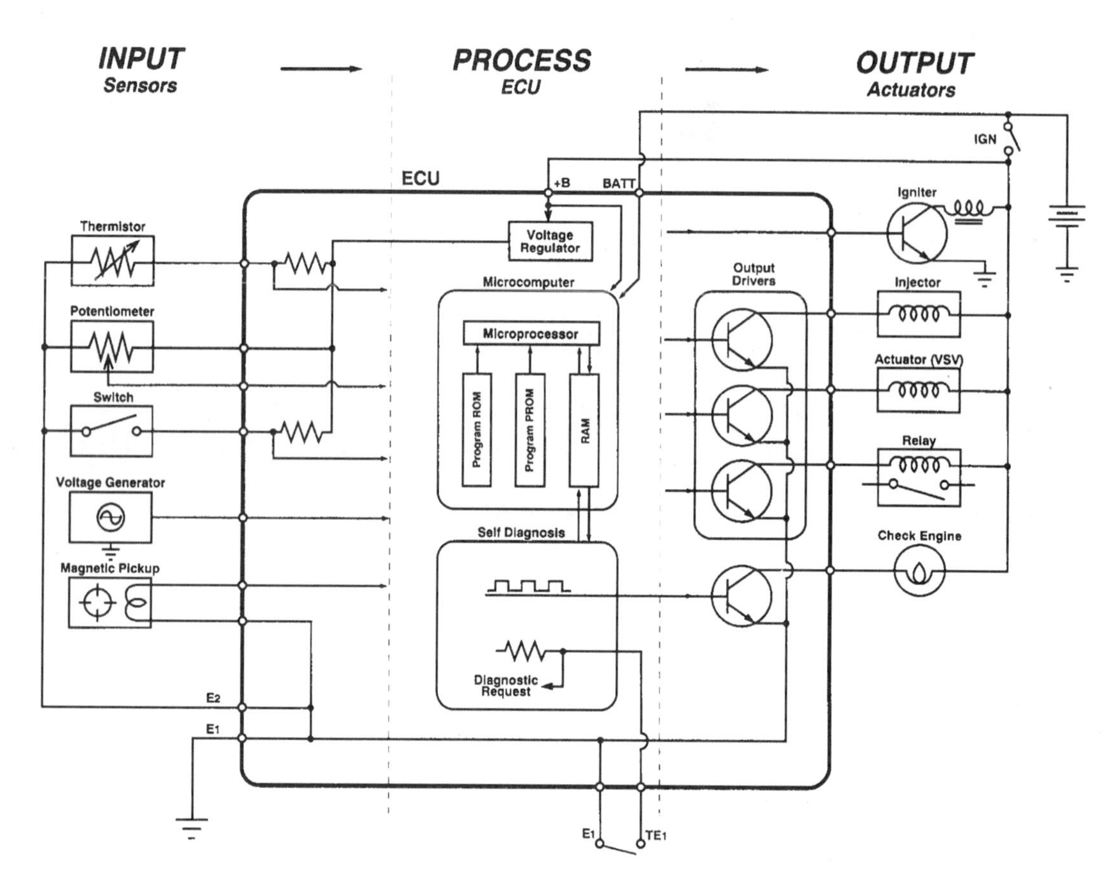

Fig. 14-1 Processing of information. (courtesy Toyota).

A BRIEF HISTORY

The first automotive application of microprocessors was back in 1968, when a computer was used to control the Bosch D-Jetronic fuel injection system on Volkswagens. In 1975, Cadillac introduced a computer-controlled electronic fuel injection system. In 1976, Chrysler introduced its "Lean Burn" computer-controlled spark system. In 1977, Oldsmobile came out with its "MISAR" (microprocessed sensing automatic regulation) spark control system. Though all of these early systems used a computer, the computer in each case had only one job to do: either regulate fuel delivery (in the case of VW and Cadillac), or to regulate spark timing (Chrysler & Oldsmobile). What's more, these early computers were all "analog" rather than digital computers.

In 1980 many California cars were equipped with a new generation of digital computerized engine controls that integrated spark timing, fuel management, emissions control, and drivetrain functions for the first time. The new generation of computerized engine controls appeared in California to meet that state's emissions regulations (which preceded the new federal emissions standards by one year), and to give the auto makers another year to further develop the new technology before it went nationwide.

In 1981, computerized engine controls became standard on most domestic automobiles. The imports soon followed suit, and for over a decade now, all new cars and trucks have been computerized.

ANALOG VS. DIGITAL

The early analog computers were designed to handle variable voltage signals (ones that change continuously over time). Analog signals look like wavy lines on an oscilloscope (Fig. 14-2). Like a moving needle on a speedometer or analog voltmeter, an analog signal is constantly changing and varying.

A variable resistor like a coolant sensor, throttle position sensor, or manifold pressure sensor typically produces an analog signal that changes in response to changes in internal resistance. An oxygen

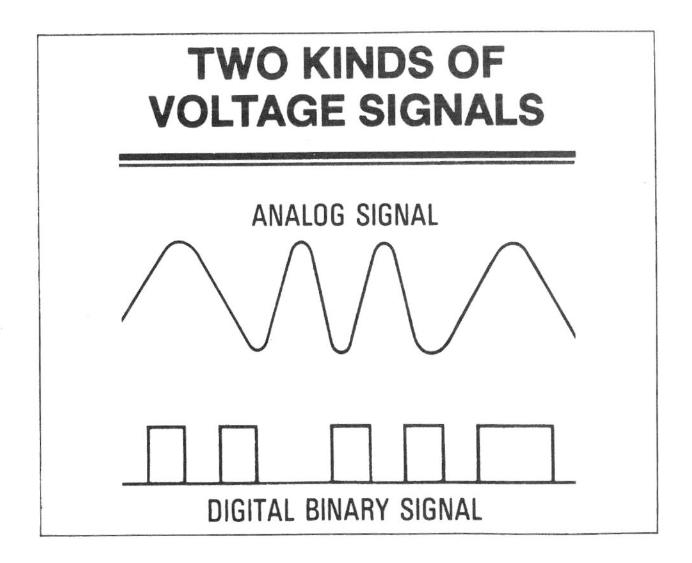

Fig. 14-2 Two types of voltage signals. (courtesy General Motors).

sensor produces a variable voltage signal that may range from 0.1 volts up to about 0.9 volts.

Analog signals are okay for some functions (such as radio and TV signals and audio cassettes), but require more complex circuitry to process accurately. What's more, analog signals are more prone to noise and errors in processing than digital signals, because of the variable nature of the signal itself. Storing an analog signal inside a computer is also more difficult because it involves charging small capacitors to hold a piece of the signal. Such a method is okay for short-term storage, but is not not a reliable means of storing information for indefinite periods of time. Thus the future for analog computers was limited because of the inherent limitations of the technology itself.

The auto makers decided that digital computers were the way to go because of their simpler circuitry, less expensive componentry, and greater reliability. So digital computers have become the industry standard and will likely remain so until something better comes along.

A digital computer is designed to process digital signals. Digital signals are on-off or square-wave signals like those produced by flipping a switch on and off. A mass airflow sensor or vehicle speed sensor typically produces a digital signal. So too does a digital EGR valve position sensor.

A digital signal delivers discreet bits of information which can be converted directly to "binary" numbers for rapid processing as well as long-term storage. This, in turn, gives digital computers a "long-term" memory capability that allows them to learn and perform more complex self-diagnostic functions.

Digital signals are also less vulnerable to noise or interference. Because the signals comes in little on-and-off bits, it's less likely to be distorted during transmission or misread when it is received. That's why digital recordings on compact disks provide superior sound fidelity to analog recordings on magnetic tape.

The circuits in a digital computer are essentially switching circuits that shunt or switch signals in one direction or another. The switches are actually transistors (see Chapter 2) photo-etched on wafer-thin chips of silicon. By combining hundreds and even thousands of such transistors on a single integrated circuit chip, a microprocessor can be created that's capable of manipulating numbers and processing information in sophisticated ways.

Most automotive digital computers up through the latter half of the 1980s have had "8-bit" microprocessors, meaning they process information in chunks or "bytes" of eight binary numbers at a time. When translated into binary code, a number or voltage value is represented by a series of eight zeros or ones (zero representing "off" and one representing "on"). Thus, the number zero in binary code would be "00000000", while the number nine would be "00001001" and so forth (Fig. 14-3). The maximum number of different combinations for a 8-bit binary code is 255.

Until recently an 8-bit microprocessor had sufficient capacity to handle most processing needs. But as vehicle complexity has increased, and the need to process even greater volumes of informa-

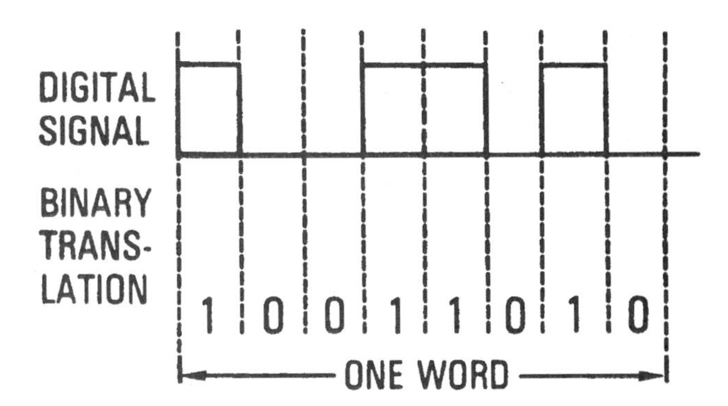

Fig. 14-3 Digital signals are used to transmit binary code (courtesy General Motors).

tion has come about, 16-bit computers have been introduced by General Motors, Chrysler, and Bosch. The next generation of computers will likely go to 32-bit microprocessors, or may even jump to 64-bit microprocessors.

COMPUTER SUBCOMPONENTS

In addition to the microprocessor itself, which is on an integrated circuit chip, the typical computer contains a number of other elements that are also essential to carry out its control functions:

Interface circuits convert analog signals from the outside world into binary code for processing (most sensors produce analog voltage signals). The interface circuits also shield the delicate circuits in the microprocessor from the higher voltages that exist in the circuits outside the computer. Interface circuits, called "drivers," convert the microprocessors' binary output signals into analog signals to drive the injectors, alter ignition timing, and so forth.

Memory circuits are provided to store information. There are three different types: Read-Only Memory (ROM), Random-Access Memory (RAM), and Programmable Read Only Memory (PROM) (Fig. 14-4).

ROM chips contain permanently stored information that tells the microprocessor how to handle information. The microprocessor can read the algorithms (mathematical instructions) from the ROM chip, but cannot modify or change the basic instructions that are there.

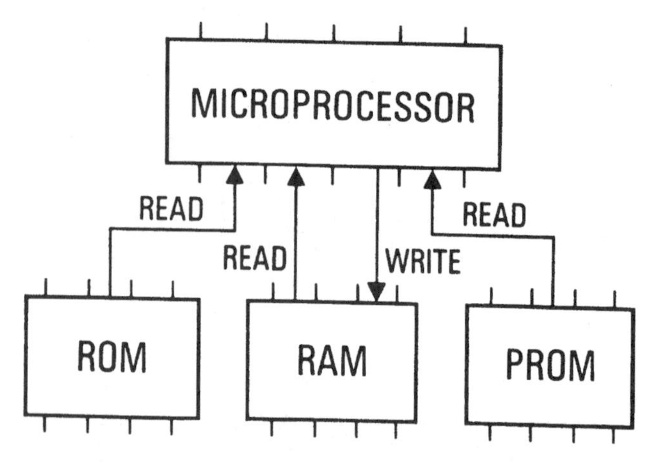

Fig. 14-4 Computer memories (courtesy General Motors).

The PROM chip (which may be an electronically eraseable or EEPROM chip) is where the computer's calibration program is stored. The PROM chip gives the microprocessor specific instructions on fuel enrichment, spark curves, when certain emission functions should occur in relation to vehicle speed, throttle position, time, and so on. This information is akin to a three-dimensional electronic "map" that guides the microprocessor as it makes its control decisions. Like ROM, the information in the PROM is "read-only" as far as the microprocessor is concerned and cannot be modified or changed—unless the PROM is replaced or reprogrammed.

To change the spark or fuel enrichment curves in a GM computer, for example, you can physically remove and replace the PROM with a different or upgraded PROM (Fig. 14-5). This is sometimes necessary to cure driveability problems and other glitches. GM has issued a number of technical service bulletins covering PROM replacement with upgraded factory PROMS. Aftermarket performance PROMS are also available that change fuel metering, spark advance, and torque converter lockup for faster acceleration. Depending on how the PROM is calibrated, it may or may not affect emissions. Emissions-legal PROMS should not affect emissions, but PROMS calibrated for all-out performance can have an adverse affect on emissions and are not legal for street-driven vehicles. Even so, these PROMS often find their way into street-driven cars and may cause a vehicle to fail an emissions test.

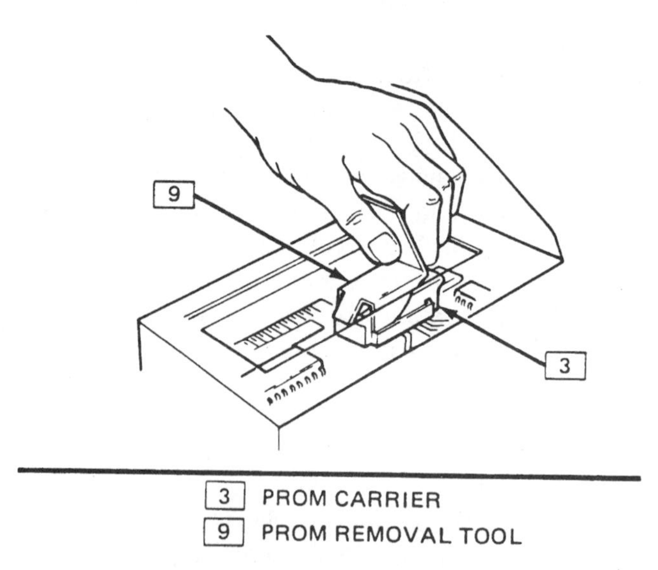

Fig. 14-5 PROM removal (courtesy General Motors).

GM is the only manufacturer who has included replaceable PROMS in its computers. The PROM chips in other manufacturers' computers are part of the circuit board and cannot be easily removed or replaced. But there are aftermarket add-on PROMS that can be installed to modify engine performance. As with the GM PROMS, only emissions-certified chips are considered legal for street-driven vehicles.

Another way to change a computer's calibration is if the computer has an EEPROM. This type of chip can be reprogrammed electronically by entering a special code number that allows new information to be written over the old information. An auto maker may issue a technical service bulletin that includes downloading new information into the computer to correct an emissions or driveability problem. Access to the information and the capability to download new programming into a vehicle's computer may be limited to new car dealers only. The new instructions may be provided through a direct modem hookup with the car maker, or via software provided on a floppy disk or compact disk.

Some computers also have E-PROMS ("flash" PROMS) that can likewise be reprogrammed. E-PROMS can be written over if ultraviolet light is flashed on a light-sensitive area of the chip. To protect the chip from unwanted erasure, the chip is usually sealed in a compartment or covered with a strip of tape. As with EEPROMS, reprogramming is limited primarily to new car dealers who have the required equipment to download new programming into the chip.

RAM chips provide short-term memory storage, and can be used for both read and write functions. When data enters the computer from the engine's various sensors, it will be temporarily recorded in the RAM chips until the microprocessor is ready to process it. Likewise, some of the microprocessor's output information may be temporarily stored in RAM, until it is ready to be passed along to the appropriate control device.

There are two kinds of RAM: volatile and nonvolatile. A volatile RAM (sometimes called "keep-alive memory" or "KAM") is a temporary type of memory that is erased when the computer's power source is disconnected. Most trouble codes, for example, are stored in volatile RAM memory so they can be erased once the code has been retrieved by a technician. The only problem with this approach is that disconnecting the battery wipes out all the information that's stored in volatile RAM. This includes trouble codes that might be needed for troubleshooting an emissions problem, short-term fuel calibration curves that the computer has learned to compensate for, variables in engine performance, etc. Preset channels in an electronic radio's volatile RAM can also be lost this way as can climate control settings and other information. So some computers have a "nonvolatile" RAM memory (NVRAM) that prevents valuable information from being lost in the event the battery is disconnected or goes dead. As with processing speed, memory requirements for today's computers are increasing too. When Ford introduced EEC-IV back in the early 1980s, it had 64K of RAM which was on par with the personal computers of the time and certainly seemed adequate for the foreseeable future.

A clock maintains an orderly flow of information into and out of the microprocessor (Fig. 14-6). A quartz crystal like the kind that's used in a digital watch provides a regular pulse signal that the microprocessor uses to coordinate and regulate the passing of signals. During the time interval between clock pulses, one bit of binary code information is passed from one part of the microprocessor to another. This, in turn, determines the rate or speed at which the microprocessor operates

The clock pulse is referred to as the "baud rate." The higher the baud rate, the faster the computer. Early GM computers had a baud rate of 160 bits per second. Later GM computers operate at a baud rate of 8,192, while Ford's run at 12,500.

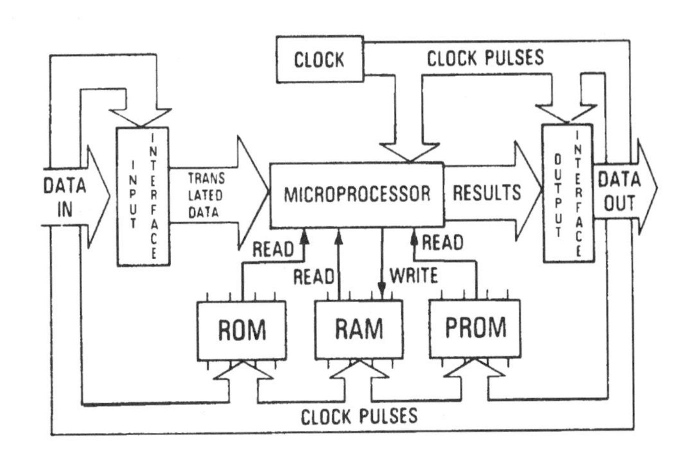

Fig. 14-6 How computers communicate internally and externally (courtesy General Motors).

FUEL LOOPS

One of the most important control functions that the computer handles is regulating the fuel mixture (which is covered in greater detail in Chapter 17). It does this by using what's called a "feedback loop" (Fig. 14-7). The primary sensor input here comes from an oxygen (O_2) sensor (called a "Lambda" sensor in import applications) located in the exhaust manifold. The O_2 sensor's voltage signal changes in response to the concentration of oxygen in the exhaust. A high concentration of oxygen is interpreted as a lean fuel mixture, while a low concentration means a rich mixture. The computer monitors the constantly fluctuating input from the O_2 sensor, and orders the fuel mixture to do exactly the opposite of what the O_2 sensor reads to compensate. In other words, a lean O_2 sensor reading makes the fuel mixture go rich, and a rich O_2 sensor reading makes the fuel mixture go lean. The constant readjusting and flip-flopping of the fuel mixture allows the computer to maintain a relatively balanced fuel mixture for maximum fuel economy and low emissions.

But all this doesn't happen the instant the engine starts. It takes a few minutes for most oxygen sensors to heat up and start producing a signal. So until there's an O_2 sensor signal, the computer stays in what's called an "open-loop" mode of operation. In this mode, the fuel mixture is slightly rich and does not change. This is necessary to maintain a good idle while the engine is warming up. Once the computer starts to receive a signal from the O_2 sensor, and/or the coolant sensor tells it the engine has warmed up enough to start leaning out the fuel mixture, the computer shifts into the "closed-loop" mode. In closed loop, it uses the signal from the O_2 sensor to constantly fine-tune the fuel mixture.

So how does the computer actually change the fuel mixture? On engines with carburetors, it does it by changing the "duty cycle" (on-time versus off-time) of a mixture control (M/C) solenoid in the carburetor's fuel-metering circuit (Fig. 14-8). The M/C solenoid is a little valve that cycles open and shut very rapidly. (You can actually hear it buzz while it's running.) Increasing the on-time allows more fuel to flow through the carburetor's jets, which richens the fuel mixture. Decreasing the on-time restricts the flow of fuel to lean the fuel mix. Mechanics can read the duty cycle of the solenoid by hooking up a dwell meter to its test lead, or by using a scan tool to monitor it through the computer itself.

On fuel-injected engines, the computer switches the injectors on and off in little pulses once every revolution of the crankshaft, or just before the intake valve opens, depending on the type of system. (The specific duty cycle varies considerably from one application to another.) The longer the injector is on, the more fuel it squirts into the engine. Thus the computer can make the fuel mixture richer or leaner by increasing or decreasing the duration of the injector pulses.

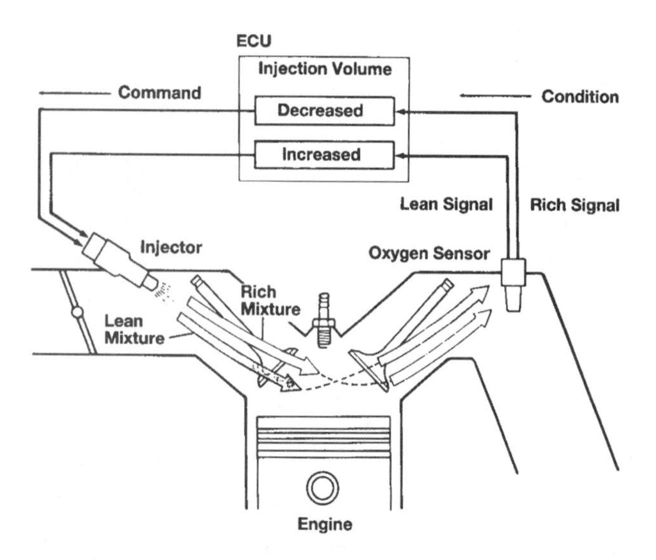

Fig. 14-7 Feedback loop (courtesy Toyota).

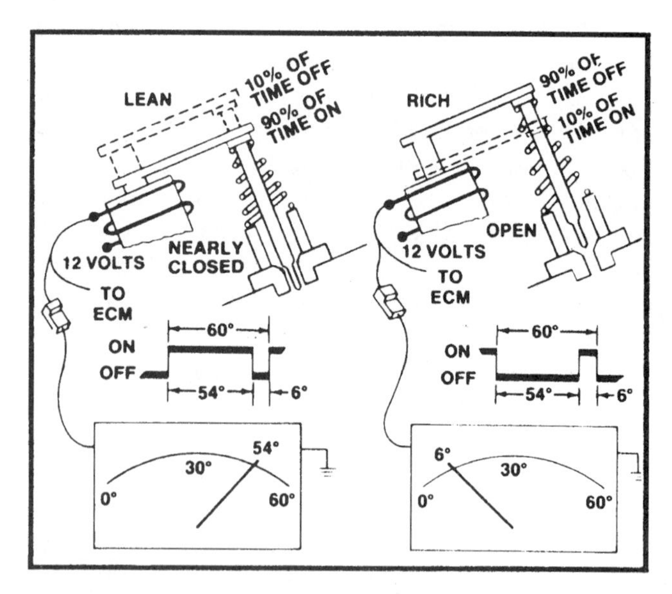

Fig. 14-8 Lean/rich dwell (duty cycle) (courtesy General Motors).

IDLE SPEED CONTROL

On many engines, the computer also regulates idle speed. It does this on carbureted engines by driving a little electric motor (the idle speed control motor) in and out to change the throttle opening (Fig. 14-9). This keeps the idle speed within the programmed rpm range. Engine speed is monitored by counting the pulses from the ignition pickup in the distributor or the signal from the crankshaft position sensor.

On fuel-injected engines, idle speed is regulated by running an idle air control (IAC) motor in and out to change the amount of air flowing through the throttle bypass circuit. Increasing airflow raises the idle speed, while decreasing airflow lowers the idle speed.

TIMING, TOO

Another vital control function handled by the computer is ignition timing. To alter the amount of spark advance, the computer monitors input from two important sensors: the throttle position sensor (TPS) and manifold absolute pressure (MAP) sensor (Fig. 14-10). The TPS tells the computer how far open the throttle is, and how quickly it is being opened. The MAP sensor reads intake vacuum, which the computer uses to determine engine load (reduced vacuum indicates increased load).

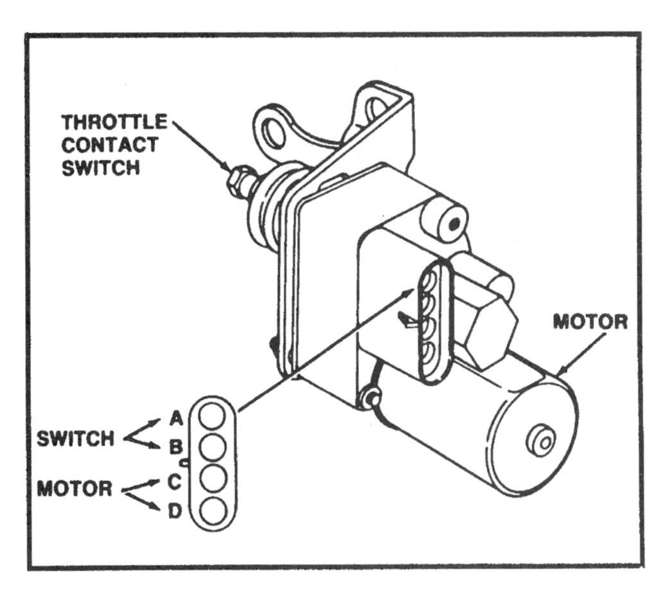

Fig. 14-9 Idle speed control motor (courtesy General Motors).

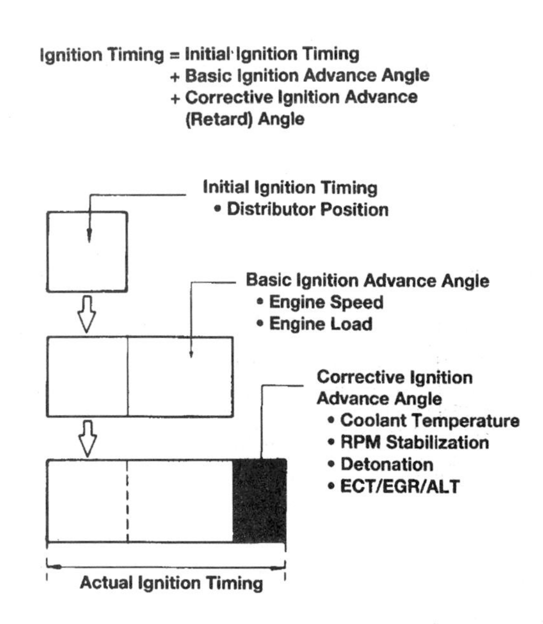

Fig. 14-10 Spark advance (retard) angle (courtesy Toyota).

When the engine is accelerating or under load, the computer backs off the amount of timing advance to prevent the engine from pinging (too much spark advance causes detonation). The computer does this by retarding (delaying) the timing signal that goes from the ignition module to the ignition coil.

The computer can also alter ignition timing to compensate for changes in altitude by using input from a barometric pressure (BARO or BP) sensor.

EMISSIONS FUNCTIONS

Computers also control a variety of related functions that affect emissions, engine cooling, and drivetrain functions. To help the catalytic converter burn the pollutants in the exhaust, the computer controls the routing of air from the air pump. When the coolant sensor tells the computer the engine is cold, air is routed to the exhaust manifolds to help reduce carbon monoxide (CO) and unburned hydrocarbons (HC) in the exhaust. But as the engine warms up, vehicles with "three-way plus oxygen" converters need the air rerouted into a middle chamber in the converter so it can reduce oxides of nitrogen (NOX).

The computer controls the opening of the purge valve on the charcoal canister so accumulated fuel vapors can be siphoned into the engine and burned. It also keeps track of the position of the EGR valve so the engine receives the correct amount of EGR at various speeds to keep NOX formation under control.

The operation of the electric cooling fan may also be under the computer's control. When the incoming signal from the coolant sensor tells the computer that the engine is getting hot, the computer grounds the relay that turns on the cooling fan. When the coolant is back within the correct range, the computer opens the relay and shuts off the fan.

The computer may also turn on the relay that operates the electric fuel pump when the ignition is switched on. It may also control the cycling of the air conditioner clutch, the charging rate of the alternator and/or the lockup of the torque converter and shifting of an electronic automatic transmission. For example, input from the vehicle speed sensor tells the computer the vehicle is going fast enough to lockup the converter for better fuel economy. The computer then grounds the lockup solenoid circuit, causing the torque converter lockup clutch to engage. When the brakes are applied, the brake pedal switch signals the computer, and the computer deenergizes the lockup solenoid.

WHEN THINGS GO AMISS

One thing all computers need to function properly is good information. Most computers have what's called a "limp-in" or "fail-safe" mode that takes over when a vital sensor signal is lost. When this happens, the computer substitutes an approximate value for the missing signal. The engine will still run, but not very well because the computer doesn't have the right information to keep things in proper balance.

Computers also have a certain amount of built-in self-diagnostic capability (Fig. 14-11). This ability is mostly limited to detecting faults within the computer system itself, not in components that are "outside" the system. A computer can recognize a problem when it loses a sensor signal, or a signal is out of normal range or doesn't make sense (engine idling, but the TPS says the throttle is wide open, for example). Most computers can't detect a fouled spark plug, leaky exhaust valve, worn timing chain, or other such problems that could potentially affect emissions but not trigger a fault code. But the next generation of vehicles with "onboard diagnostics II" (OBDII) will be able to detect such things as intermittent misfires and similar problems that affect emissions. (See Chapter 19 for more information about onboard diagnostics and troubleshooting computer problems.)

Because of its complexity, the computer is often blamed for many unrelated driveability problems. Consequently, a lot of computers are replaced needlessly—which often leads to comebacks when the new computer doesn't perform any better than the old one. The problem, of course, wasn't the computer. It was something else. According to Delco Electronics, a very high percentage (well in excess of 50%) of the supposedly defective computers that are returned under warranty have nothing wrong with them. That's why accurate diagnosis is so important.

REMANUFACTURED ECMS

In recent years, the aftermarket has wrestled a good chunk of the replacement computer business from the new car dealers by offering customers better prices and convenience. With most of cars on the road today computer-equipped, and many of them out-of-warranty (the computer is covered by the EPA-mandated 5-year/50,000-mile emissions warranty—which goes to 8 years and 80,000 miles in 1995), many aftermarket suppliers now offer remanufactured ECMs in their product lines. But what exactly is a remanufactured ECM? Is it like a remanufactured starter or steering rack? No, it's more like a reconditioned TV or VCR.

The quality of a "reman" computer depends on who's doing the reconditioning and how it is done. Computers don't wear out in the same respect that water pumps and starters do. Either it works or it doesn't. There seems to be no correlation between the number of miles driven and the frequency of repairs for onboard computers. Failures are often sudden and unpredictable. Many computers are, however, damaged by voltage overloads and shorts.

When a computer is remanufactured, it is

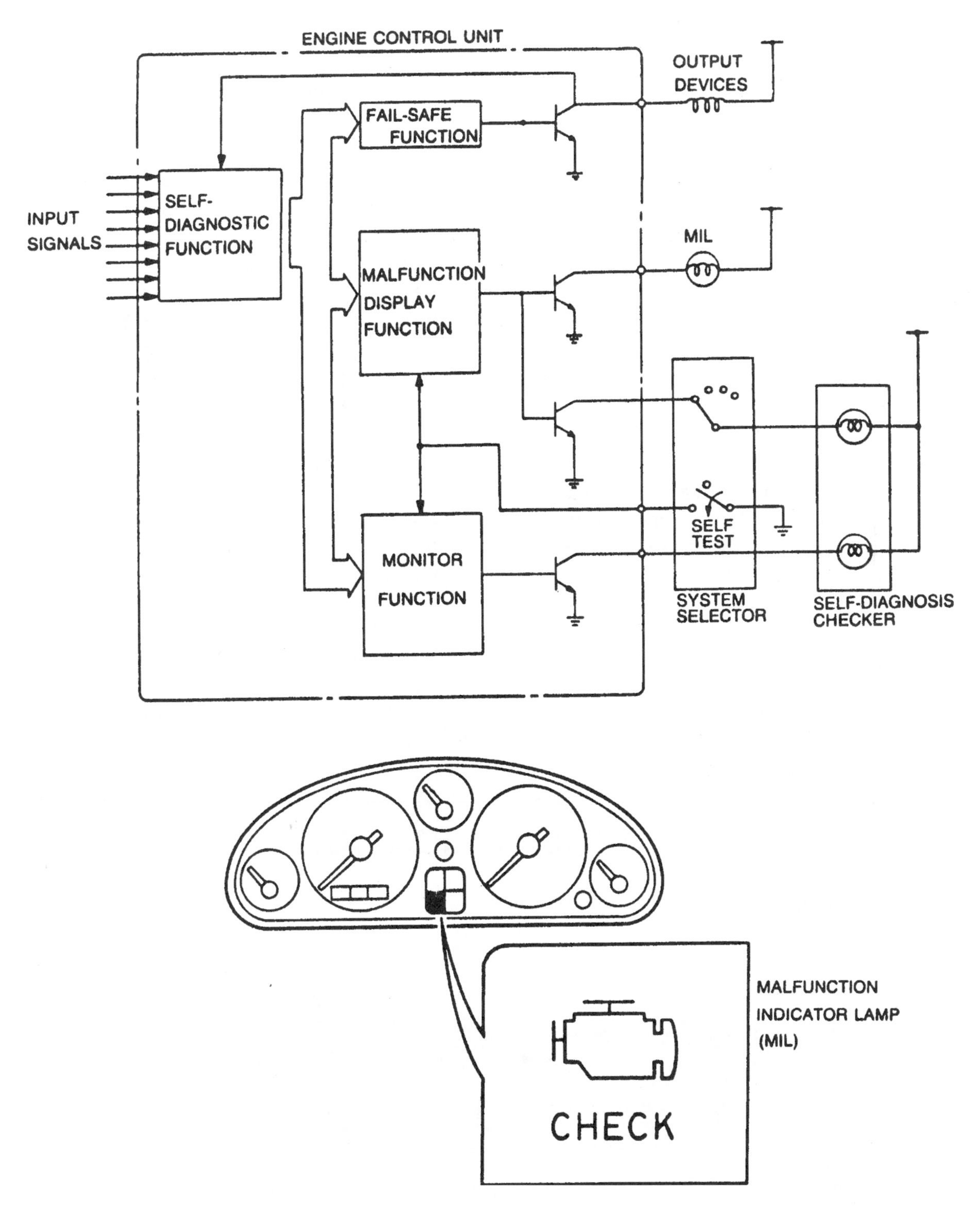

Fig. 14-11 Self-diagnosis system (courtesy Mazda).

cleaned then "exercised" on special test equipment that simulates each of the various sensor inputs. By monitoring the computer's output, faults can be identified, isolated, and repaired. Some items (like connectors or power transistors) that are prone to failure may also be replaced during the reconditioning process to improve the product's reliability. The computer may also be subjected to vibration and thermal tests to make sure it doesn't have any intermittent problems.

GENERAL SERVICE PRECAUTIONS FOR COMPUTERIZED CARS

- Never disconnect the computer's wiring harness or a sensor circuit while the ignition is on. Doing so can produce a momentary voltage spike that may damage sensitive electronic components.
- Be careful when charging the battery. More than 18 volts may damage the computer. Disconnect one or both battery cables to isolate the battery from the computer when recharging.
- If arc welding in the vicinity of the computer, disconnect the computer from its wiring harness.
- Disconnect the battery before doing any electrical repairs to avoid shorts that might damage the computer.
- Do not disconnect either battery cable while the engine is running. This too can cause a sudden voltage surge that may damage the computer and/or alternator.

HANDLE WITH CARE

Computers are valuable commodities, so handle with care! Fragile connectors and delicate electronics can be easily damaged by careless handling, so use care if you're replacing a computer.

Core identification is very important because many computers look identical on the outside (Fig. 14-12). An ID tag with the vehicle year, make, model, transmission type, and any other useful information should be completely filled out and attached to the core for proper exchange credit.

Static electricity can damage an ECM, so leave it in its packaging until it is ready to be installed. The vehicle's battery should be disconnected, and you should wear a grounding strap on your wrist to protect against static discharges. Sliding across a seat can produce a static discharge of 25,000 or more volts, which can damage a chip if you get a shock when you touch the computer housing.

On GM computers, the PROM from the old computer has to be transferred to the replacement computer. Care must be taken not to damage the PROM (use a PROM-removal tool and don't touch the PROM's pins).

The replacement computer will take 4–5 miles of driving to readjust itself to its new situation. Accelerating the car from 35 to 55 mph using progressively more throttle each time can speed up the resetting process.

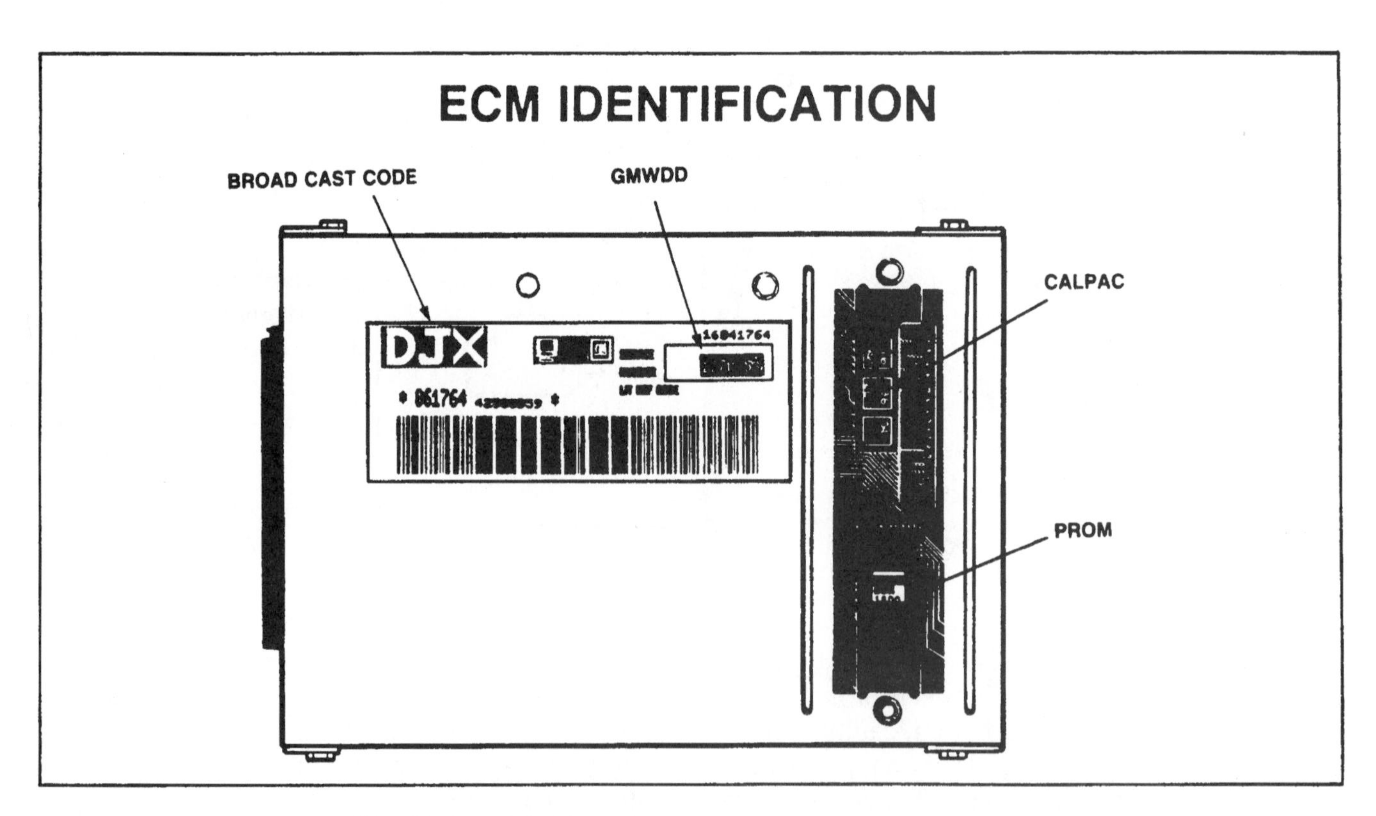

Fig. 14-12 ECM identification (courtesy General Motors).

CHAPTER 15

Sensors

OXYGEN SENSORS

The exhaust gas oxygen sensor (EGO or O_2), or Lambda sensor as the Europeans call it, is the key sensor in the engine control feedback loop. The computer uses the O_2 sensor's input to balance the fuel mixture, leaning it when the sensor reads rich and enrichening it when the sensor reads lean.

The O_2 sensor produces a voltage signal that is proportional to the amount of unburned oxygen in the exhaust. An oxygen sensor is essentially a galvanic battery that generates its own voltage. When hot (at least 600 degrees F.), the zirconium dioxide element in the sensor's tip produces a voltage that varies according to the amount of oxygen in the exhaust compared to the ambient oxygen level in the outside air. The higher the exhaust oxygen content, the lower the O_2 differential across the sensor tip, and the lower the sensor's output voltage. The voltage output ranges from 0.1 volts (lean) to 0.9 volts (rich). A perfectly balanced or "stoichiometric" fuel mixture of 14.7:1 will give a reading of around 0.5 volts.

Some O_2 sensors have three wires and an internal heating element (Fig. 15-1) to help the sensor reach operating temperature more quickly. The heater also keeps the sensor from cooling off when the engine is idling.

Driveability Symptoms

The O_2 sensor's normal life span is 30,000 to 50,000 miles. But the sensor may fail prematurely if it becomes clogged with carbon, or is contaminated by lead from leaded gasoline or solvents from the wrong type of RTV silicone sealer.

As the sensor ages, it becomes sluggish. Eventually it produces an unchanging signal or no signal at all. When this happens, the engine will experience driveability problems. It may lack power, idle rough, deliver poor fuel mileage, or flunk an emissions test.

Sometimes an apparent O_2 sensor problem is not really the sensor's fault. An air leak in the intake or exhaust manifold or even a fouled spark plug, for example, will cause the O_2 sensor to give a false lean indication. The sensor reacts only to the presence or absence of oxygen in the exhaust. It has no way of knowing where the extra oxygen came from. So keep that in mind when troubleshooting O_2 sensor problems.

O_2 SENSOR CHECKS

A good oxygen sensor should produce a fluctuating signal that changes quickly in response to changes in the oxygen level in the exhaust. To check a zirconium O_2 sensor, you'll need a 10K-ohm impedance digital voltmeter. On GM applications, you can also tap the sensor's voltage readings directly

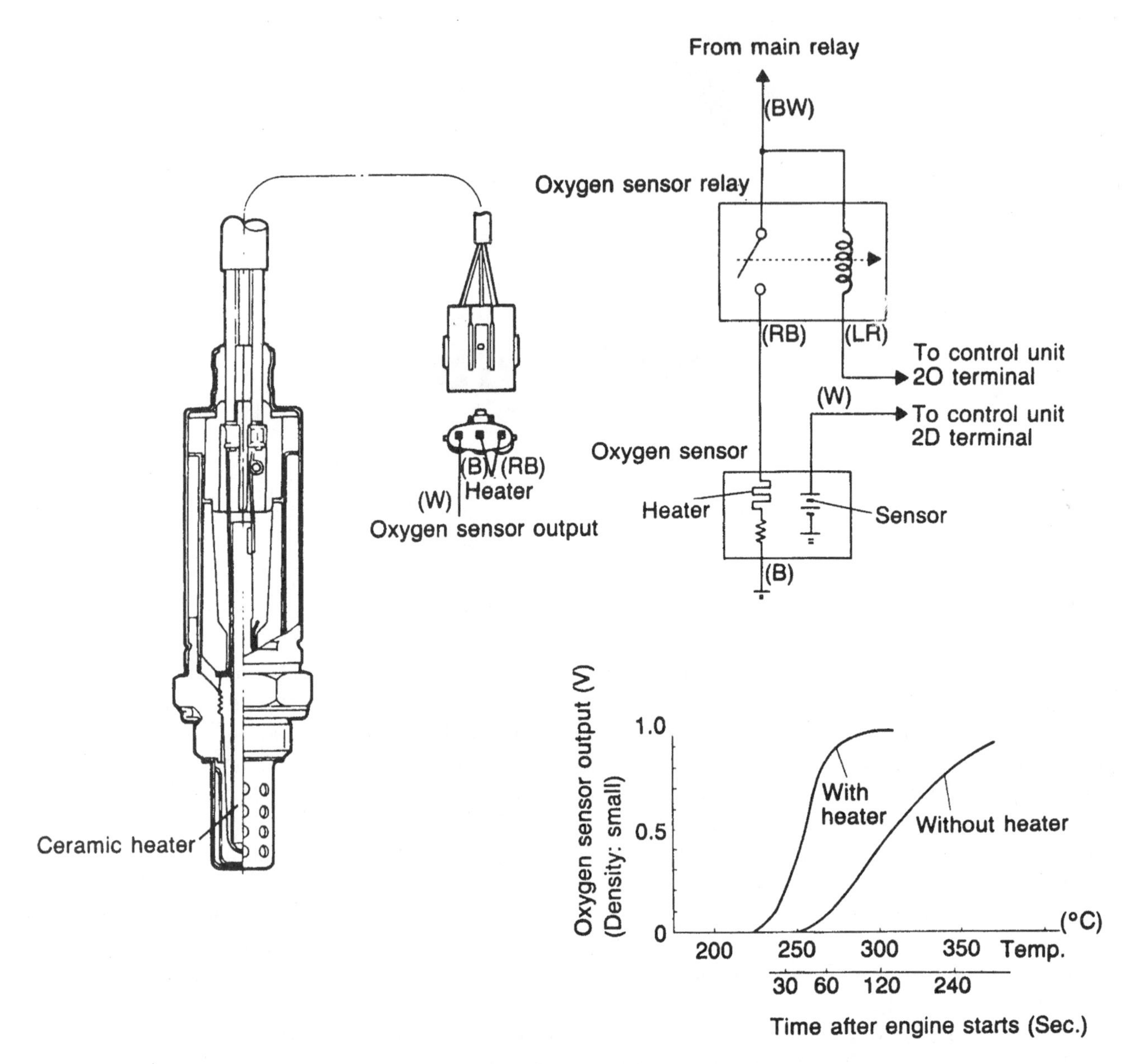

Fig. 15-1 Heated O_2 sensor (courtesy Mazda).

through the ALDL diagnostic connector using a scan tool. Ford also have self-diagnostic procedures for checking the sensor's output, but do not provide direct data stream readings.

Never use an ohmmeter in an attempt to check the sensor because doing so can damage it. And never jump or ground the sensor's leads.

To check the sensor's response to changing oxygen levels in the exhaust, first create an artificially lean condition by pulling a large vacuum line. If using a scan tool, activating the air pump solenoid valve to dump extra air into the exhaust will accomplish the same thing. When extra air is introduced into the engine, the sensor's voltage output should drop below 0.5 volts.

On carbureted engines, the sensor's rich response can be checked by creating an artificially rich mixture. Push the choke almost shut while the engine is idling. This should cause the O_2 sensor's voltage to increase above 0.5 volts.

If the sensor's output voltage fails to change in response to changes you've created in the level of oxygen in the exhaust, it needs to be replaced.

Another way to test a one- or two-wire zirconium O_2 sensor is to remove it and heat the tip with a propane torch, while monitoring the sensor's voltage output with a digital voltmeter. Connect the positive voltmeter lead to the wire coming out of the O_2 sensor, and the negative voltmeter lead to the sensor's outer shell. Then heat the tip of the sensor with the propane torch. The tip should be hot enough to turn cherry-red, and the flame must enter the opening into the sensor tip. If you get a voltmeter reading above 600 millivolts (0.6 volts), and the reading quickly changes as you move the flame back and forth over the tip, the sensor is okay. A low reading or one that is slow to change means the sensor needs to be replaced.

Another tool that can help you diagnose O_2 (and other) sensor problems is a "sensor simulator tester." The unit can provide an artificial O_2 sensor signal directly into the sensor's wiring harness (after unplugging the sensor, of course) to verify the computer's feedback response and wiring circuit. If the computer responds correctly to the simulated O_2 sensor signal, it means the wiring circuit and computer are okay. No change, however, would indicate a problem in the wiring circuit or computer.

O_2 Sensor Replacement

Like all the other engine sensors, the O_2 sensor is covered under the original equipment vehicle manufacturer's 5-year/50,000-mile emissions warranty.

Removing the sensor when the engine is cold will lessen the odds of stripping the threads in the exhaust manifold. Penetrating oil may be needed to loosen rusted threads. Once the sensor has been removed, the threads in the manifold should be cleaned before the new sensor is installed.

Most aftermarket replacement oxygen sensors are of a "universal" design which means some wire splicing may be necessary during installation. Apply graphite anti-seize compound to the sensor threads, unless the threads are precoated. The rubber boot that fits over the end of some sensors should not be pushed down further than half an inch from the sensor's base.

TITANIA O_2 SENSORS

The major applications where you'll find a titania O_2 sensor (Fig. 15-2) include 1986 and later Nissan 300ZX and Stanza 4WD wagons, 1987 and up Nissan Maxima and Sentra models, and 1986 1/2 and up Nissan D21 trucks. Chrysler also uses them on the Jeep Cherokee and Wrangler (because of the sensor's ability to handle off-road driving through water), and the Eagle Summit.

The operating principle of a titania O_2 sensor is entirely different from that of a zirconia O_2 sensor. In ways, a titania O_2 sensor works like a coolant sensor. The titania O_2 sensor changes resistance as the air/fuel ratio goes from rich to lean. But instead of a gradual change, it switches very quickly from low resistance (less than 1000 ohms) when the mixture is rich, to high resistance (over 20,000 ohms) when the mixture is lean.

The engine computer supplies a base reference voltage of approximately one volt to the titania O_2 sensor, and then reads the voltage flowing through the sensor to monitor the air/fuel ratio. (This means you can still use a digital voltmeter to read sensor output on this type of sensor, too.) When the fuel mixture is rich, therefore, resistance in a titania O_2 sensor will be low, so the sensor's voltage signal will be high (close to 1.0 volt). And when the fuel mixture is lean, resistance shoots up, and the voltage signal drops down to about 0.1 volt.

Both types of O_2 sensors (zirconia and titania) produce a high-voltage signal (1.0 volt) when the fuel mixture is rich, and a low-voltage signal (0.1 volt) when the mixture is lean. But titania sensors change resistance rather than generate their own voltage.

The three primary advantages of titania O_2 sensors are: (1) they don't need an air reference, which means there's no internal venting to the outside atmosphere to plug up; (2) they have a fast warm-up time (about 15 seconds), which means the engine goes into closed loop sooner to reduce emissions after start-up; and (3) they work at lower exhaust temperatures, which means they won't cool off at idle and they can be located further downstream from the engine or used with turbochargers (which soak up a lot of exhaust heat).

The reason why titania sensors warm up so quickly and won't cool off is because they're internally heated. Initially, this was considered a big advantage over zirconia O_2 sensors. But now that three-wire heated zirconia sensors are common, there's really little advantage of using titania versus zirconia.

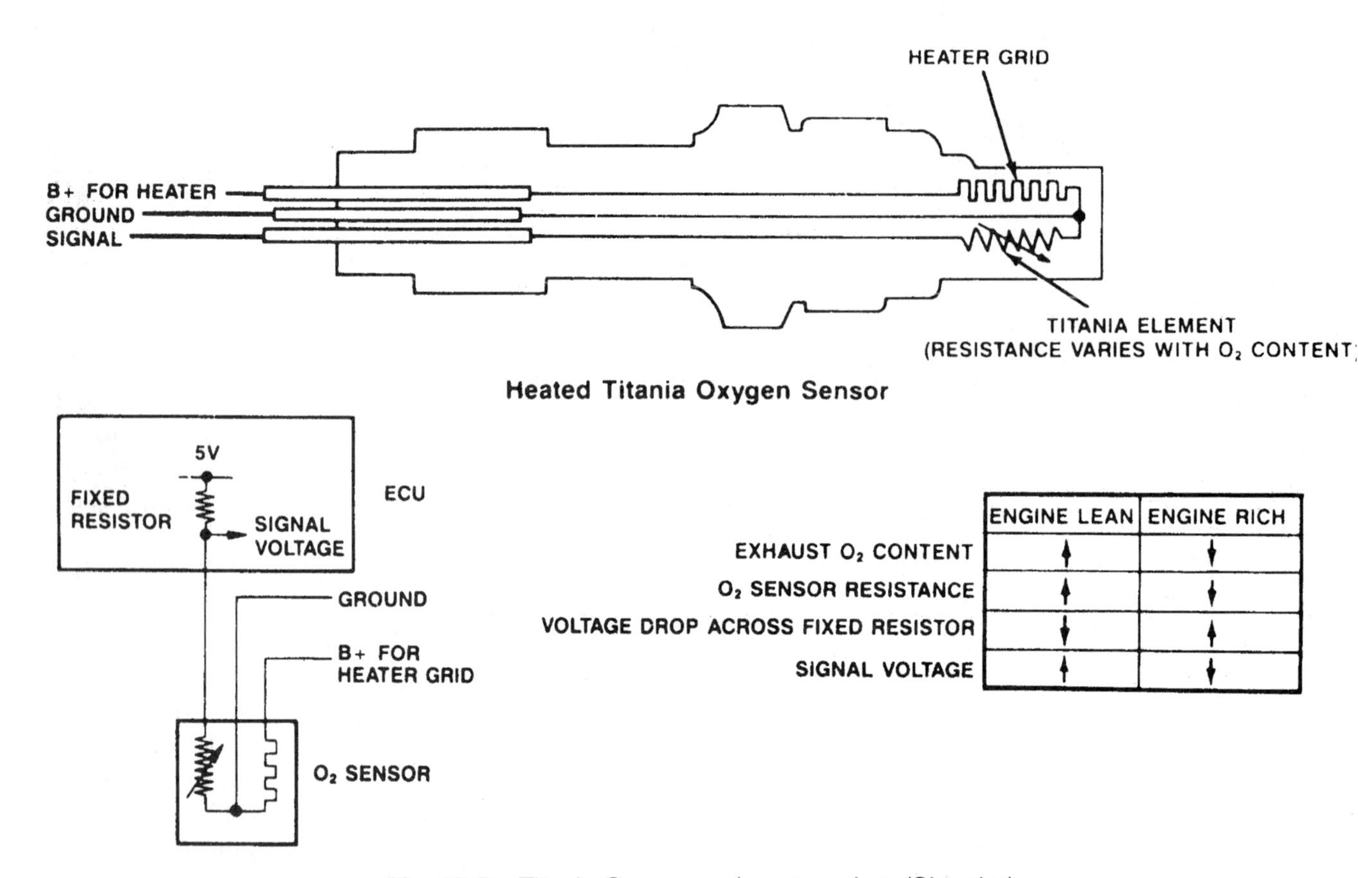

Fig. 15-2 Titania O_2 sensor. (courtesy Jeep/Chrysler)

Titania O_2 Sensor Diagnosis

A faulty titania O_2 sensor will cause the same kinds of driveability and emissions problems as a bad zirconia O_2 sensor (rich fuel mixture, failure to go into closed loop, poor fuel mileage, high CO readings, etc.). So there's nothing unusual in terms of diagnostics. The best place to start if you suspect a faulty sensor is to check the engine's computer system to see if the onboard diagnostics has detected any problems in the O_2 sensor circuit. If it has, then follow the step-by-step troubleshooting procedure in the manual to isolate the fault. A Code 33 (three red lights and three green lights) on Nissan's ECCS engine control system, for example, would indicate a problem in the O_2 sensor circuit.

Sensor output can be read with a digital voltmeter by tapping into the return signal line and watching for a change in voltage as the mixture is made artificially lean by momentarily pulling off a vacuum hose. No change would indicate the need for a new sensor.

A titania sensor can fail for all the same reasons as a zirconia O_2 sensor: carbon contamination from excessive oil burning, lead fouling from using leaded gasoline, silica fouling from RTV silicone sealer or bad fuel, or physical damage. In fact, titania sensors tend to be more fragile than zirconia sensors, so use care when handling them.

Because titania O_2 sensors work differently than the more common zirconia O_2 sensor, you can't use a "universal" type of replacement sensor. In other words, only a titania sensor can be used to replace a titania sensor.

COOLANT SENSORS

Ford calls them engine coolant temperature (ECT) sensors while Chrysler and General Motors refer to them as coolant temperature sensors (CTS). Most of the us simply call them coolant sensors.

The coolant sensor is often called the "master" sensor because the computer uses its input to regulate many other functions, including:

- Activating and deactivating the Early Fuel Evaporation (EFE) system such as the electric heating grid under carburetor or the thermactor air system.
- Open/closed-loop feedback control of the air/fuel mixture. The system won't go into closed loop until the engine is warm.
- Start-up fuel enrichment on fuel-injected engines, which the computer varies according to whether the engine is warm or cold.
- Spark advance and retard. Spark advance is often limited until the engine reaches normal operating temperature.
- EGR flow, which is blocked while the engine is cold to improve driveability.
- Canister purge, which doesn't occur until the engine is warm.
- Throttle kicker or idle speed.
- Transmission torque converter clutch lockup.

The coolant sensor is usually located on the head or intake manifold, where it screws into the water jacket. The sensors come in two basic varieties: variable resistor sensors called "thermistors," because their electrical resistance changes with temperature; and on/off switches, which work like a conventional temperature sending unit or electric cooling fan thermostat by closing or opening at a preset temperature (Fig. 15-3).

The variable resistor sensors are "smarter" than the on/off switches because they provide the computer with an more accurate indication of actual engine temperature. The computer feeds the sensor a fixed reference voltage of about 5 volts when the key is on. The resistance in the sensor is high when cold (see chart), and drops about 300 ohms for every degree Fahrenheit as the sensor warms up. This alters the return voltage signal back to the computer, which the computer then reads to determine engine temperature.

The switch-type sensor may be designed to remain closed within a certain temperature range (say between 55 and 235 degrees F., for example), or to open only when the engine is warm (above 125 degrees F.). Switch-type coolant sensors can be found on GM "T" car minimum function systems, Ford MCU, and Chrysler Lean Burn systems.

Coolant Sensor Driveability Symptoms

Because of the coolant sensor's central role in triggering so many engine functions, a faulty sensor (or

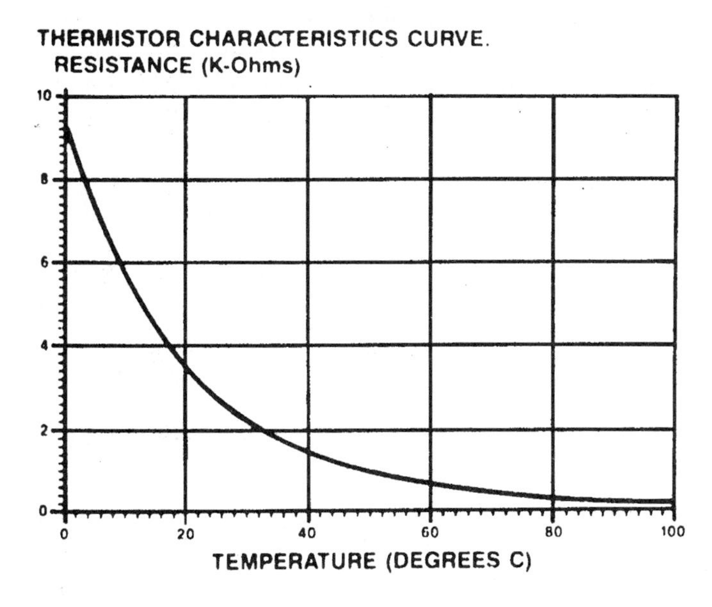

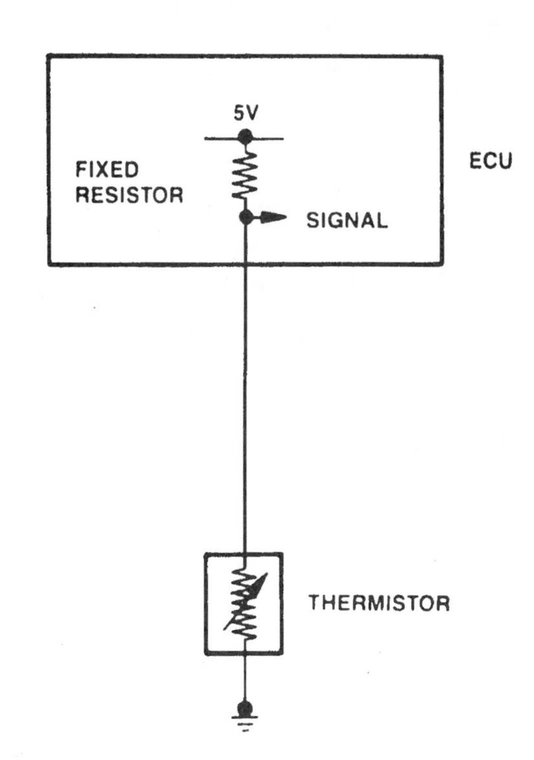

Fig. 15-3 Thermistor circuit & temperature curve. (courtesy Jeep/Chrysler)

sensor circuit) can cause a variety of cold-performance problems. The most common symptom is failure of the system to go into closed loop once the engine is warm. Other symptoms include:

- Poor cold idle (due to no EFE, heated air, or rich fuel mixture).
- Stalling (rich mixture, retarded timing, slow idle speed).
- Cold hesitation or stumble (no EFE, or EGR occurring too soon).
- Poor fuel mileage (rich mixture, open loop, spark retarded).

Keep in mind that coolant sensor problems are more often due to wiring faults and loose or corroded connectors than failure of the sensor itself. Correct operation of the sensor can also be upset by installing the wrong temperature-range thermostat.

Coolant Sensor Checks

On GM, Chrysler, and some imports systems, you can read sensor output directly through a scan tool (output will be displayed in degrees Centigrade). You can also check and compare the sensor's electrical resistance with an ohmmeter and/or its voltage reading (most read between 2.0 to 4.4 V).

Trouble codes that indicate a problem in coolant sensor circuit:

- General Motors: Code 14 (shorted) and 15 (open).
- Ford: Codes 21, 51 and 61.
- Chrysler: Code 22.

Testing the coolant sensor consists of measuring its electrical resistance with a DVOM at low and high temperatures to: (1) see that it changes, and (2) see that it matches the resistance specifications for the application (refer to a shop manual for the sensor's resistance values at various temperatures).

Coolant Sensor Replacement

Replacement is the only repair for a defective sensor. Coat the threads with sealer or teflon tape to prevent coolant leaks and do not overtighten.

THROTTLE POSITION SENSORS

Engines with feedback carburetion or electronic fuel injection have what's called a "throttle position sensor" (TPS). It informs the computer about the rate of throttle opening and relative throttle position. A separate idle switch (sometimes called a "nose" switch), and/or wide-open throttle (WOT) switch, may also be used to signal the computer when these throttle positions exist.

The throttle position sensor (Fig. 15-4) may be mounted externally on the throttle shaft, as is the case on most fuel injection throttle bodies, or internally in the carburetor, as it is in the Rochester Varajet, Dualjet, and Quadrajet.

The TPS is essentially a variable resistor that changes resistance as the throttle opens. Think of it as the electronic equivalent of a mechanical accelerator pump. By signaling the computer when the throttle opens, the computer can richen up the fuel mixture to maintain the proper air/fuel ratio.

The initial setting of the TPS is critical, because the voltage signal the computer receives back from the TPS tells the computer the exact position of the throttle. The initial adjustment, therefore, must be set as closely as possible to the factory specs. Most specs are given to the nearest hundredth of a volt! And since there is no range of "acceptable" specs given for a specific application, the TPS should be adjusted as closely as possible to those in the manual.

This is accomplished by reading the TPS voltage at a specific throttle position using a 10-megohm impedance digital voltmeter. On GM vehicles, use a

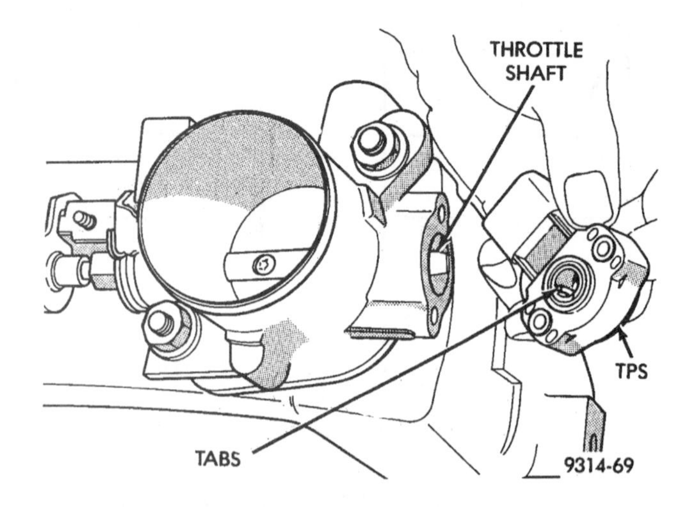

Fig. 15-4 TPS location. (courtesy Chrysler)

hand-held scan tool that plugs into the vehicle diagnostic connector.

TPS Driveability Symptoms

The classic symptom of a defective or misadjusted TPS is hesitation or stumble during acceleration. (In other words, the same symptoms a bad accelerator pump would produce.) The fuel mixture leans out because the computer doesn't receive the right signal telling it to add fuel as the throttle opens. The oxygen sensor feedback circuit will eventually provide the necessary information, but not quickly enough to prevent the engine from stumbling.

TPS Checks

There are three voltage checks to make. The first is for the presence of voltage at the TPS (with the key on). The TPS can't deliver the proper signal if it doesn't receive reference voltage from the computer. The second check is the base voltage adjustment (compare it to the specs). The third check is for the proper voltage change as the throttle opens and closes. Voltage should rise smoothly from about 1 volt to a maximum of 5 volts at wide-open throttle. No voltage rise or skips means the sensor is bad.

Trouble codes that may indicate TPS problems are:

- General Motors: 21, 22.
- Ford (EEC-IV): 23, 53, 63, 73.
- Chrysler: 24.

TPS Replacement And Adjustment

An adjustment screw is provided on the top of the Rochester carburetors, but it is covered by an anti-tampering plug which must first be removed. Most external throttle position sensors are adjusted by loosening the mounting screws and rotating the sensor slightly one way or the other until the desired voltage reading is obtained.

Under normal circumstances, the TPS does not require adjustment. But if your diagnosis reveals a problem with the TPS voltage setting, if the TPS is defective and must be replaced, or if the carburetor or TBI is replaced, then adjustment is needed.

NOTE: The TPS on most remanufactured carburetors is preset at the factory to an "average" setting for the majority of applications the carb fits. Even so, the TPS should be reset to the specific application upon which it is installed.

The basic adjustment procedures goes as follows:

1. Remove the anti-tampering plug on the Rochester carburetors by drilling a small hole (5/64 inch) in it, threading in a sheetmetal screw, and popping or prying the plug out. With an external TPS, loosen the mounting screws or remove the rivets holding it in place.
2. Refer to the electrical diagram in the manual to determine which connectors are used to make the TPS reading. On the Rochester carbs, use the TPS center terminal "B" and bottom terminal "C." If using a scan tool, just plug into the ALDL connector.
3. Turn the ignition on. Adjust the TPS with the throttle in the specified position (idle, high step of fast idle cam, or resting against the throttle stop screw with the ISC plunger fully retracted) until the proper voltage reading is obtained.

GM Feedback Carburetor Internal Throttle Position Sensors

Late-model GM vehicles with Computer Command Control and Rochester Varajet, Dualjet, and Quadrajet feedback carburetors use a throttle position sensor (TPS) that's mounted inside the carburetor. The internal TPS, which works the same as any other TPS, is a variable resistor that decreases resistance as the throttle opens. It goes from maximum resistance at idle to minimum resistance at wide-open throttle. This produces a TPS voltage return signal to the electronic control module (ECM) that goes from minimum at idle (typically from 0.26 to 0.56 volts) to maximum (up to 5.0 volts) at wide-open throttle. The ECM uses the TPS signal to modify the fuel mixture as the throttle opens, and to alter spark timing and torque converter clutch lockup.

The TPS sensor is moved up and down by a small plunger pin located under the accelerator pump lever. Watch out for this pin when working on the carburetor, because it can be easily damaged or lost. If the pin is lost, the TPS will always indicate an idle position. This will result in too lean a mixture during acceleration, causing hesitation and detonation problems. If the pin is jammed in the wide-open throttle position, the TPS signal will cause the air/fuel mixture to run too rich, resulting in poor fuel economy and rough running.

The trouble code that indicates a problem in the TPS circuit is a Code 21. A Code 21 means the ECM has noted too high a TPS signal for the engine's operating conditions. A Code 21 will be set if the engine speed is below 800 rpm and the ECM receives a TPS signal that indicates the throttle is more than 25% open for more than 10 seconds. The most likely cause of this condition is an open in the wires that run to ECM terminals 2, 21, or 22.

Because of the internal resistor inside the ECM between terminals 2 and 21, an open anywhere in the TPS circuit creates a high TPS signal (5 volts) at terminal 2 on the ECM.

The easiest way to check for an open in the TPS circuit, therefore, is to use a 10-megohm voltmeter to check the voltage at ECM terminal 2 (Fig. 15-5). If you see 5 volts, there's an open in the TPS or the TPS wiring harness. You can track it down with an ohmmeter by first checking the TPS, and then checking the continuity of the wiring harness.

To check the TPS, disconnect the wiring harness (with the key off, of course), and check resistance first between TPS terminals "A" and "B," then "A" and "C." Both should be under 20,000 ohms if the sensor is good. If either reading is over 20,000 ohms, replace the sensor.

To check the TPS base voltage adjustment, unplug the TPS connector (with the key off), and connect jumper wires across the three terminals between the connector and the plug. Then using a digital 10-megohm voltmeter, check the voltage between center terminal "B" and lower terminal "C" after turning the key on. Refer to your manual for the specified throttle position. On some applications, the TPS base voltage reading is taken with the engine off and the throttle at the curb idle position. On others, it is taken with the throttle on the high step of the fast idle cam. On those with idle speed control (ISC), it is usually taken with the ISC fully retracted.

Compare the voltage reading to the manual specs. TPS voltage values are specified to the nearest hundredth of a volt, so ballpark figures should never be used. If the base TPS voltage reading is not within .05 volts of the specified value, the TPS needs adjustment.

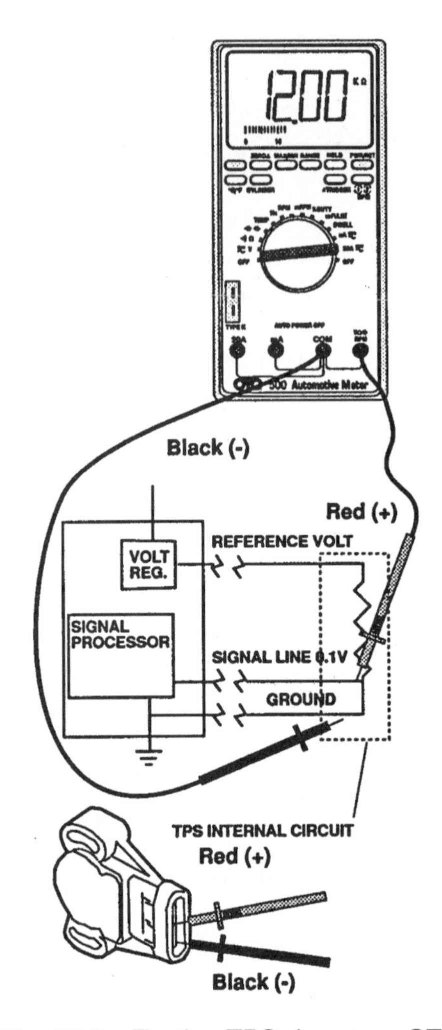

Fig. 15-5 Testing TPS. (courtesy OTC)

The TPS should also be checked through full sweep for a proper voltage response. Gradually open the throttle while noting the voltage reading. It should increase smoothly from the minimum value at idle up to a maximum of close to 5 volts when the throttle is wide open (Fig. 15-6). No

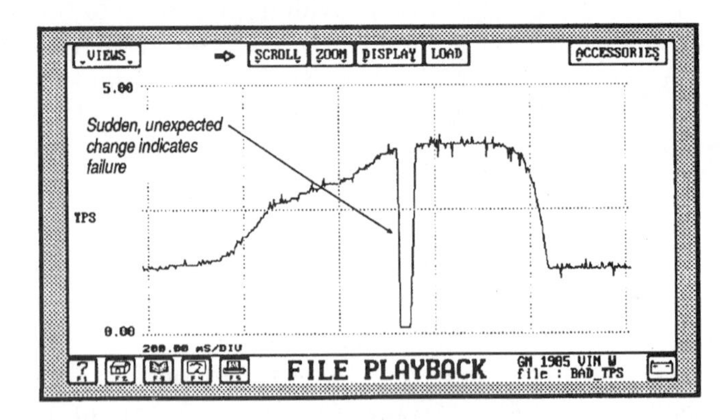

Fig. 15-6 Bad TPS signal. (courtesy Edge Diagnostic System)

change in the voltage reading, or jumps or skips in the reading indicate a new sensor is needed.

If adjustment is all that's needed, remove the anti-tamper plug that covers the TPS adjustment screw on top of the carburetor by drilling a small hole in the plug and using a self-tapping screw to pop it out. The TPS adjustment screw can now be turned until the desired voltage reading is obtained. After adjusting, replace the anti-tamper plug to seal the screw.

To replace a faulty TPS, the carburetor must be opened up by removing the air horn. The TPS is staked in place, so it must be pushed down against spring tension so the staking can be relieved by prying upward with a small screwdriver or chisel. The sensor can then be removed by pushing up on the electrical connector.

GM TBI Throttle Position Sensors

The throttle position sensor (TPS) that's used on General Motors throttle body injection (TBI) systems is mounted externally on the TBI throttle shaft. The TPS is a potentiometer (variable resistor) that changes resistance as the throttle opens. The engine control module (ECM) supplies the sensor with a 5.0-volt base reference voltage. The voltage return signal back to the ECM depends on the throttle's position and the internal resistance of the TPS.

The TPS voltage signal increases from a minimum value of around 0.5 volts at idle up to a maximum of close to 5.0 volts at wide-open throttle. Inputs from the TPS and those from the manifold absolute pressure (MAP) sensor are monitored by the ECM so changes can be made in fuel enrichment and ignition timing according to throttle position and engine load. The throttle position sensor must therefore be adjusted accurately to a specified reference voltage because the ECM uses the minimum voltage value as a reference point for determining the throttle's position.

If the TPS is set too high or too low, the TPS signal won't be in sync with the programmed curve in the ECM, and engine performance will suffer accordingly. Driveability symptoms include hesitation, stalling, and hard starting.

If the TPS is loose, it will produce an erratic signal, leading the ECM to believe the throttle is opening and closing. The result can be an unstable idle and intermittent hesitation.

If the TPS is shorted, the computer will receive the equivalent of a wide-open throttle signal all the time. This will make the fuel mixture run rich and set a Code 21 (voltage signal too high).

If the TPS is open, the computer will think the throttle is closed all the time. The resulting fuel mixture will be too lean and a Code 22 (voltage signal too low) will be set.

The TPS voltage signal can be checked by plugging a scan tool into the ALDL connector and reading live data stream, or it can be read manually with a 10-megohm digital voltmeter.

To read the TPS voltage manually, unplug the three-wire TPS connector (key off), and insert three jumper wires between the sensor and harness connector. Turn the ignition on, and take your voltage reading between terminals "B" and "C" on single TBI applications, or between terminals "A" and "B" on dual TBI Crossfire applications.

The voltage reading should be within the range specified at idle, and increase smoothly to a maximum of at least 3.5 volts at wide-open throttle. If there's no voltage reading, check for reference voltage from the ECM at connector terminal "A" on the single TBI applications, or terminal "C" on the dual TBI Crossfire applications. If the sensor is receiving voltage from the ECM but is producing no output signal, a new TPS is needed. It should also be replaced if the reading does not change smoothly from idle to wide-open throttle, or if the sensor sticks or binds.

General Motors uses a number of different throttle position sensors on its family of Rochester TBIs, and the TPS voltage settings vary according to the application:

- The TBI 100 used on the Cadillac 6.0L V8 requires an initial voltage setting of 0.55 volts plus or minus 0.5 volts.
- The TBI 300 on 1.8L and 2.5L engines requires an initial voltage setting of 0.525 volts plus or minus 1.25 volts.
- The TBI 400 on Crossfire 5.0L and 5.7L V8 applications calls for 0.525 volts plus or minus 0.75 volts. **Note**: only the forward TBI has a throttle position sensor.
- The TBI 500 on 2.0L engines also calls for 0.525 volts plus or minus 0.75 volts.

Through 1982, all GM throttle position sensors are adjustable. But on certain late-model applications, some sensors are not adjustable. Starting in 1984, for example, the TPS on the 1.8 and 2.5L Pontiac engines is nonadjustable. Likewise, Chevy switched to a nonadjustable TPS starting in 1985 on the 2.0L engine. On the newer engines with the nonadjustable sensors, the initial adjustment of the TPS isn't as critical because the ECM uses whatever idle reading it gets from the TPS as the base voltage reference point. On the adjustable variety, however, accurate adjustment is a must.

To adjust the TPS, some TBIs must be removed from the engine and turned over so spot welds on the TPS retaining screws can be drilled out (Fig. 15-7). Use a 5/16 inch drill, and do not drill deeper than necessary to free the screws.

If the sensor is being replaced, make sure the sensor pickup is above the throttle lever tang. Install two new screws, and use locking compound on the threads.

Reinstall the TBI on the manifold, start the engine, and set the minimum air idle adjustment before making any adjustments to the TPS. This is done by disconnecting the TV cable from the throttle bracket and removing the tamper-resistant plug on the side of the TBI so you can turn the minimum air screw. Plug the vacuum port for the Thermac system, start the engine, and allow the rpm to stabilize. Then install a plug (J-33047 or equivalent) in the idle air passage of the throttle body, and adjust the idle speed to specs. Now the TPS can be adjusted to the specified voltage.

The TPS is adjusted to the specified voltage range with the engine at curb idle or at minimum idle with the ISC solenoid fully retracted. Refer to your service manual to see which position is required.

MAP SENSORS

A MAP sensor is a "manifold absolute pressure" sensor. Its function is to sense air pressure or vacuum in the intake manifold, which the engine computer uses as an indication of engine load to adjust the air/fuel mixture and spark timing. Computerized engine control systems that do not use a MAP sensor rely on throttle position and air sensor input to determine engine load.

Under low-load, high-vacuum conditions, the computer leans the fuel mixture and advances spark timing for better fuel economy. Under high-load, low-vacuum conditions, the computer richens the fuel mixture and retards timing to prevent detonation. The MAP sensor, therefore, serves as the electronic equivalent of both a vacuum advance diaphragm on a distributor and a power valve in a carburetor.

The MAP sensor reads vacuum and pressure through a hose connected to the intake manifold. A pressure-sensitive ceramic or silicon element and electronic circuit in the sensor generates a voltage signal that changes in direct proportion to pressure.

MAP sensors should not be confused with VAC (vacuum) sensors, DPS (differential pressure) sensors, or BARO or BP (barometric pressure) sensors (Fig. 15-8). A vacuum sensor (which is the same as

Fig. 15-7 Removing TPS. (courtesy AC Delco)

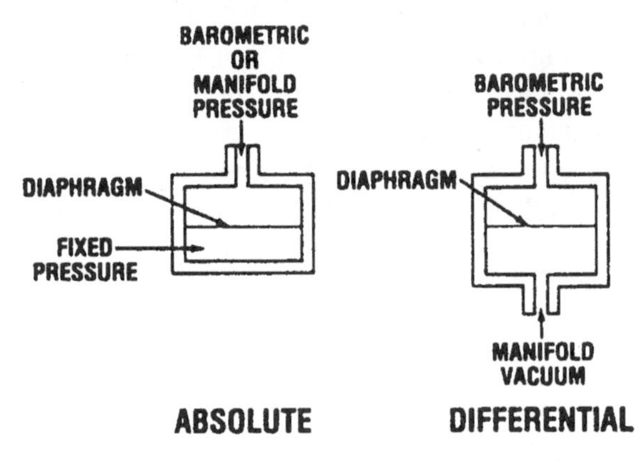

Fig. 15-8 Pressure/vacuum sensors. (reprinted, by permission, from King, *Computerized Engine Control, 4E.* Copyright 1993 by Delmar Publishers, Inc.).

a differential pressure sensor) reads the difference between manifold vacuum and atmospheric pressure. In other words, it reads the difference in air pressure above and below the throttle plate. A VAC sensor, therefore, is sometimes used instead of a MAP sensor to sense engine load.

A MAP sensor, on the other hand, measures manifold air pressure against a precalibrated absolute (reference) pressure. What's the difference? A vacuum sensor only reads the difference in pressure, not absolute pressure, so it doesn't take into account changes in barometric (atmospheric) pressure. A separate BARO sensor is usually needed with a vacuum sensor, therefore, to compensate for changes in altitude and barometric pressure.

MAP Sensor Driveability Symptoms

Anything that interferes with accurate sensor input can upset both the fuel mixture and ignition timing. This includes a problem with the MAP sensor itself, grounds or opens in the sensor wiring circuit, and/or vacuum leaks in the intake manifold.

Typical driveability symptoms include:

- Detonation due to too much spark advance and a lean fuel ratio.
- Loss of power and/or fuel economy due to retarded timing and an excessively rich fuel ratio.

A vacuum leak, for example, will cause the MAP sensor to indicate a higher-than-normal pressure (low vacuum) in the manifold, which makes the computer think the engine is under more load than it really is. Consequently, timing is retarded, and the fuel mixture is richened up.

MAP Sensor Checks

MAP sensors typically produce a voltage signal that drops with decreasing manifold pressure (rising vacuum). A VAC or DPS sensor's voltage signal does just the opposite: it rises with decreasing pressure (rising vacuum). The voltage reading should change in direct proportion to the vacuum reading.

On Gm and Chrysler systems, MAP sensor output voltage can be read directly through a scan tool plugged into the diagnostic connector. Or the voltage can be read at the sensor itself with a digital voltmeter by connecting jumpers to the wiring terminals and using a hand vacuum pump to apply vacuum to the sensor. The three terminals on a GM MAP sensor are: "A" (ground), "B" (sensor output voltage), and "C" (reference voltage, usually about 5.0 volts) (Fig. 15-9).

Test specs vary according to the application, so always refer to the voltage tables listed in the service manual. A typical MAP sensor will read 4.6 to 4.8 volts with zero inches of Hg vacuum applied to it. At 5 in. Hg vacuum, the reading should drop to

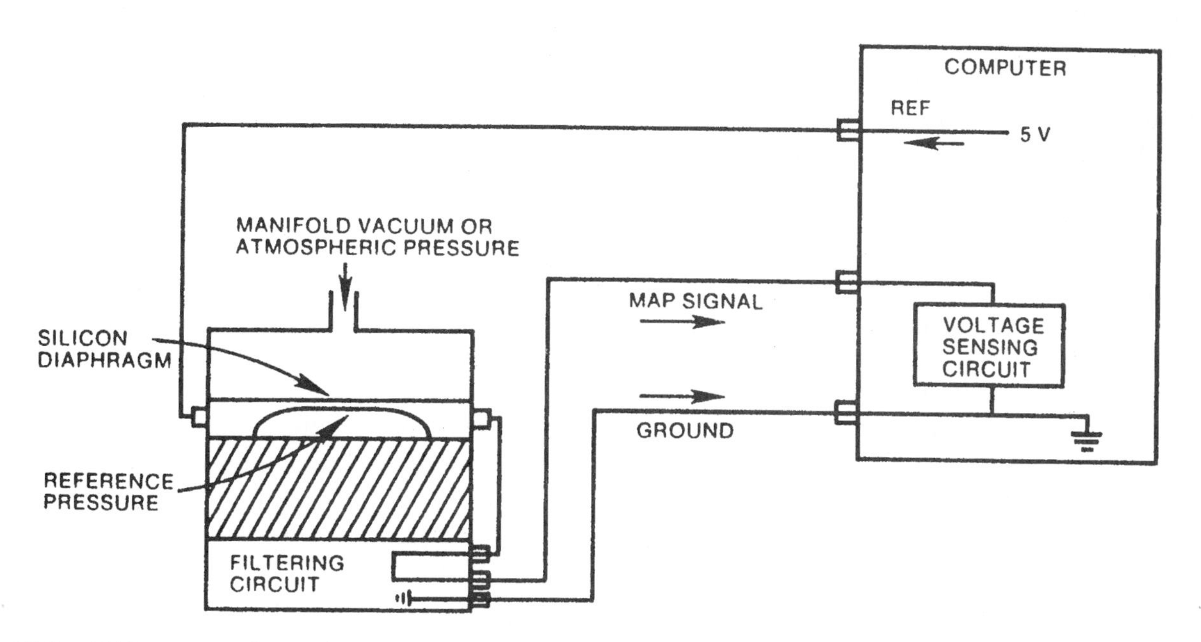

Fig. 15-9 MAP sensor circuit. (reprinted, by permission, from King, *Computerized Engine Control, 4E.* Copyright 1993 by Delmar Publishers, Inc.).

3.75 volts, then 2.85 volts at 10 in. Hg, 2.00 volts at 15 in. Hg, and finally 1.12 volts at 20 in. Hg.

Trouble codes that indicate a problem with the MAP sensor circuit:

- General Motors: Codes 34, 33, 31 (also E31, E32 and E34 on Cadillac Deville with electronic display).
- Ford: Codes 22, 72 (no change during goose test).
- Chrysler: Codes 13, 14.

When a MAP sensor trouble code is detected, follow the diagnostic procedure outlined in the manual to isolate the fault. If it turns out to be the sensor, it must be replaced since neither adjustment or repair is possible.

BAROMETRIC PRESSURE SENSORS

Barometric pressure (BARO, BP, or BAP) sensors are used on some electronic engine control systems to monitor changes in air density that occur with altitude or weather. The computer needs this information so it can adjust the air/fuel ratio and ignition timing advance curve to compensate for changes in air density. On some Ford systems, input from the sensor is also used to modify engine idle speed and the opening of the EGR valve.

Barometric pressure is an indication of air density. The pressure that the atmosphere exerts downward decreases with altitude, so the air becomes thinner (less dense) the higher up you go. Consequently, the amount of fuel that's required to maintain a constant air/fuel ratio also decreases with increasing elevation.

As one drives up a mountain, for example, the air becomes gradually thinner causing the fuel mixture to become progressively richer unless changes are made to compensate. Likewise, if a carburetor is calibrated for high-altitude operation, but the vehicle is then driven at a lower altitude, the fuel mixture will run too lean because the air is thicker (more dense), which requires more fuel.

A barometric pressure sensor measures air pressure electronically and converts the information into an electronic signal which the computer then uses to make the necessary adjustments. On GM C3 systems, the computer checks the BARO reading when the ignition is first turned on. But it won't make any subsequent adjustments based on air pressure changes until the ignition is cycled off and on again. If a vehicle is being driven up a mountain like Pikes Peak where there is a considerable change in elevation, therefore, it may be necessary to stop the car momentarily and turn the ignition off and on to "reset" the computer.

The barometric pressure sensor may be located on the firewall, inner fender, or under the dash. On GM applications, it is sometimes identified by a blue dot or blue connector. Most BARO sensors work like a MAP sensor and produce a voltage signal that decreases with pressure (altitude). One exception is the BAP sensor Ford uses on some 2.3L engines. This generates a frequency signal rather than a voltage signal (do not try to check this type of sensor with a voltmeter or ohmmeter). Another exception is the "altitude compensator" used on General Motors J-cars that works like an on-off switch to give a high or low altitude indication.

Another variation is the combination BMAP barometric pressure/MAP sensor Ford uses on some applications that combines both functions.

On most General Motors C3 systems that have a BARO sensor, you can read the sensor's output voltage with a hand-held scan tool. Some scan tools also display the actual pressure reading in kilopascals (kPa) which can be converted to inches of mercury (Hg). There are a few GM applications which do not have a BARO sensor, but will show a fixed BARO voltage reading anyway.

Baro Sensor Checks

A faulty sensor, or a grounded, open, or shorted sensor circuit can alter the air/fuel ratio and timing curve (also the idle speed and operation of the EGR valve on the Ford 2.3L engine). Typical symptoms may include poor performance at high altitude and/or spark knock.

On GM applications, the sensor should produce the same voltage reading whether the engine is running or not when the ignition key is on. Compare the sensor voltage reading to the specs. If it's above the range listed for a given altitude, check for a short in the sensor circuit. No short? Then replace the sensor. A low sensor reading could indicate a problem with the sensor, the wiring circuit, or the computer. Follow the diagnostic chart in the manual to isolate the cause. If a sensor is reading within range, you can confirm its accuracy by applying vacuum with a hand pump and watching for a voltage change.

Trouble codes that indicate a problem in the barometric pressure sensor circuit:

- General Motors: Code 32.
- Ford: Code 22 during engine running test (2.3L engine only).
- Chrysler: Code 37 (Turbo II only), Code 41, 51 (others).

Always follow the diagnostic procedure outlined in the manual to isolate the fault when a sensor appears to be misbehaving. If it turns out to be a bad sensor, replacing it is all you can do because no adjustments or repairs are possible.

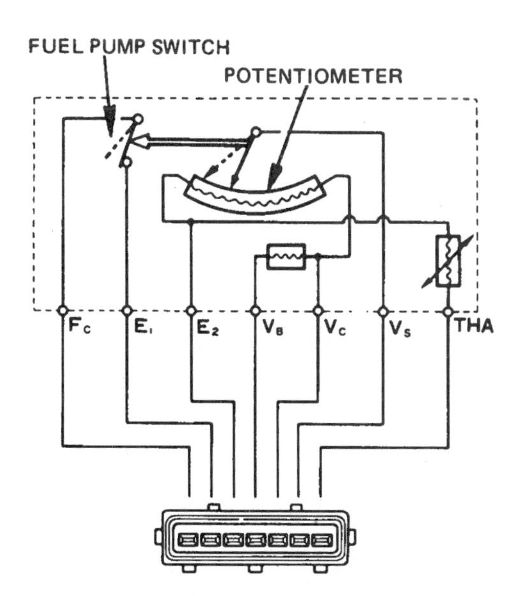

Fig. 15-10 VAF circuit. (courtesy Toyota)

VANE AIRFLOW SENSORS

Vane airflow sensors (also called airflow meters) are used mostly on German imports equipped with Bosch L-Jetronic fuel injection, Japanese imports equipped with Nippondenso multiport electronic fuel injection (made under Bosch license), and Ford vehicles equipped with the Bosch multiport EFI (Escort/Lynx, Turbo T-Bird, and Mustang with the 2.3L turbo engine and the 1989 Probe).

A vane airflow sensor (VAF) is located ahead of the throttle and monitors the volume of air entering the engine by means of a spring-loaded mechanical flap. The flap is pushed open by an amount that's proportional to the volume of air entering the engine (Fig. 15-10). The flap has a wiper arm that rotates against a sealed potentiometer (variable resistor or rheostat), allowing the sensor's resistance and output voltage to change according to airflow. The greater the airflow, the further the flap is forced open. This lowers the potentiometer's resistance and increases the voltage return signal to the computer. Thus a vane airflow sensor measures airflow directly, enabling the computer to calculate how much air is entering the engine independent of throttle opening or intake vacuum. The computer then uses the information to adjust injector duration for a balanced fuel mixture.

The vane airflow sensor sometimes also contains a safety switch for the electric fuel pump relay. Airflow into the engine activates the pump. So if the engine won't start because the fuel pump won't kick in, check the airflow sensor. The easiest to do so is to turn the key on and push the flap open. If you don't hear the fuel pump come on, the contact inside the sensor is defective.

A sealed idle mixture screw is also located on the airflow sensor. This controls the amount of air that bypasses the flap, and consequently the richness or leanness of the fuel mixture.

Airflow sensors can't tolerate air leaks. A vacuum leak downstream of the sensor allows "unmetered" or "false" air to enter the engine. The extra air can lean out the fuel mixture, causing a variety of driveability problems. On systems that have an oxygen sensor, the oxygen sensor can compensate for small air leaks once the engine warms up and goes into closed loop, but not large air leaks. The oxygen sensor is also an "after-the-fact" sensor, which means air leaks can cause hesitation or stumbling when the throttle is suddenly opened.

Vane airflow sensors are also vulnerable to dirt. Unfiltered air passing through a torn or poor-fitting air filter can allow dirt to build up on the flap shaft, causing the flap to bind or stick. You can test the operation of the flap by gently pushing it open with your finger. It should open and close smoothly with even resistance. If it binds or sticks, a shot of carburetor cleaner may loosen it up. Otherwise the sensor will have to be replaced.

Backfiring in the intake manifold can force the

flap backwards violently, often bending or breaking the flap. Some sensors have a "backfire" valve built into the flap that's supposed to protect the flap in case of a backfire by venting the explosion. But the anti-backfire valve itself can become a source of trouble if it leaks. A leaky backfire valve will cause the sensor to read low and the engine to run rich.

VAF Sensor Checks

You don't need the special Bosch tester to check out the airflow sensor. All you need is a voltmeter or analog ohmmeter to probe the appropriate sensor terminals. (Refer to your shop manual for specific terminal references and resistance specs.)

On the Ford EFI systems, you can use a breakout box and voltmeter to check voltage readings. Pushing the flap open should cause a steady and even increase in the sensor's output to a maximum of 5.0 volts. Ford also gives you trouble codes to help diagnose sensor problems: Code 56 indicates sensor input too high, Code 66 is sensor input too low, and Code 76 indicates no sensor change during goose test.

On the import systems, unplug the sensor and measure the potentiometer's resistance with an analog ohmmeter as you push the flap open. Resistance should decrease smoothly and steadily as the flap is opened. If the needle on your ohmmeter skips or jerks as you open the flap, the potentiometer in the sensor has bad spots and the sensor needs to be replaced.

You can also check static resistance between the various terminals. Here are some sample specs for the Bosch units:

Terminals 6 and 9	200–400 ohms
Terminals 6 and 8	130–260 ohms
Terminals 8 and 9	70–140 ohms
Terminals 6 and 7	40–300 ohms
Terminals 7 and 8	100–500 ohms
Terminals 6 and 27	max 2800 ohms @ 68 degrees F.

The vane airflow sensor is considered a sealed unit, preset at the factory with nothing that can be replaced or adjusted except the idle mixture screw. Opening up a VAF that's still under warranty voids the warranty. So if the unit is defective in any way, it must be replaced.

The potentiometer is set and sealed at the factory, so the only adjustment that's required when a sensor is replaced is to set the idle mixture screw. This should be done using an exhaust analyzer to obtain the proper carbon monoxide readings.

MASS AIRFLOW SENSORS

Mass airflow sensors (MAF), which are used on a variety of multiport fuel injection systems, come in two basic varieties: hot wire and hot film. Though slightly different in design, both types of sensors work on essentially the same principle. They measure the volume and density of the air entering the engine so the computer can calculate how much fuel is needed to maintain the correct fuel mixture.

Mass airflow sensors have no moving parts (Fig. 15-11). Unlike a vane airflow meter that uses a spring-loaded flap, mass airflow sensors use electrical current to measure airflow. The sensing element, which is either a platinum wire (hot wire) or nickel foil grid (hot film), is heated electrically to keep it a certain number of degrees hotter than the incoming air. In the case of hot film MAFs, the grid is heated to 75 degrees C. above incoming ambient air temperature. With the hot wire sensors, the wire is heated to 100 degrees C. above ambient temperature. As air flows past the sensing element, it cools the element and increases the current needed to keep the element hot. Because the cooling effect varies directly with the temperature, density, and humidity of the incoming air, the amount of current needed to keep the element hot is directly proportional to the air "mass" entering the engine.

MAF sensor output to the computer depends on the type of sensor used. The hot wire version, which Bosch introduced back in 1979 on its LH-Jetronic fuel injection systems, is used on a number of multiport systems including GM's 5.0L and 5.7L tuned port injection (TPI) engines. It generates a voltage signal that varies from 0 to 5 volts. Output at idle is usually 0.4 to 0.8 volts increasing up to 4.5 to 5.0 volts at wide-open throttle.

The hot film MAFs, which AC Delco introduced in 1984 on the Buick turbo V6 and has since been used on the 2.8, 3.0, and 3.8L V6 engines, produces a variable frequency output rather than a voltage output. The frequency range varies from 30 to 150 Hz, with 30 Hz being average for idle and 150 Hz for wide-open throttle.

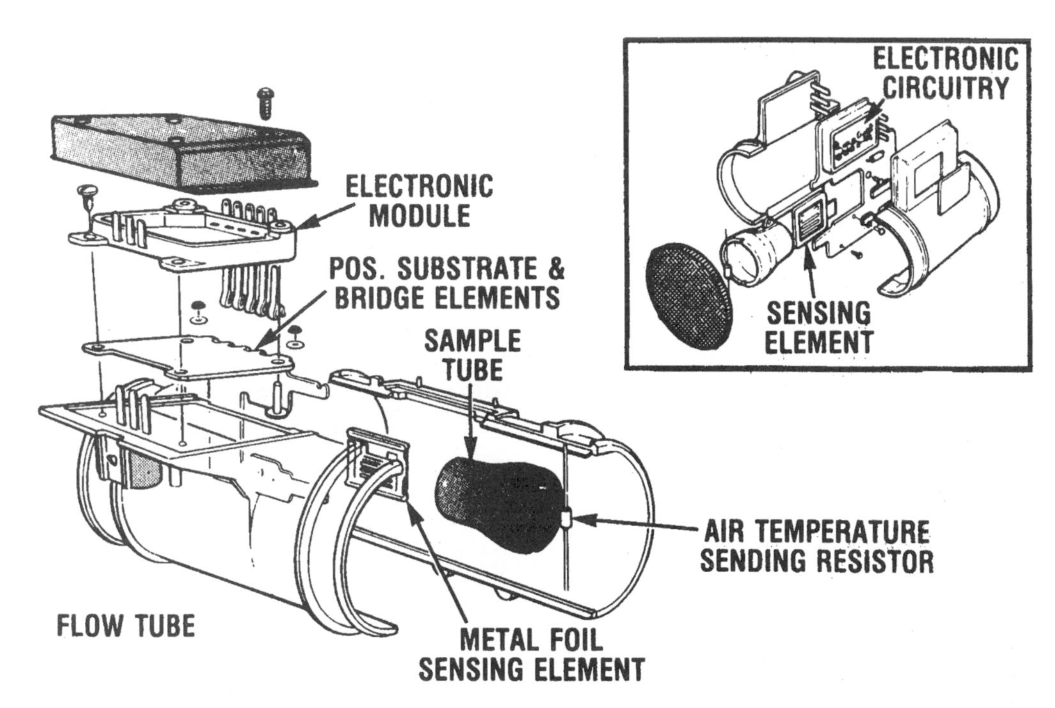

Fig. 15-11 Mass air flow. (reprinted, by permission, from King, *Computerized Engine Control, 4E.* Copyright 1993 by Delmar Publishers, Inc.).

Another difference between the hot wire and hot film sensors is that the Bosch hot wire units have a self-cleaning cycle where the platinum wire is heated to 1,000 degrees C. for one second after the engine is shut down (Fig. 15-12). The momentary surge in current, controlled by the onboard computer through a relay, burns off contaminants that might otherwise foul the wire and interfere with the sensor's ability to read incoming air mass accurately.

MAF Diagnosis

Unlike vane airflow meters with their moveable flaps, MAFs have no moving parts. So the only way to know if the unit is functioning properly is to look

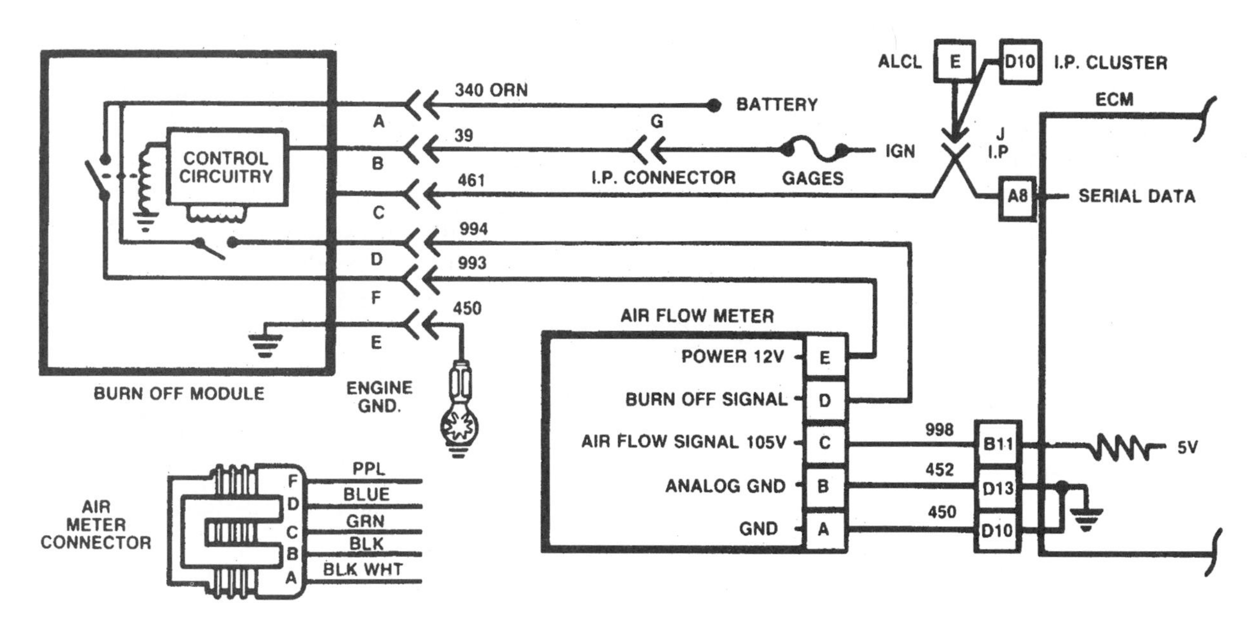

Fig. 15-12 Bosch MAF circuit and burn-off module (reprinted, by permission, from King, *Computerized Engine Control, 4E.* Copyright 1993 by Delmar Publishers, Inc.).

at the sensor's output, or its effect on injector timing.

With the Bosch hot wire sensors, sensor voltage output can be read directly with a voltmeter by probing the appropriate terminals. (Refer to your shop manual for the specific check points and reference voltages.) If the voltage readings are out of range, or if the sensor's voltage output fails to increase when the throttle is opened with the engine running, the sensor is defective and needs to be replaced. A dirty wire (which may be the result of a defective self-cleaning circuit or external contamination of the wire) can make the sensor slow to respond to changes in airflow. A broken or burned-out wire would obviously prevent the sensor from working at all. Power to the MAF sensor is provided through a pair of relays (one for power, one for the burn-off cleaning cycle), so check the relays first if the MAF sensor appears to be dead or sluggish.

On the GM hot film MAFs, you can't read frequency directly but you can tap into the ALDL data stream and read the sensor's output in "grams per second" (GPS) which corresponds to frequency. At idle, you should see something like a reading of 5 to 8 GPS on a scan tool that increases as the throttle is opened with the engine running.

Another way to check MAF sensor output is to see what effect it is has on injector timing. Using an oscilloscope or multimeter that reads milliseconds, connect the test probe to any injector ground terminal. (One injector terminal is the supply voltage and the other is the ground circuit to the computer that controls timing.) Then look at the duration of the injector pulses at idle (or while cranking the engine if the engine won't start). Injector timing varies depending on the application, but if the mass airflow sensor is not producing a signal, injector timing will be about four times longer than normal (possibly making the fuel mixture too rich to start). You can also use millisecond readings to confirm fuel enrichment when the throttle is opened during acceleration, fuel leaning during light-load cruising, and injector shut-down during deceleration. Under light-load cruise, for example, you should see about 2.5 to 2.8 Ms duration.

Trouble codes that may indicate a problem with the mass airflow sensor include:

- GM—Code 33 (too high frequency) and Code 34 (too low frequency) on engines with multiport fuel injection only, and Code 36 on 5.0L and 5.7L engines that use the Bosch hot wire MAF, if the burn-off cycle after shut-down fails to occur.
- Ford—Code 26 (MAF out of range), Code 56 (MAF output too high), Code 66 (MAF output too low), and Code 76 (no MAF change during "goose" test).

Because there are no moving parts and the sensor is all solid-state electronics, any defects require sensor replacement. No adjustment is possible or needed.

SPEED SENSORS

Vehicle speed sensors (VSS) are used in late-model vehicles for a variety of purposes. One such purpose is to monitor how fast the vehicle is traveling so the computer can lockup the torque converter clutch (TCC) at the appropriate speed (usually somewhere between 27 to 45 mph, depending on the application).

Input from a speed sensor may also be used by the computer to determine when the vehicle is moving, at rest, or cruising for various emissions control functions. On some GM EFI-equipped engines, input from the VSS is used by the computer to reset the idle air control motor as well as canister purge and TCC lockup. Such engines should not be driven without a speed sensor, as idle quality may be adversely affected.

Speed sensors are also used with variable assist power steering (Honda, Mazda, Lincoln and Ford Probe to name a few). Sensor input is used by the control electronics to vary the amount of power assist according to speed. The amount of power assist is greater at low speeds for easier parking and maneuvering, but is reduced at higher speeds for more road feel.

Speed sensors are also used in conjunction with electronically adjustable shocks (Mazda and Ford Probe, for example) to change shock valving according to speed. The ride control system in the Probe automatically switches the shocks to the "firm" setting when 50 mph is exceeded in the "auto" mode, and "extra firm" when driving in the "sport" mode. In the Mitsubishi Galant, input from a speed sensor is used to automatically lower ride height to reduce aerodynamic drag at speeds above 56 mph. The computer vents air from the air springs to lower ride height approximately one inch. Normal ride height is then restored when the vehicle's speed drops back below 43 mph.

Another use of vehicle speed sensors is to replace a mechanical speedometer cable for supplying input to an electronic speedometer or electronic cruise control system.

Types OF Speed Sensors

There are essentially two types of vehicle speed sensors. One is the LED type used by GM (Fig. 15-13). Here the VSS consists of a light emitting diode and a photo transistor located in the back of the speedometer housing. The rotation of the mechanical speedometer cable turns a shutter that causes the LED to wink at the photo-receptor. This generates a pulsating electrical signal which is proportional to the vehicle's speed. The computer then uses the signal to calculate the exact mph.

The other type of speed sensor is a permanent magnet (PM) type usually located in the transaxle or transmission. Like a crank position sensor or magnetic ignition pickup, it generates an alternating current signal as it reads gear teeth on the sensor ring. This type of sensor may have a "buffer" box between it and the computer to convert the AC pulses into a direct current (DC) signal that can be used by an electronic speedometer and/or the engine control.

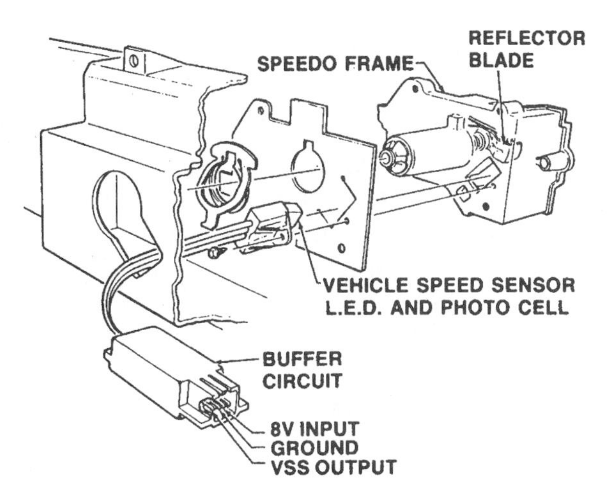

Fig. 15-13 Vehicle speed sensor (reprinted, by permission, from King, *Computerized Engine Control, 4E.* Copyright 1993 by Delmar Publishers, Inc.).

On General Motors light trucks with rear-wheel antilock brakes (RWAL), for example, the AC signal from the speed sensor is routed to a "digital ratio adapter controller" (DRAC) module that converts the AC signal into three separate DC signals: a 4,000 pulse per mile frequency signal for the electronic cruise control and instrumentation modules, a 2,000 pulse per mile frequency signal for the engine control module, and a 128,000 pulse per mile frequency signal for the antilock brake module.

VSS Driveability Symptoms

The driveability symptoms that result from a faulty vehicle speed sensor depend on what type of control functions require an accurate speed input. In an application where the computer needs to know how fast the vehicle is traveling before locking up the torque converter clutch, premature lockup or no lockup at all may result if the sensor is bad. On those where speed input is needed for variable assist power steering, the steering feel may not change as speed increases. On some GM EFI-equipped engines, idle quality may be affected by a bad sensor. On those where the sensor supplies input to electronic instrumentation, fluctuating, unstable, or obviously inaccurate speedometer readings may result from sensor problems.

VSS Sensor Checks

The fastest way to check a suspicious vehicle speed sensor on a GM application is to plug in a scan tool and read the mph output directly while spinning the drive wheels off the ground. You should see a speed indication whenever the drive wheels are turning faster than 3 mph.

You'll also find a Code 24 on GM applications if the system detects no VSS sensor input when the car is in gear, engine speed is between 1400 and 3600 rpm, and the throttle position sensor and airflow sensors indicate the engine in under light load. Be warned, however, that a false Code 24 can sometimes be generated when the engine is raced in neutral in vehicles with manual transmissions.

One of the steps in the GM diagnostic chart is to unplug the VSS sensor and connect a "signal generator" (J33431-B) to simulate sensor input. If you still can't read a mph signal through your scan tool, you've found a wiring or computer problem.

Chrysler's trouble code for a vehicle speed sensor problem is 15. On older models you can't read mph directly on older Chrysler applications, but you can get an indication of the sensor's status by looking for a change in the switch status display. A reading that changes from 88 to 00 and back means the sensor is producing an input signal.

On Ford systems where a speed sensor is used to generate a speed signal for electronic instrumentation, the sensor can be checked two ways: by looking for intermittent resistance at the sensor while the wheels are turning, and by checking sensor resistance (it should be between 200 and 230 ohms). The electronic speedometer also has a self-test capability which can be initiated by holding the trip reset button while turning the ignition from off to run. If the speedometer checks out, but the display is unstable at higher speeds, fluctuates, or decreases when accelerating, the speed sensor is faulty and needs to be replaced. Broken or missing gear teeth on the transmission output shaft can also affect the accuracy of the sensor's reading.

The speed sensor is a sealed assembly and is not repairable. If defective, it must be replaced. No adjustment is needed.

KNOCK SENSORS

The knock sensor is an auxiliary sensor used to detect the onset of detonation. The sensor has no affect on fuel or emissions and is actually part of the electronic spark control circuit. Consequently, it affects ignition timing only.

When the knock sensor detects the characteristic pinging or knocking vibrations produced by detonation, it signals the computer to momentarily retard timing. The computer then backs off the timing a fixed number of degrees or in increments, depending on how it is programmed, until the detonation stops. Then timing returns to normal.

The sensor (Fig. 15-14), which is mounted on the intake manifold or engine, generates a voltage signal when engine vibrations between 6-8 kHz are detected. The sensor works on a "wall vibration" principle, wherein a vibrating plate inside the sensor presses against a piezo-ceramic crystal to generate a voltage signal. When the plate oscillates at the right frequency, the knock sensor signals the computer to retard timing.

Fig. 15-14 Knock (detonation) sensor. (reprinted, by permission, from King, *Computerized Engine Control, 4E.* Copyright 1993 by Delmar Publishers, Inc.).

The location of the sensor on the engine is critical. It must be positioned so it can detect vibrations from the most detonation-prone cylinders. On some in-line 5- or 6-cylinder engines, two sensors are needed to pickup detonation at both ends of the engine.

Knock Driveability Symptoms

If the knock sensor circuit fails, the computer won't retard timing to prevent detonation. The result will be an audible pinging or knocking from the engine during acceleration or under load. Light detonation usually causes no harm, but heavy detonation over time can crack pistons and rings, flatten rod bearings, and cause head gaskets to fail.

Knock sensors can sometimes be fooled by other sounds in the engine, causing the timing to retard unnecessarily. A bad rod bearing or piston slap in a high-mileage engine, for example, may trigger the sensor. So too can a worn timing chain or mechanical fuel pump. A drop in fuel economy or performance would result from retarded timing.

Knock Sensor Checks

Just because an engine detonates doesn't mean the knock sensor is defective. The causes of detonation include:

- Defective EGR valve (stuck shut or inoperative).
- Too much compression due to accumulated carbon in cylinders.
- Overadvanced timing.
- Lean fuel mixture (or a vacuum leak).
- Overheated engine.
- Low octane fuel.

The knock sensor can be ruled out on most applications by rapping on the intake manifold near the sensor with a wrench (never on the sensor itself!) while the engine is running off-idle, and observing the timing with a timing light. The sound of the wrench should simulate the vibrations produced by detonation, causing the knock sensor to signal the computer to back off the timing. You should see a corresponding decrease in timing advance of usually 6 to 8 degrees. If nothing happens, the sensor, wiring circuit, or computer may be faulty.

Use a scan tool to read the knock sensor status directly. Some systems give a yes/no or on/off indication, while other show the actual number of degrees of spark retard. If the sensor gives an indication of knock retard when you rap on the engine, the sensor and its wiring harness are okay. But if the timing fails to retard, there's a problem in the computer spark control circuit.

A knock indication that fails to change when you rap on the engine, or one that shows constant retard at idle, would indicate a faulty sensor or wiring circuit (or a false retard caused by a mechanical problem in the engine as described earlier).

Trouble codes that indicate a problem with the knock sensor circuit:

- General Motors: Code 43.
- Ford: Code 25.
- Chrysler: Code 17.

The knock sensor is a sealed unit, so if defective it must be replaced. No adjustment or repair is possible.

AIR TEMPERATURE SENSORS

Air temperature sensors go by a variety of names. Ford has air charge temperature (ACTs), vane air temperature (VAT) and manifold charging temperature (MCT) sensors. General Motors has manifold air temperature (MAT) sensors. Chrysler has charge temperature sensors (CTS). And Bosch has air temperature sensors (ATS).

Though all these sensors are basically the same in design and function as a coolant sensor, air temperature sensors are used in vehicles equipped with fuel injection for one of two purposes. On some systems, the sensor is used to monitor the temperature of the air entering the engine while the engine is cold. The sensor is mounted in the intake manifold with the tip exposed to air. It gives a much faster response to temperature changes than the coolant sensor. If the air is cold enough, the sensor triggers the cold-start injector to spray additional fuel into the manifold to richen the fuel mixture. In this way it acts something like an electronic choke while the engine is cold. On some applications, the air temp sensor is also used to delay the opening of the EGR valve until the engine warms up.

In other applications the sensor is used to continuously monitor the incoming air temperature so the engine computer can alter the fuel mixture to compensate for changes in air density. Cold air is denser than warm air. So when the sensor reads cold, injector duration is increased to maintain a balanced fuel mixture. If the sensor reads hot, then the duration of the injectors is shortened to prevent the mixture from becoming too rich.

All air temperature sensors work like coolant sensors. Their electrical resistance changes in direct response to temperature. When the computer applies a control voltage to the sensor, the amount of resistance encountered determines the return signal voltage. When the sensor is cold, resistance is high, and the return signal to the computer is low. As the temperature rises, resistance drops, and the voltage output goes up.

Air Temp Sensor Driveability Symptoms

An air temperature sensor can sometimes be damaged by backfiring in the intake manifold. Carbon and oil contamination can also affect the accuracy and responsiveness of the sensor, as can old age. Driveability symptoms for the type that are used to trigger the cold-start injector include stalling or rough idle when cold. A faulty sensor may also allow the EGR valve on some applications to open before the engine is warm, resulting in cold hesitation or detonation.

On applications where air temperature is monitored continuously to help balance the fuel mixture, a faulty sensor can contribute to cold stumbling and warm surging.

Sometimes what appears to be a fuel mixture balance problem due to a faulty air temperature sensor is in fact due to something else, like a restricted catalytic converter! A severe exhaust restriction will reduce intake vacuum and airflow, causing the sensor to read hotter than normal. Once the engine is warm, the air temperature sensor will usually read a few degrees cooler than the coolant sensor.

Trouble codes that may indicate an air temperature sensor problem are:

- General Motors: Codes 23 and 25.
- Ford: Codes 24, 54 and 64.
- Chrysler: Code 23.

Air Temp Sensor Checks

An air temperature sensor can be checked several ways. On GM applications, the easiest way is to tap into the onboard electronics with a scan tool. Plugging into the ALDL connector will allow you to get a direct temperature reading (in degrees Celsius) from the sensor. You should see a slight drop in the temperature reading when the throttle is opened while the engine is running and air rushes into the manifold. You can also compare the air temperature and coolant sensor readings to see if they're close on a warm engine.

The sensor's resistance can also be checked with an ohmmeter. Remove the sensor and check the resistance when cold. Then blow hot air at the tip with a blow dryer (never use a propane torch!), and watch for a drop in resistance. No change means a new sensor is needed.

All you can do with a defective sensor is replace it (unless the tip is coated with carbon, in which case cleaning the tip in solvent may restore its function). Be careful not to overtighten the sensor when it is installed, and use sealer on the threads to prevent vacuum leaks.

CRANKSHAFT POSITION SENSORS

The advent of distributorless ignition systems has added yet another engine sensor to the inventory of underhood electronics, the crankshaft position sensor. A crankshaft position sensor serves essentially the same purpose as the ignition pickup and trigger wheel in an electronic distributor. The only difference between the two is that the timing signal is read off the crankshaft or harmonic balancer instead of the distributor shaft. This eliminates ignition timing variations that can result from wear and backlash in the timing chain and distributor gear. It also does away with timing adjustments (or misadjustments as the case may be).

General Motors uses three different types of crankshaft position sensors. One is a Hall effect crank position sensor that reads a notched metal "interrupter" ring on the back of the harmonic balancer. This is used on the early 3.8L V6 Buick Sequential Fuel Injection (SFI) engines (and turbos) with distributorless Computer-Controlled Coil Ignition (C3I). The crank position sensor provides an on-off signal to the ECM that the ECM uses to monitor engine rpm and crank position. The system also uses a separate "cam position sensor" in place of the original distributor to inform the ECM about valve timing. This enables the ECM to determine the correct firing sequence, which it then uses to control both injector and ignition timing. Ford uses a similar setup on its 5.0L V8 with distributorless ignition.

Another type of crankshaft position sensor GM uses is the "combination sensor," which you'll find mounted on the front of the 3.0L V6. GM calls it a combination sensor because the crank position sensor contains a pair of Hall effect switches that generate two separate signals. There are two notched interrupter rings on the back of the harmonic balancer. One ring has three notches, which causes one of the Hall effect switches to generate three crank position signals every revolution. The other ring has only one notch, which causes the other Hall effect switch to generate a single "sync-pulse" signal that the ECM uses to calculate rpm and ignition timing.

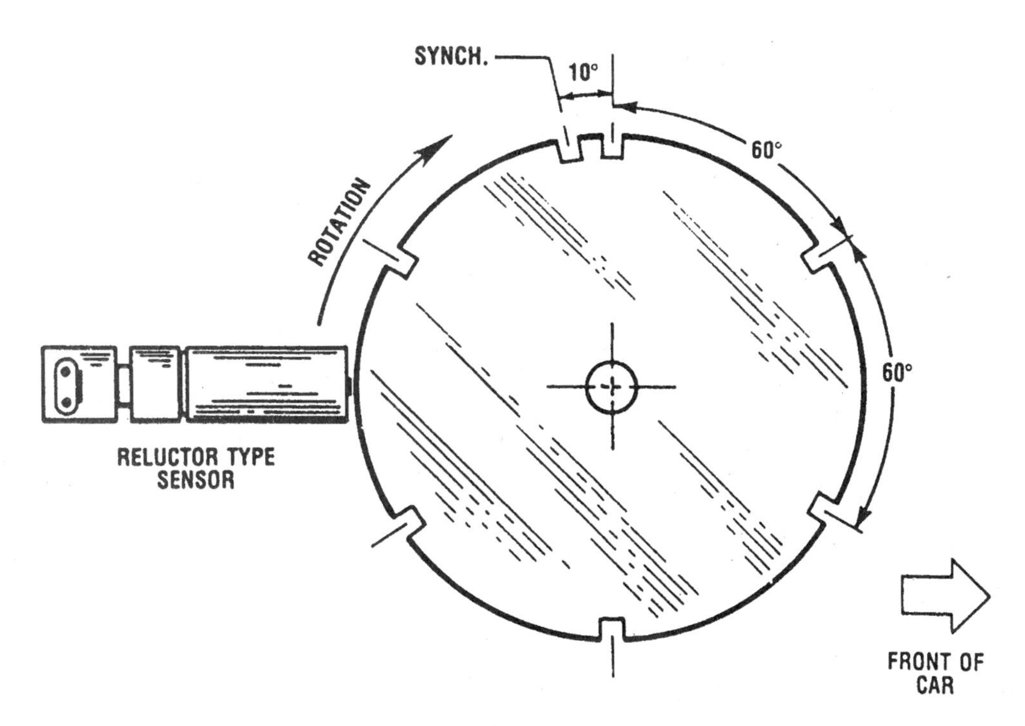

Fig. 15-15 Reluctor-type sensor. (reprinted, by permission, from King, *Computerized Engine Control, 4E.* Copyright 1993 by Delmar Publishers, Inc.).

The third type of crankshaft position sensor GM uses is a magnetic pickup that reads slots machined in a "reluctor" ring (Fig. 15-15) in the center of the crankshaft. This setup is used with the Direct Ignition Systems (DIS) on the 2.0L, 2.5L and 2.8L engines, and the Integrated Distributorless Ignition (IDI) on the 2.3L Quad 4. The crank reluctor ring has six equally spaced slots 60 degrees apart. A seventh slot is spaced 10 degrees from one of the others so the crank sensor will generate an extra "sync-pulse" every revolution. The ECM then uses the information to calculate proper ignition and injector timing. This type of sensor must be carefully positioned so the air gap is within .050 in. of the crankshaft reluctor ring.

CRANK POSITION SENSOR CHECKS

Whether a crankshaft position sensor is the magnetic type or a Hall effect switch, most problems can be traced to faults in the wiring harness. A disruption of the sensor supply voltage, ground, or return circuits can cause a loss of the all-important timing signal, resulting in an engine that cranks but won't start.

When troubleshooting a suspected crankshaft position sensor problem, you have to follow the diagnostic flow charts in your service manual to isolate the faulty component. Until you perform some basic electrical checks, there's no way to know if the problem is in the ignition module, computer, wiring harness, or crank sensor.

On GM applications, a trouble Code 12 while cranking would indicate no reference signal being generated. On Ford applications, a Code 14 would indicate a problem with the crank position sensor signal, which Ford calls a "PIP" (profile ignition pickup) signal.

Magnetic sensors can be checked by unplugging the electrical connector and checking resistance between the appropriate terminals. On the 2.3L Quad 4, for example, the sensor should read between 500 and 900 ohms.

A magnetic crank position sensor should also produce an alternating current when the engine is cranked. So a voltage output check is another test that can be performed. With the sensor connected, read the output voltage across the appropriate module terminals while cranking the engine. If you see at least 20 mV on the AC scale, the sensor is good, meaning the fault is probably in the module.

Hall effect crankshaft position sensors typically have three terminals: one for current feed, one for ground, and one for the output signal. The sensor must have voltage and ground to produce a signal, so check these terminals first with a voltmeter. Sensor output can be checked by jumping the feed terminal to battery voltage and the ground terminal to ground. Then slide a metal knife blade or feeler gauge through the sensor slot to see if it produces a signal (which you read with your voltmeter). When the conductive blade is in the slot, a Hall effect switch (Fig. 15-16) should show zero output. When the blade is pulled out, the switch should produce an 8- to 12-volt signal.

If your diagnosis reveals a faulty crank sensor, the only option is to replace it. With Hall effect sensors, the sensor must be properly aligned with the interrupter ring to generate a clean signal. Any rubbing or interference could cause idle problems as well as sensor damage. Magnetic crankshaft position sensors must be installed with the proper air gap, which is usually within .050 in. of the reluctor wheel on the crankshaft.

CAMSHAFT POSITION SENSORS

On most engines with distributorless ignition systems and sequential fuel injection, a camshaft position sensor is used to keep the engine's control module informed about the relative position of the crankshaft. Monitoring cam position allows the control module to determine when the intake and exhaust valves are opening and closing. The control module can use the cam position sensor's input along with that from the crankshaft position sensor to determine which cylinder in the engine's firing sequence is approaching top dead center. This information is then used by the engine control mod-

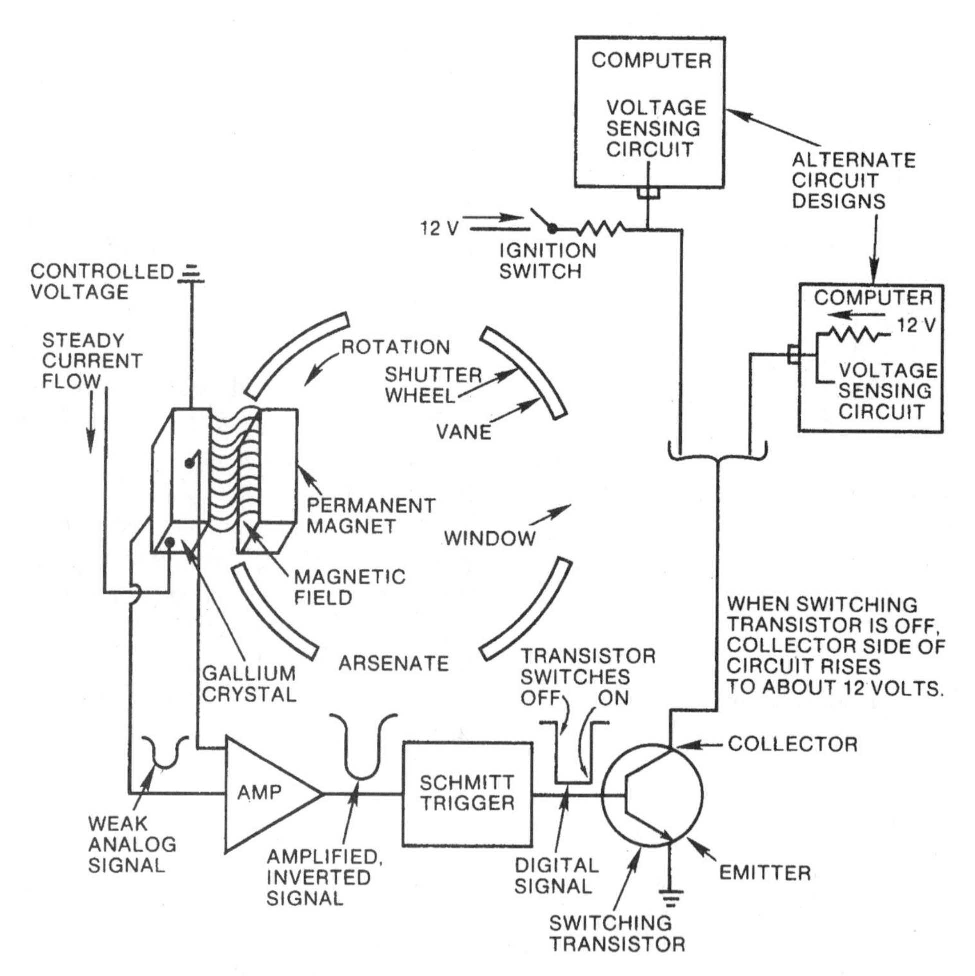

Fig. 15-16 Hall effect switch circuit (reprinted, by permission, from King, *Computerized Engine Control, 4E.* Copyright 1993 by Delmar Publishers, Inc.).

ule to synchronize the pulsing of sequential fuel injectors so they match the firing order of the engine. On some applications, input from the camshaft position sensor is also required for ignition timing.

The camshaft position sensor may be magnetic or Hall effect, and mounted on the timing cover over the camshaft gear, on the end of the cylinder head in an overhead cam application, or in a special housing that replaces the distributor (in the case of some of the GM applications). Operation and diagnosis is essentially the same as that for a crankshaft position sensor.

FORD EGR VALVE POSITION SENSORS

The EGR valve position (EVP) sensor is used primarily on Ford EEC-IV V6 and V8 engine applications to monitor the position of the EGR valve. By keeping the EEC-IV computer informed about the EGR valve's position, the computer can tell when the valve is closed, when it's open and by how much. From this, the computer can calculate the optimum exhaust recirculation flow rate that's needed to minimize NOX emissions. The operation of the EGR valve is then regulated by opening and closing two vacuum solenoids. One is the EGR control (EGRC) solenoid and the other is the EGR vent (EGRV) solenoid.

The EVP sensor itself is a linear potentiometer that works much like a throttle position sensor. Its electrical resistance changes in direct proportion to the movement of the EGR valve stem. When the EGR valve is closed, the EVP sensor registers maximum resistance. As the valve opens, resistance decreases until it reaches a minimum when the EGR valve is fully open.

Ford says the EVP sensor should have no more than 5,500 ohms resistance when the EGR valve is closed, and no less than 100 ohms when the valve is fully open.

The EVP sensor has a three-terminal electrical connector. One terminal is for the 4- to 6-volt voltage reference (VREF) circuit from the computer. Another is for the signal return (SIG RTN) ground circuit. The third terminal is for the EVP sensor output signal, which will vary from close to zero volts when the EGR valve is closed, to a maximum voltage nearly equal to the reference signal when the valve is fully open.

EVP DRIVEABILITY SYMPTOMS

If something happens to the EVP sensor circuit to disrupt the EVP signal back to the computer, the computer has no way of monitoring or confirming the exact position of the EGR valve. Even so, it will still attempt to regulate the EGR valve based on other inputs such as throttle position and manifold absolute pressure. However, control will not be as accurate and some driveability problems may result.

If the engine is experiencing hesitation during acceleration, rough idle, and/or hard starting, the problem is probably the EGR valve itself. Any of these symptoms could be caused by an EGR valve that's stuck in the open position, probably due to a buildup of carbon at the valve's base. Removing the valve and removing the accumulated carbon should solve the problem.

When detonation is the primary driveability symptom, it means the EGR valve is failing to open. The loss of exhaust gas recirculation allows combustion temperatures to soar, resulting in spark knock during acceleration or when running under load. In this instance, the problem may be a faulty EGR valve (failed vacuum diaphragm), or a loss of vacuum to the valve (vacuum leak, faulty EGRC or EGRV vacuum solenoid, or an electrical problem in either of the two solenoid's control circuits).

Problems with the EVP sensor itself can be caused by electrical shorts or opens in the VREF, SIG RTN, or EVP circuits. Bent or corroded pins in the electrical connectors account for most such faults. Another would be wear or sticking in the EVP sensor that would affect the accuracy of the EVP signal.

Ford lists a number of trouble codes that are related to the EVP sensor. These include:

- Code 31—EVP out of limits. This code means the computer is not receiving a proper signal from the sensor. Disconnect and plug the vacuum line to the EGR valve and perform an "Engine Running Quick Test" to see if the code repeats. If it does, there's an electrical problem at the sensor. Use a digital ohm meter on the 200,000 scale to measure the sensor resistance between the EVP SIG and VREF terminals, and then the SIG RTN and VREF terminals. It should be between 5,500 (EGR valve closed) and 100 ohms (EGR valve open), and change smoothly and gradually as the valve is opened with a hand vacuum pump. You can also do this manually by removing the EVP sensor from the EGR valve and pushing the EVP shaft. If the sensor's resistance is out of specs, fails to change smoothly, or change at all, replace the EVP sensor.

If the sensor itself checks out, use a digital voltmeter on the 20 V scale to measure the VREF to SIG RTN voltage, and the VREF to EVP signal voltage at the EVP harness connector (key on, engine off). You should see between 4-6 V. If not, there's an electrical short or open in the wiring. Turn the key off, unplug the computer, and use a breakout box to check wiring continuity. If no wiring faults can be found, replace the computer.

- Code 32—EGR not controlling.
- Code 33—EVP not closing.
- Code 34—No EGR flow.

Any of these three codes indicates a problem with the EGR valve itself, the EGRC or EGRV vacuum solenoids, or a vacuum leak. A Code 83 (EGRC circuit fault) or Code 84 (EGRV circuit fault) would indicate an electrical problem in one of the solenoid circuits. The solenoids should have between 30 and 70 ohms resistance.

If the problem is not electrical, you'll have to use a vacuum gauge to figure out why vacuum isn't getting through. You should see the vacuum cycle on and off in less than 2 seconds at the EGR valve when the engine is at normal operating temperature and the throttle is opened and closed.

COOLING FAN TEMPERATURE SWITCHES

Since the advent of front-wheel drive, electric cooling fans have taken over the job of radiator cooling. On the older cars, fan operation is controlled by a temperature switch located in the radiator or engine. When the temperature of the coolant exceeds the switch's temperature rating (typically 220 to 230 degrees F.), the switch closes and energizes a relay circuit that supplies voltage to run the fan. The fan then continues to run until the temperature of the coolant drops back below the opening point of the switch (usually 210 to 215 degrees F.).

Fan voltage is routed through a relay rather than the coolant switch itself, because relays can handle higher voltage loads. How the fan is wired depends on the application. Some are wired to turn on only when the ignition is on, while others can come on at any time. The latter variety provides continued cooling after a hot engine is turned off to help prevent the formation of steam pockets that might damage a radiator with plastic end-tanks. With this type of fan, you have to be careful, because the fan can come on unexpectedly if the engine is hot even though it's off. So unplug the fan beforehand to save your fingers.

Switch-controlled electric cooling fans can succumb to three types of problems: a temperature switch failure, a relay failure, or a motor failure—any of which will result in loss of cooling. The problem may not be apparent when driving at highway speeds because there is usually enough airflow through the radiator to keep the coolant from overheating. But in slow-moving traffic or during hot weather, the loss of the cooling fan can result in instant boilover.

Troubleshooting Fan Switches

Troubleshooting switch-controlled fans usually involves isolating the bad component by a process of elimination. Step one is to check the fan by unplugging it and jumping it to the battery. If the fan works, that leaves the relay and the temperature switch. Step two is to reconnect the fan and "hot-wire" the relay. You'll have to look up the relay wiring diagram in your shop manual for this step to see which terminal to jump to simulate the coolant switch closing. If the fan runs, you can rule out the relay. That leaves the temperature switch as the culprit—unless the fuse is blown. But if the fan doesn't run when you jump the relay, it's the relay that needs to be replaced.

A temperature switch can also be checked with an ohmmeter or self-powered test light. Remove the switch and dip the tip of the switch into a pan of boiling coolant (50% antifreeze and 50% water). You can't use straight water because it boils at 212 degrees F., which isn't hot enough to close many temperature switches. If the switch fails to close, it's defective and needs to be replaced.

If you want to determine the switch's exact opening and closing points, use a digital pyrometer or thermometer to check the temperature of the coolant while monitoring the switch with an ohmmeter or self-powered test light. If the switch's open and close points don't match the specs listed in the manual, someone may have installed the wrong switch for the application. This can cause either overheating or low heater output.

Most electric cooling fans are wired to come on when the A/C is on (unless a separate fan is used for the A/C condenser). Some are wired parallel to the A/C compressor clutch, while others are wired into a pressure-sensing switch in the A/C high pressure line. So another way to quick-check the system is to simply turn on the A/C to see if the fan also comes on (it should).

If the fan seems to function normally when the A/C is off, but the engine suffers from overheating when the A/C is on, a faulty pressure switch may be the cause.

COMPUTER-CONTROLLED FANS

In most newer vehicles with computerized engine controls, the operation of the cooling fan has been turned over to the computer. Input from the engine coolant sensor, and in many cases the vehicle speed sensor too, is used to determine when the fan needs to be on. This allows more precise control of engine temperature, and also complicates the diagnostic process because the computer itself is now involved.

To illustrate how complicated some of these systems can be, consider late-model Ford front-wheel drive vehicles as an example. Those with automatic transaxles and a 2.5L, 3.0L or 3.8L engine have a two-speed "electro-drive" cooling fan (manual transaxles have a single speed fan). The fan only runs when the ignition is on, and is controlled by the EEC-IV computer through an "integrated control assembly" module. The fan comes on at low speed if the coolant sensor reads above 215 degrees F. (102 degrees C.), and turns off when the coolant temperature drops back below 210 degrees F. (99 degrees C.). The fan normally runs on low speed, unless the coolant temperature is higher than normal and low speed isn't doing the job. The fan also comes on when the A/C is on, but only if the vehicle is traveling slower than 43 mph. Above 48 mph, there's sufficient airflow through the radiator that the fan isn't needed. GM uses a similar setup in certain applications, but uses two separate relays for high and low fan speeds.

Failures in computerized systems can include the coolant sensor, the wiring harness, the computer, the fan relay, or the fan motor. If an engine has been overheating and the electric cooling fan appears to be inoperative (doesn't come on when the A/C is turned on), check the computer for any trouble codes related to the coolant sensor circuit. On GM applications, this would be a Code 14 or 15; on Chrysler a Code 88, 12, 35, or 55; and on Ford Codes 21, 51, or 61.

If a coolant sensor-related trouble code is present, refer to the step-by-step diagnostic procedures in the manual to isolate the problem. On a Ford, for example, you unplug the computer and hook up a breakout box to perform pinpoint voltage checks. If the fan and relay work when the proper pins are jumped, and the coolant sensor checks out, it means computer is bad and needs to be replaced.

Coolant sensors are variable resistors that change resistance according to temperature. The higher the temperature, the lower the sensor's resistance. So if the sensor's resistance doesn't compare with the manual specs when you check it with an ohmmeter, replace it.

CARBURETOR IDLE AND WOT THROTTLE SWITCHES

Though throttle position is monitored by a variable resistor throttle position sensor (TPS) on many computerized engine control systems, contact switches may also be used to keep track of the throttle's position in some applications. An idle switch and/or a wide-open throttle (WOT) switch may be mounted on the throttle linkage to signal the computer when the throttle is at either of these two positions. The computer can then make the necessary adjustments in the fuel mixture and other control functions as needed.

The idle switch (Fig. 15-17) tells the computer when the throttle is closed. The switch may be normally open or closed depending on the application.

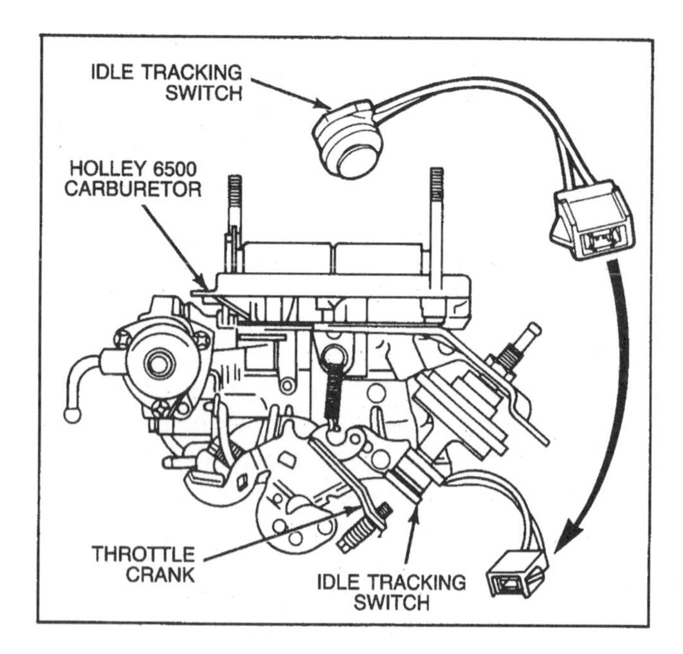

Fig. 15-17 Idle tracking switch. (courtesy Motorcraft - Ford)

A change in switch status signals the computer that the engine is either at idle or off-idle. An idle indication is typically used to actuate idle speed control (ISC), prevent operation of the EGR valve and/or canister purge, and to change the routing of air from the air pump.

The wide-open throttle switch tells the computer when the throttle is full-open. Since maximum power is needed at WOT, a WOT signal typically causes the computer to enrichen the fuel mixture. On some vehicles it also cuts out the air conditioning compressor and prevents the EGR valve from opening.

To better understand these two types of switches, let's look at some typical applications.

Switches/Chrysler

On Chrysler 2.2L engines with Holley 5220/6520 feedback carburetors, a "carburetor switch" is mounted on the throttle kicker solenoid to signal the computer when the throttle is at idle. The switch is normally open, and closes (grounds) when the throttle closes at idle.

Chrysler doesn't have a specific trouble code for the carburetor idle switch, but symptoms that might indicate a faulty switch include idle problems and stalling.

Switch continuity can be checked with a test light or ohmmeter. The switch should show continuity when the throttle is closed. If not, check the alignment and contact of the switch against the linkage.

Switch voltage can be checked by placing a thin piece of cardboard between the throttle linkage and switch to hold the switch open. Switch voltage should be within one volt of battery voltage with the key on. If it isn't, turn the key off and remove the 10-pin connector from the spark control computer. Then check cavity #2 for battery voltage. No voltage indicates a problem in the wiring. If voltage is present at cavity #2, check continuity between the switch and cavity #7 in the 10-way connector. Lack of continuity indicates a wiring problem.

Switches/General Motors

General Motors Computer Command Control (C-3) systems with feedback carburetors and idle speed control have a throttle contact switch (called a "nose" switch) mounted on the tip of the ISC motor. The switch closes when the throttle linkage contacts the ISC motor plunger at idle. This activates the ISC circuit so the ISC motor can regulate idle speed. When the throttle moves off-idle and the switch opens, the ISC motor is deactivated, and the ISC plunger cannot retract.

The status of the nose switch can be monitored with a hand-held scan tool that plugs into the ALDL connector. A trouble Code 35 would indicate a possible short circuit in the ISC control switch.

To check the nose switch, remove the connector from the back of the ISC motor, and use an ohmmeter or test light to check the switch. Insert one probe of your ohmmeter into terminal "A" on the ISC motor assembly and the other into terminal "B." If using a test light, connect it between battery positive (+) and terminal "B," and ground terminal "A". When the throttle is not touching the ISC plunger, there should be no continuity (nose switch open). Closing the throttle or pushing in on the ISC plunger should close the switch and give you continuity. If the switch fails to open and close as it should, the entire ISC motor assembly must be replaced.

On certain 1979 and 1980 C-4 applications, a WOT switch is used by the computer to provide a fixed dwell signal (5 to 10 degrees) for maximum fuel enrichment when the throttle is floored.

Switches/Ford

Ford uses an "idle tracking switch" (ITS) on 2.3L MCU and EEC-IV applications so the computer can modify fuel enrichment and Thermactor operation as needed. A WOT vacuum switch (Fig. 15-17) is also used on the 2.3L and 4.9L applications for fuel enrichment at full throttle.

The idle tracking switch is mounted on the throttle linkage and is used to monitor both engine idling and prolonged deceleration when the throttle is closed. The Ford ITS switch is normally closed, and opens when the throttle is at idle (just the opposite of GM's nose switch).

The WOT vacuum switch senses engine vacuum to signal the computer when the throttle is wide open. The switch is normally closed. When the engine is warmed up, a temperature vacuum switch (TVS) opens to allow vacuum to enter the WOT switch and keep it open. As the throttle begins to open wide and vacuum drops, the WOT switch closes depending on the "set point" at which it is calibrated. For automatic transmissions, the WOT switch closes when manifold vacuum drops below 3 inches Hg. For automatics, the set point is 5 inches Hg.

MCU trouble codes that indicate possible problems with the ITS or WOT switches include Codes 52, 53, 62, and 63. EEC-IV trouble codes that indicate possible switch trouble include Codes 58, 68, and 86.

CHAPTER 16

Actuators

Actuators are typically electric solenoids used to provide motion to a valve, plunger, or other device. When the solenoid coil is energized by an electrical current, it creates a magnetic field that pulls a plunger to create motion. Actuators include fuel injectors (which are covered in Chapter 9), throttle position kickers, idle solenoids, airflow and vacuum control valves, and canister purge valves.

Another type of actuator are small electric "stepper" motors that provide rotary motion in response to computer control. These include idle speed control motors, "liner" EGR valves (covered in Chapter 10), and the motors that change hydraulic valving in electronic shock absorbers and struts.

THROTTLE POSITION KICKERS AND IDLE SOLENOIDS

Many carburetors and some fuel injection throttle bodies have what's called a throttle kicker solenoid (TKS), throttle solenoid positioner (TSP), throttle position kicker, idle solenoid, or a vacuum-operated throttle modulator (VOTM) on the throttle linkage. These devices are used to open the throttle slightly when the engine is cold for better idle quality, to increase idle speed when an engine with an automatic transmission is idling in drive, to boost idle speed when the air conditioning compressor is engaged, and/or to regulate idle speed.

These devices should not be confused with an idle stop solenoid, which is usually a separate device used to prevent engine run-on when the ignition is turned off. An idle stop solenoid is typically energized whenever the ignition is on, to hold the throttle open to its curb idle setting. When the ignition is turned off, the solenoid retracts allowing the throttle to close and kill the engine.

Some carburetors are equipped with a fuel cutoff solenoid (Fig. 16-1) to prevent engine run-on. When the ignition is turned off, the solenoid retracts and blocks the flow of fuel through the idle circuit.

When used to improve cold-idle quality, an idle kicker vacuum diaphragm receives vacuum through a temperature vacuum switch (TVS). Once the engine warms up, the vacuum supply is cut off, and the throttle returns to its normal curb idle position.

On applications where an idle kicker solenoid is used to keep the engine from lugging down when idling in drive, a switch on the transmission shift linkage energizes the solenoid when the transmission is in drive.

When used to compensate for the added load placed on the engine by an A/C compressor, an idle kicker solenoid or vacuum diaphragm opens the throttle slightly when the A/C is used. On A/C systems where the compressor cycles on and off (as opposed to those that run continuously), the idle kicker solenoid may cycle on and off with the compressor, or it may stay on as long as the heater control is set to A/C. This can sometimes cause a

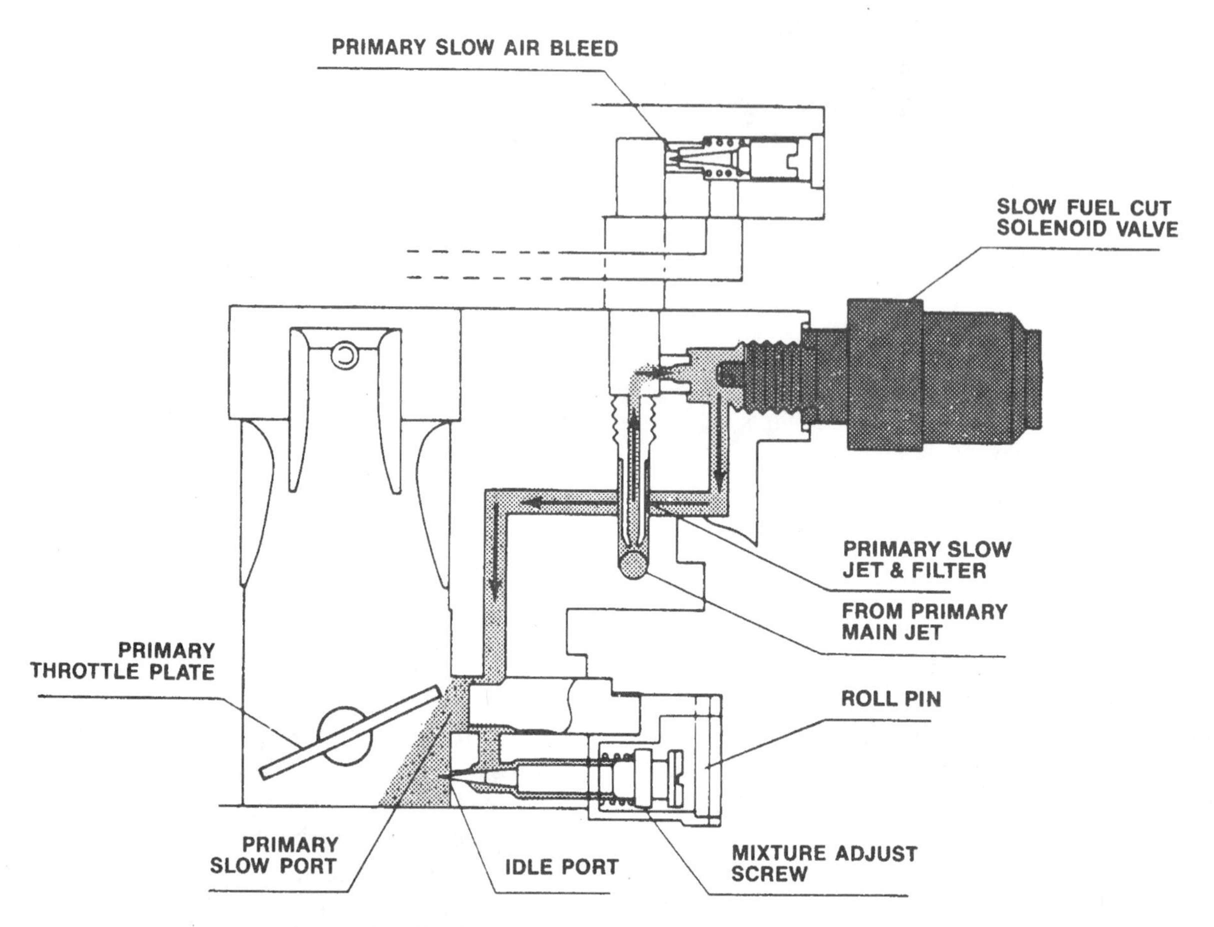

Fig. 16-1 Fuel cut solenoid (courtesy Motorcraft - Ford).

noticeable fluctuation in the idle speed, which is normal for this type of arrangement. Where problems arise is when the idle kicker fails to work, and the engine lugs down or stalls when the A/C is switched on.

When used to regulate idle speed, intake vacuum causes the plunger to move in and out in response to changing engine load. High vacuum (low load) reduces idle speed, while low vacuum (high load) increases idle speed (Fig. 16-2).

Kicker Checks

To check the performance of the idle kicker solenoid or vacuum diaphragm, switch the A/C on momentarily to check the effect on idle speed. A separate idle speed adjustment is usually required to set the idle kicker, "A/C-ON" idle speed. The idle kicker spec may be listed on the emissions decal or in the manual (usually 50 to 200 rpm higher than the curb idle speed). Readjust the "kicker-on" idle speed if it is not within 25 rpm of specs. The adjustment is made by turning a screw on the unit, or by loosening an adjustment nut on the kicker bracket and repositioning the kicker in or out with respect to the throttle linkage.

If the idle kicker solenoid or vacuum diaphragm plunger fails to move when the A/C is switched on, sometimes the pressure of the return spring is enough to prevent the solenoid or plunger from kicking out. Try opening the throttle slightly to relieve pressure. If the kicker still doesn't move, check the electrical connector for the presence of voltage with a voltmeter or test lamp, or check the vacuum connection for the presence of vacuum. You should see battery voltage (or vacuum) if the control circuit is functioning properly. If voltage (or vacuum) is reaching the idle kicker, but nothing is happening (no clicking or extension of the plunger), you can assume the kicker is kaput and needs to be replaced. If you don't get voltage (or vacuum) at the kicker when the A/C is on, then the problem is in

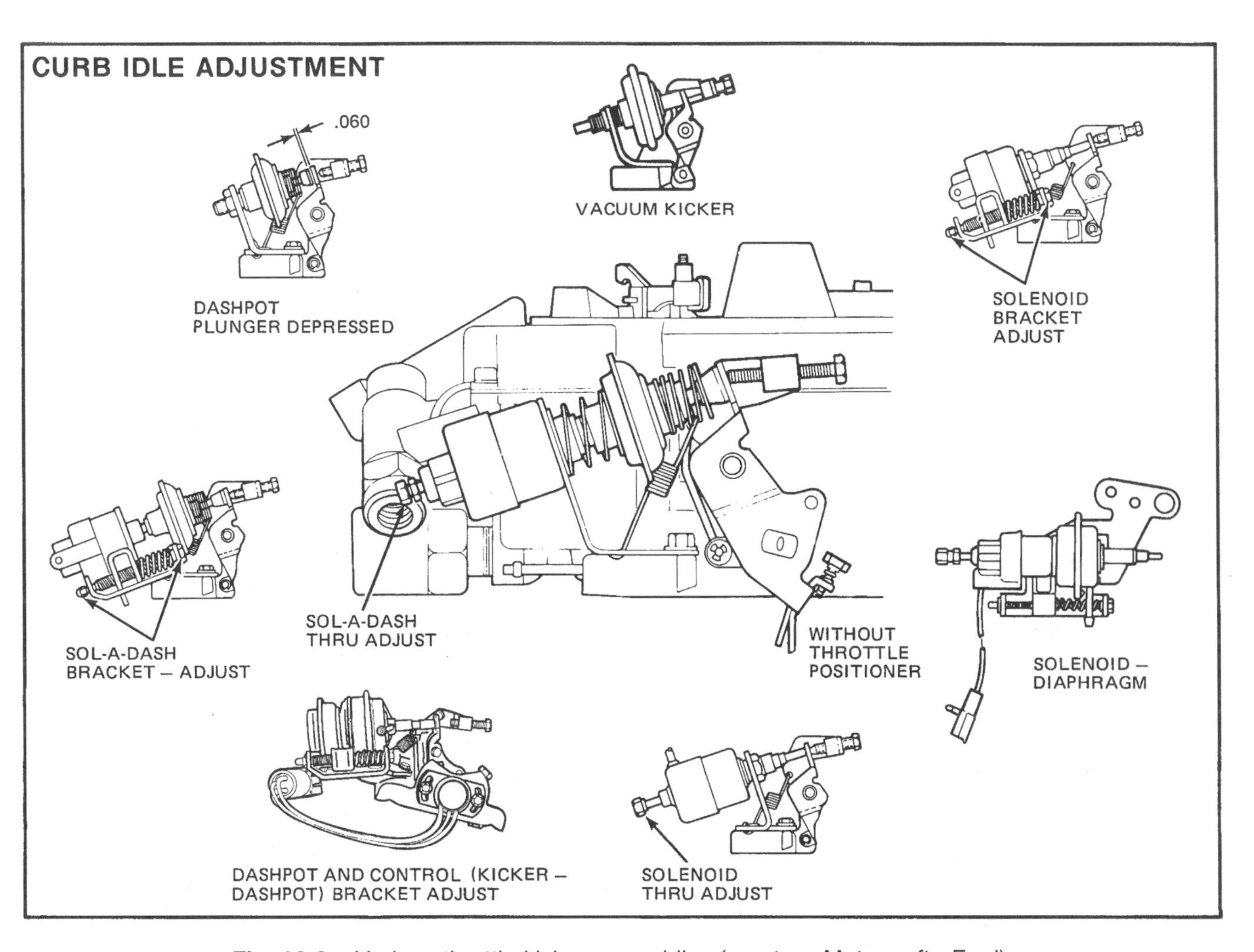

Fig. 16-2 Various throttle kicker assemblies (courtesy Motorcraft - Ford).

the wiring harness (or vacuum plumbing or vacuum switch).

Ford VOTM and Vacuum Idle Speed Control

Some Fords (typically 1980 and later models) use a VOTM (vacuum operated throttle modulator) to regulate idle speed. Others use a slightly different system called mechanical vacuum idle speed control (Fig. 16-2). Both have a vacuum diaphragm with a screw adjustment and plunger that rests against the throttle linkage.

The basic idle speed adjustment procedure for carburetors with a VOTM that retracts when vacuum is applied is to disconnect and plug the VOTM vacuum hose. Then apply full vacuum to the unit with a slave hose from an intake vacuum source or with a hand-held pump. Vacuum pulls the throttle plunger away from the linkage so the basic idle speed adjustment can be made by turning the screw on the top of the VOTM.

NOTE: if the VOTM can't hold vacuum, the diaphragm is leaking and the unit needs to be replaced.

On other Ford applications (1980-1982 140 CID 4-cylinder engines, for example), a different style of VOTM is used. On engines without air conditioning, there is no VOTM but there is an idle tracking switch on the throttle linkage to inform the engine computer when the engine is at idle. On all engines with air conditioning, a "reverse" type of VOTM is used, wherein the plunger extends and pushes against the throttle linkage to increase idle speed when vacuum is applied. The VOTM is supposed to receive vacuum when the air conditioner is running.

This increases the throttle opening to maintain idle speed and to compensate for the extra load of the A/C compressor to prevent stalling. When adjusting the curb idle speed on carbs with the reverse-style VOTM, make sure the A/C is off.

An easy way to tell the difference between the two types of VOTMs is to apply vacuum and see if the throttle linkage plunger retracts or extends. If it retracts, then it is the type that needs to be fully retracted with full vacuum before adjusting curb idle. If the VOTM plunger extends when vacuum is applied, then it must not receive vacuum while the idle speed adjustment is made.

GM Idle Speed Control Motors

On General Motors carburetors that have idle speed control (ISC) motors (Fig. 16-3), a small electric stepper motor is connected to the throttle linkage to regulate idle speed. The ISC motor receives its commands from the engine control module (ECM), which monitors engine rpm through distributor pickup reference pulses. The ECM then maintains the desired idle speed by moving the ISC motor plunger in and out to change the throttle position. The ECM is also programmed to increase the throttle opening when the air conditioning compressor clutch is engaged and when the transmission is in drive.

The ISC control circuit is only activated at idle when the throttle linkage closes a contact switch in the ISC motor housing (called the nose switch).

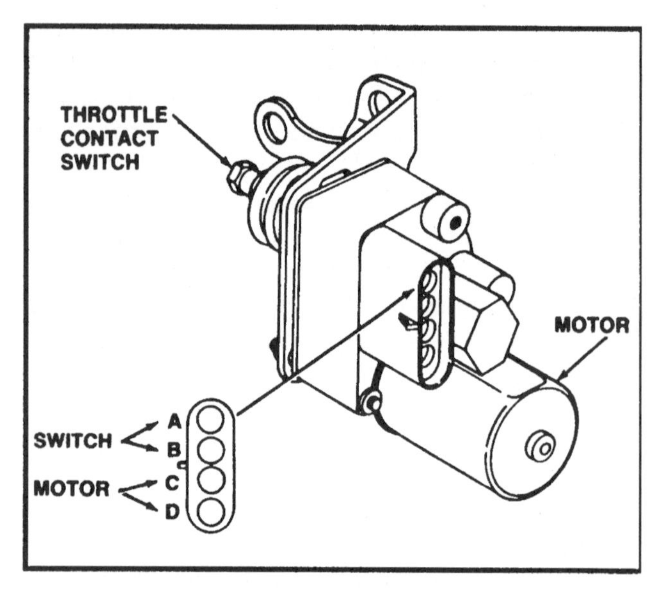

Fig. 16-3 ISC motor (courtesy General Motors).

ISC Diagnosis

Problems such as stalling, low idle speed, fast idle, or an erratic idle can be symptoms of trouble in the ISC motor or control circuit. But be sure to rule out other possibilities first, such as a leaky EGR valve or vacuum leak.

The operation of the ISC motor can be checked by switching on the A/C while the engine is idling and watching for the ISC motor to increase the throttle opening. When the ignition is turned off, the ISC plunger should also move to the fully extended position on most applications.

If the ISC plunger doesn't move, the nose switch may not be closing and activating the ISC control circuit. You can check the status of the nose switch with a scan tool by watching for an "on" signal when the throttle closes. No change in switch status when the throttle opens and closes means the switch is defective, and the ISC motor must be replaced.

If the nose switch is working but the ISC plunger isn't moving, you can check the motor by turning the key off, unplugging the connector on the back of the ISC motor, and using a pair of fused jumper wires to power the motor (which will require readjusting the idle speed after you're through) (Fig. 16-4).

Grounding terminal "D" on the back of the ISC motor and applying battery voltage to terminal "C" should cause the plunger to fully retract. Reversing the connections (grounding "C" and supplying voltage to "D") should run the plunger all the way out. If nothing happens when you jump the ISC motor, the motor is defective and needs to be replaced. But if it works, the problem is in the wiring harness or ECM.

Do not jump the ISC motor for more than 4 seconds after the plunger reaches the fully extended or retracted position, because doing so may overload and damage the motor. Many ISC motors burn out because the plunger is running at the fully extended or retracted position. This can be the result of a vacuum leak, fuel problem, overadvanced or retarded ignition timing, or plunger misadjustment. The computer continues to supply voltage to the ISC motor in an attempt to regulate the idle speed, but because the motor is all the way in or out, it can't go any further, overheats, and burns out.

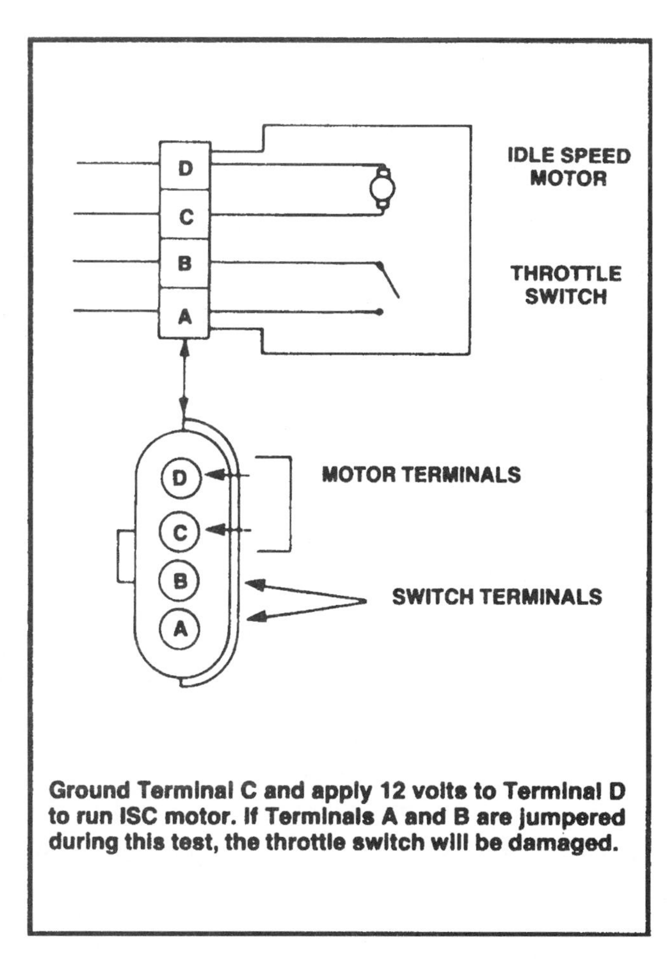

Fig. 16-4 Retracting/expanding ISC motor (courtesy General Motors).

Also, do not connect voltage across ISC motor terminals "A" and "B," as doing so can also damage the nose switch.

A shorted ISC control circuit will set a Code 35 in the computer's memory. This code can also be triggered on 1985 models by a shorted nose switch. A "false" Code 35 can sometimes be set by a misadjusted or sticking throttle position switch (TPS). This tricks the onboard diagnostics into thinking the throttle is open when it really isn't (over 50% throttle for over 2 seconds). An air leak in the hose to the vacuum sensor on some 2.5L engines can also do the same thing.

If you get a Code 35, turn the ignition off and unplug the connector on the back of the ISC motor. Connect an ohmmeter between terminals "A" and "B" (the top two terminals), then open the throttle and note the resistance. Less than 10 ohms means the motor is shorted and needs to be replaced. If the reading is over 10 ohms, check for voltage between terminals "A" and "B" in the harness connector. "A" is ground and "B" is the computer control circuit. If the reading is at least 6 volts, check the operation and adjustment of the TPS. If the TPS is okay, the fault is in the ECM, and it needs to be replaced.

If the reading between connector terminals "A" and B" is less than 6 volts, check for continuity between terminal "A" and ground, then for continuity between terminal "B" and terminal #8 at the ECM. If the wiring is okay, the problem is in the ECM.

Idle Speed Adjustment

Idle speed adjustments should not be needed unless the ISC motor has been replaced—or someone has played around with the motor or plunger. The basic adjustment procedure is to unplug the ISC motor (with the ignition off to prevent electrical damage to the motor). Then, set a minimum idle speed with the ISC plunger fully retracted, and then a maximum idle speed with the plunger fully extended. The minimum and maximum speed settings determine the range of idle adjustments that can be made by the ISC system.

Before the ISC adjustment is made, the engine must be at normal operating temperature and in closed loop (fluctuating M/C dwell signal). Ignition timing must be within specs as should the adjustment of the throttle position sensor. The ISC adjustment is then made as follows:

1. Turn the ignition, A/C, and outside temperature buttons off.
2. Unplug the connector from the back of the ISC motor and run the plunger all the way in by grounding terminal "D" on the back of the ISC motor and supplying battery voltage to terminal "C". Set the minimum idle speed to specs (usually 450 to 500 rpm) by turning the throttle stop screw.
3. Now reverse the connections (ground terminal "C," and supply battery voltage to terminal "D") and run the plunger all the way out. Start the engine and turn the ISC plunger to set the maximum idle speed to specs (1200 to 1800 rpm, depending on the application) by turning the ISC plunger. On automatic transmissions, the transmission must be in drive.
4. Turn the engine off, and reconnect the ISC connector.

5. Clear the trouble code memory by pulling the ECM fuse for 10 seconds. Disconnecting the ISC motor sets a Code 35 and causes the "Check Engine" light to come on. When the ISC motor is reconnected, the light will go out but the fault code will remain in memory unless it's cleared.

IDLE AIR CONTROL VALVES

Idle air control (IAC) valves are used on both throttle body and multipoint fuel-injected engines to regulate idle speed. Chrysler calls theirs an automatic idle speed (AIS) motor while Ford refers to theirs as an idle speed control (ISC) solenoid.

The IAC valve opens a small bypass circuit that allows air to flow around the throttle. Increasing the volume of air that's allowed to bypass the throttle increases idle speed, while reducing the bypass airflow decreases idle speed.

The IAC valve is controlled by the engine's computer. The computer monitors idle speed by counting ignition pulses from the ignition module in the distributor when the throttle position sensor or throttle switch signals the computer that the throttle is closed. On engines with distributorless ignition, idle speed is tracked by counting pulses from the crankshaft position sensor.

When the engine's idle speed is above or below the preset range in the computer's program, the computer commands the IAC valve to either increase or decrease the bypass airflow. Additional sensor inputs from the coolant sensor, brake switch, and speed sensor may also be used by the computer to regulate idle speed according to various operating conditions.

Idle speed may also be increased when the A/C compressor is engaged, the alternator is charging above a certain voltage, and/or the automatic transmission is in gear to prevent the engine from lugging down.

Idle Air Diagnosis

A common condition is an idle air bypass motor that's run all the way closed (fully extended). This is often a symptom of an air leak downstream of the throttle such as a leaky throttle body base gasket, intake manifold gasket, vacuum plumbing, injector O-rings, etc. The computer has shut the bypass circuit in an attempt to compensate for the unmetered air leak that is affecting idle speed.

A Code 11 on a General Motors application indicates a problem in the idle air control circuit. The diagnostic procedure involves disconnecting IAC motor, then starting the engine to see if the idle speed increases (it should). Turn the engine off, reconnect IAC and start the engine again. This time the idle speed should return to normal. If it does, the problem is not in the IAC circuit or motor. Check for vacuum leaks or other problems that would affect idle speed.

If the idle speed does not change when the IAC is unplugged, and/or does not return to normal after reconnecting the unit, use a test light to check the AIC wiring circuits while the key is on. The test light should flash on, and/or go from bright to dim on all four circuits, if the ECM and wiring are okay (indicating the problem is in the IAC motor). If the test light fails to flash on one or more circuits, the fault is in the wiring or ECM.

Ford doesn't use idle air bypass to regulate idle speed on its throttle body (CFI) applications, but uses a solenoid or vacuum diaphragm instead to open the throttle linkage. Idle air bypass (Fig. 16-5) is used only on multipoint injection applications. Codes 12, 13, 16, 17, and 19 all indicate idle speed is out of spec (too high or too low). Codes 47 and 48 indicate a fuel mixture problem which could be caused by an air leak. The diagnostic procedure when any of these codes are found is to turn the engine off, unplug the ISC bypass air solenoid connector, then restart the engine to see if the idle rpm drops (it should if the ISC solenoid is working). No change would indicate a problem in the motor or wiring.

The ISC solenoid can be checked by measuring its resistance. With the positive lead of a digital volt/ohm meter on the VPWR pin and the negative lead on the ISC pin, you should see 7.0 to 13.0 ohms. If it's out of specs, the ISC solenoid is bad. Also check for shorts between both ISC solenoid terminals and the case.

If the ISC checks out okay, check for battery voltage between the ISC connector terminals while the key is on. Voltage should also vary when the engine is running. No voltage indicates a wiring or computer problem.

With Chrysler, a Code 25 means there's a prob-

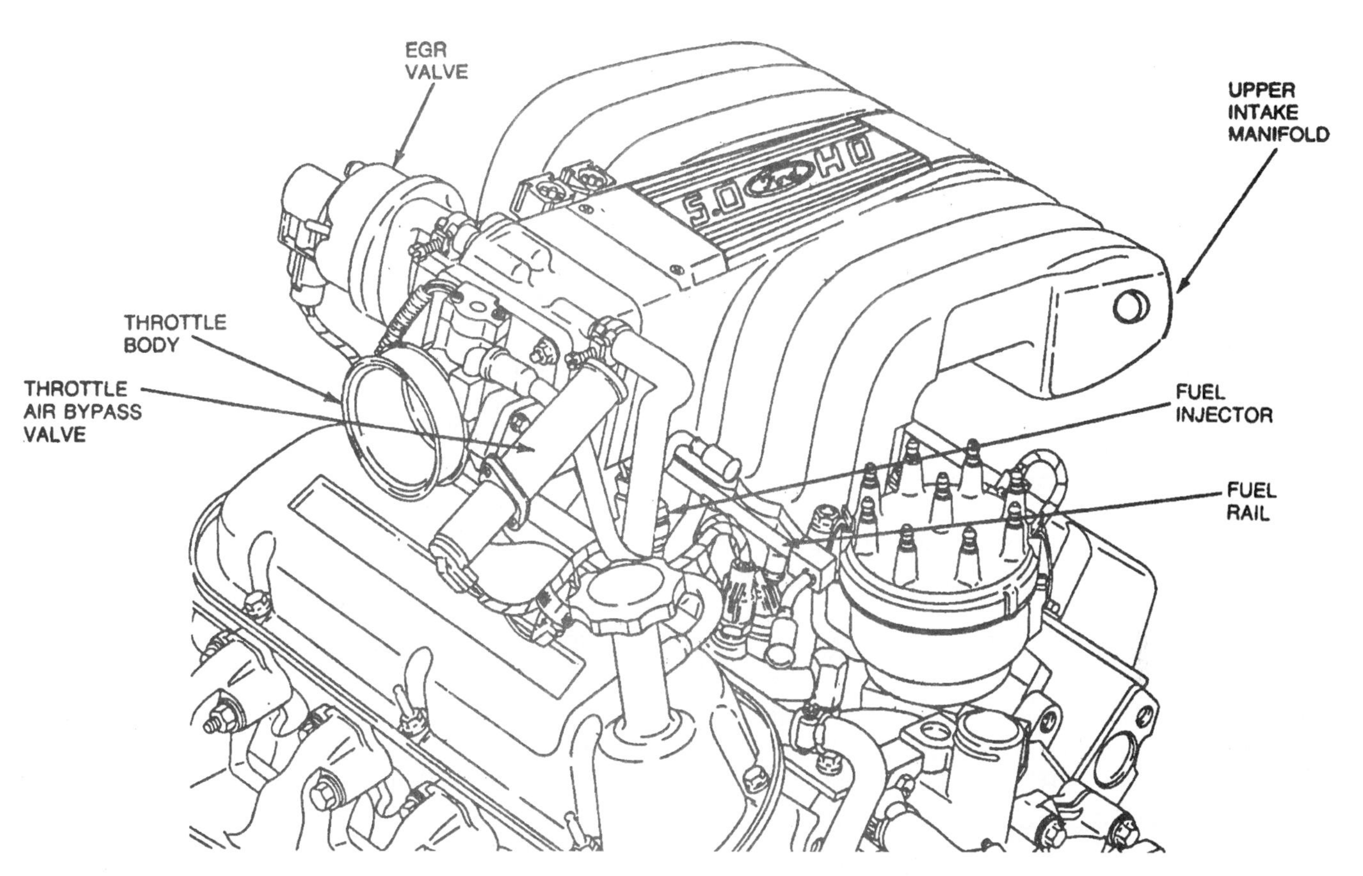

Fig. 16-5 Air bypass solenoid. (courtesy Ford).

lem in the AIS motor driver circuit. The AIS driver circuit can be checked in the self-diagnostic actuator tests. When the actuator display shows a Code 03, the AIS motor should cycle one step open and one step closed every 4 seconds. You can remove the AIS from the throttle body to see if the valve pintle is moving in and out, or simply listen for the motor to buzz.

In the engine running test mode #70, which checks the throttle body minimum air flow, depressing and holding the proper button on a handheld scan tool should close the AIS bypass circuit. At the same time, ignition timing and fuel mixture are fixed. Idle speed should increase to about 1300 to 1500 rpm. If it doesn't match the specs, the minimum air flow through the throttle body is incorrect.

When installing a new GM IAC or Chrysler AIS motor (Fig. 16-6), the pintle must not extend more than a certain distance from the motor housing. The specs vary so check the manual. Chrysler says 1 inch (26 mm) is the limit, while some GM allows up to 28 mm on some units and 32 mm on others. If the pintle is overextended, it can be retracted by either pushing it in (GM), or by connecting it to its wiring harness and using actuator test 03 to move it in (Chrysler).

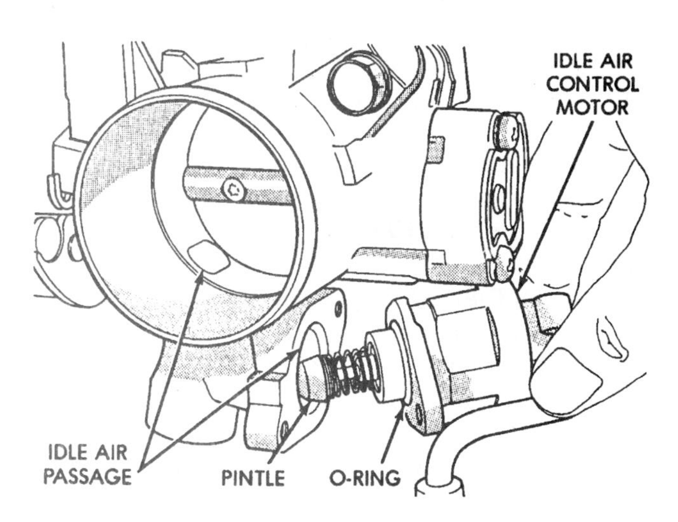

Fig. 16-6 Idle air control motor (courtesy Chrysler).

FORD TAB/TAD THERMACTOR AIR CONTROL SOLENOIDS

Ford's Thermactor air control system uses an air control valve to route air from the air pump to either the exhaust manifold, the catalytic converter, or back into the atmosphere, depending on operating conditions. When the engine is cold, the air control valve routes air to the exhaust manifold to reduce hydrocarbon (HC) and carbon monoxide (CO) emissions. When the engine is warm, the air control valve routes air to the three-way catalytic converter, so nitrous oxides (NOX) can be controlled as well as HC and CO. When the engine is subjected to prolonged idling, the air control valve dumps the air pump's output back into the atmosphere so the converter doesn't overheat.

The operation of the air control valve is regulated by the engine control computer via two vacuum solenoid valves, the Thermactor air bypass (TAB) valve and the Thermactor air diverter (TAD) valve. The TAB and TAD valves are connected to a manifold vacuum source, and pass vacuum to the appropriate section of the air control valve when energized. The TAB and TAD valves are normally closed, and vent vacuum when deenergized.

When the TAB valve receives battery voltage from the computer, it opens and passes manifold vacuum to the air control valve bypass circuit. This opens the valve in the bypass circuit allowing air from the air pump to be vented back into the atmosphere.

When the TAD valve is energized, it passes manifold vacuum to the diverter valve portion of the air control valve. This opens the diverter valve causing it to reroute air from the exhaust manifold to the catalytic converter.

TAB/TAD Diagnosis

A failure of the TAB/TAD valves may prevent the catalytic converter from receiving extra oxygen, which will upset the operation of the Thermactor system, causing an increase in HC/CO/NOX emissions.

When the onboard diagnostics detect a problem with the operation of the Thermactor system or the TAB/TAD valves, it will trigger one of three trouble codes. A Code 44 signals the Thermactor system is inoperative. A Code 45 indicates trouble in the TAD/diverter valve portion of the system, while a Code 46 means trouble with the TAB/bypass circuit. When a trouble code is found, follow the step-by-step diagnostic procedure in the manual to isolate the cause.

The following quick checks can be made to speed your diagnosis:

- With the key off and the TAB/TAD valves disconnected from their wiring harness, an ohmmeter check should find 50 to 110 ohms of resistance if the valves are okay. More or less resistance means the solenoid valve is defective and needs to be replaced.
- With the key on, engine off, the TAB/TAD valves should both be receiving battery voltage (minimum 10.5 volts). If not, check the wiring harness for continuity, grounds, or shorts using a 60-pin breakout box at the computer. Continuity checks can be made using pin 51 for the TAB valve and pin 11 for the TAD valve. Check for shorts between the TAB/TAD circuits and ground using pins 40, 46, and 60. Hot shorts can be checked by checking between the TAB/TAD circuits and pins 37 and 57.
- The operation of the TAB/TAD valves can be checked by connecting a vacuum pump to the valve's inlet vacuum (manifold) port and plugging the outlet (air control valve side) port. The TAB/TAD valves should bleed (not hold) vacuum when deenergized. When energized using a jumper wire, the TAB/TAD valves should open to pass vacuum. The valves should now hold vacuum assuming the outlet port is plugged tightly.
- Check the vacuum supply at each valve with the engine idling. The TAB/TAD valves should both see a minimum of 10 in. Hg of manifold vacuum at idle. If there's no vacuum or the vacuum reading is low, check the vacuum hose connections for leaks, and compare the routing to the emissions decal.
- When the diagnostic self-test is initiated, a vacuum gauge connected to the TAB/TAD valves should pulsate above and below 5 in. Hg. No pulsation indicates a wiring problem, bad solenoid, or vacuum leak.

The TAB/TAD valves are sealed assemblies, so if you find a problem with one, the only repair option is to replace it. On some applications, the TAB/TAD valves are separate (2.3L engines, for example), while on others (V6 & V8 engines), the TAB/TAD valves are paired and must be replaced as a unit.

CANISTER PURGE SOLENOIDS

One of the physical characteristics of gasoline that makes it such a good engine fuel is that it evaporates easily. Unfortunately, this also creates evaporative emissions from the fuel tank and carburetor that contribute to air pollution. This type of pollution is especially bad because it continues even when the engine is not running.

Evaporative emissions have been eliminated by sealing the fuel tank and carburetor bowl. The gas tank is sealed with a cap that contains a spring-loaded pressure relief valve. The tank itself is also vented to the "charcoal canister" that traps and stores fuel vapor until the engine is started so the vapors can be siphoned into the engine and burned. The carburetor bowl is likewise vented to the charcoal canister, so when the engine is off, fuel vapors cannot escape into the atmosphere. On some carburetors, a solenoid is used to shut off the bowl vent when venting is not required.

The charcoal canister is filled with activated charcoal that absorbs fuel vapors like a sponge soaks up water. When the engine starts, engine vacuum opens a "purge valve," which may be on the canister, in the vacuum line, or mounted remotely. The purge valve allows air to be drawn through the canister to pull out the fuel vapors. The vapors are then siphoned into the engine and burned.

Purging may be delayed by a thermal vacuum switch or by a computer-controlled solenoid in the vacuum line until the engine is warm, running above idle, and/or above a certain speed to prevent an excessively rich fuel mixture.

Depending on the design of the canister, purging may be controlled in different ways. Those that use ported vacuum to open the purge valve employ a technique called "constant and demand purge." The purge valve allows constant purging at a restricted rate through an orifice, until a certain level of vacuum exists at the canister outlet. When ported vacuum is applied to the purge control valve, it allows a higher purging rate.

The reason for having constant and demand purging is because the engine cannot handle a large flow of air through the canister at idle or slow speeds. The additional air and fuel vapor usually makes the air/fuel mixture go rich, causing a rough idle and increased CO emissions. At higher rpms, however, the engine can ingest the canister's vapors without affecting the fuel ratio significantly. A vapor feed rate of around 12 cfm is typical for a small V8 cruising at highways speeds.

Canister Purge Problems

Most charcoal canisters are sealed and draw fresh air in through a separate hose that connects to the air cleaner. But on some older GM canisters, the bottom of the canister is open and contains a filter through which fresh air can enter to flush the fuel vapors from the canister. The filter requires periodic replacement, but is frequently neglected.

The canister can become saturated with fuel, and/or cannot be purged because of a faulty purge valve, vacuum leak in the purge hose, or misrouted vacuum connections. Fuel vapors can then "back up," causing a strong odor of gasoline when the engine is not running.

If the purge valve sticks open or the hoses are misrouted so the canister is purged at all times, the engine will run rich, idle poorly, and show elevated CO exhaust emissions.

The purge valve can be checked by applying vacuum to the "control port" with a hand-held pump. The purge valve should hold about 15 in. Hg. vacuum without leaking. If the valve leaks, it needs to be replaced.

The vacuum "purge port" (which is connected to the PCV plumbing or another source of manifold vacuum) should also hold vacuum when no vacuum is applied to the control port. Applying vacuum to the control port should open the purge valve (which can be checked with a second vacuum gauge or pump), allowing fuel vapors to pass from the canister to the engine. If the purge port leaks or fails to open when vacuum is applied to the control port, the purge valve needs to be replaced.

On engines with solenoid-controlled purge valves, the solenoid prevents vacuum from purging the canister until a certain engine temperature, throttle position, and/or vehicle speed is reached. If the valve doesn't hold vacuum when the engine is cold, and the ignition is on, either the solenoid is defective or there's a problem in the engine computer (look for a trouble Code 31 on Chrysler, or a Code 85 on Ford).

GM Purge Solenoids

On late-model GM applications with a computer-controlled solenoid purge valve, the computer not only opens and closes the solenoid to control purging, but also cycles it on and off (changes the duty cycle) to vary the rate of purging (Fig. 16-7). The solenoid is energized (on) when the engine is cold to block purging, and deenergized (off) to allow purging when the engine is above 80 degrees C., has been running for at least 3 minutes, and the vehicle speed is above 15 mph. For variable-rate purging, the computer monitors airflow and engine load through the MAP sensor.

You can check the status of the purge solenoid through the computer system with a scan tool to see if its status and duty cycle matches the engine's operating conditions. (Use the "canister purge" or "idle learn" mode, depending on the application.)

With the ignition on and the engine off, the purge solenoid should be energized and hold vacuum (apply to the port that connects to the throttle body). If it fails to hold vacuum, use a test light to check the supply voltage and ground at the solenoid. If there's no electrical problem, and the solenoid won't hold vacuum, the solenoid is defective and needs to be replaced.

The solenoid deenergizes when the ALDL diagnostic connector is grounded. So another way to test it is to apply vacuum, ground the diagnostic connector, and see if vacuum drops (it should). No change means the solenoid is stuck and needs to be replaced (Fig 16-8).

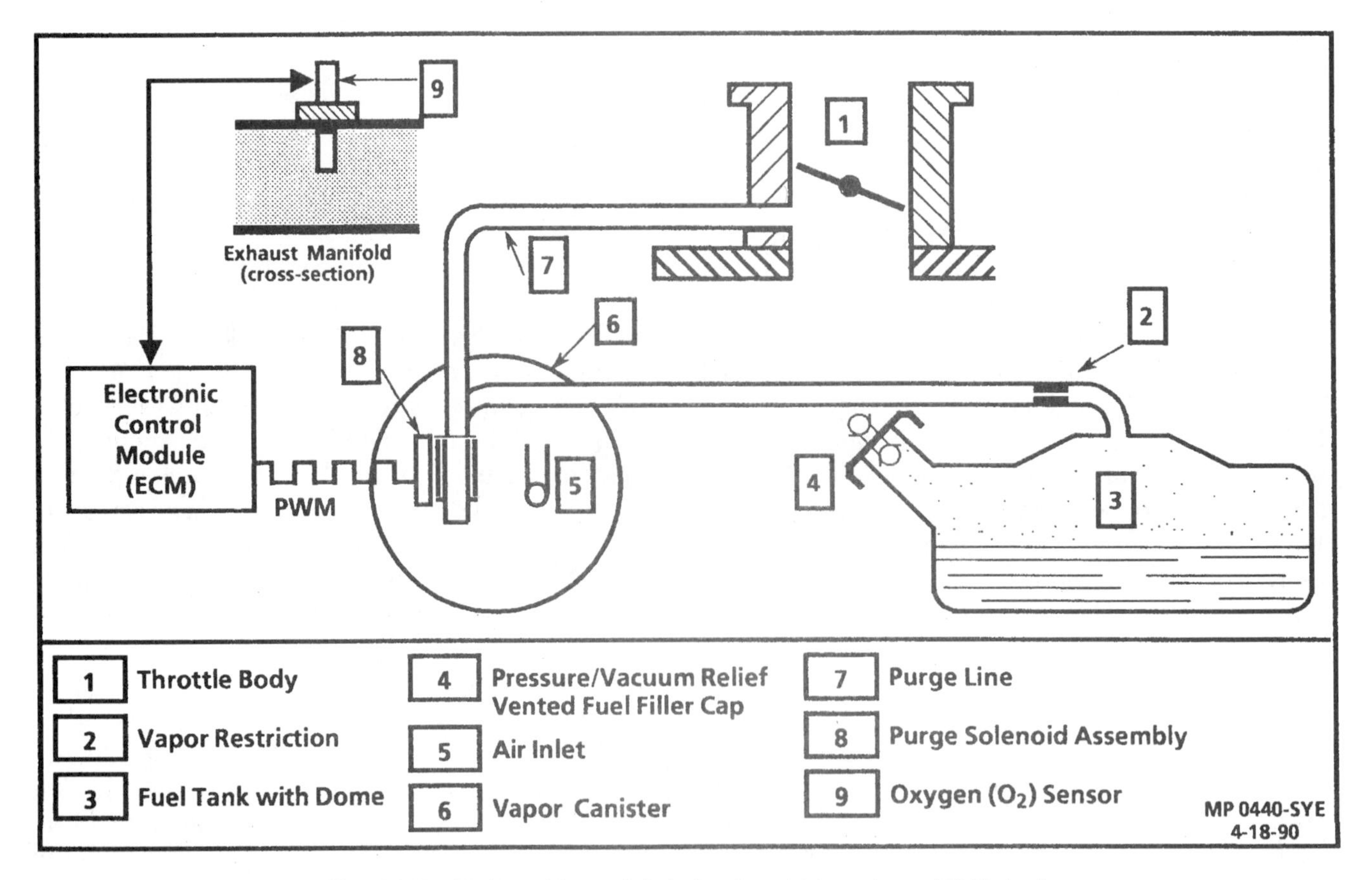

Fig. 16-7 Pulse with modulated solenoid (courtesy AC Delco).

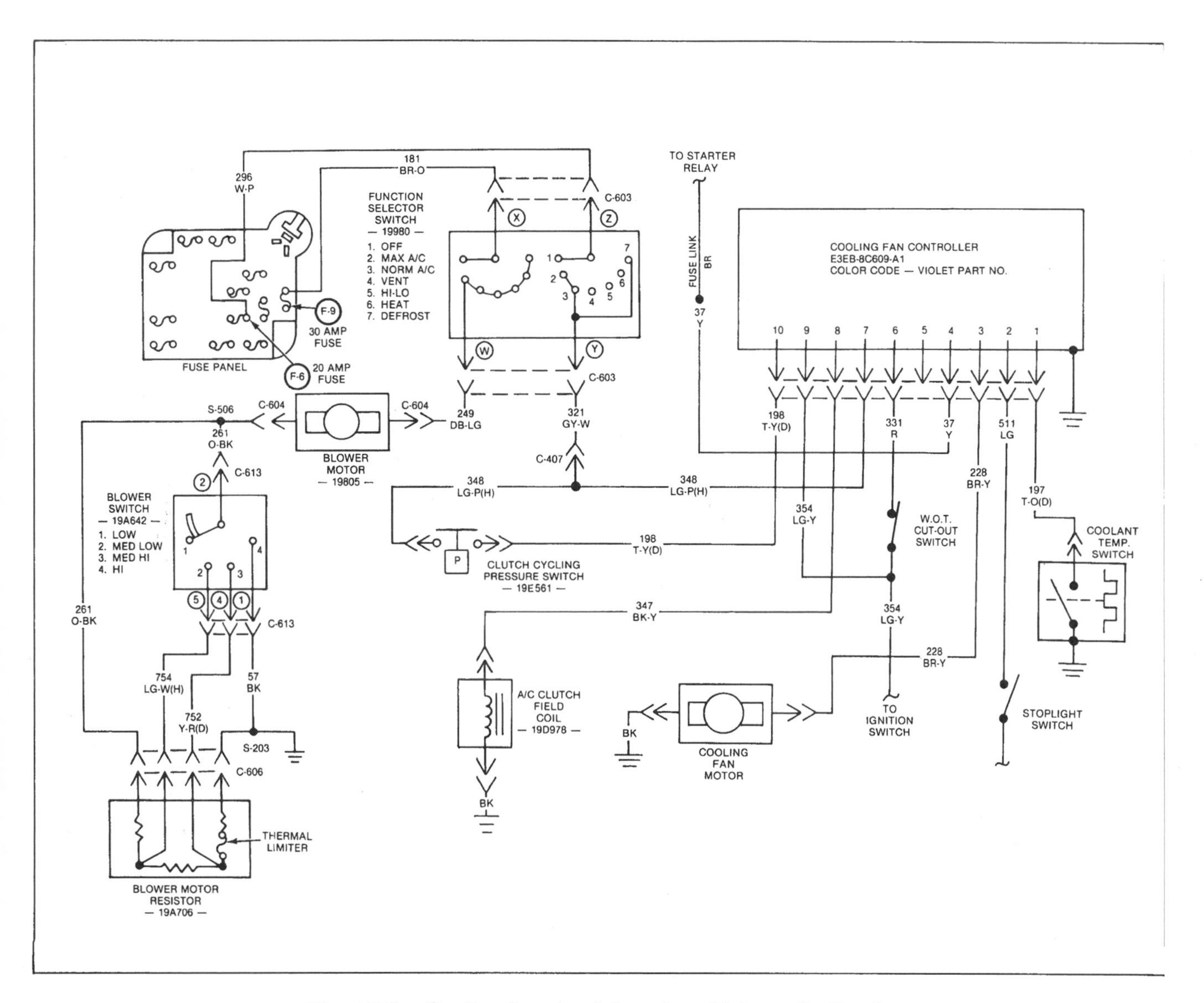

Fig. 16-8 Cooling fan circuit (courtesy Motorcraft - Ford).

CHAPTER 17

Computerized Management

FEEDBACK CARBURETION

Beginning in 1980 for California emissions-certified cars, and in 1981 for the remaining 49 states, computerized engine controls with "feedback" carburetion were added to reduce emissions and improve fuel economy and driveability. The feedback principle is based upon a "loop" (Fig. 17-1) that begins at the oxygen (O_2) sensor in the exhaust manifold, goes to the engine control computer, and then to the carburetor.

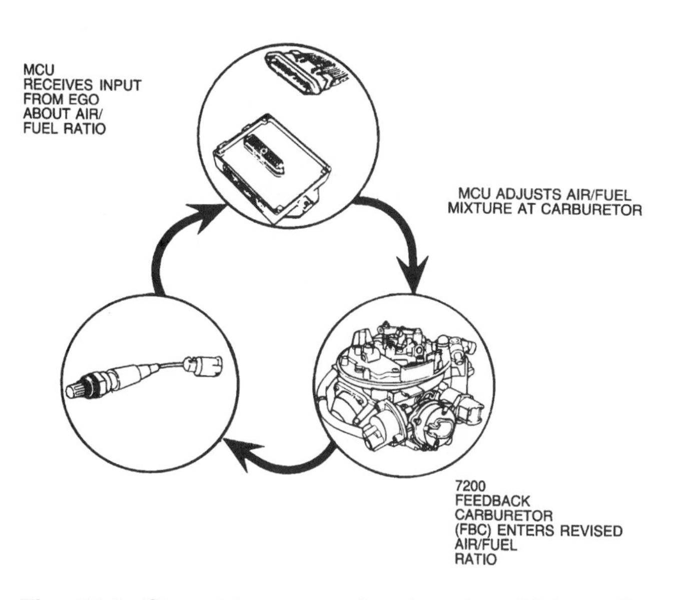

Fig. 17-1 Closed-loop operation (courtesy Motorcraft-Ford).

Feedback carburetion involves monitoring the amount of unburned oxygen in the exhaust to determine if the fuel mixture is running too rich or too lean. The engine computer uses the oxygen sensor's input to constantly readjust the air/fuel mixture. This maintains the optimum balance for low emissions, good fuel economy, and performance.

If something happens in the feedback circuit that prevents the computer from correctly regulating the fuel mixture, the system will lose its ability to maintain a balanced air/fuel mixture, causing emissions, fuel economy, and performance to suffer. The important thing to remember is that many things can cause feedback problems other than a faulty oxygen sensor.

An air leak in the intake or exhaust manifold or even a fouled spark plug, for example, can cause an O_2 sensor to give a false lean indication. The sensor doesn't know the difference between oxygen in the fuel mixture oxygen and oxygen from an air leak. It only reacts to the presence or absence of oxygen in the exhaust. It can't tell the computer where the oxygen came from. When the computer tries to compensate for what it thinks is a lean mixture by increasing the richness of the air/fuel mixture, it makes the engine run rich. This results in elevated carbon monoxide emissions, poor fuel mileage, and possibly a rough idle or surging condition.

The key sensor in the feedback circuit is obviously the oxygen sensor. The coolant sensor is also important because it tells the computer when the engine is warm enough to go into "closed-loop" operation and to start using the O_2 sensor signal to control the fuel mixture.

The O_2 sensor produces a voltage signal (Fig. 17-2) that varies according to how much oxygen is in the exhaust. The sensor produces a low-voltage signal when there's a lot of unburned oxygen in the exhaust (which the computer interprets as a lean mixture). And it produces a high-voltage signal when there's little unburned oxygen in the exhaust (which the computer interprets as a rich mixture).

The computer then orders the fuel delivery system to do the exact opposite of what the oxygen sensor is telling it. If the O_2 sensor reads lean, the computer orders the fuel system to go rich. If the O_2 sensor reads rich, the computer orders the fuel system to go lean to compensate. As the air/fuel mixture comes back into balance, it tends to overshoot the mark which causes the system to oscillate back and forth in close proximity to an ideal mixture.

M/C DWELL

On carbureted engines, the fuel mixture is leaned by increasing the "dwell" ("on" time) of the carburetor mixture control (M/C) solenoid (Fig. 17-3). Decreasing the dwell enrichens the mixture. On GM carburetors, the M/C solenoid uses a stepped

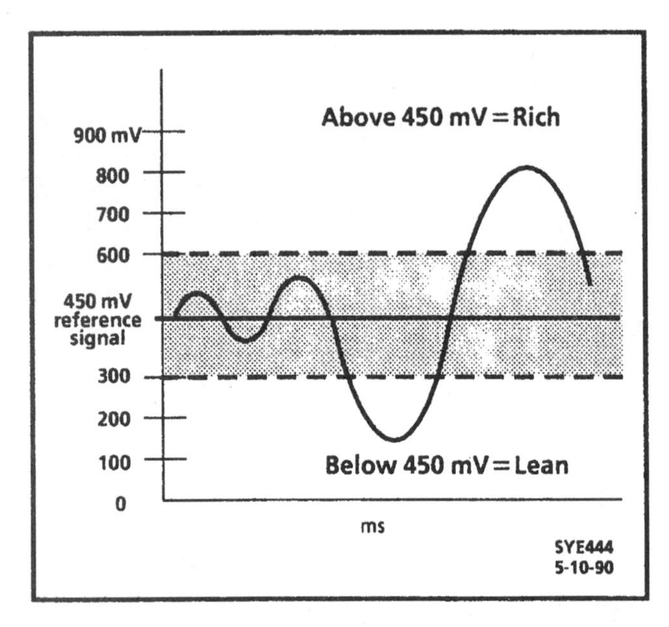

Fig. 17-2 Oxygen (O_2) sensor signal (courtesy AC Delco).

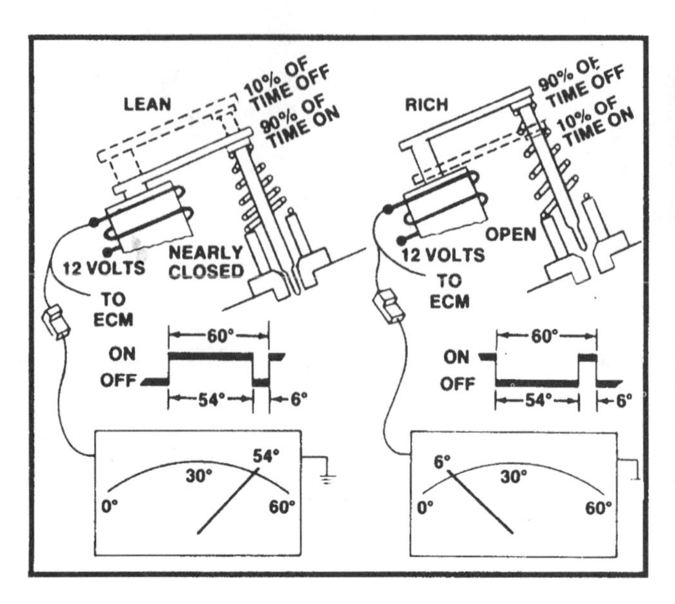

Fig. 17-3 Carburetor dwell readings (courtesy General Motors).

metering rod to vary the amount of fuel that flows through the main jet. When the solenoid is energized, the metering rod moves down into the full-lean position. When deenergized, it raises to the full-rich position. The M/C solenoid cycles on and off 10 times a second to regulate the relative richness or leanness of the fuel mixture (Fig. 17-4).

There are basically three types of feedback mixture control:

1. Vacuum Regulator Control. With this system, the computer controls a vacuum solenoid regulator (Fig. 17-5) which supplies a controlled amount of vacuum to a diaphragm connected to the carburetor fuel metering rods and air bleeds. Examples include the Holley 6500 and 6510C used by General Motors and early Chryslers.

2. Electrically Controlled Main Metering. The computer controls a mixture solenoid which changes the amount of air passing through the idle air bleed circuit or main metering circuit. Examples include the Rochester E2ME, E2MC, E4ME, E4MC, and E2SE. On Chrysler carburetors (Holley 5220 and 6280 two-barrel, and 6145 one-barrel), the "duty cycle" solenoid allows fuel to flow through a circuit parallel to the main jet to enrichen the mixture.
3. Electrically Controlled Air Bleed. The computer controls a stepper motor or solenoid which con-

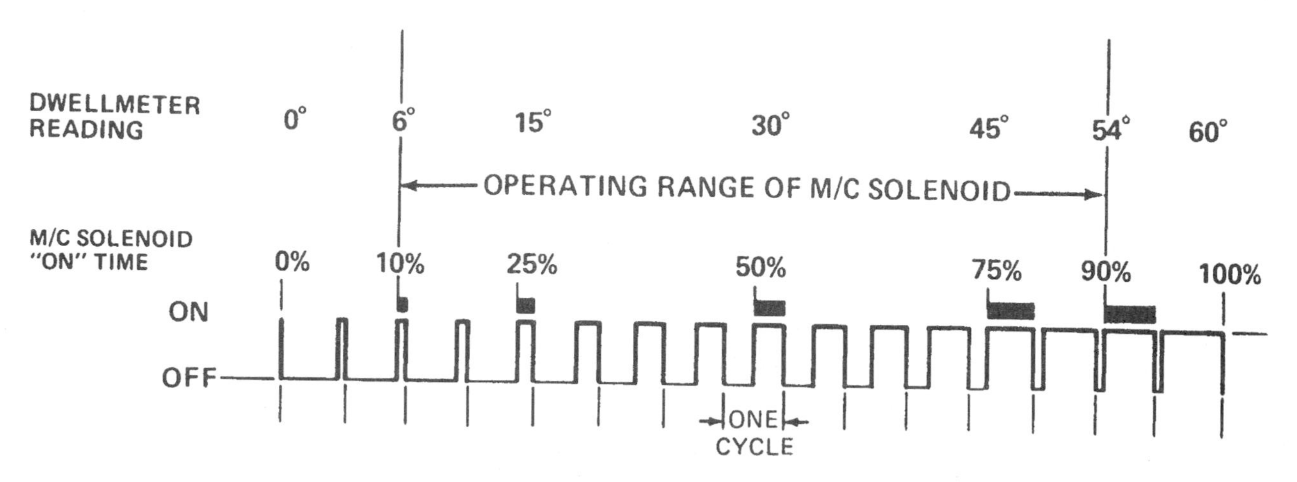

Fig. 17-4 Duty cycle vs. dwell meter (courtesy General Motors).

trols the idle and main metering air bleeds, not the fuel metering rods. Examples here (Fig. 17-6) include the Ford 7200VV, Carter YFA1, and Carter BBD two-barrel with the stepper motor as used by Jeep and AMC.

Reading M/C Dwell

Fuel mixture control dwell can be read directly on General Motors systems by using a scan tool or by finding the green carburetor M/C dwell test lead

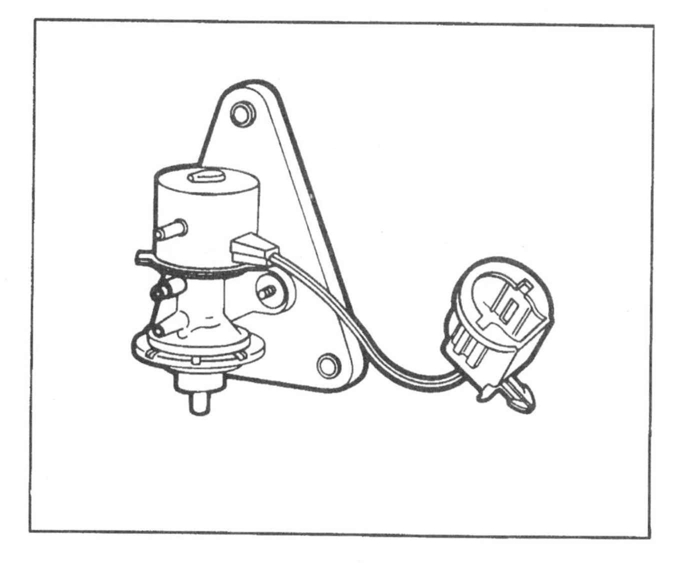

Fig. 17-5 Vacuum regulator control (remote mounted) (courtesy Motorcraft-Ford).

and connecting an ordinary dwell meter to it. With the dwell meter set to read on the 6-cylinder scale, the M/C carburetor dwell readings will vary from zero to 60 degrees (Fig. 17-3).

Remember, a high-dwell reading means the carburetor is running lean, and a low-dwell reading means it is running rich. But what does this really mean? It means the carburetor is doing exactly the opposite of what the oxygen sensor is reacting to in the exhaust. A low-dwell number, therefore, really means the oxygen sensor is seeing a lean condition in the exhaust. The feedback system thus sends a rich command (shorter dwell) to the M/C solenoid. A high-dwell reading indicates a rich exhaust condition, which produces a lean command to the mixture control solenoid. The dwell signal to the M/C solenoid is always opposite of what the oxygen sensor sees in the exhaust:

- *Rich fuel mixture = Rich exhaust condition = Lean command to M/C solenoid.*
- *Lean fuel mixture = Lean exhaust condition = Rich command to M/C solenoid.*

The dwell readings remain fixed while the engine is warming up in the "open-loop" mode of operation. Dwell then begins to fluctuate once the engine is warm and the oxygen sensor is producing a signal. The computer then goes into "closed-loop" operation and starts to vary the dwell signal in response to input from the O_2 sensor.

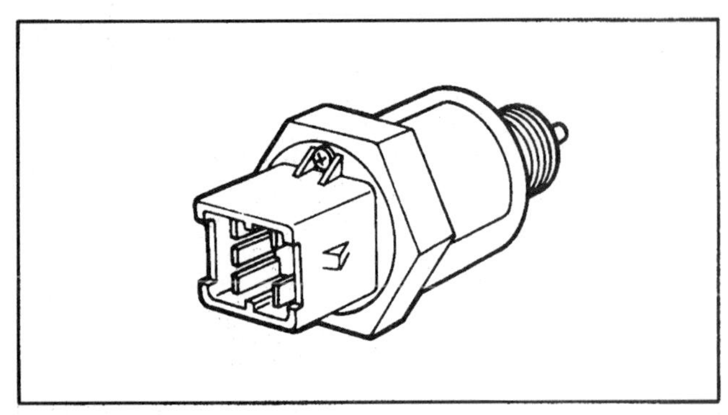

Feedback Control Motor

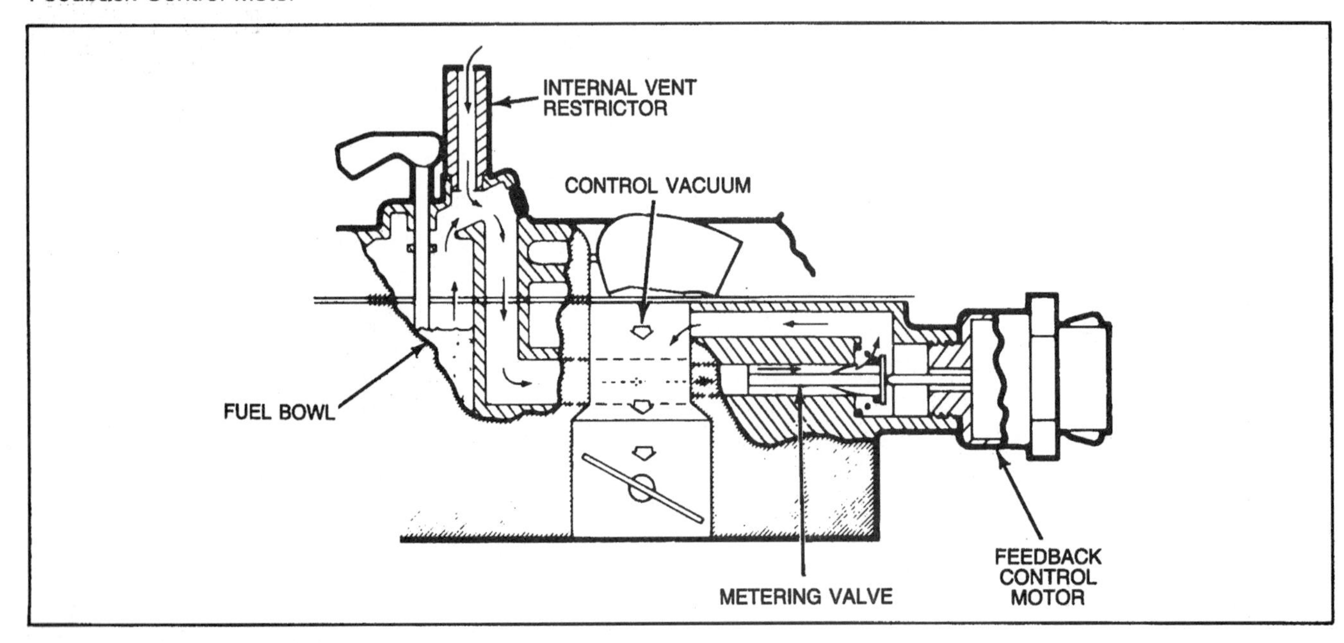

Fig. 17-6 Controlled air bleed (courtesy Motorcraft - Ford).

While cruising at highway speeds, the M/C dwell should hover around 35 degrees. At wide-open throttle, the computer will automatically enrichen the mixture, and the dwell reading will momentarily drop to a fixed 6 degrees.

EFI FUEL ENRICHMENT

On fuel-injected engines, the computer enrichens the fuel mixture by lengthening the duration of the injector pulses. A longer pulse delivers more fuel and richens the mixture. A shorter pulse duration leans the mixture. As with carbureted engines, the duration of the injector pulses will be fixed until the O_2 sensor reaches operating temperature and starts producing a signal. At that point, the system will go into closed-loop operation with the feedback circuit controlling the fuel mixture.

On engines with multipoint fuel injection where all the injectors are pulsed simultaneously, the engine computer maintains an "average" fuel mixture for all the cylinders. In other words, one or two individual cylinders may be running slightly rich, while another one or two may be slightly lean. But because the oxygen sensor in the exhaust reads the average oxygen content for the exhaust from all the cylinders, the best the system can do is average out the mixture for all the cylinders.

On an engine that has sequential fuel injection, each injector is driven separately and pulsed in a sequence that corresponds to the ignition firing order. In this type of application, the computer may have the ability to increase or decrease the duration of individual injectors as needed to maintain a balanced fuel mixture.

Injector dwell can be viewed on an oscilloscope using the millisecond sweep function. The duration should vary with engine speed and load if the feedback loop is functioning correctly (Fig. 17-7).

THE O_2 SENSOR: THE MAIN FEEDBACK CONTROL

Oxygen sensors are covered in Chapter 15, but to summarize briefly: an oxygen sensor is a kind of miniature generator. The zirconium dioxide element in the sensor's tip produces a voltage that varies according to the amount of oxygen in the exhaust. The greater the amount of oxygen, the lower the sensor's output voltage. Sensor output ranges from 0.1 volts (lean) to 1.0 volts (rich). A perfectly balanced or "stoichiometric" fuel mixture 14.7:1 gives a mid-range sensor reading of about 0.5 volts. To produce this signal, the sensor must be hot (over 600 degrees F.). Sometimes the sensor will cool off enough at idle to stop producing a signal, so the engine may have to be revved to bring it back up to operating temperature.

Many O_2 sensors have three wires and an internal heating element to help the sensor reach operating temperature more quickly, and to maintain that temperature should the exhaust cool (as when idling).

Some O_2 sensors have a titania element that changes resistance in response to exhaust oxygen content, which causes the input voltage signal to change.

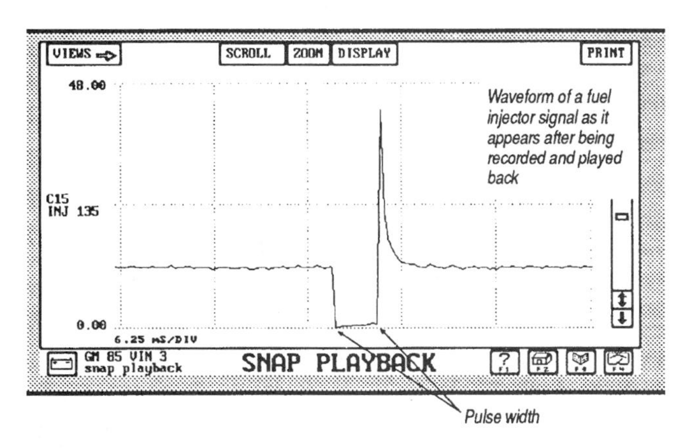

Fig. 17-7 Fuel injector waveform (courtesy Edge Diagnostic Systems)

Regardless of the type of oxygen sensor, a good one will produce a fluctuating signal that changes almost immediately in response to the oxygen level in the exhaust. Using a scan tool to divert air from the air pump through the air solenoid into the exhaust manifold should cause the sensor's voltage to momentarily drop (read lean). Artificially enrichening the fuel mixture by holding the choke shut or restricting the air intake should cause the O_2 sensor's voltage to go to maximum (rich).

The oxygen sensor's response time deteriorates with age. Normal life span is 30,000 to 50,000 miles. The sensor may fail prematurely, however, if it becomes clogged with carbon, contaminated by lead, or contaminated by silicone vapor drawn through the PCV system from using the wrong type of RTV silicone gasket compound. If the sensor is getting old or goes dead, it will usually trigger the "Check Engine" light, set a trouble code in the computer memory, and cause the system to run rich in open loop.

Testing the Feedback Loop

There are two ways to test the response of an O_2 sensor in the vehicle. One is to check its voltage output in response to changing lean/rich fuel conditions. If the sensor's output fails to change or is sluggish, then it needs to be replaced. The other method is to check the onboard computer for an O_2 sensor trouble code, then to follow the step-by-step diagnostic procedure as outlined in the vehicle manufacturer's shop manual.

The best way to read the sensor's voltage is through the onboard computer system with the scan tool. Another alternative is to use an O_2 sensor tester that uses LED lights to give a rich-lean indication. The other alternative is to use a 10-megohm impedance digital voltmeter to read the sensor's output voltage at the sensor's electrical connector.

Though you can read the O_2 sensor's output voltage directly on most later-model applications, there are some you can't. On older GM T-car systems, for example, you can't get a voltage reading, but only a number that corresponds to the number of times the sensor goes from rich to lean. GM calls this "cross counts," and it should be under 40 if the sensor is good. On some older Chrysler systems, all you can get is a relative rich or lean indication from the sensor. You also can't read the O_2 sensor voltage directly on Ford MCU or EEC systems because Ford does not provide a data output stream. The same for most imports—unless you tap into the computer's wiring harness with an interface tool that allows you to read the system's live data stream.

An analyzer or scope can also be used to view the O_2 sensor's output voltage directly (Fig. 17-8). But unless the information is provided through the vehicle's diagnostic connector, you'll have to tap into the computer harness to get the live data stream readings.

When viewing the oxygen sensor's output, make sure you run the engine at fast idle (1500 to 2000 rpm) for several minutes to burn off any accumulated carbon on the sensor tip and to bring the sensor up to operating temperature.

To test the sensor's response to oxygen, introduce a vacuum leak by pulling off a large vacuum hose (the vacuum hose to the power brake booster works well). The O_2 sensor readings should change rapidly and drop below 500 millivolts (0.5 volts) if the sensor is good. Replace the hose and create an artificially rich mixture by partially closing the choke (on fuel-injected cars, partially block the air intake). The voltage readings should now jump back above 500 millivolts.

If you have an analyzer or scope that can also display the carburetor mixture control dwell or injector duration in milliseconds at the same time, note how these values change in response to changes in output from the oxygen sensor. When the oxygen sensor reads rich, you should see a corresponding decrease in carburetor or injector dwell, as the computer attempts to compensate by leaning the fuel mixture. And when the oxygen sensor reads lean, you should see an increase in carburetor or injector dwell, as the computer compensates by increasing fuel flow.

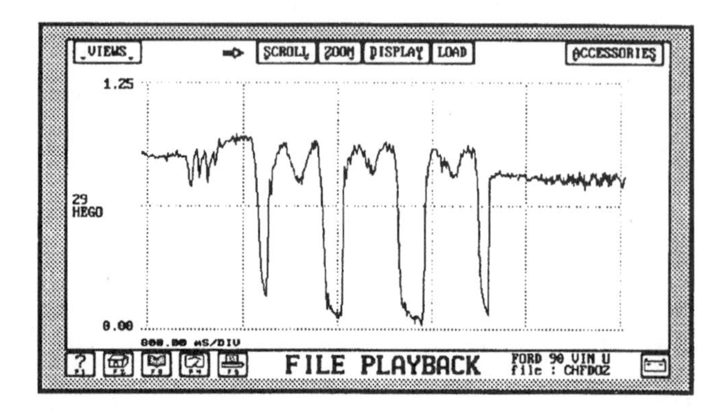

Fig. 17-8 O_2 sensor signal waveform (courtesy Edge Diagnostic Systems)

Another method for testing the feedback system is to use a special tester that "simulates" a good oxygen sensor signal (a sensor simulator). The tester is connected to the oxygen sensor's wiring connector and used to provide either a rich or lean signal to the computer. If sending a rich signal to the computer doesn't lean out the fuel mixture (or a lean signal doesn't cause a corresponding increase in fuel delivery), then it indicates a problem in the wiring between the oxygen sensor and computer, or the computer itself. If the system responds normally, but there's either a fault code indicating a problem in the oxygen sensor circuit or an emissions problem, then the problem is in the sensor not the wiring or computer.

ADDITIONAL INPUTS

In addition to the basic feedback loop between the oxygen sensor, computer, and carburetor or injectors, other inputs into the system can also affect the relative richness or leanness of the fuel mixture. Airflow, engine load, rpm, throttle position, and whether the vehicle is starting, accelerating, or decelerating can all influence how much fuel is needed.

When the throttle is suddenly opened, for example, a somewhat richer mixture is needed to compensate for the increased airflow. With a carburetor, the extra shot of fuel is provided by the accelerator pump. But with fuel injection there is no accelerator pump. So the computer has to decide when extra fuel is needed and how much, based on its various sensor inputs.

During deceleration, many fuel injection system are programmed to temporarily cut off fuel delivery as a means of increasing fuel economy. The computer has to be able to determine when the vehicle is decelerating by monitoring inputs from the throttle position sensor, MAP sensor, vehicle speed sensor, and/or airflow sensor. When conditions are "right" and the computer determines the vehicle is decelerating, it turns off the injectors, until it determines fuel is again needed to keep the engine from stalling.

When an engine is pulling a heavy load or accelerating, extra fuel may be needed to improve performance and to resist detonation. Again, the computer will rely on other inputs to determine if and when the fuel mixture needs adjusting to compensate for changing driving conditions.

Bosch L and LH-Jetronic Fuel Injection

To better understand how all these inputs affect the basic feedback loop and the computer's ability to manage fuel delivery, let's look at a couple of fuel injection systems starting with Bosch L and LH-Jetronic, which are found on many domestic as well as import applications.

In the basic L-Jetronic system, airflow is measured mechanically with a spring-loaded flap and potentiometer. The flap rotates when incoming air pushes against it. The amount of deflection is proportional to the volume of airflow. The rotation of the flap slides an electrical wiper contact across a resistor grid (the potentiometer). This changes a voltage signal to the injection system control box. The greater the airflow, the greater the deflection of the flap, and the higher the resistance created by the potentiometer. A dropping voltage signal tells the control box more fuel is needed. The computer then increases fuel delivery by lengthening the "on time" or duration of the injector pulses. The further the flap in the airflow meter is pushed open, the more the duration of the injector pulses are lengthened.

The LH-Jetronic systems use an air mass sensor that works on a different principle. Here, airflow is measured electronically. What's more, the sensor reads air mass rather than volume. Mass takes into account both volume and density. Air density changes with temperature (cold air is denser than warm air) and barometric pressure (air density thins with increasing altitude). A reference voltage is applied to a thin wire inside the sensor that heats it to 100 degrees C. hotter than ambient air temperature. As air flows through the sensor and past the hot wire, it carries away heat and cools the wire. The electrical control circuit for the wire is designed to maintain a constant temperature differential. So the amount of extra voltage that's required to offset the cooling effect and keep the wire hot tells the control box how much air is entering the engine. Injector duration can then be increased the appropriate amount to keep the air/fuel mixture in balance.

The accurate measurement of airflow or air mass is absolutely critical with fuel injection. The control electronics must know precisely how much air is entering the engine in order to adjust injector duration to maintain a balanced fuel mixture. Obviously any air leaks downstream of the airflow meter or air mass sensor will allow "unmetered" air to enter the engine. The extra air will lean out the fuel mixture, which can adversely affect starting, idle, emissions, and performance. A leaky vacuum hose, loose air hose clamps, a leaky manifold gasket, or even leaks around the injector O-rings can all be sources of "false" air.

The airflow meter flap in older versions of L-Jetronic incorporate a spring-loaded backfire valve to protect the flap against damage should the engine hiccup. Even with this protection, a severe backfire can bend or deform the flap causing a severe imbalance in the air/fuel mixture. In some cases, the backfire valve itself can be a source of air leakage if it doesn't seat tightly. On later versions, the backfire valve is replaced by a redesigned flap that allows the pressure of a backfire to blow past it.

The flap-type airflow meters should always be inspected by pushing the flap with your finger. There should be no binding when the flap is pushed open, and spring pressure should return it to its closed position. On the earlier L-Jetronic systems, the fuel pump contact switch is also located in the airflow meter. When the ignition key is on, the pump is energized when the air flap is pushed open.

The L-Jetronic airflow meter can also be checked out with an ohmmeter by unplugging the wiring harness and checking resistance between the various numbered contacts (Fig. 17-9). LH-Jetronic air mass sensors can be checked by reading voltages and comparing the readings against the specs listed in the manual.

Compensation Inputs

Additional "compensation inputs" are needed to fine-tune the fuel mixture for changes in temperature, engine load, and speed. In the L-Jetronic applications, changes in air density are measured by an inlet air temperature sensor (usually referred to as "Temperature Sensor I" in Bosch service literature). The resistance of the sensor decreases as air

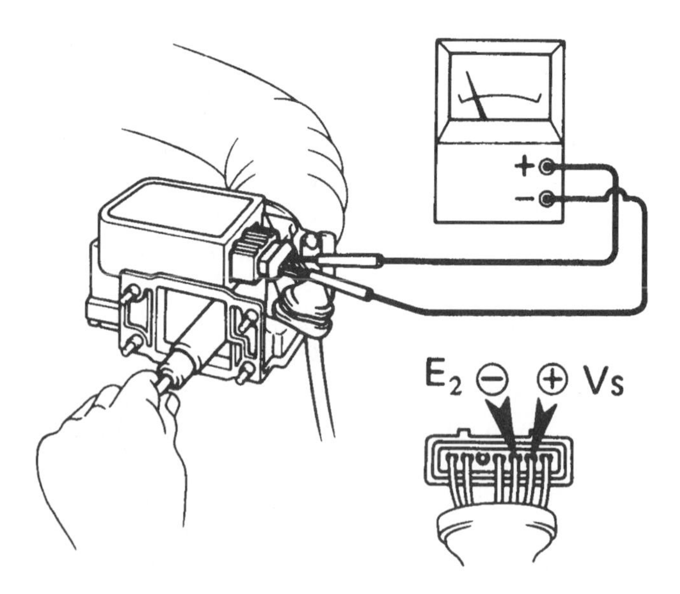

Fig. 17-9 VAF resistance check (courtesy Toyota).

temperature rises. Cold air, therefore, triggers an increase in injector duration time.

The engine temperature or coolant sensor ("Temperature Sensor II") tells the computer when the engine is cold. The engine requires a richer fuel mixture while it warms up. The resistance of the coolant sensor also drops as the engine warms up. Resistance should read 2,000 to 3,000 ohms when cold (up to 65 degrees F.), and 270 to 390 ohms when hot (176 degrees F.).

A "throttle valve switch" signals the control box when the throttle is at idle (closed) and at wide open. The engine requires additional fuel during acceleration, so the throttle valve switch helps the control box determine when extra fuel is needed.

The throttle valve switch also helps low fuel consumption and emissions during deceleration. When the control box receives a closed-throttle signal above a certain engine rpm, it assumes the engine is coasting down. The injectors are momentarily switched off and no fuel is injected until the throttle is reopened.

Cold Starting

Because a cold engine needs a rich fuel mixture to start, many L-Jetronic and some early LH-Jetronic systems have an additional fuel injector in the intake manifold called a "cold-start valve" or CVS. The cold-start valve sprays extra fuel into the engine for a short period of time (usually 2 to 8 seconds) as determined by a "thermo-time switch."

The thermo-time switch is yet another temperature sensor, but one that is independent of the control computer in such applications. It is energized when the ignition switch is turned on, and completes the circuit to the cold-start valve. The number of seconds the circuit remains closed to provide cold-start fuel enrichment depends on how quickly the electrically heated bimetal contact inside the thermo-time switch heats up. The warmer the engine, the more quickly the switch opens and cuts off the cold-start fuel enrichment valve.

Starting problems can often be traced to the cold-start valve and thermo-time switch. Checking for battery voltage with a 12-volt test light or voltmeter when the engine is started will tell you if the cold-start valve and thermo-time switch are being energized, and for how long. Sometimes a cold-start valve will leak. If squeezing off the valve's fuel supply hose smooths out a rough idle, the valve is leaking an needs to be replaced. Clogging can also be a problem. The valve can be checked by removing it from the manifold to see if it sprays fuel when energized. Position the valve in a suitable (nonbreakable) container and energize by jumping it off the positive ignition coil terminal with a fused jumper wire (Fig. 17-10). Make the final jumper connection at the coil to keep any sparks away from the fuel mist! If the valve fails to spray, or drips when off, it needs to be replaced.

In later versions of LH-Jetronic, troubles with the cold-start valve and thermo-time switch have been eliminated by incorporating the start-up fuel enrichment into the control electronics. Extra fuel

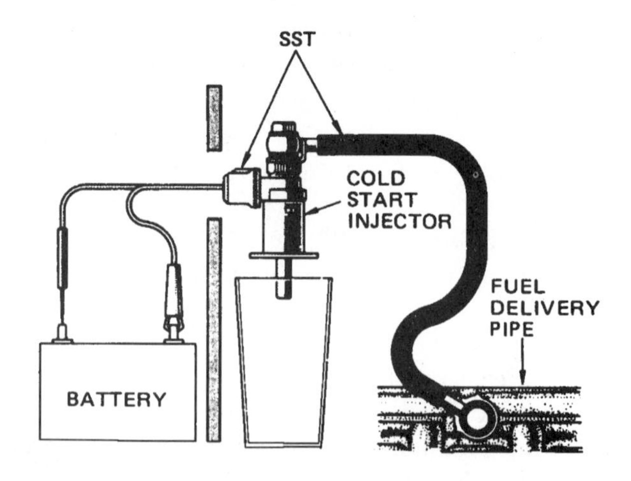

Fig. 17-10 Testing cold-start injector (courtesy Toyota).

is provided during cranking and warm-up by lengthening the duration of the injector pulses.

MECHANICAL FACTORS THAT INFLUENCE COMPUTER CONTROL OF FUEL DELIVERY

To accurately control the amount of fuel sprayed each time an injector opens, a "fuel pressure regulator" is used on the fuel rail to maintain a consistent pressure differential across the tip of the injectors. The regulator (which is a purely mechanical device and is not linked to the computer) has a spring-loaded diaphragm and valve that reacts to changing manifold vacuum by bleeding off fuel pressure. When manifold vacuum is high, less pressure is needed to spray the same amount of fuel into the manifold as when vacuum is low (Fig. 17-11). So the valve in the regulator is pulled open under vacuum to return excess fuel pressure to the fuel tank through a separate return line. When the throttle is floored, more fuel pressure is needed and the regulator closes, allowing fuel pressure to rise. The computer has no direct control over the operation of the regulator. But, it's fuel calibration programming takes the pressure changes into account when it calculates how much fuel is needed under various operating conditions.

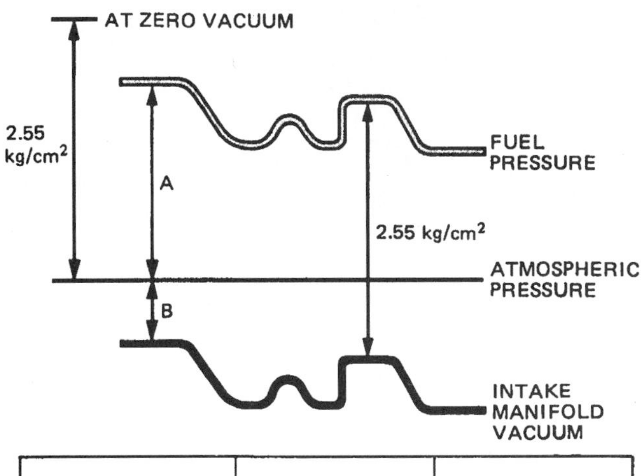

FUEL PRESSURE	LOW	HIGH
INTAKE MANIFOLD VACUUM	HIGH (LOW PRESSURE)	LOW (HIGH PRESSURE)
INJECTION VOLUME	(SAME)	(SAME)

Fig. 17-11 Presure vs. vacuum (courtesy Toyota).

Injector problems can also influence the fuel mixture, which in turn can cause the feedback loop to make adjustments in an attempt to compensate for too much or too little fuel. Dirty injectors won't deliver as much fuel as clean ones, causing the affected cylinders to run lean. This, in turn, can cause the computer to increase injector dwell in an attempt to get more fuel into the engine. Leaky injectors, on the other hand, will create a rich condition by allowing extra fuel to enter the engine. The oxygen sensor will read the rich condition, and the computer will try to compensate by leaning out the mixture.

Bosch KE-Jetronic Fuel Injection

The Bosch KE-Jetronic system is a third-generation version of it's K-Jetronic system. It is used on 1984 and newer Volkswagen, Audi and Mercedes models. Unlike L and LH-Jetronic, KE-Jetronic is a continuous injection system that uses a fuel distributor and mechanical injectors rather than electronic injectors. But KE-Jetronic has an electronic rather than mechanical airflow sensor as well as electronic control over the fuel distributor.

The air/fuel mixture is controlled by a "differential pressure regulator" (also called an "electro-hydraulic pressure actuator"). This device (Fig. 17-12) takes the place of the "frequency valve" that was used in the earlier K-Lambda Jetronic systems. The pressure regulator is a small plastic box mounted on the side of the fuel distributor. It varies the control pressure in the fuel distributor to make the fuel mixture richer or leaner. Inside are two electromagnets that play tug-o-war to deflect a valve plate. The deflection of the plate varies fuel pressure between the upper and lower chambers inside the fuel distributor, which varies the rate of fuel delivery to the injectors. But before we go any further, let's take a look at how fuel is managed in the KE-Jetronic system.

To manage fuel delivery, the KE-Jetronic electronic control module (ECU) needs information on the relative richness or leanness of the fuel mixture so it can maintain a balanced mixture for efficient combustion. This vital bit of information is provided by the "Lambda sensor" (oxygen sensor) in the ex-

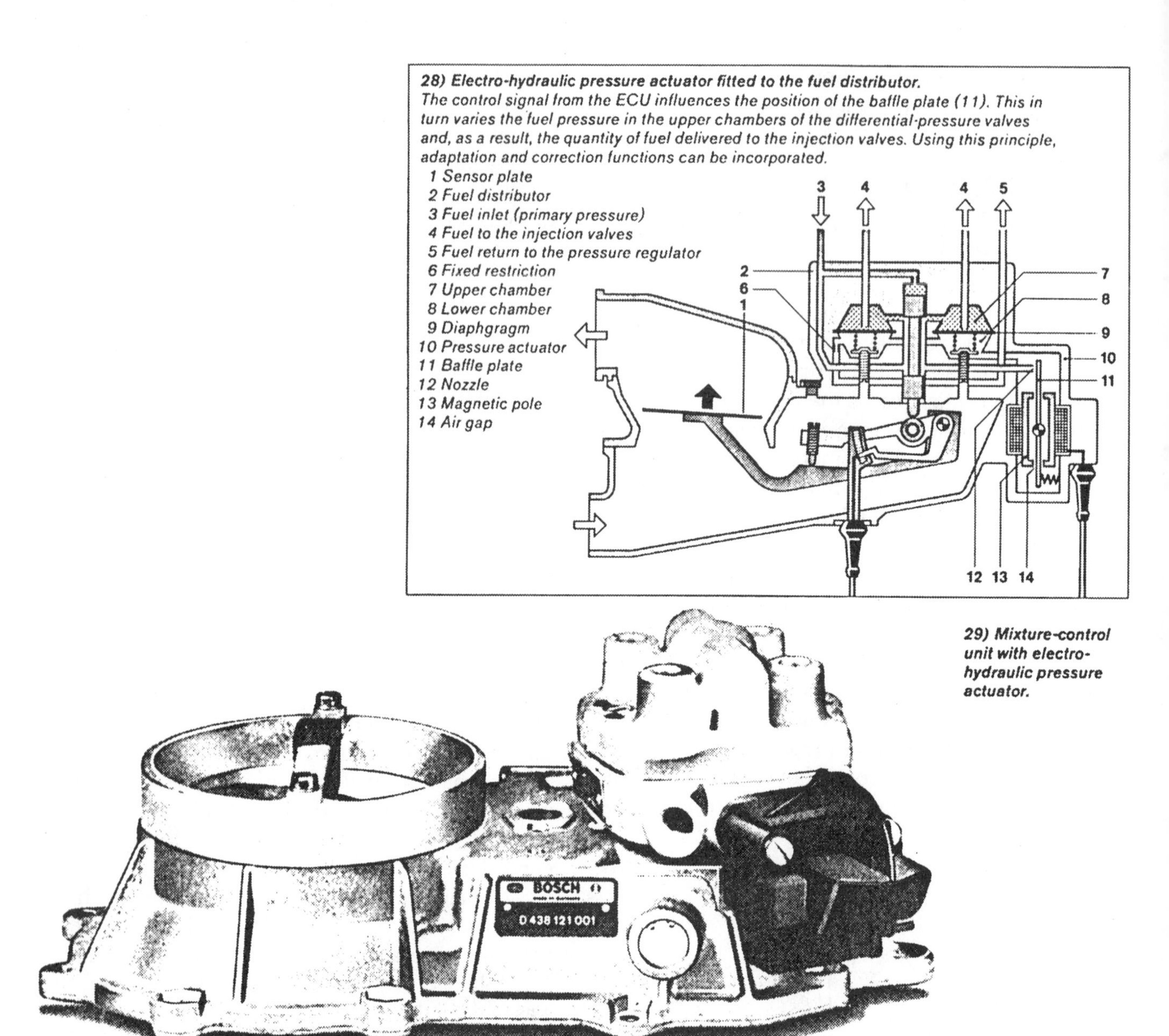

Fig. 17-12 Pressure actuator (courtesy Robert Bosch).

haust manifold. (Lambda, by the way, is the Greek letter that's used to represent a "stoichiometric" or balanced air/fuel mixture of 14.7:1.)

The Lambda sensor works the same as any other oxygen sensor in this application, generating a voltage signal that is proportional to the amount of unburned oxygen in the exhaust. This gives the computer an indication of the relative richness or leanness of the fuel mixture so corrections can be made to maintain a closely balanced fuel mixture. Corrections are made when the ECU receives a rich or lean indication from the oxygen sensor. The ECU changes its current signal to the differential pressure regulator, which in turn changes the pressure inside the lower chamber of the fuel distributor to alter fuel flow to the injectors.

If the oxygen sensor reads lean, for example, the ECU increases current to the differential pressure regulator. This increases the differential pressure in the fuel distributor, which increases fuel flow to richen the mixture. The opposite happens when the oxygen sensor reads rich. The ECU lowers and/or reverses the current to the pressure regulator, which lowers the differential pressure to decrease fuel flow to the injectors.

The ECU constantly alters the relative richness

or leanness of the fuel mixture by changing both the strength and direction of the current to the differential pressure regulator.

Temperature Compensation

By constantly adjusting the fuel mixture, the ECU delivers the right amount of fuel under a wide variety of operating conditions. A cold engine, for example, idles best with a rich fuel mixture. But a rich mixture when the engine is warm increases emissions and can make it idle rough. Most KE-Jetronic systems have a cold-start valve for initial fuel enrichment when the engine is cranked, but there's no warm-up regulator. Temperature compensation is provided via the "coolant temperature sensor."

When the coolant sensor gives a cold indication to the ECU, the ECU responds by making the fuel mixture richer for improved driveability. And as the engine warms up, the ECU keeps tabs on the engine's temperature and gradually leans out the mixture to reduce emissions and improve fuel economy.

The coolant sensor is the type that loses resistance as it heats up. Maximum resistance (up to 20K ohms) occurs when the sensor is cold, and minimum resistance (as low as 80 ohms) occurs when it is hot. At 32 degrees F. (zero degrees C.), for example, the coolant sensor will read around 2,500 ohms. But at 190 degrees F. (88 degrees C.), it will read around 200 ohms. The exact reading will vary depending on the vehicle application, so refer to a manual for the specifics.

CHAPTER 18

Computerized Spark Management

Computerized spark management is an important element of computerized engine control as is fuel management, emissions control, and other drive-train control functions. Prior to the advent of integrated computerized engine controls that combine the overseeing of all of these functions electronically, spark timing was regulated mechanically by the distributor's centrifugal advance mechanism and vacuum advance diaphragm. Centrifugal advance used weights and springs to advance timing as engine speed increased. Vacuum advance used intake vacuum to advance timing under light load to improve fuel economy. By changing the weights and springs in the centrifugal advance mechanism, the rate at which the ignition timing advanced (the "spark curve") could be recalibrated on a distributor machine to improve performance or reduce emissions. But a centrifugal advance mechanism could only produce a linear spark curve. It could be modified to increase timing advance at a faster or slower rate, but it couldn't change spark timing suddenly or independent of engine speed. It had to move in lock step up or down with the rpms of the engine. Vacuum advance, on the other hand, had more flexibility. The rate at which vacuum advance (or retard) was applied could be modified by varying the vacuum source (intake, ported, or venturi vacuum), by adding various delay valves, coolant temperature switches, transmission gear switches, etc., and/or by using dual diaphragm (advance and retard) distributors. But even with all these "tricks," there was only so much that could be done to manipulate spark timing to reduce emissions. A better means of controlling spark timing was needed (Fig. 18-1).

EARLY ELECTRONIC SPARK TIMING

In 1976, Chrysler introduced its Lean Burn system (Fig. 18-2) with a dedicated spark control computer hung on the side of the engine's air cleaner. For the first time, an electronic control module could modify spark timing based on various engine inputs independent of engine speed and load. The Lean Burn computer used a vacuum transducer to sense engine load, eliminating the need for a vacuum advance diaphragm in the distributor. No centrifugal advance mechanism was needed because the computer added the equivalent of centrifugal advance electronically as engine speed increased. Additional inputs came from an air temperature sensor, throttle position sensor, carburetor idle switch, coolant sensor, and dual pickups in the distributor (a start pickup and a run pickup). This allowed the Lean Burn system to optimize spark timing in ways that were never before possible.

For example (Fig. 18-3), the computer could advance ignition timing 2 degrees one millisecond, and jump it 10 degrees the next without passing through any intermediate steps in between. When the throttle was floored, the computer would rec-

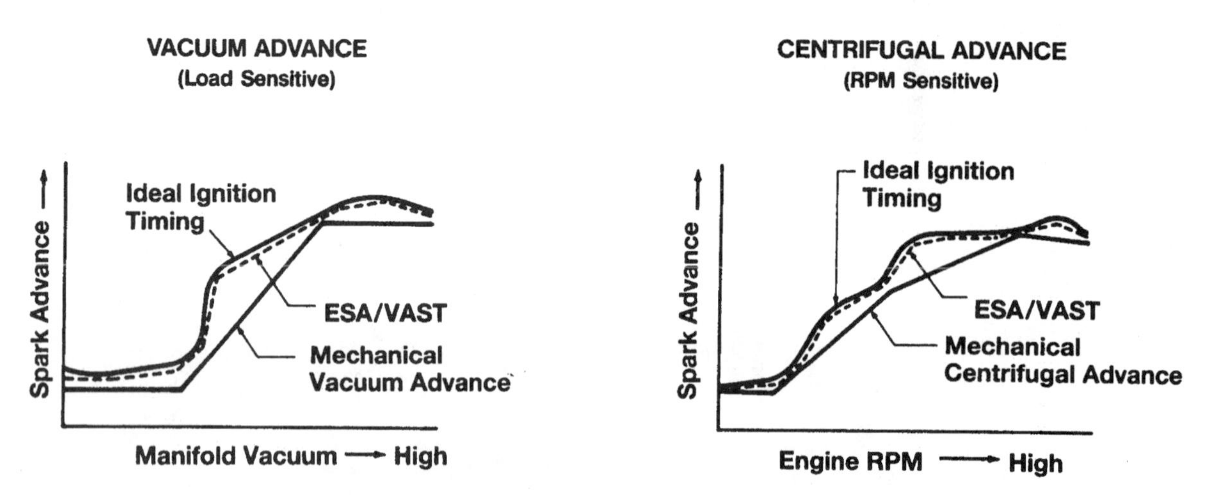

Fig. 18-1 Conventional & vacuum advance vs electronic spark advance (courtesy Toyota).

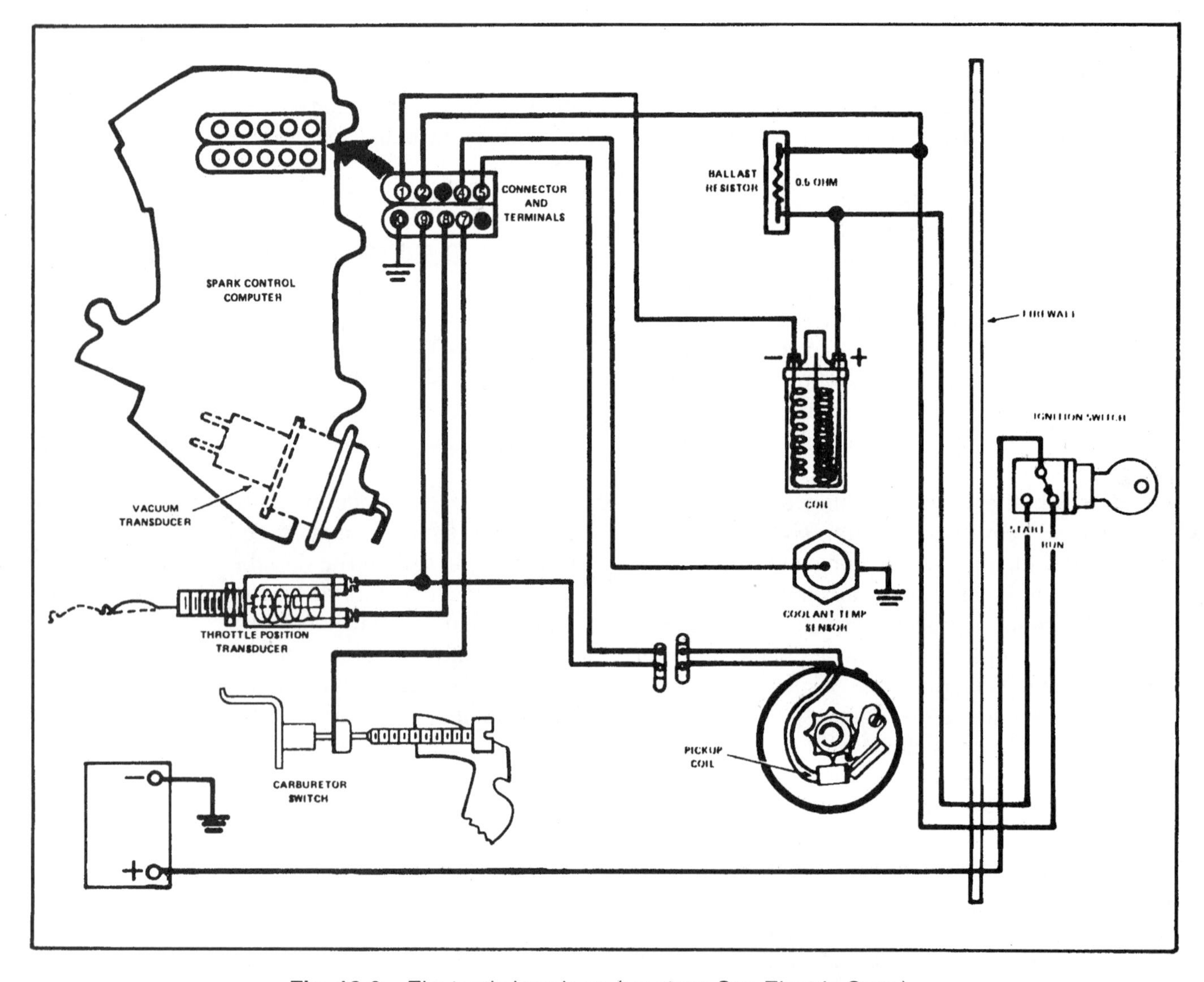

Fig. 18-2 Electronic lean burn (courtesy Sun Electric Corp.).

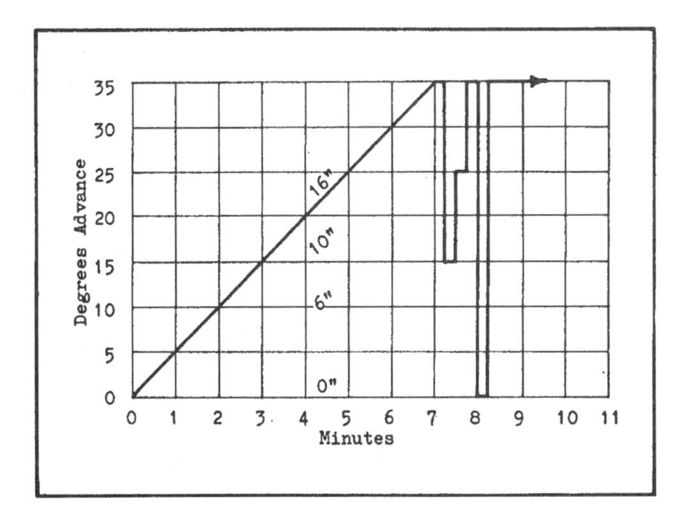

Fig. 18-3 Vacuum advance at three different engine loads (courtesy Sun Electric Corp.).

ognize the sudden change in throttle position and immediately add up to 15 degrees of additional timing advance for crisper throttle response—a trick no mechanical distributor could ever do. In fact, flooring the throttle with a mechanical distributor typically results in a momentary timing retard as the drop in intake vacuum causes the advance diaphragm to back off timing. The computer could also add more than the usual amount of timing advance when cruising under light load to improve fuel economy, then take it away if the engine came under increased load to prevent detonation.

The early Lean Burn spark control computers were troublesome due to the design of the circuit boards and the computer's rather vulnerable location directly atop the engine. But, they opened the door to computerized spark management by demonstrating the advantages of "programmed" timing over mechanical systems. Thus the die was cast, and soon other systems began to appear. In 1977, Oldsmobile introduced its MISAR spark control system. Then in 1980, technology advanced to the next plateau, and spark control was integrated with fuel management and other emissions and drivetrain control functions in the first computerized engine control systems. These were introduced on General Motors, Ford, and Chrysler-built cars in the California market. The following year, computerized engine controls went nationwide, and the rest, as they say, is history. The technology has continued to evolve with each passing model-year, with the computer gradually taking over more and more control functions. With the advent of distributor less ignition systems, the distributor was eliminated altogether, along with its maintenance requirements (cap and rotor replacement) and potential for trouble.

THE BASICS OF ELECTRONIC SPARK TIMING

Being able to manipulate spark timing electronically independent of engine speed (centrifugal advance) or load (vacuum advance) allows the engine to be "fine tuned" at any given instant in time for optimum performance and emissions. By using inputs from various sensors, a computer can determine the exact number of degrees of timing advance (or retard) that's needed for any given set of operating circumstances—and can change it in a split second if need be to keep it optimized, a feat that is impossible with mechanical controls or add-ons.

Take detonation, for example (Fig. 18-4). When an engine experiences detonation (spark knock), hydrocarbon emissions increase because the colliding flame fronts create turbulence in the combustion chamber that leaves small pockets of unburned fuel. An engine with mechanical spark advance controls has no means of coping with detonation. Overall timing can be retarded by adjusting the engine's base timing setting, or by adjusting an "octane rod" (if one is provided on the distributor). But retarding the entire timing curve to prevent detonation under severe load conditions hurts performance and fuel economy when the engine is not

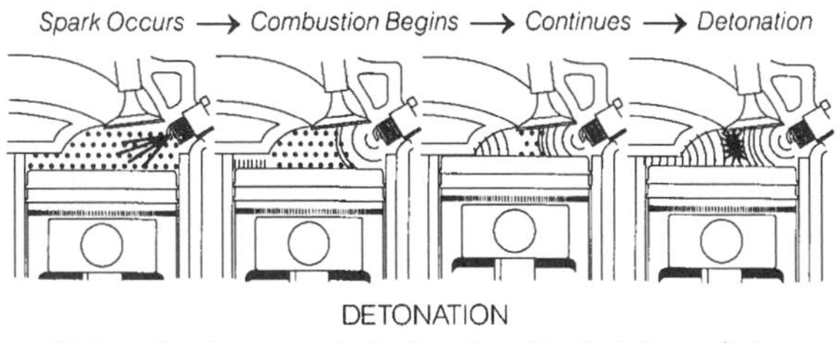

Fig. 18-4 Cylinder detonation—colliding flame fronts (courtesy Champion).

under load. Detonation typically occurs when the engine is lugging under load and combustion temperatures and pressures are highest. The vacuum advance diaphragm on a distributor will reduce timing advance when intake vacuum drops—but it may not be enough to totally prevent detonation from occurring. Consequently, the engine may still ping. So it should be obvious that mechanical controls offer little flexibility in timing control. Not so with computerized spark management.

The ability to retard ignition timing electronically when the engine is experiencing detonation (Fig. 18-5) is just one of the features that computerized spark management makes possible. Knock retard relies on input from a knock sensor. When vibrations of the right frequency are detected by the knock sensor, it signals the computer that detonation is occurring. The computer responds by momentarily retarding timing in incremental steps until the knocking stops. Then it adds back the timing unless the engine starts to ping again.

In addition to knock retard, electronic spark management allows the entire spark curve to be changed infinitely and most instantly to accommodate most any operating condition. It can be varied by: engine speed, engine temperature, engine load, turbo boost, throttle position, barometric pressure (which compensates for changes in altitude), other emissions control functions that are occurring (such as EGR and/or canister purge), vehicle speed (without the need for add-on solenoids, timers, delay valves, etc.), and/or gear position or the status of the torque converter lockup clutch.

To achieve all this, the computer receives inputs from the various sensors and status switches. Information about engine speed or rpm is provided by the distributor pickup or a crankshaft position sensor. Input on engine load is provided by the manifold pressure sensor. A separate barometric pressure sensor may provide air pressure data for altitude compensation. Throttle position is monitored via the throttle position sensor, while engine temperature is watched by the coolant sensor. The knock sensor signals the computer if detonation occurs. Communication with the powertrain control module (if there is one) provides data about what's happening inside an electronically controlled automatic transmission. From these various inputs, the computer can determine if the engine is being cranked, idling, accelerating, or decelerating. The computer processes all of this information and compares the data against its programmed map to determine the exact amount of spark advance needed (Fig. 18-6). It then makes the necessary corrections by either advancing or delaying the firing of the coil.

On some applications, timing is advanced or retarded electronically by modifying the ignition module's switching of the coil on and off. A special circuit between the computer and control module allows the computer to delay or advance the switching of the coil's primary voltage. On some applications, the module is only used to control the coil when the

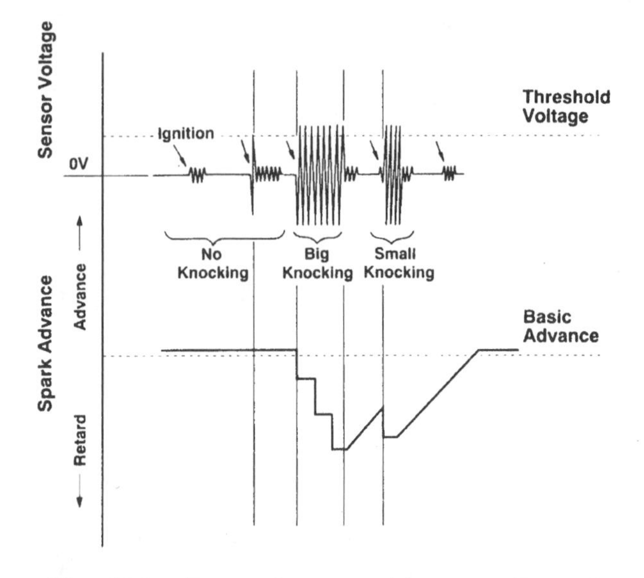

Fig. 18-5 Detonation control (courtesy Toyota).

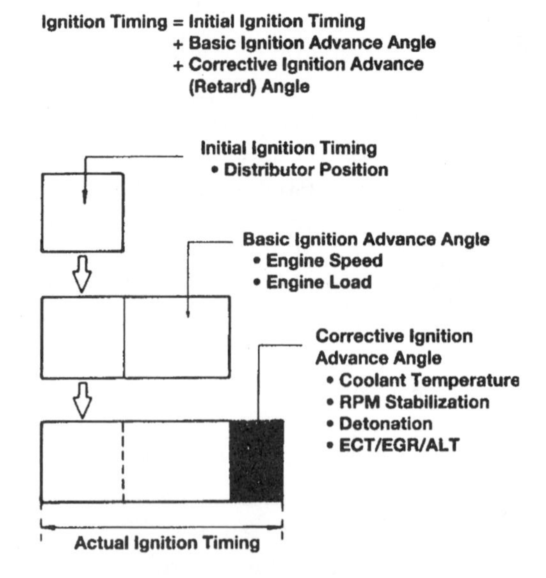

Fig. 18-6 ECU spark advance strategy while running (courtesy Toyota).

engine is being cranked. Once the engine starts, the computer bypasses the module and takes over direct control of the ignition coil. On other applications, there is no ignition module and the ignition coil is controlled directly by the computer's electronic spark timing circuitry.

GM Spark Controls

In early General Motors HEI ignitions with C3 (computer command control) and electronic spark timing (EST) or electronic spark control (ESC), the ignition module in the distributor served primarily as a signal converter for the electronic control module (ECM). The module converted the distributor's pickup voltage signal from AC to DC, which then served as the reference or "REF" signal for the ECM to regulate timing. Later versions also included a Hall effect switch to generate the main timing signal once the engine was running. The Hall effect switch was added to improve timing accuracy.

As the engine was being cranked, the module would switch the power transistor on and off to fire the ignition coil. When the computer saw a REF pulse of 200 rpm or higher for 5 seconds, it meant the engine had started, and it was time to take control of timing. The ECM would then send a 5-volt signal to the module's bypass relay to bypass the module's control circuitry so the ECM could switch the power transistor on and off directly.

On GM's distributorless ignition systems such as the computer-controlled coil ignition (C3I), the reference signal for determining engine rpm comes from a Hall effect crankshaft sensor mounted behind the crankshaft balancer (Fig. 18-7). The balancer has three sets of vanes which face the engine. As the balancer rotates, the vanes alternately pass through the Hall effect switch, producing an rpm signal that is fed to both the ignition module in the coil pack assembly and the ECM. Crankshaft position on Type I versions of the C3I system on 3.0L engines is determined by a second Hall effect switch also mounted behind the balancer. On 3.8L turbo engines with the Type I or Type II version of the system, a separate Hall effect camshaft position sensor is mounted where the distributor used to be. On nonturbo versions of the 3.8L engine, the camshaft position sensor is mounted on the camshaft timing cover and reads a vane on the cam gear. The camshaft position sensor is necessary so the computer can figure out when the number one piston is at top dead center so it can keep the firing sequence in sync with the engine. The later Type III version of the C3I system (sometimes called the "Fast Start" system) has a third Hall effect sensor mounted behind the balancer with the crankshaft sensor (Fig. 18-8). The additional sensor (called the "18X" sen-

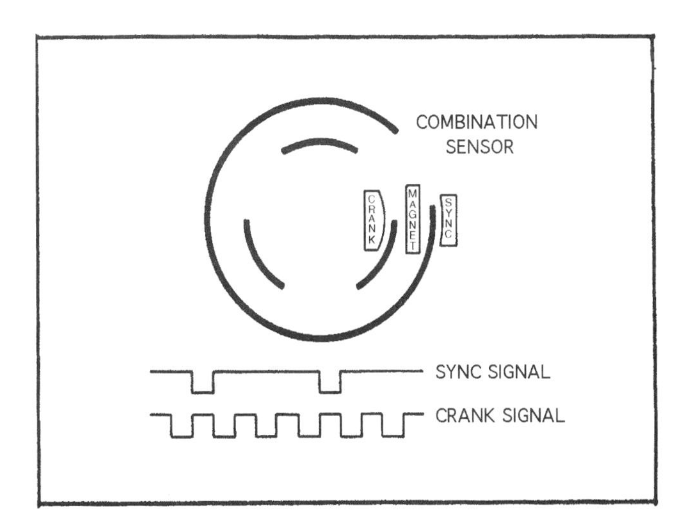

Fig. 18-7 3.0L dual crankshaft sensor (courtesy Sun Electric Corp.).

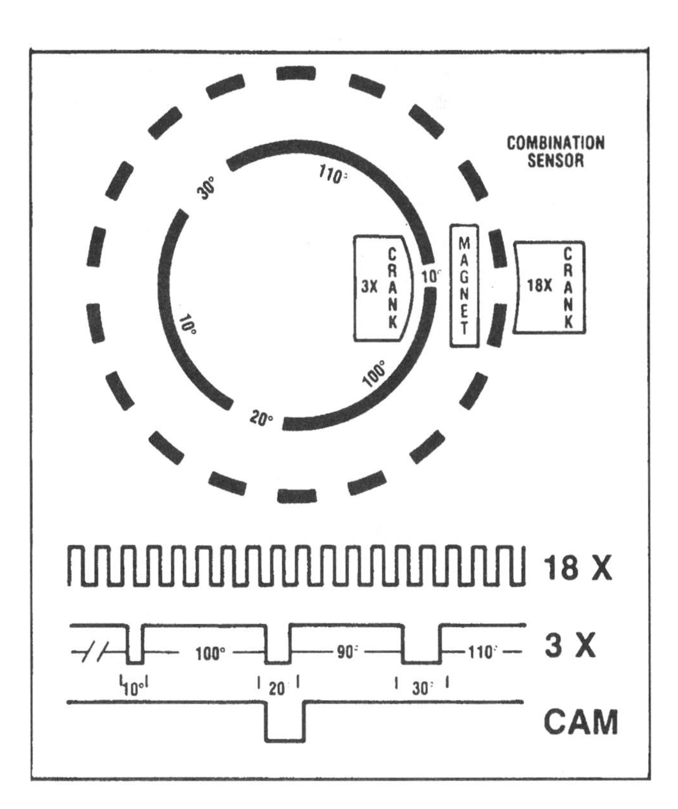

Fig. 18-8 Buick 3800 "Fast Start" crankshaft sensor (courtesy Sun Electric Corp.).

sor) allows the ECM to fire a cylinder's coil within the first 120 degrees of crankshaft rotation to provide faster starting, more accurate timing (especially at low speed), and greater reliability.

On GM's direct ignition system (DIS), a magnetic crankshaft position sensor provides the engine speed reference signal. The sensor, which is located in the side of the engine block, reads notches cut into a disk in the center of the crankshaft. The disk has 7 notches cut in it and is referred to as the "reluctor." Six of the notches are evenly spaced 60 degrees apart, while the seventh notch is located between notches 1 and 6, offset by only 10 degrees from notch 6. The odd position of the seventh notch, plus its close proximity to notch number 6, produces an usually high reference pulse (which GM calls the "cylinder event" pulse) once every revolution. This serves as a reference signal for the coil module so it can synchronize the coil's firing sequence with the position of the crankshaft.

One point worth noting about GM's DIS system on 4-cylinder engines is that the system doesn't fire every time a notch passes by the crankshaft position sensor. Because the notches are spaced 60 degrees apart, the coils only fire on every third notch. The spark control module skips the signal from the notch 6 and 1, then fires the 2-3 coil on notch 2, then skips notches 3 and 4, and fires the 1-4 coil on notch 5. On 6-cylinder applications, the module skips every other notch signal. It ignores notch 1, fires coil 2-5 on notch 2, then skips notch 3 and fires coil 3-6 on notch 4, then skips notch 5 and fires coil 1-4 on notch 6.

GM's integrated direct ignition (IDI) system, which is found on the Quad Four engine, uses a similar control setup as the DIS system, with a magnetic crankshaft position sensor and a seven-notched crankshaft reluctor. But the IDI system has a coil and module assembly that sits on top of the engine directly over the spark plugs. There are no spark plug wires because the IDI assembly attaches directly to the spark plugs.

GM's Opti-Spark ignition system, which was introduced in 1992 on the Corvette, uses light-emitting diodes (LEDs) in the distributor and a slotted disk to generate the timing reference and sync signals. Electronic spark timing is controlled in much the same way as on the other GM systems. The ECM looks at the signals from the distributor along with its other sensor inputs, figures out the correct timing that's needed, then switches the coil on and off as needed to advance or retard timing.

Ford Spark Control

Ford's electronic engine control IV (EEC-IV) system uses a Hall effect switch in the distributor (which Ford refers to as the profile ignition pickup or "PIP" sensor) to generate a reference signal for both engine speed and crankshaft position. One of the vanes that passes through the PIP sensor in the distributor is narrower than the others. This produces a shorter signal (called the "signature signal"), which the computer can use as a sync signal to determine crankshaft position. The signature signal is actually the second cylinder in the firing order, but the computer is able to synchronize ignition timing because it knows where the signature signal falls in the firing order.

On Ford engines with EEC-IV and a thick film integrated (TFI-IV) ignition module (so named because of the type of chip on which the module circuitry is mounted), the TFI module on the base of the distributor handles the job of switching the coil on and off—but with input from the engine's computer. If the spark command fails to arrive from the computer, the TFI module will continue to open and close the coil's primary voltage in response to the PIP signal, but with no timing advance.

The computer receives the PIP signal from the Hall effect switch in the distributor by way of the TFI module. From this signal it determines engine rpm and crankshaft position. The computer then calculates the amount of timing advance needed according to the PIP input and other sensor inputs, and sends a "spark output" (SPOUT) command signal to the TFI module to control timing. The SPOUT signal opens the primary coil circuit and fires the coil.

In most TFI-IV ignition systems (Fig. 18-9), ignition dwell (the duration of coil on time) is con-

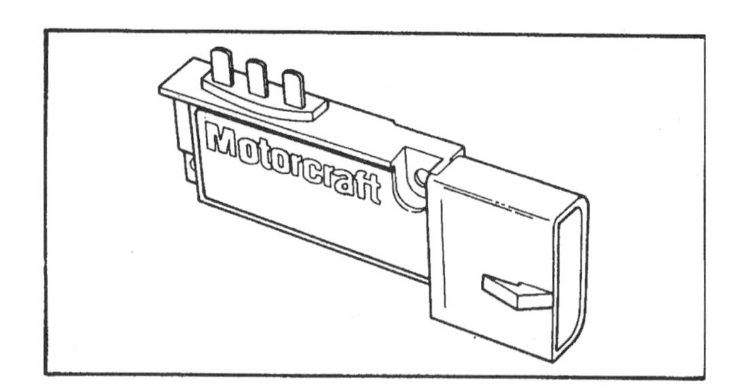

Fig. 18-9 Thick film ignition (TFI-IV) module (courtesy Motorcraft-Ford).

trolled by the TFI module. The TFI module decides when to turn the coil's primary voltage on. The computer signals the TFI module with its SPOUT signal when to turn the coil's primary voltage off (firing the coil). The TFI module increases dwell as engine speed increases. Beginning in 1989, however, some applications changed to computer-controlled dwell. In these applications, the trailing edge of the SPOUT signal tells the TFI module when to close the primary circuit, and the leading edge of the SPOUT signal tells the TFI module when to open the primary circuit.

In 1989, Ford introduced its distributorless ignition system (DIS) on the 3.8L supercharged Thunderbird engine, the 3.0L Taurus SHO engine, and 2.3L Ranger engine. The DIS system has one ignition coil for each pair of cylinders. When the coil fires, it produces a spark when one cylinder is at the top of its compression stroke and a "waste spark" when the other cylinder is at the top of its exhaust stroke. The computer determines how much spark advance is needed and tells the DIS module when to fire the coil, and the DIS module decides which coil to fire. But to choose the correct coil, the DIS module has to know the position of the crankshaft so it can determine which cylinder is approaching TDC.

On the DIS applications, the PIP sensor (Fig. 18-10) is located on the front of the engine just behind the crankshaft pulley and hub assembly. A vaned cup on the back of the pulley triggers the Hall effect switch. Because the cup is mounted on the crankshaft, only one vane is used for each pair of cylinders. As the vanes pass through the Hall effect PIP sensor, a voltage signal is sent to the ignition module and computer. The vanes are positioned on most engines so that the PIP signal will go high each time a piston reaches 10 degrees BTDC. The PIP signal serves as the rpm reference signal, but does not tell the DIS module about the firing position of the crankshaft. That job is handled by a separate cylinder identification (CID) sensor. The CID sensor is mounted on the right end of the rear head, and is driven by one of the overhead cams on the 3.0L SHO engine. On the 3.8L applications, the CID sensor is mounted where the distributor used to be, and is driven by the camshaft. On the 2.3L engine, though, the CID sensoı is combined with the PIP sensor, and is mounted behind the crank pulley and nub assembly.

On 2.3L engines with DIS and two spark plugs per cylinder, the system has two separate coil packs: one on the right-side of the engine and one on the left side. The right coil pack contains coils 1 and 2. Coil 1 fires the right-side spark plugs in cylinders 1 and 4. Coil 2 fires the right-side spark plugs in cylinder 2 and 3. Coil 3 on the left side of the engine fires spark plugs in cylinders 1 and 4, while coil 4 fires the plugs in cylinders 2 and 3. When the engine is cranked, only the right-side plugs fire. But once the engine starts, both plugs in each cylinder are fired simultaneously. To disable the left coil pack when cranking, the computer sends a 12-volt signal to pin 6 of the DIS module. This is called the "dual plug inhibit signal." When pin 6 has 12 volts, the DIS module does not ground coils 3 and 4. But once the engine starts, the computer removes the 12-volt signal to pin 6 and the left coil pack starts to fire along with the right coil pack.

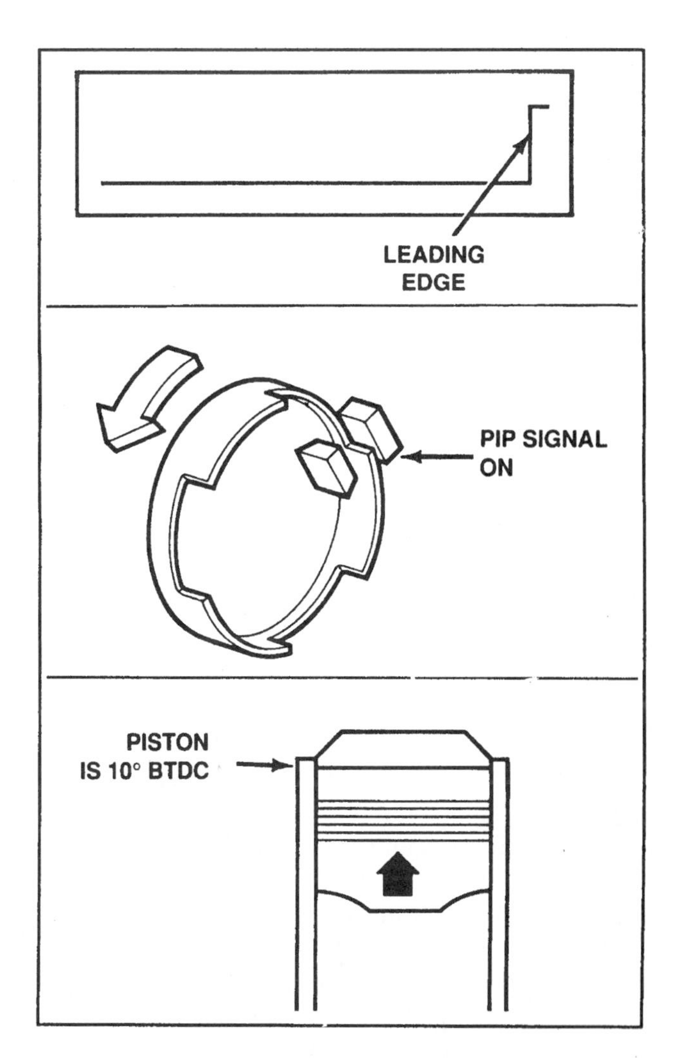

Fig. 18-10 PIP leading edge 10 BTDC (courtesy Motorcraft-Ford).

Chrysler Spark Control

Chrysler's electronic fuel injection systems from 1984 up through 1987-88 were controlled by two separate modules: a "logic" module in the passenger compartment that contained the main microprocessor, and a "power" module in the engine compartment which handled the job of driving the injectors and switching the ignition coil on and off. In mid-1987, Chrysler introduced its single module engine control (SMEC) system where the logic and power modules were combined.

A Hall effect switch in the distributor generates the basic distributor reference signal that the logic module or SMEC computer needs to determine engine speed. The REF pickup, as it is know, has four blades spaced 90 degrees apart. A second Hall effect pickup in the distributor and beneath the reference pickup provides a sync signal for determining crankshaft position. The "SYNC" pickup has only a single vane that extends for 180 degrees of rotation. The leading and trailing edges of the SYNC pickup provide a signal to both the logic module and power module for timing and injector control.

The 3.0L V6 Mitsubishi engine that was introduced in 1987 in Dodge and Plymouth minivans has an optical distributor similar to Nissan's ECCS system and GM's Opti-Spark system. The distributor's reference signal and sync signals are produced by a pair of light-emitting diodes (LEDs) and a slotted disk. As light from the LEDs flashes through two sets of slits in the disk, it generates both rpm and crankshaft position signals that the computer uses for spark timing.

Chrysler's direct ignition system (DIS) (Fig. 18-11) appeared in 1990 on their 3.3L V6 engine. Chrysler's DIS system is similar to GM and Ford's systems, and uses a Hall effect sensor to monitor the positions of the crankshaft and camshaft. The coil pack contains its own ignition module, but the module varies timing according to commands from the computer. The crankshaft sensor is mounted on the bell housing and reads notches on the torque converter plate. There are four notches for each pair of cylinders. The computer uses the crankshaft sensor signal to determine both engine speed and crankshaft position. The camshaft sensor, which reads notches on one of the engine's overhead cam gears, produces a signal which is used primarily to synchronize the injectors. But it also allows faster starting by telling the computer where the engine's firing order begins, so the plugs can start to fire within one revolution when cranking.

WHEN TROUBLE STRIKES

Great as it is, things occasionally go wrong with computerized spark control systems. The computer needs accurate input data in order to control spark timing and dwell properly. So garbage in equals garbage out (GIGO). In most instances, so-called computer problems are due to faulty sensor inputs because of wiring problems or sensor failures. Most vehicle manufacturers insist that a high percentage of the computers that are returned under warranty have nothing wrong with them. Accurate diagnosis, therefore, is the key to pinpointing problems with computerized spark timing systems.

It's important to remember that the secondary side of the ignition system is still the same whether an engine has a computer or not. A bad rotor, cracks in the distributor cap, weak coil, bad plug wires, or fouled spark plugs will still result in the same kind of driveability and emissions problems. The primary side of the system, which includes the distributor pickup, ignition module, and ignition switch circuit, likewise will function (or malfunction) in the same manner whether or not a computer is telling it what to do.

The only thing the computer does is modify the instant at which the ignition coil's primary voltage is switched on and off. On applications where the coil is fired by an ignition module, the system is often capable of running without a timing signal input from the computer. It won't run well, but it will at least run. If the ignition system is misfiring, therefore, look for traditional problems in the secondary components (fouled plugs, bad plug wires, defective cap or rotor, weak coil, arcing, etc.). If the engine won't start because there's no spark, check the primary components (distributor pickup or crankshaft sensor, ignition module, coil, ignition switch, ignition circuit voltage) before condemning the computer. If the timing appears to be under- or over-advanced, resulting in spark knock, hard starting, poor mileage, etc., check base ignition timing, and the engine's sensors (manifold pressure, coolant, knock, barometric pressure, and/or throttle position sensors) to rule out "GIGO."

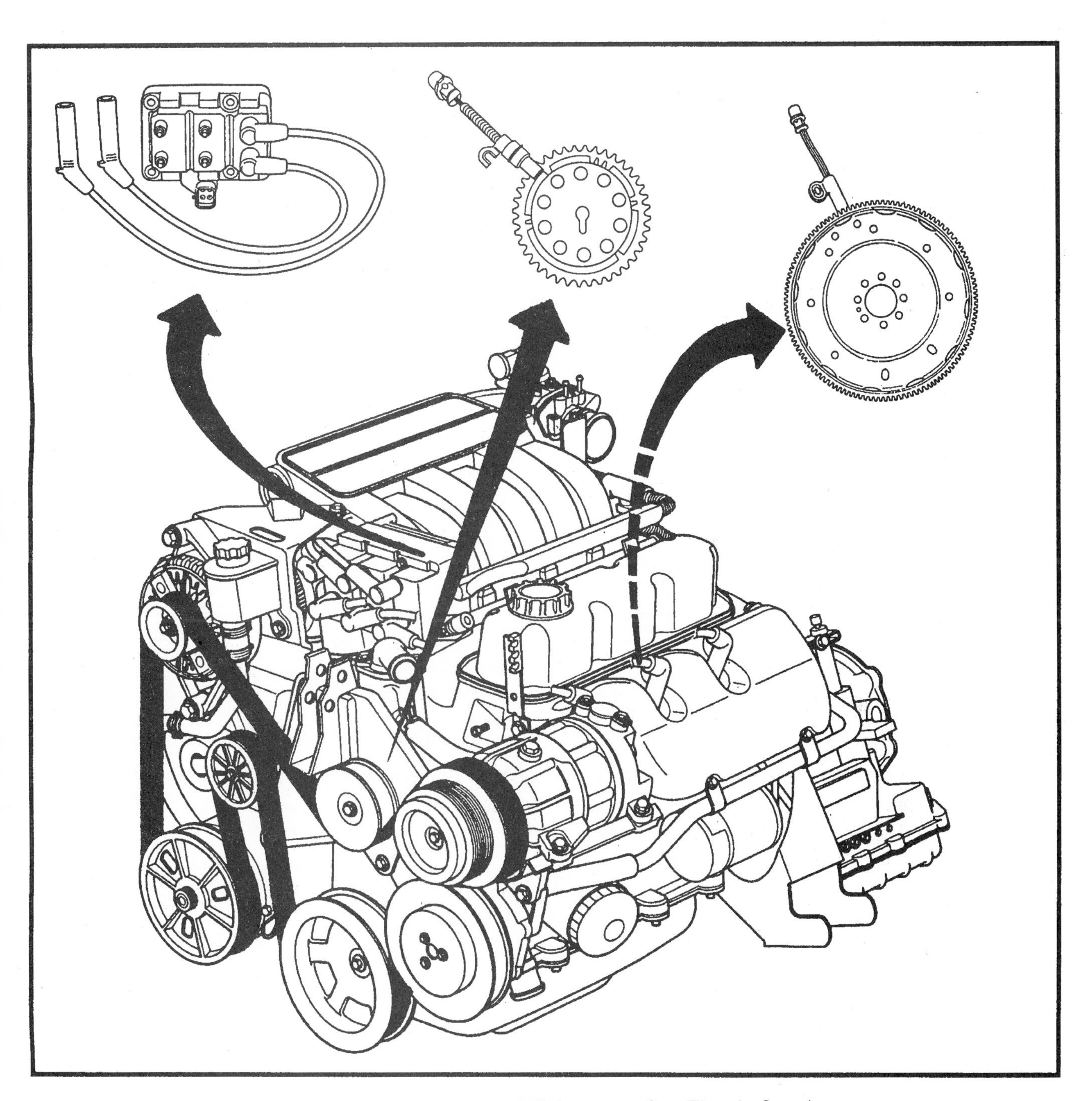

Fig. 18-11 Chrysler 3.3L DIS (courtesy Sun Electric Corp.).

CHAPTER 19

Drivability Diagnosis

When computerized engine controls were introduced, it was obvious to the people who designed them that some type of self-diagnostic capability would be necessary for troubleshooting system faults. With fuel management, spark timing, emissions control, and other powertrain functions interwoven electronically, some means of identifying specific faults within the system was absolutely necessary. Thus onboard diagnostics were born, and with it a whole new way of diagnosing and troubleshooting engine performance and emissions problems.

The basic idea behind onboard self-diagnostics is fairly simple. The vehicle's computer (or computers in applications where there are separate control modules for the transmission, antilock brake system, etc.) monitors the constant flow of inputs from its various sensors. These inputs are necessary to carry out the computer's control functions, so the computer makes sure that: (1) it is receiving data from its various inputs, (2) that the data it does receive is within acceptable norms or limits, (3) that the data received from any given sensor agrees with that from other sensors where a direct relationship exists, and (4) that its commands are being carried out via various feedback mechanisms.

As long as everything "fits" so to speak, there are no apparent problems, and the system continues to function normally. But when a fault arises such as loss of input from a vital sensor, input data that is out of range (too high or too low), input that doesn't make sense (a wide-open throttle signal with the engine idling, for example), or loss of a return signal from an actuator, it triggers the self-diagnostics to react. What happens next depends on the nature of the problem, the diagnostic capabilities of the system, and the vehicle application. In most cases where input from a sensor circuit is lost or disrupted, the computer will generate and/or store a fault code and illuminate the malfunction indicator lamp (the new "generic" term for the "Check Engine" or "Power Loss" warning light). If the sensor's input is essential for continued engine operation, the computer may go into a "limp-in" or "fail-safe" mode, where artificial data is substituted for the missing sensor data to keep the engine running. The engine may not run properly, but it will at least run.

For example, say the computer loses the signal from the throttle position sensor (TPS) or manifold absolute pressure (MAP) sensor. Without this data, the computer has no way of knowing if the engine is under load or not (a condition which typically requires a somewhat richer fuel mixture and changes in spark timing). So to compensate, the computer may substitute bogus data until the problem can be fixed. The engine, meanwhile, may hesitate under sudden acceleration or lack power under load, but it will continue to run.

Loss of input from either the coolant sensor (Fig. 19-1) or oxygen sensor will cause most systems to revert to an open-loop mode of operation. In

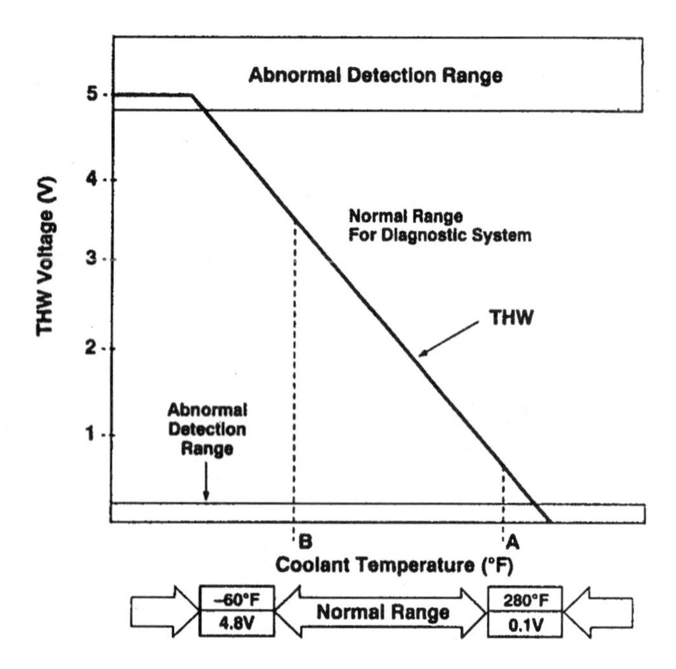

Fig. 19-1 Coolant sensor fault detection range (courtesy Toyota).

open loop, the fuel mixture will be fixed (unchanging) and will be richer than normal, causing an increase in emissions and reduced fuel economy. As before, the engine will continue to run, but won't perform as well as it should.

If the distributor reference signal or crankshaft sensor signal is lost, however, there's no way for the system to substitute data for the missing information. The distributor reference signal or crankshaft sensor signal (on engines with distributorless ignition systems) are absolutely essential for switching the ignition coil on and off (ignition firing), and injector pulsing on most applications with multiport fuel injection. Without the key rpm signal, the engine won't start or run.

Sometimes a fault occurs when the data from a particular sensor is outside the normal range of values or doesn't change or react appropriately in response to other changes that are occurring. The oxygen sensor, for example, should normally produce a signal that fluctuates between 0.1 and 0.9 volts. If the computer suddenly sees a signal that is outside this range, it will interpret it as a fault. Likewise, if the oxygen sensor signal fails to change and remains fixed at a constant high or low voltage when the engine is at normal operating temperature, the computer will interpret the data as indicating a problem with the sensor or its wiring circuit.

The exact conditions under which a fault code might be set are spelled out in many shop manuals. Referring to this information (Fig. 19-2) may help you better understand why a particular code might have been set, and what might have caused it.

That's the essence of how self-diagnostics works. There's no magic or hocus-pocus involved. The computer simply monitors the incoming data and compares the data to a list of "acceptable" values and conditions under which the data is supposed to change. If things don't match up, it logs a fault code or fault message that corresponds to the problem.

RETRIEVING CODES

Some fault codes are "hard" codes, meaning the computer stores a numeric code in its memory that corresponds to the problem. Most hard codes will cause the malfunction indicator lamp to come on and remain on, but some types of problems may only cause the lamp to come on while the problem is actually occurring. It depends on the application and how the self-diagnostics of the system is programmed. In instances where there is an intermittent problem and the lamp goes out, a code will remain in memory for later retrieval. But on many applications, the code will be erased if the problem doesn't reoccur after 50 ignition cycles. On others, though, the code will remain in the computer's memory indefinitely until it is erased by disconnecting the computer's power supply (which doesn't work if the computer has a nonvolatile memory), or signaling the computer with the appropriate input code to clear its memory.

Fault codes or messages can be read by putting the vehicle's computer into a special diagnostic mode. The exact procedure varies somewhat from application to application, but as a rule it involves grounding or jumping terminals on the system's diagnostic connector, hooking up a scan tool to the diagnostic connector, hooking up an engine analyzer with scan capabilities to the diagnostic connector, or pressing certain buttons on the vehicle's automatic climate control panel or driver information center in a specific sequence. Always refer to a shop manual if you're not familiar with the exact procedure that's required for the vehicle you're servicing.

Once the system is in the diagnostic mode, stored codes can be retrieved, various tests can be

Code	Type	Power Loss/ Limit Lamp	Circuit	When Monitored By The Engine Control Module	When Put Into Memory
37	Fault	No	Transmission Lockup Solenoid	All the time when the ignition switch is on.	If the lockup solenoid does not turn on and off when it should.
41	Fault	No	Alternator Field Control (Charging System)	All the time when the ignition switch is on.	If the field control fails to switch properly.
42	Fault	No	Auto Shutdown	All the time when the ignition switch is on.	If the relay does not turn on and off when it should.
43	Fault	No	Spark Control	During cranking only.	If the spark control interface fails to switch properly.
44	Fault	No	Fused J2	All the time when the ignition key is on.	If fused J2 is not present in the logic board of the engine control module.
46	Fault	Yes	Battery Voltage Sensing (Charging System)	All the time when the engine is running.	If the battery sense voltage is more than 1 volt above the desired control voltage for more than 20 seconds.
47	Fault	No	Battery Voltage Sensing (Charging System)	When the engine has been running for more than 6 minutes, engine temperature above 160°F and engine rpm above 1,500	If the battery sense voltage is less than 1 volt below the desired control voltage for more than 20 seconds.
51	Fault	No	Oxygen Feedback System	During all closed loop conditions.	If the O_2 sensor indicates a lean condition for more than 2 minutes.
52	Fault	No	Oxygen Feedback System	During all closed loop conditions.	If the O_2 sensor indicates a rich condition for more than 2 minutes.
53	Fault	No	Engine control module.	When the ignition key is initially turned to the on position.	If the logic board fails.
54	Fault	No	Distributor Sync. Pickup	All the time when the engine is running.	If there is not a distributor sync. pickup signal.
55	Indication	No			Indicates end of diagnostic mode.
88	Indication	No			Indicates start of diagnostic mode. **NOTE:** This code must appear first in the diagnostic mode or fault codes will be inaccurate.
0	Indication	No	Indicates oxygen feedback system is lean with the engine running.		
1	Indication	No			Indicates oxygen feedback system is rich with the engine running.

Fig. 19-2 Setting codes into memory and activating "Power loss/limit" light (courtesy Chrysler).

performed (including dynamic tests and actuator tests depending on the application), and live sensor data can be read, provided the system has serial data information, and you have the proper scan tool.

Stored codes are usually displayed in numeric sequence. On many vehicles, the fault codes are proceeded by an "indicator" code that tells you to get ready because the system is going to read out any fault codes that are present or stored in memory. The fault codes may then be followed by a second indicator code that tells you "end of message." The codes may then repeat for a fixed number of times or continuously depending on the application.

On vehicles that provide "manual" code retrieval, the check engine, power loss or malfunction indicator warning light on the instrument panel may be used to display "flash" codes. With this method, the light blinks on and off in a series of flashes when the system is in the diagnostic mode. You decipher the code by counting flashes. But to interpret what the numbers mean, you have to refer to a service manual.

On many import applications (Nissan, Subaru, Honda), special diagnostic LED lights on the engine control module itself are illuminated to display code information (Fig. 19-3). To read the LEDs, however, you first have to uncover the module which may be tucked away under a seat, behind a kick panel, or under the dash. You then have to refer to a service manual to interpret what the LED light pattern or sequence means.

On older Fords and early Toyota Cressidas and Supras, an analog voltmeter could be used to manually read codes. By attaching the voltmeter's leads

CODE	L.E.D. display		Malfunctioning area	Items retained in memory
	Red	Green		
11	●	●	Crank angle sensor circuit	X
12	●	●●	Air flow meter circuit	X
13	●	●●●	Cylinder head temperature circuit	X
14	●	●●●●	Vehicle speed sensor (VG30ET engine)	–
21	●●	●	Ignition signal missing in primary coil	X
22	●●	●●	Fuel pump circuit	X
23	●●	●●●	Throttle valve switch (Idle switch) circuit	–
24	●●	●●●●	Neutral/Park switch (VG30ET engine)	–
31	●●●	●	Load signal circuit (Power steering oil pressure switch, Headlamp switch, Radiator fan switch, Rear defogger switch, Heater/air conditioner switch)	–
32	●●●	●●	Starter signal circuit	–
34	●●●	●●●●	Detonation sensor (VG30ET engine)	X
41	●●●●	●	Fuel temperature sensor circuit	X
44	●●●●	●●●●	No malfunctioning in the above circuit (Check other electrical systems.)	–

X: Yes –: No

Fig. 19-3 Self-diagnosis chart (courtesy Nissan).

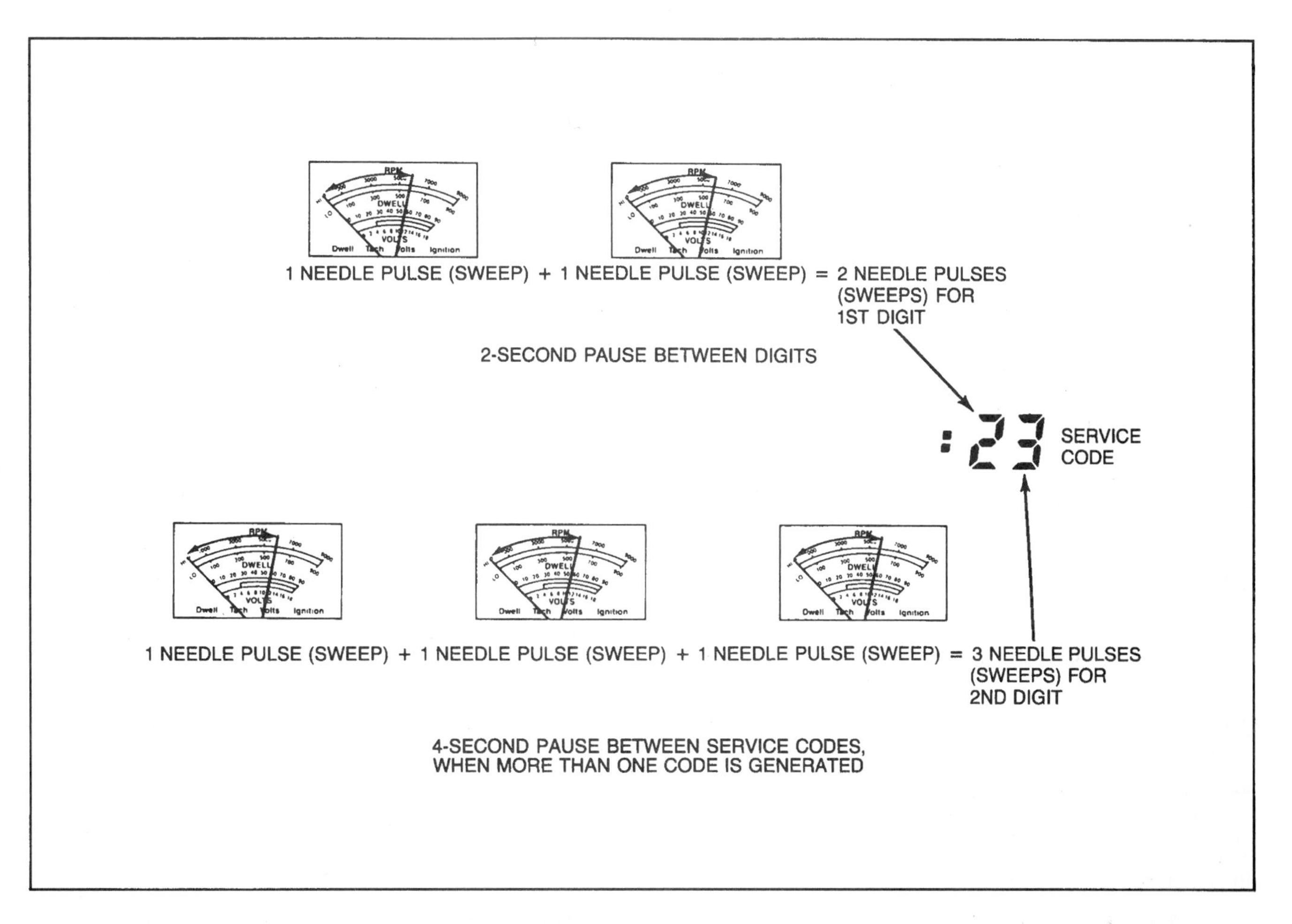

Fig. 19-4 Reading service codes with an analog voltmeter (courtesy Motorcraft - Ford).

to certain terminals on the diagnostic connector and putting the system into the diagnostic mode, the needle on the voltmeter would jump as the codes fed out. By counting the sweeps (Fig. 19-4) of the meter needle, you interpreted the codes.

Some vehicles have the equivalent of a built-in scan tool that allows you to read out codes digitally on the speedometer, climate control display, or driver information center. On these vehicles, pressing certain buttons in a specified sequence causes the system to display fault codes as well as other useful diagnostic information (Fig. 19-5).

Most technicians, however, prefer to read fault codes and other diagnostic information electronically with a scan tool or computerized engine analyzer that has scan capabilities. Reading numbers on a digital display is much easier than trying to count flashes of light. You're less apt to make a mistake, and you're usually able to extract more useful information from the system. Computerized engine analyzers are an especially powerful tool, because many can do a complete "sweep" of all the system's sensors in addition to displaying any fault codes that might be present.

On vehicles that provide live serial data through the diagnostic connector (most GM, newer Ford and Chrysler, all Lexus, Geo Metro and Spectrum, most Mazda, Mitsubishi, Hyundai and Chrysler imports, and some Isuzu, Nissan, Subaru, Suzuki and Toyota), a scan tool opens the window to the inner workings of the system. In addition to retrieving fault codes, you can read various live sensor inputs to determine whether or not a sensor is responding correctly in a given situation. Some applications (Chrysler, for example) also allow you to perform various actuator tests to see if the injectors, canister purge solenoid, and other relays are responding to computer commands.

Most older import applications as well as older Fords and some carbureted Chryslers, however, do

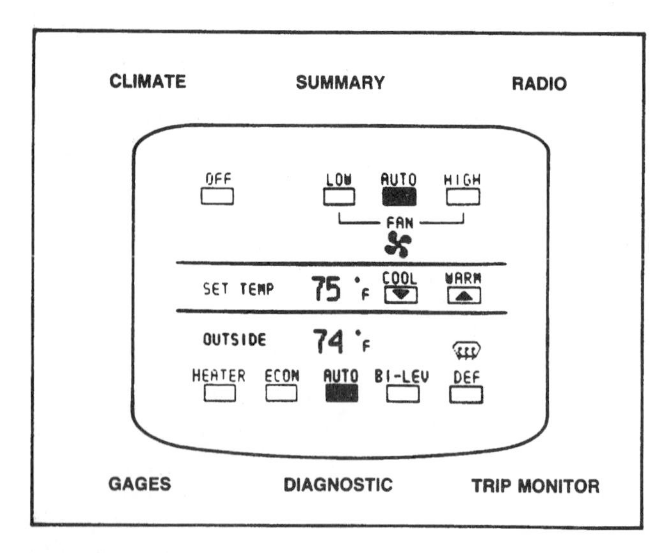

Fig. 19-5 Buick CRT display. Codes and data can be accessed by touching off/warm simultaneously (courtesy General Motors).

not provide live serial data through the diagnostic connector. Even so, there are aftermarket scan tools and analyzers available that can be plugged directly into the computer's harness to tap the flow of data within the system (Fig. 19-6). Some of the more sophisticated analyzers allow you to display sensor data as scope patterns so you can "see" what's going on. Displaying several related sensors simultaneously as scope patterns on a screen allows you to catch problems that would be difficult, if not impossible, to detect reading digital information alone. These types of analyzers are certainly more expensive than simple hand-held scan tools, but well worth the price when you consider their increased diagnostic capabilities.

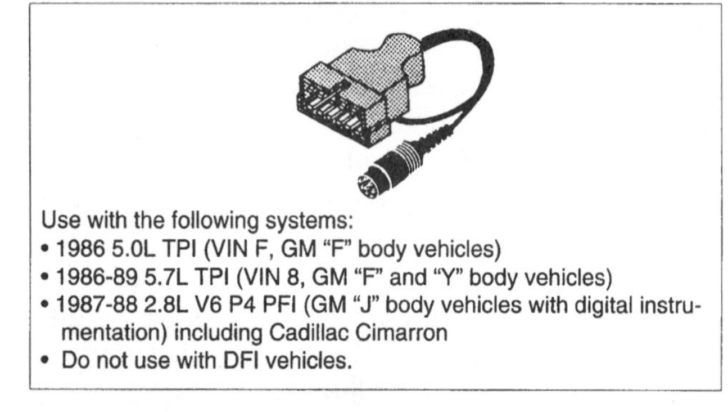

Fig. 19-6 Scan tool to diagnostic connector cable (courtesy OTC).

UNDERSTANDING FAULT CODES

The most important thing to remember about fault codes is that codes by themselves are NOT the final diagnosis. They are only a starting point to begin further diagnosis (see Table 19-1). In other words, codes only tell you where to start, not what to replace. When you find a fault code using either a manual retrieval procedure or electronically with a scan tool or analyzer, therefore, do not replace anything until you've done further testing to determine what's actually causing the problem. The only thing the code tells you is that something abnormal has occurred in a particular circuit. So until you do additional testing, there's no way to know for sure which component is bad.

To isolate a fault you must usually proceed through a detailed step-by-step diagnostic procedure using a digital multimeter and a service manual as your guide. Some diagnostic charts (Fig. 19-7) can be quite lengthy and involved, but all of the steps are usually necessary to rule out all other possibilities and arrive at the correct conclusion. Skipping steps to save time or failing to perform the required tests correctly won't help you nail down the problem. If anything, shortcuts will lead you astray and you'll end up replacing the wrong part. This puts you right back where you started, because you failed to fix the problem.

Isolating faults is usually the most time-consuming part of the job. Is it the sensor, a bad wiring connector, a wiring open or short, or a problem within the computer itself? You have no way of knowing until you start performing the various tests. By a process of elimination, you should be able to narrow down the list of possibilities until you arrive at some sort of conclusion. This is supposed to eliminate much of the guesswork and trial-and-error diagnosis that runs up unnecessary repair bills—in theory, that is. The catch is fault codes don't always reveal the true nature of a problem. And there are all kinds of problems that are beyond the detection capabilities of most onboard self-diagnostic systems (problems in the secondary side of the ignition system or mechanical problems, for example, like loss of compression or a bad timing chain). The latest generation of onboard diagnostics (OBD-II) is doing more to address problems, such as intermittent misfires, etc., that can affect emissions. But on older vehicles you're stuck with the limitations of the system.

CODE 34
FUEL INJECTION
MAP SENSOR
(SIGNAL VOLTAGE TOO LOW)

- IGNITION "OFF", CLEAR CODES.
- DISCONNECT MAP SENSOR AND JUMPER HARNESS CONNECTOR PINS "B" TO "C".
- DIAGNOSTIC TERMINAL NOT GROUNDED.
- START ENGINE AND RUN FOR 1 MINUTE OR UNTIL "CHECK ENGINE" LIGHT COMES ON.
- IGNITION "ON", ENGINE STOPPED.
- GROUND DIAGNOSTIC TERMINAL AND NOTE CODE.

CODE 33

REPLACE SENSOR

CODE 34

- REMOVE JUMPER FROM PINS "B" TO "C".
- CHECK VOLTAGE BETWEEN HARNESS CONNECTOR PINS "A" AND "C" USING VOLTMETER J-29125.

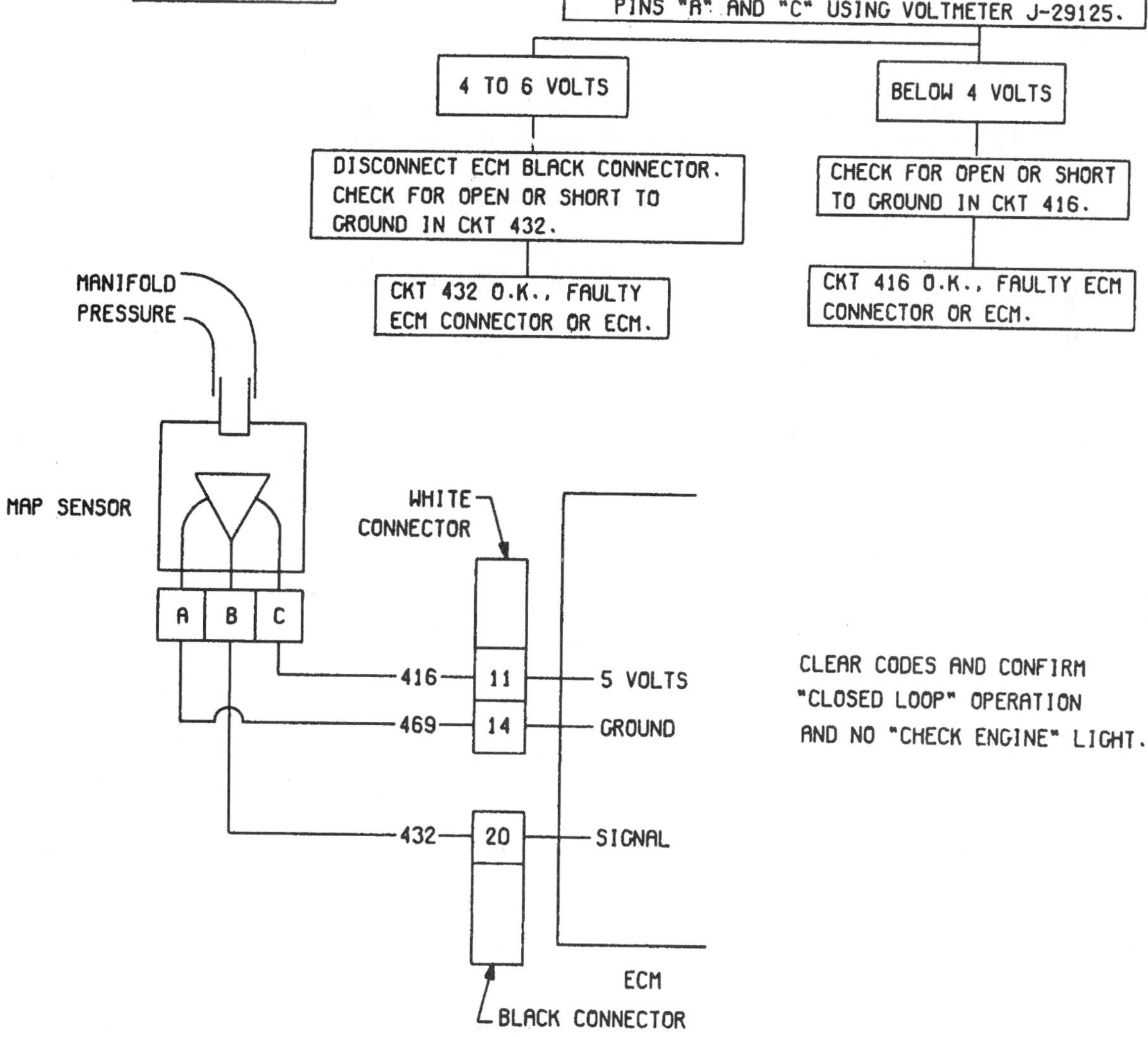

Fig. 19-7 Diagnostic trouble tree (courtesy General Motors).

USING A SENSOR SIMULATOR

One way to speed up the diagnostic process once a fault code has been found is to use a "sensor simulator" tool in conjunction with a scan tool to perform a quick check of a sensor circuit. A sensor simulator can simulate various voltage, frequency, and resistance inputs to help you determine if a problem is in the sensor or the wiring. Such a tool can simulate a high or low oxygen sensor voltage signal, a frequency signal from a mass airflow sensor, MAP sensor, knock sensor or vehicle speed sensor, or the variable resistance of a coolant sensor or throttle position sensor.

If substituting a simulated signal for the real thing either at the sensor connector or through a breakout box solves the problem, it tells you the wiring and module are okay and that the problem is in the sensor. Bingo, you've found the problem and eliminated the need to make a lot of time-consuming and unnecessary circuit checks.

If there's no change in performance or operation with a simulated signal, then the signal is not getting through to the computer (a wiring problem requiring further continuity checks), or that the signal is not being processed correctly (a bad control module).

For example, let's say an engine is running rich, the "Check Engine" light is on, and there's a fault code for the O_2 sensor. Is the O_2 sensor bad or is it something else? To find out, disconnect the O_2 sensor, set the sensor simulator to produce an O_2 sensor signal (which you can vary from a fixed lean reading to a rich signal), connect the test leads to the O_2 sensor's wiring connector, and start the engine and see what happens. If there's no change in the fuel mixture when you switch the input signal back and forth from lean to rich, the problem isn't a bad O_2 sensor. It's in the wiring or ECM.

A sensor simulator can also be used to alter the operation of a computerized engine control system when performing other diagnostic tests. By simulating a hot- or cold-input signal from the coolant sensor, you can change the operation of the computer from closed loop to open loop. This type of substitution may be helpful when performing a power balance test on some vehicles. Running in open loop prevents the computer from changing the idle speed when cylinders are deactivated.

Some other uses of this type of diagnostic tool include:

- Simulating a throttle position (TPS) sensor signal to vary the idle fuel mixture (changes the duration of the injector pulses).
- Simulating a barometric pressure sensor or MAP sensor signal to switch the fuel mixture and timing between "high altitude" and "normal" settings.
- Simulating a manifold absolute pressure (MAP) sensor signal to check for changes in ignition timing and fuel enrichment.
- Simulating a mass airflow (MAF) sensor signal to monitor changes in fuel enrichment.

FAULT OR NO FAULT?

Great as onboard self-diagnostics are, sometimes they're not much help. An intermittent warning light is one such problem that can drive a technician crazy. With some onboard self-diagnostic programs, certain intermittent faults cause the warning light to come on only when the problem is occurring. The rest of the time, the light remains off. The vehicle has not fixed itself, however. The problem is still there. So check for codes even if there's no light.

Sometimes you'll find an intermittent warning light but no code. Is there a problem or isn't there? One way to find out is to test-drive the vehicle with a portable "flight recorder" device attached to the onboard electronics. The recorder captures the live data stream information so a "snap shot" of the data can be recorded when the problem occurs. By analyzing the data later, you may then be able to determine if indeed there's a problem. But this technique only works on vehicles that provide access to the data stream. And it only works if somebody presses the record button to capture a window of data when the problem occurs. One of the limitations of flight recorders is that they don't read data continuously, but rather take samples of data every little bit. If a momentary glitch occurs between data samples, the recorder may not catch it. That's why some technicians prefer to display and watch live sensor data on a scope, so they can see intermittent problems that don't show up well on most scan tools or flight recorders.

Intermittent faults are the worst to nail down. The pass/fail tests that are part of most diagnostic procedures may not catch an intermittent problem if the problem isn't acting up when a particular test is performed. Consequently, you continue along the wrong path and end up making the wrong diagnosis and replacing the wrong part. Or you reach the end of the diagnostic tree (and your rope), only to find yourself no closer to solving the problem than when you started. Everything apparently checks out okay, but the problem is still there. Now what?

The only way to find an intermittent fault is to try to duplicate the operating conditions that are present when the fault occurs. Heat, cold, vibration, and moisture are four variables that may play a role in causing an intermittent fault. Anything that causes a momentary short or open in a circuit will cause a change in the circuit's voltage. A loose wiring connector that breaks contact momentarily when it gets hot, wet, or vibrates is a classic cause of many intermittent faults.

Test-driving a vehicle until the fault appears is one way to get an intermittent fault to repeat. But test-driving can eat up a lot of time. And there's no guarantee the fault will reoccur. Nor is there any guarantee that a flight recorder will catch it either. A better approach might be to try duplicating in the service bay what happens during a test-drive. Start the engine and allow it to reach normal operating temperature. With a scan tool or computerized analyzer connected to the vehicle, see if you can create a momentary glitch by wiggling, blowing hot air, spraying water, etc. on the suspected wiring connectors or components. Sometimes this approach works, and sometimes it doesn't.

FALSE CODES

"False" codes are another glitch that can sometimes create ghosts in a system. A voltage spike or transient electrical surge may trigger a code for a problem that doesn't exist. An arcing plug wire, an open diode, electromagnetic interference from a spark plug wire that's routed too near a sensor circuit, etc., may all be potential causes of false codes.

On GM C3 systems, for example, arcing between the ignition coil and ground or at the spark plug wires can sometimes cause an intermittent "Check Engine" light but no trouble code. So can an open diode across the air conditioner compressor clutch, intermittent shorts, or grounds in the wiring harness (grounding terminals "A" and "U" or applying battery voltage to terminals "C" and "R").

False codes can also be triggered while performing certain test procedures or making repairs. Sometimes you'll find old codes that have been left in memory from a previous problem that has since been repaired (a good reason for always erasing codes after repairs have been completed).

The only way to be really sure whether or not there is a real problem is to: (1) read and record all stored codes, then (2) erase them with a scan tool or by removing the module's fuse, and (3) test-drive the vehicle or run a self-test (Ford) to see if the same code or codes reappear. Don't disconnect the battery to erase stored codes because you'll also erase all the settings in the electronic radio and climate control system. Also do not disconnect anything electrical when the key is on or while the motor is running. Doing so may create a voltage spike that could damage sensitive electronic components.

For some kinds of false code problems, a manufacturer may issue a technical service bulletin describing the nature of the problem and how to fix it. In some cases the cure may involve rerouting some wiring or replacing a part that seems unrelated to the problem. Having access to these bulletins either through a new car dealer or an independent CD-ROM-based information system such as those provided by All-Data, Chilton, and Mitchell can help you solve problems that may be impossible to solve without the appropriate bulletin.

READING LIVE SERIAL DATA

Tapping into the vehicle's serial data stream with a scan tool or computerized engine analyzer to read sensor voltages and other system inputs and outputs is a quick way to pinpoint problems. Reading the live data stream allows you to see what the control module thinks is occurring, and to quickly determine if something is out of whack. Reading the serial data with a scan tool can also eliminate the need to make dozens of individual sensor and circuit tests with a volt-ohmmeter or to use a breakout box. It also allows different sensor and circuit inputs to be displayed simultaneously for comparison. This is an extremely useful feature to have when you're trying to nail down the cause of an intermittent driveability problem.

Computerized test equipment that performs a complete sensor sweep typically compares the vehicle's readings against those for "known good sensors." In some cases the range for acceptable readings is determined by the vehicle manufac-

TABLE 19-1
SENSOR TROUBLE CODES

SENSOR	TROUBLE CODES GM	FORD	CHRYSLER
OXYGEN (EGO)	13,44,45,55	43,91,92,93	21,51,52
THROTTLE POSITION (TPS)	21,22	23,53,63,73	24
ENGINE VACUUM (MAP)	31,33,34	22,72	13,14
BAROMETRIC PRESSURE (BARO)	32	—	37
COOLANT (ECT)	14,15	21,51,61	17,22
KNOCK	42,43	25	17 (some only)
VEHICLE SPEED (VSS)	24	—	15
AIR TEMPERATURE (MAT, VAT, ACT)	23,25	24,54,64	23
AIRFLOW (VAF, MAF)	33,34,44,45	26,56,66,76	—
EGR VALVE (EGR, EVP)	—	31,32,33,34,31 83,84	

turer, while in other cases it may be established by looking at real world applications.

When an analyzer performs a sensor sweep, it will typically red-flag or highlight any sensor readings that are out of range. This does not necessarily mean there's a problem with the sensor, however. A sensor may be "borderline," reading a little too high or too low compared to the norm, yet be producing no noticeable emissions or driveability problem. In such cases you'll have to use a certain amount of judgment to decide whether or not an "apparent" problem is in fact a real problem that requires further investigation.

MISDIAGNOSIS

On Ford's with EEC-IV, the most common mistake that's made during diagnosis, according to Ford, is not following the specific diagnostic procedure. Ford does not allow any deviation from the factory procedure. If the manual says to open and close the car door twice, you'd better do exactly what it says.

When troubleshooting GM systems, some people try to read more into the data stream than what is actually there. GM passes a tremendous amount of information out the ALDL connector. Some of it is nothing more than complicated gibberish with little significance. But because it's there and can be read, some technicians think they have to use every bit of information. What's more important is comparing related sensor readings to see if they make sense, like watching the idle rpm and idle air control motor signals on a fuel-injected engine.

Another mistake that's often made is that of jumping to conclusions too quickly. Just because the last three Oldsmobiles that you worked on had bad coolant sensors doesn't mean the one you're working on now has the same problem. Faulty components can't be accurately identified without going through some type of testing procedure to rule out other possibilities.

All too often a technician will blame the computer or a sensor for a problem, only to find that replacing the part doesn't fix the problem. That's because he overlooked something basic while zeroing in on the computer system. Things like secondary ignition performance, battery voltage, intake vacuum, and engine compression also need to be checked when diagnosing performance problems.

According to one of the leading scan tool manufacturers, the single most common problem users of their products have is setting up the tool correctly for the vehicle being tested. Because of the tremendous variety in data stream formats from vehicle to vehicle, punching the wrong entry code into a scan tool, or misreading the vehicle identification number (VIN), can result in a "no data" display. It's essential that the exact vehicle data be entered so the tester can read the data stream correctly. Pontiac, for example, doesn't use the same C3 system on all it's 2.5-liter engines. Don't assume computer systems are the same just because two engines are the same. What's more, some GM vehicles which are totally different (a Pontiac 1.6-liter LeMans and a Chevrolet 2.5-liter S10 pickup) may have identical computer systems in terms of how the data stream is formatted and read.

Glossary of Emissions-Related Terms

acid rain—Corrosive rain formed when sulfur emissions from motor vehicles, industrial plants, or electric generating stations combine with hydrogen and oxygen in the atmosphere to form sulfuric acid (H_2SO_4). This is not only generally corrosive to anything it may come into contact with, it also raises the acidity of lakes and ponds, often to the point that fish and other aquatic life cannot survive.

air—The combination of gases that makes up the earth's atmosphere. Composed of nitrogen (76-78%), oxygen (18-21%), carbon dioxide, argon, and other gases. When air is drawn into an engine, the oxygen combines with the fuel during combustion, producing carbon dioxide and water vapor. If there is too much fuel for the available oxygen, carbon monoxide and unburned hydrocarbons are produced in the exhaust. What is more, at temperatures above 2,500 degrees F, oxygen can combine with nitrogen to form NOX another harmful pollutant.

AIR—*see* Air Injection Reactor.

air/fuel ratio—The relative proportions of air and fuel entering an engine's cylinders as produced by the carburetor or fuel injection system. The ideal or "stoichiometric" ratio for gasoline is 14.7:1 air to fuel by weight. A higher ratio would contain more air and less fuel, and would be considered a "lean" mixture. A lower ratio with more fuel and less air would be a "rich" mixture. The air/fuel ratio is determined by the orifice size of the main jets inside a carburetor, the dwell duration of the mixture control solenoid inside a feedback carburetor, or the orifice opening and fuel pulse duration of a fuel injector.

air injection system—Supplies fresh air to the exhaust stream, which helps oxidize HC and CO, and, on models equipped with a catalytic converter, gives the catalyst the extra air it needs to oxidize those pollutants.

Air Injection Reactor (AIR)—The GM name for the air injection system, which comprises a vane pump, a diverter valve, and a check valve.

air pump—A belt-driven vane pump that supplies the flow of air needed for most air injection systems.

altitude—The distance of a point above sea level. Important to automotive emissions control because the higher the altitude, the fewer oxygen molecules per given volume of air, which alters the effective compression and air/fuel ratios.

ambient temperature—The temperature of the air surrounding a vehicle.

anti-percolation valve—A carburetion system device used to prevent fuel evaporation from the fuel bowl while the engine is running. It is connected to the throttle linkage, so it is closed when the throttle is open, and open when the throttle is closed. With the engine off, hot fuel vapors boil out through the vent line and into the charcoal canister.

ASE—An abbreviation for the National Institute for Automotive Service Excellence. This organization certifies professional mechanics in emissions repair and other specialties.

aspirator—A one-way valve attached to the exhaust system of an engine that admits air during periods of vacuum between exhaust pressure pulses. Used to help oxidize HC and CO, and to supply additional air which the catalytic converter may require. Can be used instead of a belt-driven air injection pump in some applications. Called "Pulse-Air" in GM systems.

backfire—1. An explosion in the exhaust system of a motor vehicle caused when unburned air/fuel mixture is ignited, usually upon deceleration. 2. An explosion of the air/fuel mixture in the intake manifold, which is evident at the carburetor or throttle body, and may be caused by improper ignition timing,crossed spark plug wires, an intake valve that is stuck open, etc.

back pressure—Resistance of an exhaust system to the passage of exhaust gases. This can have an adverse effect on performance, fuel economy, and emissions. Excessive back pressure may be caused by a clogged catalytic converter, or a dented or crimped pipe.

back pressure EGR—Some emissions control systems use a back-pressure sensor or diaphragm to monitor backpressure so that exhaust gas recirculation (EGR) flow can be increased when the engine is under maximum load (and producing maximum back pressure).

barometric pressure—The pressure exerted by the weight of the earth's atmosphere, equal to one bar, 100 kilopascals, or 14.7 psi (often rounded off to 15 psi) at sea level. Barometric pressure changes with the weather and with altitude. Since it affects the density of the air entering the engine and ultimately the air/fuel ratio, some computerized emissions control systems use a barometric pressure sensor so that the spark advance and EGR flow can be regulated to control emissions more precisely.

blowby—Byproducts of combustion that leak out of the combustion chamber past the piston rings into the crankcase. Mostly hydrocarbons, this represents approximately 20% of the air pollution a pre-emissions controlled engine produces. In modern vehicles, the blowby vapors are drawn into the intake stream through the positive crankcase ventilation system (PVC) to be burned in the cylinders.

British Thermal Unit (Btu)—The amount of heat required to raise one pound of water one degree Fahrenheit. The heat value of various motor fuels are often compared in Btu's per gallon or per pound.

California Air Resources Board (CARB)—An agency of the state of California responsible for regulations intended to reduce air pollution, especially that created by motor vehicles.

camshaft duration—the amount of time as measured in degrees of crankshaft rotation that an intake or exhaust valve is held open. Duration specs are typically measured at .050 inch of valve lift.

camshaft lift—how far a valve is pushed open. The amount of lift is determined by the height of the cam lobe and the geometry of the rocker arms (on a pushrod engine) or the cam followers (on an overhead cam engine).

carbon dioxide (CO_2)—A harmless, odorless gas composed of carbon and oxygen. In automotive science, the product of complete combustion.

carbon monoxide (CO)—An odorless gas composed of carbon and hydrogen. A major air pollutant and potentially lethal if breathed in small doses. CO is formed by the incomplete combustion of any fuel containing carbon (gasoline, diesel fuel, alcohol, coal, wood, etc.). If additional oxygen is provided (as by an air pump) and allowed to combine with the CO (as happens inside a catalytic converter), carbon monoxide is transformed into harmless carbon dioxide (CO_2). Excessive CO emissions in vehicles are caused by overly rich air/fuel ratios. Possible causes include restrictions in the air intake, a clogged air filter, a partially closed choke, a fuel-saturated carburetor float, too high a float level, leaky needle valve and seat, oversized carburetor jets, internal carburetor fuel leaks, defective oxygen sensor, or leaky injectors.

catalysis—The action of a catalyst.

catalyst—A substance that accelerates or enhances a chemical reaction without being changed itself. When used in a catalytic converter, it can reduce the level of harmful pollutants in the exhaust. Catalysts which are commonly used include platinum, palladium, and rhodium, all of which are quite expensive. To minimize costs, a thin layer of catalyst is all that's needed on a ceramic honeycomb, ceramic pellets, or a metallic substrate inside the converter. Catalysts are quite sensitive to lead, however. If leaded gasoline is used, the lead coats the catalyst and renders it useless.

catalytic converter—An automotive exhaust system component containing a catalytic element to reduce NOX and/or HC and CO tailpipe emissions. When hot exhaust gases pass through the converter, the catalyst allows oxygen from the air pump or aspirator to "reburn" the pollutants lowering the concentrations of HC and CO to almost zero. This is called an "oxidation" reaction. It creates a great deal of heat (1,200-1,600 degrees F) so the converter is made of stainless steel surrounded with a heat shield. In three-way converters, a separate "reduction" chamber mounted just ahead of the oxidation chamber contains a different mixture of catalysts to reduce NOX emissions.

Central Port Injection (CPI)—An AC Rochester fuel injection system originally installed on the 4.3L Chevrolet Vortec V6 that produces port injection performance at a low price by using one TBI-style injector to pulse fuel directly to individual nozzles at the intake ports.

centrifugal advance—A mechanical means of advancing spark timing with flyweights and springs. These weights are located inside the distributor (except on engines with computerized engine controls), under the rotor (GM window-type distributors), or under the point or pickup base plate. The size of the weights, the amount of spring tension, and the engine rpm rating determines the rate of advance.

charcoal canister—The basic component of evaporative emissions control systems, this is a small cylindrical or rectangular container that contains activated charcoal particles. The charcoal traps gasoline vapors from a vehicle's sealed fuel system.

charge temperature sensor—On computer-controlled engines, a sensor which sends a signal to the computer that varies with the temperature of the intake stream.

check valve—A valve which permits the passage of a gas or fluid in one direction, but not in the other. For example, the check valve between the air pump and exhaust manifold in an air injection system allows air to flow to the manifold, but stops exhaust gas from entering the air pump in the event that the pump belt breaks.

choke—A manually or thermostatically actuated plate mounted on a shaft at the mouth of a carburetor which is closed when the engine is cold to greatly increase the amount of gasoline in the air fuel mixture, thus aiding starting when fuel evaporation is low. Choke problems are the primary cause of hard starting. If the choke is not closed when the engine is started, the air/fuel ratio will be too lean while the engine is cold. This results in a slow idle and stalling. If the choke does not open up at the correct rate as the engine warms up, the air/fuel ratio becomes too rich. This causes a rough idle, poor fuel economy, and excessive HC and CO emissions. The choke is controlled by a coiled bimetallic spring that reacts to temperature changes. An electric heating element inside the choke housing, or warm exhaust gases or engine coolant routed nearby are used to speed up the rate at which the choke opens.

choke heater—A device which warms the thermostatic coil of an automatic choke, causing it to open quickly. A typical late-model version comprises an electrical heating element and a timer circuit.

choloroflurocarbons (CFCs)—A family of chemicals that includes R12 automotive air conditioning refrigerant. CFCs have been blamed for a deterioration of the Earth's protective ozone layer. CFCs are being phased out of production by international agreement.

Clean Air Act—Originally passed in 1970 by the U.S. Congress, and updated in 1990, this legislation created today's auto emissions laws and established the Environmental Protection Agency as the watchdog over our nation's air quality.

closed loop—The basic principle of electronic engine management in which input from an oxygen sensor allows the engine control computer to determine and maintain a nearly perfect air-fuel ratio.

Cold Weather Modulator—A component of certain Ford heated air intake systems, this is a thermostatically controlled check valve that traps vacuum in the vacuum motor when the car is accelerated hard at temperatures below 55 degrees F. This eliminates hesitation by allowing heated air to enter the engine in spite of the drop in vacuum that naturally occurs when the throttle is opened wide.

Combined Emissions Control system (CEC)—An early GM transmission controlled spark system that uses the solenoid valve's plunger as an auxiliary throttle stop. During high gear deceleration, it holds the throttle open a predetermined amount, leaning the mixture in that normally rich, high-vacuum mode.

compression ratio—The relationship between the piston cylinder volume from bottom dead center to top dead center. Higher compression ratios improve combustion efficiency but also require higher-octane fuels. Pre-emission control engines often had compression ratios as high as 11.5:1, whereas most of today's engines are between 8.5:1 and 9.5:1. Diesel engines have very high compression ratios, from 18:1 to 22:1.

computerized engine controls—A microprocessor-based engine management systems that utilizes various sensor inputs to regulate spark timing, fuel mixture, emissions, and other functions. Most have a certain amount of self-diagnostic capability, and store or generate fault codes to help a technician diagnose system problems.

coolant temperature override switch (CTO)—A vacuum controlling device that prevents the overheating associated with late ignition. At normal operating temperature, the CTO allows ported vacuum to reach the distributor. But when overheating begins to develop at idle, it routes the strong manifold signal to the advance unit, which makes the spark occur earlier, cooling combustion and increasing idle speed.

coolant temperature sensor (CTS)—In computerized engine control systems, a thermistor which informs the ECU as to the temperature of the coolant. In the PTC (Positive Temperature Coefficient) type, ohms go up with temperature. In the more common NTC (Negative Temperature Coefficient) type, resistance goes down as heat goes up.

crankcase emissions—*see* blowby.

decel valve—A device which reduces exhaust emissions during vehicle deceleration by keeping rpm up and vacuum down. Found on some Ford engines such as the Pinto 2.0L, it works by opening an extra intake air passage under high-vacuum conditions.

Digital Electronic Fuel Injection (DEFI)—An early Cadillac electronic fuel injection system.

detonation—A phenomenon of internal combustion wherein the compressed air/fuel charge explodes violently instead of burning smoothly, usually due to the creation of a second flame front in the chamber away from the spark plug. This collides with the spark-ignited flame front, causing a noise known as "pinging" or "spark knock," and potentially damaging to the engine. Detonation can be caused by excess spark advance, low-octane fuel, lean air/fuel mixtures, and/or overheating. Carbon buildup inside the combustion chambers and on the piston face can also increase compression sufficiently to cause detonation. Mild detonation is not harmful, but heavy detonation can damage pistons, rings, and rod bearings. It can also cause elevated HC emissions.

detonation sensor—An electrical device mounted on an engine block, cylinder head, or intake manifold that generates a small voltage signal when it encounters the vibration frequency associated with detonation. This signal is sent to an electronic control unit that retards ignition timing until detonation stops.

dieseling—A condition in which a carbureted engine continues to run after the ignition is shut off. Caused by hot spots inside the combustion chambers, spontaneous fuel ignition keeps the engine running (see also pre-ignition). This does not occur in fuel-injected engines, however, because the injectors stop the flow of fuel when the ignition is shut off. To prevent dieseling in carbureted engines, an idle stop solenoid or fuel cut-off solenoid may be used.

direct-acting thermostatic air cleaner—The basic component of certain heated air intake systems, which uses a thermostatic bulb connected to the rod that operates the flapper valve in the air cleaner snorkel.

diverter valve—In an air injection system, a vacuum-operated valve that directs air pump output to the atmosphere during high-vacuum deceleration to eliminate backfiring.

dual diaphragm distributor—A distributor incorporating a vacuum advance unit which contains two separate diaphragms and vacuum chambers. Usually, one is connected to ported vacuum, which advances the spark during acceleration and cruising, and the other to intake manifold vacuum, which retards the spark at idle and deceleration.

duty solenoid—On a feedback carburetor, a solenoid that cycles many times per second to control a metering rod, hence the air/fuel mixture.

dynamometer—A machine used to simulate loaded driving conditions for emissions and diagnostic purposes. A vehicle's drive wheels are placed on a pair of rollers so the vehicle can be driven in place at various speeds while being subjected to changing loads. Engine performance and emissions are monitored with additional test equipment to determine emissions compliance and/or to diagnose emissions or performance problems.

Early Fuel Evaporation (EFE)—A GM system that promotes the evaporation of gasoline in the intake manifold through the application of heat from exhaust or an electrical heating element, thus improving the engine's cold running characteristics and reducing emissions.

EEPROM—Electronically Eraseable Program Read Only Memory, a special type of computer calibration chip that can be reprogrammed electronically by a service technician using the proper equipment and codes.

EGR valve—The exhaust gas recirculation valve meters a small amount of exhaust gas into the intake manifold to dilute the air/fuel mixture. This keeps combustion temperatures below 2,500 degrees F to reduce the formation of NOX. The amount of exhaust gas recirculated into the engine is only a few percent. The EGR valve is mounted on the intake manifold. A vacuum-operated diaphragm or small electric motor lifts a small plunger-type valve that uncovers an opening to the intake manifold. Exhaust gases then flow through plumbing from the exhaust manifold past the EGR valve and into the engine.

EGR valve position sensor—A potentiometer that keeps the engine control computer informed as to how far the EGR valve is open.

electronic control unit (ECU)—A digital computer, especially one that controls engine and sometimes transmission functions. Also known as an electronic control module (ECM), or electronic control assembly (ECA).

electronic fuel injection (EFI)—Any type of fuel injection system that uses electronic rather than mechanical controls. *See also* Central Port Injection (CPI), multiport fuel injection (MFI), K-Jetronic, L-Jetronic, sequential fuel injection (SFI), throttle body injection (TBI), and Tuned Port Injection (TPI).

electronic spark control (ESC)—The process whereby spark advance is controlled electronically rather than by mechanical means. A microprocessor calculates the necessary amount of spark advance based on engine operating conditions and what has been programmed into its memory. The traditional vacuum and centrifugal advance mechanisms are not used in the distributor.

emissions controls—The components directly or indirectly responsible for reducing air pollution, including crankcase emissions, evaporative emissions, and tailpipe emissions.

emissions —Unwanted, harmful chemicals and chemical compounds that are released into the atmosphere from a vehicle, especially from the tailpipe, crankcase, and fuel tank. These include unburned hydrocarbons, carbon monoxide, oxides of nitrogen, particulates (soot), and sulfur.

emissions warranty—A federally mandated 5-year/50,000-mile performance and defect warranty that covers all emissions-related components on all new vehicles built since 1981. Parts covered include the PCV system, EGR system, catalytic converter, exhaust manifolds, head pipe, air pump and related plumbing, charcoal canister and fuel vapor recovery system, heated air intake system, fuel injection system (except pump & filter), engine sensors, and computer. Spark plugs and other normal maintenance items are not covered after 2 years or 24,000 miles. On 1995 and later vehicles, the emissions warranty is reduced to 2 years/24,000 miles on all emissions-related components except the computer and catalytic converter which are extended to 8 years/80,000 miles. If a covered component fails during the warranty period, you can return it to a new car dealer for free replacement—but you are under no obligation to do so.

Environmental Protection Agency (EPA)—A regulatory agency of the U.S. federal government responsible for creating and enforcing regulations concerning the protection of the environment from various forms of pollution, including that which is generated by motor vehicles.

evaporative emissions—Hydrocarbons from fuel which evaporate from a vehicle's fuel tank and carburetor. They are eliminated by sealing the fuel system and using a charcoal canister to trap vapors from the fuel tank and carburetor. Some states also require that filling stations include evaporative emissions recovery devices on gasoline pumps to catch gasoline fumes when the fuel tanks are being filled.

exhaust analyzer—An automotive test and service device which uses a process involving infrared energy to determine and display the composition of an engine's exhaust gases. The two-gas type measures hydrocarbons and carbon monoxide content, while the four-gas type also measures oxygen and carbon dioxide content.

exhaust gas oxygen (EGO)—The amount of oxygen present in the exhaust stream, as measured by an oxygen sensor and reported to the computer in closed-loop, feedback systems. The computer uses this information along with signals from other sensors to control the air/fuel mixture.

exhaust gas recirculation (EGR)—An emissions control system which reduces an engine's production of oxides of nitrogen by diluting the air/fuel mixture with exhaust gas so that peak combustion temperatures (those which cause nitrogen and oxygen to form these harmful compounds) in the cylinders are lowered. A malfunction which prevents EGR will cause elevated NOX emissions and possibly detonation under load.

exhaust emissions—Pollutants identified by clean air legislation as being harmful or undesirable. These include lead, unburned hydrocarbons (HC), carbon monoxide (CO), oxides of nitrogen (NOX). Permissible levels for these pollutants are specified in grams per mile or percent of volume.

fast idle—The higher speed at which an engine idles during warm-up. When first started, a cold engine needs more throttle opening to idle properly. On carbureted engines without computer idle speed control, a set of cam lobes on the choke linkage provides a fast idle speed of 1100 to 1500 rpm during engine warm-up.

feedback—1. A principle of fuel system design wherein a signal from an oxygen sensor in the exhaust system is used to give a computer the input it needs to properly regulate the carburetor or fuel injection system in order to maintain a nearly perfect air/fuel ratio. 2. A signal to a computer that reports on the position of a component, as an EGR valve. Typically, the feedback device is a variable resistor.

feedback carburetor—A carburetor that controls the air/fuel mixture according to commands from the engine control computer, typically through the operation of a duty solenoid.

fuel injection—A system that uses no carburetor but sprays fuel under pressure into the intake manifold or directly into the cylinder intake ports. The advantage of fuel injection over carburetion is that it allows more precise control of the air/fuel mixture for improved performance, fuel economy, and reduced exhaust emissions.

grams per mile (GPM)—A measurement of the amount of emissions a vehicle produces.

gulp valve—In an air injection system, a valve that opens to admit extra air into the intake manifold upon deceleration, thus leaning out the mixture to prevent backfiring.

Hall effect—A phenomenon in which voltage is generated by the action of a magnetic field acting on a thin conducting material, commonly used to control the primary circuit of an electronic ignition system. Named for the American scientist, Edwin Hall (1855-1938).

HC-CO meter—Uses an infrared sensing device to measure the amount of HC and CO in vehicle exhaust (*see also* infrared analyzer). A probe is inserted into the tailpipe of the vehicle being checked so that samples of the exhaust gases can be drawn into the machine. HC is read in parts per million, and CO is read in percent.

heated air intake system—A system that maintains intake air at a more or less constant temperature by blending outside or underhood air with heated air picked up from a shroud over the exhaust manifold. A typical version uses a vacuum motor to power a door in the air cleaner snorkel, and a thermostatic bleed valve to control the signal to the vacuum motor. Also called Thermostatic Air Cleaner (TAC). A system malfunction that prevents the door from closing can cause hesitation and stumbling when the engine is cold. An air temperature control flap stuck shut will overheat the air/fuel mixture, possibly causing detonation and elevated CO levels (due to a rich air/fuel ratio, as warm air is less dense than cold air).

heat riser—A channel in an intake manifold through which exhaust gas flows in order to heat the manifold, thus aiding in fuel vaporization.

heat riser valve—A control valve between the exhaust manifold and exhaust pipe on one side of a V8 or V6 engine that restricts the flow of exhaust causing it to flow back through the heat riser channel under the intake manifold. This aids fuel evaporation and speeds engine warm-up. A heat riser valve stuck open will slow engine warm-up and may cause hesitation and stalling when the engine is cold. A valve stuck in the closed position will greatly restrict the exhaust system and cause a noticeable lack of power and drop in fuel economy.

heated exhaust gas oxygen sensor (HEGO)—An oxygen sensor which is heated electrically as well as by engine exhaust so that it warms up to normal operating temperature more quickly, thus allowing the engine to enter closed-loop operation sooner than with a non-heated sensor.

High-Swirl Combustion (HSC)—Ford's name for a cylinder head and valve design that promotes turbulence in the combustion chamber during the power stroke, thus contributing to complete, efficient burning of the air/fuel charge.

Honda CVCC—(Controlled Vortex Combustion Chamber) An efficient combustion chamber design that uses a small auxiliary combustion chamber (containing the spark plug) that receives a rich mixture to get an overall lean mixture in the cylinder to fire dependably.

hot idle compensator—A temperature-sensitive carburetor valve that opens when the inlet air temperature exceeds a certain level. This allows additional air to enter the intake manifold to prevent overly rich air/fuel ratios.

humidity—The amount of water vapor in the air. The amount of water air can hold before it becomes saturated depends on temperature. Warm air can hold more moisture than cold air. At 100% humidity, the air is completely saturated with moisture. Humidity affects engine performance because it tends to boost the effective octane rating of the air/fuel mixture. Engines can therefore tolerate more spark advance during humid weather than during dry weather.

hydrocarbon (HC)—Any chemical compound composed chiefly of hydrogen and carbon, especially petroleum. As an automotive pollutant, HC is simply unburned fuel and lubricating oil, and may be found in the crankcase as blowby, evaporating from the gas tank, and escaping from the tailpipe.

hydrogen—A flammable gas with the symbol H, and usually occurring as H_2. The most common element in nature.

Idle Air Control valve (IAC)—The GM name for an electrically operated valve which, according to commands from the engine control computer, varies the size of an air passage that bypasses the throttle plate of an electronic fuel injection system, thus controlling idle speed.

idle limiter cap—A plastic device pressed over a carburetor's idle mixture screw which limits the amount of adjustment available during service. Adopted to help eliminate excessive air pollution that may be caused if the idle mixture is set too rich.

idle adjustment—Refers to either the idle speed adjustment or the idle mixture adjustment.

idle mixture—the air/fuel ratio at idle.

idle mixture adjustment—The idle air/fuel mixture adjustment is usually sealed to prevent tampering. It is adjusted on carbureted engines by removing removing the anti-tampering plugs (which may require carburetor removal first) and turning the idle mixture screws until the proper idle mixture is achieved. Turning the screw out richens the idle mixture, while turning it in leans the mixture. Using the idle drop technique, the mixture is set by adjusting the mixture for smoothest idle at the slowest rpm rating. Setting the idle mixture on an emissions-controlled engine is tricky and may require monitoring exhaust CO or using a special procedure called "propane enrichment."

idle mixture screw—A screw with a tapered point which is threaded into the body of a carburetor and can be turned to vary the amount of fuel which can pass through the idle circuit.

idle speed—The speed at which an engine idles specified in revolutions per minute (rpm). Idle speed is an important value because it can affect emissions, idle quality, and driveability. For most engines, the idle speed is usually between 600 and 850 rpm. The exact idle speed is specified on the underhood emissions decal. If the idle speed is too low, the engine may stall, especially when the transmission is in drive and/or electrical accessories or the air conditioner are on. If too high, idle emissions will be increased, the vehicle may be hard to hold at a stop when in drive, and the jolt created when shifting into gear may damage the transmission or driveshaft joints.

idle speed adjustment—The idle speed on carbureted engines without computer idle speed control is set by turning a screw that opens the throttle plates. The screw is turned until the desired idle rpm level is achieved. On fuel-injected engines without computer idle speed control, idle speed is set by turning an idle air bypass screw that allows air to bypass the throttle plates. Idle speed is not adjustable on engines with computer idle speed control.

idle stop solenoid—An electromagnetic device mounted on carburetor linkage that maintains the proper throttle opening for specified idle speed while the ignition is on, but allows the throttle to close farther when the ignition is switched off, thus reducing the amount of air that can enter the engine and reducing the likelihood of dieseling.

inspection/maintenance (I/M)—The periodic inspection and maintenance of a vehicle's ignition, fuel, and emissions control systems. By maintaining the engine's various subsystems, emissions are kept to a minimum while fuel economy and performance are enhanced. A well-maintained engine is an efficient-running and clean engine.

I/M 240—An "enhanced" emissions inspection program that uses loaded-mode testing to check vehicle emissions. It is called I/M 240 because it is a 240-second test that simulates the federal urban test cycle for certifying new vehicle emissions performance. Loaded-mode testing requires a dyno and special emissions analysis equipment that can determine tailpipe emissions in grams per mile and measure oxides of nitrogen (NOX) emissions.

inches of mercury—(in. Hg.) A measurement of vacuum, related to the height that atmospheric pressure can push a column of mercury within a tube. Standard atmospheric pressure at sea level is 14.7 in. Hg, commonly rounded off to 15.

integrated circuit (IC)—A miniaturized electronic circuit having all necessary components, such as transistors, resistors, capacitors, etc., integrated into a silicon chip.

K-Jetronic—A Robert Bosch mechanical port fuel injection system which injects gasoline continuously (the "K" stands for the German word for "continuous"). Also known as CIS (Continuous Injection System). Late-model variations include KE-Jetronic and K-Jetronic with Lambda, both of which employ oxygen and other sensors to keep the air/fuel mixture within a stricter range.

knock sensor—A sensor that signals the engine control computer when detonation is detected. When the computer receives a knock signal from the knock sensor, it momentarily retards ignition timing until detonation ceases. Knock sensors react to engine vibrations in a specific frequency band, but can sometimes be fooled by other engine sounds such as those produced by worn rod bearings.

lambda—The eleventh letter of the Greek alphabet used by engineers to represent the air/fuel ratio. European auto makers typically refer to the oxygen sensor as the lambda sensor.

Lambda Sond—The first closed-loop fuel injection system to appear in production, developed jointly by Robert Bosch and SAAB.

Lean Burn— A Chrysler Corporation electronic engine control system that appeared in 1976. It used precise control of spark timing to allow a very lean mixture to burn properly, thus reducing emissions. Although it used an analog computer, which is relatively unsophisticated compared to a digital computer, it was the first mass production application of an engine control computer.

leaner and later—Refers to early calibration strategies for air/fuel mixture and ignition timing that reduced HC and CO formation.

lean misfire—A condition caused by an air/fuel mixture that is too lean to sustain combustion. Lean misfire causes one or more cylinders to pass unburned fuel into the exhaust system, causing a big increase in hydrocarbon (HC) emissions. Symptoms include a rough idle and hesitation or stumble on acceleration. Lean misfire is often caused by vacuum leaks or an EGR valve that's stuck open.

light-off mini-oxidation catalytic converter—A smaller catalyst mounted just behind the exhaust manifold that gets hot very quickly after the engine is started, so it begins working in time to neutralize much of the extra pollution that is produced during cold running. Also called a "pre-cat" or "pup" converter.

Linear EGR—An AC Rochester EGR system using a linear motor to move the valve's pintle in small steps, which provides precise control of recirculation.

liquid/vapor separator—An evaporative emissions system component mounted above the fuel tank that prevents liquid gasoline from entering the vent lines.

L-Jetronic—A Robert Bosch port EFI system which uses input on the volume of intake air to calculate fuel delivery (the "L" stands for *Luftmengenmessung,* German for "airflow management").

magnetic timing—A procedure for checking or adjusting an engine's ignition timing using a mag timing meter rather than a timing light. A magnetic probe is inserted into a receptacle near the crankshaft harmonic balancer or flywheel. The probe picks up a small notch in the balancer or flywheel with every revolution of the engine. An inductive pickup that clamps onto the number one plug wire tells the meter when the plug fires. The meter then displays the degrees of timing advance.

manifold absolute pressure sensor (MAP)—A variable resistor used as a sensor to inform an engine control computer as to the vacuum conditions in the intake manifold.

manifold air temperature (MAT)—The temperature of the intake stream in the intake manifold, as increased by a heat riser or EFE system, and/or converted to engine control computer input by a MAT sensor.

manifold vacuum—The vacuum available at an engine's intake manifold generated by the engine's pumping action, measured in inches or millimeters of mercury.

mass airflow sensor (MAF)—A device used in EFI systems which supplies the computer with input as to the volume of air entering the manifold. Found at the mouth of the intake manifold, it uses the temperature differential between a heated platinum wire or a plastic film and the passing air to generate a signal of varying voltage.

Mitsubishi Jet Valve—A tiny third valve that admits only air to churn up the air/fuel charge and promote lean running and a complete burn.

monolithic catalytic converter—A catalytic converter which has its catalytic materials coating a ceramic honeycomb, as distinguished from the pellet bed converter.

multipoint fuel injection (MFI)—A type of fuel injection system that has a separate fuel injector for each of the engine's cylinders. Multipoint injection systems deliver better performance and lower emissions than throttle body injection (TBI) systems, but are more costly and complex.

negative backpressure EGR valve—An EGR valve incorporating a bleed hole that is normally closed. When backpressure drops, indicating reduced load, the bleed opens reducing vacuum above the diaphragm and cutting EGR flow.

nitrogen—A gaseous, nearly inert element given the symbol N, which comprises approximately 80% of the Earth's atmosphere. Naturally occurring as N_2.

noble metal—A relatively rare and very expensive metal, such as the platinum, palladium, and rhodium found in catalytic converters.

NOX—*see* oxides of nitrogen.

octane— Refers to gasoline's ability to resist detonation. The higher the octane number, the greater the fuel's resistance to detonation (spark knock or pinging under load). Most of today's gasolines have a pump octane rating of 87 to 92.

onboard diagnostics (OBD)—A term for special diagnostic software and hardware that detects performance problems that adversely affect emissions. OBD rules require a standardized diagnostic connector and fault codes for emissions troubleshooting.

open loop—In engines with a computer and oxygen sensor control system, a mode of operation during which the computer ignores the signal from the oxygen sensor, typically before the engine reaches normal operating temperature.

Orifice Spark Advance Control (OSAC)—A Chrysler Corporation emissions control system which slows vacuum advance of ignition timing by means of an orifice in a component mounted on the air cleaner.

OSAC valve—An abbreviation for "orifice spark advance control valve," a device used on some older Chrysler engines to limit NOX formation. The valve delays vacuum to the distributor vacuum advance between idle and part-throttle operation.

Otto cycle—The basic principle of operation of the common four-stroke piston engine, involving intake, compression, ignition/power, and exhaust. Named for Nikolaus Otto, German inventor (1832-1891).

oxidation—Any reaction in which a chemical joins with oxygen, as rusting or combustion.

oxidation catalyst—A two-way catalytic converter which promotes the oxidation of HC and CO in an engine's exhaust stream, as distinguished from a three-way or reduction catalyst

oxides of nitrogen—Harmful gaseous emissions of an engine composed of compounds of nitrogen and varying amounts of oxygen which are formed at the highest temperatures of combustion. With other gases and in the presence of sunlight, an ingredient of photo-chemical smog.

oxygen—A gaseous element given the chemical symbol O, and occurring as O_2, which makes up approximately 20% of the earth's atmosphere.

oxygen sensor—A device, usually threaded into the exhaust manifold, which uses platinum inner and outer electrodes and a zirconium electrolyte to generate a small voltage, the strength of which is dependent upon the amount of oxygen present in the engine's exhaust stream. This voltage is used as a signal to the engine control computer, which takes the signal into consideration when determining the amount of fuel necessary to maintain a proper air/fuel ratio.

ozone—A molecule of oxygen chemically represented as O_3 which is formed by exposure of O_2 to an electrical discharge, and has a pungent odor and a strong oxidizing effect.

particulate emissions—Solid particles, such as carbon and lead, found in vehicle exhaust; soot. A problem especially in diesels.

particulate trap—An emissions control device in the exhaust system of a diesel engine which is used to capture particulates before they can enter the atmosphere.

parts per million (PPM)—A measurement of the emissions of a motor vehicle given as the number of parts of a particular chemical within one million parts of exhaust gas.

pellet bed catalytic converter—A GM converter design comprising a stainless steel shell and a bed of catalyst-coated ceramic pellets. Unlike monolithic converters, if the pellets become contaminated or otherwise rendered inoperative, they can be dumped and replaced with a new load, although this job (which requires special vibrator/aspirator equipment) is not actually done very often in the real world.

photo-chemical smog—A noxious, unhealthful gaseous compound in the atmosphere formed by the interaction of various chemicals such as the pollutants hydrocarbons and oxides of nitrogen in the presence of sunlight.

PCV valve—A steel or plastic valve used in positive crankcase ventilation systems to meter blowby into the intake stream. It is commonly mounted in a grommet in the valve cover and connected to a hose that goes to a spacer under the carburetor, or to an intake manifold port with fuel-injected engines. Inside the housing is a movable plunger and a light coil spring that bears against it.

piezo-electricity—Voltage generated by a dielectric crystal under mechanical stress. The principle is used in knock sensors.

platinum—A rare, valuable metallic element given the symbol Pt, which is highly resistant to corrosion, and is used as a catalytic agent in automotive catalytic converters of the oxidizing type.

ported vacuum—Engine vacuum as available above the throttle plates of a carburetor, used to advance ignition timing when the throttle is opened above its idle position.

ported vacuum switch—A valve which permits or stops the passage of ported vacuum to a vacuum-operated component, such as a distributor advance mechanism. It may be thermostatically operated, or controlled by electric current or the movement of a mechanical component.

positive crankcase ventilation (PCV)—An engine emissions control system which picks up crankcase gases such as blowby and meters them into the intake stream to be burned. Clogging of the PCV system may cause oil leaks as blowby increases crankcase pressure. It can also lead to rapid sludge buildup in the crankcase and possible engine damage.

positive backpressure EGR valve—A common type of EGR valve which uses exhaust system backpressure to sense engine load, thus more accurately metering the amount of exhaust recycled.

pre-ignition—The ignition of the air/fuel mixture in the combustion chamber by means other than the spark; the same as dieseling. It is usually caused by hot spots in the combustion chamber (sharp edges, carbon accumulation, or spark plugs with too hot a heat range). Pre-ignition can burn holes in pistons and contribute to detonation.

Programmed Read Only Memory (PROM)—A computer component which contains values and programming which is not lost when the power supply to the computer is shut off or interrupted, its memory being non-volatile. Used to determine the basic parameters of operation in an engine control computer system.

propane enrichment—A service procedure common in the 1970s which was used to set idle mixture. A metered amount of propane gas is added to the intake stream and the resulting rpm increase is observed.

pulse width—In EFI systems, the length of time the injectors are energized and held open ("on" time), which determines the amount of fuel injected. Measured in milliseconds, a range of one-thousandth; to seven-thousandths of a second is common.

quench area—Any internal portion of a combustion chamber which causes combustion to cease because of the temperature drop in the air/fuel charge where it meets this area.

R12—A type of refrigerant used in automotive air conditioning systems that contains ozone-damaging choloroflurocarbons (CFCs). R12 is being phased out and replaced with a new refrigerant R134a. Professional technicians are required by law to recover and recycle R12 when performing any type of A/C service or repair work.

R134a—An "ozone-safe" refrigerant that is used on most 1993 and later vehicles. It is not a direct drop-in replacement for R12, and may require various modifications to a vehicle's A/C system before it can be substituted for R12.

reference voltage—In computerized engine management systems, a five-volt signal sent out from the ECU to a variable resistance sensor such as a TPS. The computer then reads the voltage value of the return signal. Called "V-ref" by GM.

reduction catalyst—The section of a three-way catalytic converter that breaks NOX down into harmless nitrogen and oxygen through a reduction reaction.

remote backpressure transducer—In an EGR system, an exhaust backpressure sensing device mounted in the vacuum line leading to the EGR valve rather than on the valve itself. At idle or light loads, it bleeds off the vacuum signal to prevent recirculation.

resistance—The quality of reducing the flow of electrons in a circuit.

retard—To cause ignition spark to occur later in an engine's cycle.

rich mixture—An air/fuel mixture with insufficient air or excessive fuel. The ideal mixture for gasoline and air is 14.7:1 by weight. A richer mixture is needed during engine warm-up and during acceleration, but a rich mixture at other times causes elevated carbon monoxide (CO) emissions.

road-draft tube—A pre-emission control-era device for ventilating the crankcase, essentially a pipe routed under the chassis at an angle that produced a small amount of vacuum as the vehicle traveled forward. Fresh air was drawn through a mesh filter in the oil filler cap, circulated around inside the crankcase, and exhausted through the road-draft tube carrying blowby with it.

run-on—*see* dieseling.

self-diagnostics—In automotive computers, especially those for engine control, a program which assesses the condition of the system, including the sensors and the computer itself, and communicates its findings to the technician by means of trouble codes.

sensor—An electrical device used to provide a computer with input as to temperature, rpm, vacuum, etc.

sequential fuel injection (SFI)—A type of multiport injection system where the individual fuel injectors are pulsed sequentially one after another in the same firing order as the spark plugs, rather than being pulsed simultaneously. This allows more precise fuel control for lower emissions and better performance, but also requires more complex controls.

smog—A general term for air pollution, especially the photo-chemical variety. Smog forms when sunlight causes chemical reactions in air pollutants resulting in the formation of ozone and other compounds.

smog pump—A slang term for an air injection system pump.

spark advance curve—The rate at which ignition timing advances. If plotted on a graph, the line resembles a curve. It rises from some initial amount of advance and levels off at the maximum advance. On older ignition systems with mechanical and vacuum advance controls, the curve depended on engine rpm and intake vacuum. On later ignition systems with electronic spark advance, various sensor inputs are used to calculate the amount of advance needed.

spark decel valve—A vacuum valve located in the line between the distributor and carburetor. The valve advances spark advance during deceleration to reduce emissions.

spark delay valve—A vacuum valve used in the line between the distributor and carburetor to delay vacuum timing advance under certain driving conditions to reduce NOX emissions. The valve acts like a restriction in the vacuum line so that vacuum builds up and changes very slowly.

spark knock—*see* detonation.

stoichiometric—Referring to the ideal air/fuel ratio, 14.7:1 by weight, in which all the oxygen is consumed in the burning of all the fuel.

stoichiometry—The state of having a stoichiometric air/fuel mixture.

Sub-EGR Control Valve—A vacuum valve used on Chrysler/Mitsubishi 2.6L engines which is operated mechanically by means of the throttle linkage, so it varies the signal to the EGR valve according to the position of the accelerator pedal.

tetraethyl lead—A lead compound that increases gasoline's octane rating. Unfortunately, lead damages catalytic converters and oxygen sensors and therefore cannot be used in vehicles designed to operate on unleaded fuel.

Thermactor—The Ford name for an air pump or air aspirator air injection system.

thermal reactor—An emissions control device comprising a large, heavy exhaust manifold in which hydrocarbons and carbon monoxide that escape from the cylinders are oxidized. Extra air is provided by a pump or an aspirator valve to promote this reaction. It is now obsolete, as it is not as efficient as the catalytic converter.

thermistor—A resistor the value of which changes according to its temperature. Used as a sensor for a gauge or computer system.

Thermostatic Air Cleaner (TAC)—The GM name for an engine air cleaner assembly that controls the temperature of the intake air by blending relatively cool underhood or outside air with relatively hot air picked up from a shroud over the exhaust manifold.

thermostatic vacuum switch (TVS)—A valve that controls the passage of vacuum according to temperature. In a basic EGR system, for instance, it blocks vacuum until a certain coolant temperature is reached, at which point it opens, allowing vacuum to reach the EGR valve.

three-way catalyst (TWC)—A catalytic converter that oxidizes hydrocarbons and carbon monoxide, and also reduces oxides of nitrogen emissions. Usually, it has separate chambers, the one upstream handling reduction, and the one downstream handling oxidation. The noble metals used as the catalytic agents are platinum, palladium, and, for reduction, rhodium.

throttle body injection (TBI)— A fuel injection system that has fuel injectors located in a common throttle body. The throttle body resembles a carburetor from the outside, and sits in the usual position on the intake manifold, but instead of having a fuel bowl, float, and venturis it has one or two fuel injectors. TBI provides many of the advantages of fuel injection (easier starting, lower emissions, etc.) without the cost and complexity of a multiport injection system.

throttle position sensor (TPS)—In computerized engine control systems, the variable resistor-type sensor which informs the ECU of throttle position. A strip of carbon provides resistance, and a brush or wiper moves along its face as the throttle is opened. It is a three-wire device with terminals for reference voltage, output back to the ECU, and ground.

transistor—An electronic component using a semiconductor to amplify or switch current.

transmission-controlled spark (TCS)—An emissions control system which prevents distributor vacuum advance from occurring at normal operating temperature until the transmission has shifted into high gear. An electrical switch mounted on the transmission controls a solenoid-actuated vacuum valve in the hose between the vacuum port above the throttle plate of the carburetor and the distributor vacuum advance unit.

trouble code—A number generated by a computer to indicate a failure in a sensor, circuit, or the computer itself. The number may be communicated to the technician by the flashing of a dash light when the diagnostic mode is entered.

Tuned Port Injection (TPI)—A General Motors multiport fuel injection system used on 5.0L and 5.7L V8 engines featuring tuned intake runners from a common plenum.

two-way catalyst—A catalytic converter that oxidizes hydrocarbons and carbon monoxide, but has little effect on oxides of nitrogen. The noble metals used to promote this reaction are platinum and palladium. Also called an "oxidation catalyst."

vacuum—A condition of pressure which is less than that of the atmosphere; negative pressure.

vacuum advance—The principle of using the vacuum generated by an engine to advance ignition timing, accomplished through the use of a mechanism attached to the distributor which moves the breaker point or pickup coil plate when it receives vacuum.

vacuum delay valve (VDV)—An orifice-controlled valve which delays a vacuum signal to a diaphragm, such as in the distributor vacuum advance unit.

vacuum motor—A chamber containing a vacuum diaphragm used to create movement in a component, such as a heated air intake blend door, when engine vacuum is routed to the chamber.

valve overlap—The amount of time that the closing of the exhaust valve overlaps the opening of the intake valve. The amount of valve overlap depends on the spacing of the lobe centers on the camshaft and the cam's duration.

vapor-liquid separator—Part of the evaporative emissions control system that prevents liquid gasoline from flowing through the vent line to the charcoal canister.

vapor lock—A condition in carbureted engines where excessive heat has caused the fuel in the fuel line or fuel pump to boil. The bubbles block the flow of fuel to the carburetor and prevent the engine from starting.

vapor recovery system—Part of the evaporative emissions control system that prevents gasoline vapors from escaping to the atmosphere. Vapors are trapped in the charcoal canister and then drawn into the engine and burned when the engine is started. Vapor recovery can also refer to preventing gasoline vapors from entering the atmosphere when a vehicle is being refueled.

variable resistor—An electrical component which reduces the flow of current in a circuit variably according to its position, such as a TPS. Also called a "potentiometer," or "pot."

venturi—The narrow part of the carburetor throat. When air passes this point, the restriction causes an increase in velocity and a drop in pressure that siphons fuel from the fuel bowl into the airstream.

venturi vacuum amplifier—In certain EGR systems, a device that uses the weak venturi vacuum signal to regulate the application of strong manifold vacuum to the EGR valve. It usually includes a reservoir that supplies sufficient vacuum when the engine is producing too little to operate the EGR valve.

Acronyms

The advent of emissions control spawned an alphabet soup of acronyms to label everything from emissions control components to engine management systems. It seems that every time something new is introduced, another acronym enters the automotive jargon. What follows is a list of acronyms (many of which you've probably never seen before) along with the vehicle manufacturers who coined them:

EMISSIONS RELATED

AAV	anti-afterburn valve (Mazda)
AIR	Air Injection Reaction (GM)
AIS	Air Injection System (Chrysler)
CC	catalytic converter
CCP	controlled canister purge (GM)
CCV	canister control valve
CEC	Crankcase Emissions Control System (Honda)
CO	carbon monoxide
CO2	carbon dioxide
CP	canister purge (GM)
ECS	Evaporation Control System (Chrysler)
EEC	Evaporative Emissions Controls (Ford)
EECS	Evaporative Emissions Control System (GM)
EFE	Early Fuel Evaporation system (GM)
EGR	exhaust gas recirculation
EGR-SV	EGR solenoid valve (Mazda)
EGRTV	EGR thermo valve (Chrysler)
EVRV	electronic vacuum regulator valve for EGR (GM)
HAIS	Heated Air Intake System (Chrysler)
HC	hydrocarbons
NOX	nitrogen oxides
OC	oxidation converter (GM)
ORC	oxidation reduction catalyst (GM)
PAFS	Pulse Air Feeder System (Chrysler)
PAIR	Pulsed Secondary Air Injection system (GM)
PCV	positive crankcase ventilation
PVS	ported vacuum switch
SO2	sulfur dioxide
TAC	thermostatic air cleaner (GM)
TAD	Thermactor air diverter valve (Ford)
TWC	three-way catalyst
TVS	thermal vacuum switch
TVV	thermal vacuum valve (GM)

COMPUTERIZED ENGINE CONTROL SYSTEMS

C3	Computer Command Control system (GM)
C4	Computer-Controlled Catalytic Converter system (GM)
CAS	Clean Air System (Chrysler)
CCC	Computer Command Control system (GM)
CVCC	Compound Vortex-Controlled Combustion system (Honda)
ECCS	Electronic Concentrated Control System (Nissan)
EEC	Electronic Engine Control (Ford)
ELB	Electronic Lean Burn (Chrysler)
MCU	Microprocessor-Controlled Unit (Ford)
MISAR	Microprocessed Sensing and Automatic Regulation (GM)
PGM-FI	Programmed Gas Management Fuel Injection (Honda)
SCC	Spark Control Computer (Chrysler)
SMEC	Single Module Engine Control (Chrysler)
TCCS	Toyota Computer Controlled System

SENSORS

ACTS air charge temperature sensor (Ford)
AFS airflow sensor (Mitsubishi)
AFM airflow meter
APS absolute pressure sensor (GM)
APS atmospheric pressure sensor (Mazda)
ATS air temperature sensor (Chrysler)
BARO barometric pressure sensor (GM)
BMAP barometric/manifold absolute pressure sensor (Ford)
BP backpressure sensor (Ford)
BPS barometric pressure sensor (Ford & Nissan)
BPT backpressure transducer
CAS crank angle sensor
CESS cold engine sensor switch
CID cylinder identification sensor (Ford)
CMP camshaft position sensor (GM)
CP crankshaft position sensor (Ford)
CTS charge temperature switch (Chrysler)
CTS coolant temperature sensor (GM)
CTVS choke thermal vacuum switch
ECT engine coolant temperature (Ford & GM)
EGO exhaust gas oxygen sensor (Ford)
EGRPS EGR valve position sensor (Mazda)
EOS exhaust oxygen sensor
EPOS EGR valve position sensor (Ford)
EVP EGR valve position sensor (Ford)
FLS fluid level sensor (GM)
HEGO heated exhaust gas oxygen sensor
IAT inlet air temperature sensor (Ford)
IATS intake air temperature sensor (Mazda)
KS knock sensor
MAF mass airflow sensor
MAP manifold absolute pressure
MAT manifold air temperature
MCT manifold charging temperature (Ford)
OS oxygen sensor
PA pressure air (Honda)
PIP profile ignition pickup (Ford)
SS speed sensor (Honda)
TA temperature air (Honda)
TP throttle position sensor (Ford)
TPP throttle position potentiometer
TPS throttle position sensor
TPT throttle position transducer (Chrysler)
TVS thermal vacuum switch (GM)
VAF vane airflow sensor
VSS vehicle speed sensor
WOT wide-open throttle switch (GM)
WSS wheel speed sensor

ELECTRONIC COMPONENTS

ALCL assembly line communications link (GM)
ALDL assembly line data link (GM)
ASD automatic shutdown relay (Chrysler)
BCM body control module (GM)
CALPAK calibration pack
CECU central electronic control unit (Nissan)
CPU central processing unit
DERM diagnostic energy reserve module (GM)
DLC data link connector
EACV electronic air control valve (Honda)
EBM electronic body module (GM)
EBCM electronic brake control module (GM)
ECA electronic control assembly
ECM electronic control module (GM)
ECU electronic control unit (Ford, Honda, and Toyota)
EEPROM electronically erasable programmable read only memory chip
E-PROM erasable programmable read only memory chip
EI electronic ignition (GM)
IC integrated circuit
ICS idle control solenoid (GM)
ISC idle speed control (GM)
LCD liquid crystal display
LED light-emitting diode
MIL malfunction indicator lamp
PCM powertrain control module
PROM program read only memory computer chip
SES service engine soon indicator (GM)
TCC torque converter clutch (GM)
VCC viscous converter clutch (GM)

ELECTRONIC IGNITIONS

BID	Breakerless Inductive Discharge (AMC)
CDI	Capacitor Discharge Ignition (AMC)
C3I	Computer-Controlled Coil Ignition (GM)
CSSA	Cold-Start Spark Advance (Ford)
CSSH	Cold-Start Spark Hold (Ford)
DIS	Distributorless Ignition System (Ford)
DIS	Direct Ignition System (GM)
EDIS	Electronic Distributorless Ignition System (Ford)
EIS	Electronic Ignition System (generic)
HEI	High Energy Ignition (GM)
IDI	Integrated Direct Ignition (GM)
ITCS	Ignition Timing Control System (Honda)
SDI	Saab Direct Ignition
SSI	Solid State Ignition (Ford)
TFI	Thick Film Integrated (Ford)

SPARK CONTROL SYSTEMS AND DEVICES

CCEV	Coolant-Controlled Engine Vacuum switch (Chrysler)
CSC	Coolant Spark Control (Ford)
CSSA	Cold-Start Spark Advance (Ford)
CTAV	Cold Temperature-Actuated Vacuum (Ford)
CTO	Coolant Temperature Override switch (AMC)
DRCV	distributor retard control valve
DSSA	Dual Signal Spark Advance (Ford)
DVDSV	differential vacuum delay and separator valve
DVDV	distributor vacuum delay valve
ESA	Electronic Spark Advance (Chrysler)
ESC	Electronic Spark Control (GM)
ESS	Electronic Spark Selection (Cadillac)
EST	Electronic Spark Timing (GM)
OSAC	Orifice Spark Advance Control (Chrysler)
PVA	ported vacuum advance
PVS	ported vacuum switch
SAVM	spark advance vacuum modulator
SPOUT	Spark Output signal (Ford)
SRDV	spark retard delay valve
TAV	temperature actuated vacuum
TCS	Transmission-Controlled Spark (GM)
TIC	thermal ignition control (Chrysler)
TRS	Transmission Regulated Spark (Ford)
VDV	vacuum delay valve

FUEL SYSTEM

AIS	automatic idle speed motor (Chrysler)
ABSV	air bypass solenoid valve (Mazda)
ASD	automatic shutdown relay (Chrysler)
CANP	canister purge solenoid valve (Ford)
CCEI	Coolant-Controlled Idle Enrichment (Chrysler)
CEAB	cold engine air bleed
CER	cold enrichment rod (Ford)
CIS	Continuous Injection System (Bosch)
CPI	Central Port Injection (GM)
CVR	control vacuum regulator (Ford)
DEFI	Digital Electronic Fuel Injection (Cadillac)
DFS	deceleration fuel shutoff (Ford)
EFC	electronic fuel control
EFC	electronic feedback carburetor (Chrysler)
EFCA	electronic fuel control assembly (Ford)
EFI	electronic fuel injection
FBC	feedback carburetor system (Ford and Mitsubishi)
FBCA	feedback carburetor actuator (Ford)
FCA	fuel control assembly (Chrysler)
FCS	fuel control solenoid (Ford)
FDC	fuel deceleration valve (Ford)
FI	fuel injection
IAC	idle air control (GM)
ISC	idle speed control (GM)
ITS	idle tracking switch (Ford)
JAS	Jet Air System (Mitsubishi)
MCS	mixture control solenoid (GM)
MFI	multiport fuel injection
MPFI	multipoint fuel injection
MPI	multiport injection
PECV	power enrichment control valve
PFI	port fuel injection (GM)
SFI	Sequential Fuel Injection (GM)

SMPI	Sequential Multiport Fuel Injection (Chrysler)
TABPV	throttle air bypass valve (Ford)
TBI	throttle body injection
TIV	Thermactor idle vacuum valve (Ford)
TPI	Tuned Port Injection (GM)
TKS	throttle kicker solenoid (Ford)
TSP	throttle solenoid positioner (Ford)
TV	throttle valve

GENERAL

A/C	air conditioning
AC	alternating current
A/F	air/fuel ratio
A/T	automatic transmission
ATC	after top center
ATDC	after top dead center
ATF	automatic transmission fluid
AWD	all-wheel drive
BAT	battery
BHP	brake horsepower
BTC	before top center
BTDC	before top dead center
Btu	British thermal units
C	Celsius
cc	cubic centimeters
cfm	cubic feet per minute
CID	cubic inch displacement
dB	decibels
DC	direct current
DOHC	dual overhead cams
DVOM	digital volt-ohmmeter
EMI	electromagnetic interference
F	Fahrenheit
ft. lb.	foot-pound
FWD	front-wheel drive
gal	gallon
GND	ground
GPM	grams per mile
Hg	mercury
hp	horsepower
IGN	ignition
ID	inside diameter
I/P	instrument panel
kHz	kilohertz
kPa	kilopascals
km	kilometers
kV	kilovolts
L	liters
lb. ft.	pound-feet
mm	millimeters
ms	millisecond
mV	millivolts
mpg	miles per gallon
mph	miles per hour
Nm	Newton meters
OD	outside diameter
OE	original equipment
OEM	original equipment manufacture
OHC	overhead cam
P/B	power brakes
P/N	part number
PPM	parts per million
PS	power steering
psi	pounds per square inch
pt.	pint
qt.	quart
RFI	radio frequency interference
rpm	revolutions per minute
RPO	regular production option
RWD	rear-wheel drive
SOHC	single overhead cam
TACH	tachometer
TDC	top dead center
V	volts
VAC	volts alternating current
VDC	volts direct current
VIN	vehicle identification number
WOT	wide-open throttle

MISCELLANEOUS

ASE	Automotive Service Excellence
CAFE	corporate average fuel economy
CARB	California Air Resources Board
DOT	Department of Transportation
DRBII	Diagnostic Readout Box (Chrysler)
EPA	Environmental Protection Agency
FMVSS	Federal Motor Vehicle Safety Standards
IM240	inspection/maintenance 240 program
ISO	International Standards Organization
NHTSA	National Highway Traffic Safety Administration
OBD	onboard diagnostics
SAE	Society of Automotive Engineers

Index

Note: Page Numbers in **Bold type** reference non-text material.

D

F

G

H

I

K

L

M

N

O

P

R

S

T

NOTES

NOTES

NOTES

DATE DUE

JUN 3 0 1999

Printed in USA

HIGHSMITH #45230